Concepts and Applications of Biochemistry

Concepts and Applications of Biochemistry

Edited by **Oliver Stone**

SYRAWOOD
PUBLISHING HOUSE
New York

Published by Syrawood Publishing House,
750 Third Avenue, 9th Floor,
New York, NY 10017, USA
www.syrawoodpublishinghouse.com

Concepts and Applications of Biochemistry
Edited by Oliver Stone

International Standard Book Number: 978-1-68286-250-6 (Hardback)

Printed in China.

Contents

Preface

This book was inspired by the evolution of our times; to answer the curiosity of inquisitive minds. Many developments have occurred across the globe in the recent past which has transformed the progress in the field.

The field of biochemistry aims to study the chemical processes taking place in living organisms on a cellular level. The chapters included in this book are a compilation of updated information on structure and interaction of proteins and nucleic acids, structure of cells, applied biochemistry, molecular biology, etc. Some of the core concepts of biochemistry like functions of biomolecules and metabolic processes have been explained in this book. It aims to serve as a resource guide for students and experts alike and contribute to the growth of the discipline.

This book was developed from a mere concept to drafts to chapters and finally compiled together as a complete text to benefit the readers across all nations. To ensure the quality of the content we instilled two significant steps in our procedure. The first was to appoint an editorial team that would verify the data and statistics provided in the book and also select the most appropriate and valuable contributions from the plentiful contributions we received from authors worldwide. The next step was to appoint an expert of the topic as the Editor-in-Chief, who would head the project and finally make the necessary amendments and modifications to make the text reader-friendly. I was then commissioned to examine all the material to present the topics in the most comprehensible and productive format.

I would like to take this opportunity to thank all the contributing authors who were supportive enough to contribute their time and knowledge to this project. I also wish to convey my regards to my family who have been extremely supportive during the entire project.

Editor

1

Polyamines, peroxidase and proteins involved in the senescence process

Marcos Ribeiro da Silva Vieira[1] Fernando Jose Freire[2], Damiana Cleuma de Medeiros[3]Fabiana Bezerra Moura[1], Thialla Laranjeira Amorim[1] and Adriano Nascimento dos Simões[1]

[1]Universidade Federal Rural de Pernambuco, Unidade Acadêmica de Serra Talhada, CEP: 59909-460, Serra Talhada, PE, Brasil.
[2]Departamento de Agronomia, CEP: 52171-900, Recife, PE, Brasil'
[3]Instituição: UFRN - Unidade Acadêmica Especializada em Ciências Agrárias, Departamento de Agropecuária, Macaíba, RN, Brasil.

Senescence is the natural aging process at the cellular level or range of phenomena associated with this process. The objective of this review was to show the involvement of substances that may be related to senescence in plants, such as polyamines, peroxidase and proteins. These substances were related with the terminal stages of development of various plant organs (leaves, flowers, roots).

Key words: Flowers, putrescine, spermidine, spermine, reactive oxygen species.

The plants are generally classified as highly products perishable, the ephemeral nature of the different tissues that form, the high activity respiratory and the low-carbohydrate reserves (Nowak and Rudnicki, 1990; Skutnik et al., 2001). After harvesting occur flowers' biochemical, structural and physiological change processes leads to tissue disorganization and disintegration of organs which promote senescence. The polyamines [putrescine (PUT), spermidine (SPD) and spermine SPM)], classified as plant growth regulators, are related to different responses physiologically, such as senescence and stress. In plants, the diamine and putrescine synthesizes from arginine and ornithine. The putrescine is converted to spermidine and spermine by successive transfers of one or two aminopropyl groups via SAM (S-adenosyl methionine) (Lima et al., 1999).

Other substances related to senescence such as peroxidases, are enzymes that typically catalyze reactions using oxi-reduction hydrogen peroxide (H_2O_2) as electron acceptor to catalyze different oxidative reactions (Blokhina et al., 2003) and proteins (Sugawara et al., 2002; Azeez et al., 2007), whose synthesis is characterized by high and rapid adaptability molecular and physiologically functional with respect to the medium (Larcher et al., 2000).

POLYAMINES

The polyamines (PAs) are aliphatic molecules of low molecular weight present in all organisms. The main PAs found in the higher plants are PUT, SPD and SPM occurring in free form or conjugated to phenolic acids and low molecular weight molecules (Bouchereau et al., 1999;

Kuznetsov et al., 2006). In addition to biosynthesis and conjugation, the oxidative degradation is one way of regulating the PAs levels (Kusano et al., 2008). In plants, the PAs are located not only in the cytosol but also in the organelles, such as mitochondria, chloroplasts and vacuoles (Kumar et al., 1997). Its synthesis is initiated by two different routes: the PUT is formed directly from ornithine or indirectly by arginine and the formation of SPM occurs by adding a decarboxylated S-adenosyl methionine (SAM) group added to PUT. The formation of SPM occurs by adding another decarboxylated SAM (Coruzzi and Last, 2000).

The presence of positive charges on the molecules of PAs allows their electrostatic binding to cellular macromolecules, including DNA, RNA, proteins, chromatin and may cause stabilization or destabilization of these macromolecules (Kusano et al., 2008). Thus, they are involved in fundamental cellular processes including gene expression regulation, signal modulation and cell proliferation and membrane stabilization (Tabor and Tabor, 1984; Cohen, 1998; Igarashi and Kashiwagi, 2000).

Recent studies show that PAs have function in modulating many physiological processes in plants, such as cell division and differentiation, organogenesis, embryogenesis and tolerance to biotic and abiotic stress (Bouchereau et al., 1999; Crozier et al., 2001; Martin-Tanguy, 2001; Baisand Ravishankar, 2002; Silveira et al., 2004; Silveira et al., 2006; Santa-Catarina et al., 2006; Groppa and Benavides, 2008).

Some plants use the mechanism called osmotic adjustment to tolerate the effects of abiotic stresses, which allows the cell to maintain its metabolic functions even in adverse environmental conditions, thus promoting tolerance to stress and maintain relatively high osmotic potential (Bayuelo-Jiménez et al., 2002). This mechanism enables the cell to accumulate substances called compatible osmolytes, which preserve cell integrity resulting in the continuation of vital activities to plant growth and development (Bray et al., 2001). There is evidence that the PAs stabilize the membrane and retard senescence (Smith, 1985). SAM can be turned successively into aminopropyl-carboxylic acid and ethylene (Slocum, 1984), once the PAs and ethylene compete for the same precursor (Bouchereau et al., 1999; Pandey et al., 2000).

The PAs inhibit the ethylene production by regulating the activity of synthase and oxidase of 1-aminocyclopropane-1-carboxylic acid (ACC) (Lee et al., 1997), while ethylene alters the PAs formation by reducing the arginine decarboxylase activity (ADC) and SAM decarboxylase (SAMDC) (Roustam et al., 1994). Changes in PAs levels and ethylene during senescence have been reported in some plants such as plum (de Dios et al., 2006) and *Hibiscus syriacus* L. (Seo et al., 2007). Li and Wang (2004) observed that between ethylene and polyamines, there is metabolic competition only under high stress conditions.

The PAs concentration may vary by plant organ, maturity and post-harvest treatment (Carbonell et al., 2000; Chattopadhayay et al., 2002; Nayyar and Chander, 2004). Pomegranates show higher levels of PAs during cold storage; however, the main change occurred for SPM (Mirdehghan et al., 2007). PUT accumulation was also noted in pepper, cucumber, zucchini and citrus (orange and lemon) during exposure to cold chamber (Serrano et al., 1998; Martínez-Romero et al., 2003). In Faroe chrysanthemum stems Vieira et al. (2010) observed decreases in the levels of PAs throughout the cold storage. In several species, the PAs increase was correlated with the reduction of injuries caused by such stress (Kramer and Wang, 1989; Wang and Ji, 1989). It has been suggested that the accumulation of polyamines in tissues can confer more tolerance to various types of stresses due to these amines act in the removal of reactive oxygen species and also assist the stabilization of cell membranes (Larher et al., 2003; Groppa and Benavides, 2008).

High temperature also causes reduction in cell division and this effect may be linked to its action on the Pas. Under high temperatures would occur reduction in the concentration of these polyamines, affecting cell division (Poljakoff-Mayber and Lerner, 1994). In post-harvest of cloves kept at 21°C, Serrano et al. (1998) observed increase in PUT content and decrease in SPD. In addition to PAs, other substances may be related to senescence in plants such as peroxidase and proteins.

PEROXIDASE

Plants produce reactive oxygen species (ROS) in various metabolic processes, when they suffer some kind of stress (Jin et al., 2006). Thus, ROS may be generated during the post-harvest phase of flowers, which according to Shigeokaet al. (2002) would cause oxidative damage in plants. The accumulation of these ROS could promote damage to lipids, proteins, among others, forming toxic products (Pennycooke et al., 2005). Plants have antioxidative enzymes such as peroxidase (POD) that diminish the damage caused by excess peroxides (Scandalios, 1993). The POD's (EC 1.11.1.7) contain a heme prosthetic group (ferriprotoporphyrin IX) and in the catalytic process transiently oxidize the ferric ion (Fe^{3+}) to the higher valence states (FE^{4+} or FE^{5+}). In the reaction involving the POD, the electron donor can be ascorbate, amines and other organic compounds. The oxidation product show in many cases, intense staining (Richardson and Hyslop, 2000).

The POD can be considered a stress enzyme stimulated by low temperatures (El-Hilari et al., 2003). In baroapotato, Menolli et al. (2008) observed increased activity of the enzyme POD until the 7th day of storage at 5°C. Martinez-Téllez and Lafluente (1993) and El-Hilari et al. (2003) analyzing orange fruitsof Navelina cultivar stored at 1, 2.5, 5 and 10°C for 60 days and Mandarin' 'Fortune" stored at 4 and 8°C for 4 weeks, verified changes in the

POD activity. Zauberman et al. (1985) did not detect variation in the POD activity in avocados 'Fuerte' stored at 0, 2 and 5°C for up to 18 days.According to Yang and Hoffman (1984), storage at low temperatures causes induction of various enzymes such as POD, which can convert the acid 1-aminocyclopropane-1-carboxylic (ACC) into ethylene. At higher temperatures, Costa (2009) did not observe a standard trend for POD activity in stems of *Heliconia Bilhai* at 12 and 19°C. Responses of this nature were also obtained by Vieira et al. (2010) in Faroe chrysanthemum stems at 10and 25.2°C.

The POD is also involved in various reactions such as oxidation of phenols, oxidation of indole-3-acetic acid (IAA), bonds of polysaccharides, wound healing, defense against pathogens in the hypersensitive response (HR), lignin and suberin synthesis to the cell wall thickening, regulation of cell elongation and senescence (Quiroga et al., 2000; Silva, 2000; Kao, 2003; Campos et al., 2004). Hossain et al. (2006) report that increased levels of peroxides could be a programmed regulation of the activity of POD enzymes which seems to be a prerequisite for the onset of petals senescence. Bartoli et al. (1997) observed increased POD activity in *Chrysanthemum morifolium* Ram petals during senescence and related to defense against oxidant molecules that promote membrane damage.

The POD can occur in cells programmed for death and in flowers the deterioration is certainly programmed, it is not reversible and inevitably leads cells to death (Rogers, 2006). The substances deposited on the cut surface of flowers, as well as tyloses, can migrate into the xylem vessels, serving as barrier for microorganisms at the time they cause obstruction, preventing the water absorption. The formation of such substances, according to Vaslier and Van Doorn (2003), probably is involved in the ethylene synthesis and action of enzymes such as POD.

The POD enzyme is involved in vascular blockage of some species of flowers by oxidation of p-coumaryl alcohol, coniferyl and sinapil, which are precursors of lignin. Lignin is a compound that is part of the secondary metabolism of plants and, although providing support and structuring to water transport through the xylem, it may function as a protective mechanism against pathogen attack in case of stress by depositing on the cut surface preventing also the entry of water into vases (Boerjan et al., 2003). Loubaud and Van Doorn (2004) concluded that the blockage in *Astilbe* stems, since once developed in flowers kept in humid and dry storage, however, in roses cv. Red One and *Viburnum opulus*the blockage observed in these stems was apparently related to the presence of bacteria in the xylem.

PROTEINS

Proteins are linear polymers of amino acids (macromolecules) with molar mass ranging from hundreds to thousands of Daltons (1 Da = 1.661 × 10^{24} g). Only about 20 amino acids are found in animal and plant proteins, and these are combined in myriad ways to form a variety of different proteins (Magalhães, 2008). An amino acid has a carboxylic group and an amino group in protein molecules and they are responsible for peptide bonds formation, except the N and C terminals of the protein. Each amino acid has two ionization constants, which represent the ionization constants of 1-amino and 1-carboxylic acid groups of the free amino acid. However, some have more than two pKa values, once they have other ionizable groups in their chains. The net charge of a protein molecule in aqueous solution depends on the ionization constant of the constituent amino acid side chains and the solution pH (Melvin, 1987).

The proteins are divided into a plant cell among various organelles and have specific functions. As an example, a small amount of proteins may be found in the cell wall and part of them consists of enzymes that initiate formation reactions, remodeling or breakdown of the wall structure. The chloroplast has a double membrane and there are transport proteins in the outer layer of this membrane (Newton et al., 2004). The synthesis of proteins is the central function of all cells. In its absence, growth and maintenance of organs cease and this represents a limiting factor to the growth rate of plants (Porter and Lawlor, 1991). According to Larchjer et al. (2000), the protein synthesis is characterized by high and rapid capability of molecular, functional and physiologic adaptation relative to the medium.

Changes in the levels of proteins have been associated in part as a result of synthesis of "new" specific proteins during senescence (Woodson and Handa, 1987), such as the ribonucleases, the β-glucosidases and proteases (Sood et al., 2006).The knowledge of the synthesis and the nature of modifications that proteins regulate have important implications for successful handling of longevity of the flowers (Woodson and Handa, 1987). The percentage of total protein is a data that evaluates the plants conditions in the field and post-harvest. Sood et al. (2006) noted that the protein content was higher in young seedlings and lower in the stages of maximum development. Drevdahl and Thimann (1977) observed decrease in protein contents during development in oat plants. In *Sandersonia* petals (Eason et al., 2002) and *Dendrobium* cv. Khao Sanan (Lerslerwong et al., 2009), there was decreased protein content during senescence. This response was also observed in leaves and flowers of Faroe Chrysanthemum post-harvest (Vieira et al., 2010), while Laschi (2000) observed increased proteinsin rose stems. Souza (2008) reported decreases in protein levels between the third and tenth day of assessment in inflorescences of Heliconia Golden Torch. This author studied the percentage of proteins in heliconia bilhai and observed 7.0% increase from the harvest day and the four days of storage, followed by decreases until the tenth day.

The reduced proteins can be attributed to the action of

proteases. Elanchezhian and Srivastava (2001) suggested that the decreased protein contents in chrysanthemum petals during senescence was due to the synthesis inhibition and increased protein degradation by proteases, resulting in loss of functional capacity of the membranes, increased output of ions and finally senescence and death of tissues. Through studies on the programmed cell death (PCD), it was shown that proteases increased and their relationship with senescence as a possible regulator of programmed death in plant cells (Gerreiro et al., 1998; Gietl and Schimid, 2001).

Proteins play a critical role in modulating plant response to stress. Saliveit (2000) and Thomashow (2001) claim that the low temperature during storage induces changes in protein content. Ferguson et al. (1988) also mentions that high temperature can directly or indirectly injure plant proteins by inactivating enzymes, changing the conformation of peptides or disrupting membrane complexes. Within this field, important advances have been made in understanding the plant response to stress.

REFERENCES

Bais HP, Ravishankar GA (2002).Role of polyamines in the ontogeny of plants and their biotechnological applications.Plant Cell, T and O Culture. 69:1-34.

Bar|toli GC, Guaiamet JJ, Montaldi ER (1997). Ethylene Production and Responses to Exogenous Ethylene in Senescing Petals of *Chrysanthemum morifolium* RAM cv. Unsei.Plant Science. 124:15-21.

Bayuelo-Jimenez J, Debouck DG, Lynch JP (2002). Salinity tolerance in *Phaseolus* species during early vegetative growth. Crop Sci. 42:2184-2192.

Boerjan W, Ralph J, Baucher M (2003). Lignin biosynthesis. Annual R.P. Biology. 54:519-546.

Bouchereau A, Aziz A, Larher F, Martin-Tanguy J (1999). Polyamines and environmental challenges: recent development. Plant Science. 140:103-125.

Bray EA, Bailey-Serres J, Weretilnyk E (2001). Responses to abiotic stress. In: Buchanan, B.; Gruissem, W.; Jones, R. (Eds). Biochemistry & molecular biology of plants.American Society of Plant Physiologists. 22:1167-1168.

Campos AD, Ferreira GA, Hampe MMV, Antunes IF, Brancão N, Silveira EP, Osório VA, Augustin E (2004). Atividade de peroxidase e polifenoloxidase na resistência do feijão à antracnose. Pesquisa Agropecuária Brasileira.39(7):637-643.

Carbonell BA, Valero GD, Martínez RD, Serrano MM, Burló CF (2000). Polyamines: biosynthesis, metabolism, and their role in ripening and postharvest handling of fruits. Food Science and T. International. 6:85-95.

Chattopadhayay MK, Tiwari BS, Chattopadhayay G, Bose A, Sengupta DN, Ghosh B (2002). Protective role of exogenous polyamines on salinity stressed rice (*Oryza sativa*) plants. Physiologia Plantarum. 116:192-199.

Coruzzi G, Last R (2000). Amino acids. In: Buchanan, BB; Gruissem, W; Jones, RL Biochemistry and Molecular Biology of Plants. American Society of Plant Physiologists. 358-410.

Costa AS (2009). Conservação pós-colheita, sintomas e respostas fisiológicas da senescência e injúria por frio em hastes florais de *Heliconia bihai*(L). 83f. Tese (Doutorado em Botânica) - Universidade Federal Rural de Pernambuco, Recife.

Crozier A (2001). Biosynthesis of hormones and elicitor molecules. In: Buchanan, B.; Gruissem, W.; Jones, R. (Eds). Biochem. Mol. Biol. Plants. Amr. Soc. Plant Physiol. 911-915.

de Dios P, Matilla AJ, Gallardo M (2006). Flower fertilization and fruit development prompt changes in free polyamines and ethylene in damson plum (*Prunusinsititia*l.). J. Plant Physiol. 163:86-97.

Drevdahl RA, Thimann KV (1977). Proteases of senescing oat leaves I. purification and general properties. Plant Physiol. 59:1059-1063.

Eason JR, Ryan JD, Pinkney TT, O'donoghue EM (2002). Programmed cell death during flowers senescence: isolation and characterization of cysteine proteinases from *Sandersonia aurantiaca*. Funct. Plant Biol. 29(60):1055-1064.

Elanchezhian R, Srivastava GC (2001). Physiology changes during flower senescence. Biol. Plant. 44(3): 411-415.

El-Hilari F, Ail-Oubahou A, Remah A, Akhayat O (2003). Chilling injury and peroxidase activity change in "Fortune" mandarin fruit during low temperature storage. Bulgarian J. Plant Physiol. 29(1-2):44-54.

Ferguson JD, Blannchard T, Galligan DT (1988) Infertility in dairy cattle fed a high percentage of protein degradable in the rumen. J. Anim. Vet. Med. Assoc. 192:659-662.

Gietl C, Schmid M (2001). Ricinosmes: an organelle for development regulated cell death in senescence in plant tissues. Naturwissenschaften. 88:49-58.

Groppa MD, Benavides MP (2008). Polyamines and abiotic stress: recent advance. Amino Acids.34:35-45.

Guerreiro C, Calle M, Reid MS, Valpuesta V (1998). Analysis of the expression of two thioproteinase genes from daylily (*Hemerocallis*spp) during flower senescence. Plant Mol. Biol. 36: 565-571.

Hossain Z, Mandal AKA, Datta SK, Biswas AK (2006). Decline in ascorbato peroxidase activity—a prerequisite factor for tepal senescence in gladiolus. J. Plant Physiol. 163:186-194.

Igarashi K, Kashiwagi K (2000). Polyamines: mysterious modulators of cellular functions. Biochem. Biopsychol. Res. Commune 271(3):559-564.

Jin J, Shan N, Ma N, Bali J, Gao J (2006). Regulation of ascorbate peroxidase at the transcript level is involved in tolerance to postharvest water deficit stress in the cut rose (*Rosa hybrida* L.) cv. Samantha, Postharvest B Technol. 40(30):236-243.

Kao CH (2003). Differential effect of sorbitol and polyethylene glycol on antioxidant enzymes in rice leaves. Plant G. Reg. 39(2):83-89.

Kramer GF, Wang Y (1989). Correlation of reduced chilling injury with increased spermine and spermidine levels in zucchini squash. Journal of Plant Physiology, 76(3):479-484.

Kumar A, Taylor M, Altabella T, Tiburcio A (1997). Recent advances in polyamine research. Trend Plant Sci. 2(4):124-130.

Kusano T, Berberich T, Tateda C, Takahashi Y (2008). Polyamine: essential factors for growth and survival. 228:367-381.

Kuznetsov V, Shorina M, Aronova E, Stetsenko L, Rakitin V, Shevyakova N (2006). NaCl-and ethylene-dependent cadaverine accumulation and its possible protective role in the adaptation of the common ice plant to salt stress.Plant Science.172:363-370.

Larchjer W (2000). Ecofisiologia vegetal.São Carlos: RiMa, p.531.

Larher FR, Aziz A, Gibon Y, Trotel-Aziz P, Sulpice R, Bouchereau A (2003). An assessment of the physiological properties of the so-called compatible solutes using in vitro experiments with leaf discs. Plant Physiol. Biochem. 41(6-7):657-666.

Laschi D (2000). Fisiologia de hastes cortadas de rosa (Rosa sp) cv. Grand Gala. 110 f. Tese (Doutorado em Agronomia/Horticultura) - Instituto de Biociência de Botucatu, Universidade Estadual Paulista.

Lerslerwong L, Ketsa S, van Doorn WG (2009). Protein degradation and peptidase activity petal senescence in *Dendrobium* cv. Khao Sanan. Postharvest Band Technol. 52:84-90.

Lee MM, Lee SH, Park KY (1997). Effects of spermine on ethylene biosynthesis in cut carnation (*Diathuscaryophyllus* L.) flowers during senescence. J. Plant Physiol. 151:68-73.

Li C, Jiao J, Wang G (2004). The important roles of reactive oxigen species in the relationship between ethylene and polyamines in leaves of spring wheat seedlings under root osmotic stress. Plant Sci. 166:303-315.

Loubaud M, Van Doorn G (2004). Wound-induced and bacteria-induced xylem blockage in roses, *Astible*, and *Viburnum*. Postharvest B Technol.32:281-288.

Magalhães CS (2008). Avaliação comparativa de procedimentos de extração de proteínas em plantas medicinais e fitoterápicos e quantificação de metais associados a essas proteínas. 151 f. Tese (Doutorado em Ciências)-Universidade Estadual de Campinas.

Martinez-Tellez MA, Lafuente MT (1993). Chilling-induced changes in phenylalanine ammonia-lyase, peroxidase, and polyphenol oxidase activities n citrus flavedo tissue. Acta Horticult. 3(43):257-263.

Martínez-Romero D, Serrano M, Valero D (2003). Physiological changes in pepino (*Solanummuricatum* Ait.) fruit stored at chilling and nonchilling temperatures. Postharvest B Technol. 30:177-186.

Martin-Tanguy J (2001). Metabolism and function of polyamines in plants: recent development (new approaches). Plant Growth Reg. 34:135-148.

Melvin M (1987). Electrophoresis: analytical chemistry by open learning. London: John Wiley p. 127.

Menolli LN, Finger FL, Puiatti M, Barbosa MJ, Barros RS (2008). Atuação das enzimas oxidativas no escureci-mento causado pela injúria por frio em raízes de batata-baroa. Acta Scientiarum Agronomy.30(1):57-63.

Mirdehghan SH, Rahemi M, Martínez-Romero D, Guillé F, Valverde JM, Zapata PJ, Serrano M, Valero D (2007). Reduction of pomegranate chilling injury during storage after heat treatment: Role of polyamines. Postharvest B. Technol. 44: 19-25.

Nayyar H, Chander S (2004). Protective effects of polyamines against oxidative stree induced by water and cold stress in chickpea. J. Agron. Crop Sci. 190:355-365.

Newton RP, Brenton AG, Smith CJ, Dudley E (2004). Plant proteome analysis by mass spectrometry: princi-ples. Problems, pitfalls and recent developments, Phyto-chemistry. 65(2):1449-1453.

Pandey RK, Maranville JW, Chetima MM (2000). Deficit irrigation and nitrogen effects on maize in a Sahelian environment. II. Shoot growth, nitrogen uptake and water extraction. Agric. Water Mangt.46:1-13.

Pennycooke JC, Cox S, Stushnoff C (2005). Relationship of cold acclimation, total phenolic content and antioxidant capacity with chilling tolerance in petunia (Petunia x hybrid). Environ. Expt. Bot. 53:225-232.

Poljakoff-Mayber A, Lerner HR (1994). Plants in saline environments. In- Handbook of plant and crop stress. Pessarakli, m. (Ed.). Marcel Dekker. New York. 65-96.

Porter JR, Lawlor DW (1991). Plant Growth: interactions with nutrition and environment Cambridge: University of Cambridge Press, 284 p.

Quiroga C, Guerreiro C, Botella MA, Barceló A, Amaya I, Medina MI, Alonso FJ, Forchetti SM, Tigier H, Valpuesta V (2000). A tomato peroxidase involved in the synthesis of lignin and suberin. Plant Physiol. 122:1119-1127.

Richardson T, Hyslop DB (2000). Enzimas. In: Fennema, OR (Dir.) Química de los alimentos. Acríbia. 501-503.

Rogers HJ (2006). Programmed cell death in floral organs: How and why do flowers dye? Ann. Bot. 97:309-315.

Roustan JP, Chraibi KM, Latche A, Fallot J (1993). Relationship between ethylene and polyamine synthesis in plant regeneration. In: Pech, JC (Ed.) Cellular and molecular aspects of the plant hormone ethylene. Kluwer Acad. Pub. 365-366.

Saliviet ME (2001). Discovery of chilling injury. In: Kung, SD; Yang, SF (Eds.) Discoveries in Plant Biology. Singapore: World Sci. Pub. 423-448.

Scandalios JG (1993).Oxygen stress and superoxide dismutases. Plant Physiol. 101:7-12.

Seo SG, Shim IS, Usui K, Fujihara S (2007). Analysis of polyamines, 1-aminocyclopropane-1-carboxylic acid and their conjugated forms in floral organs of *Hibiscus syriacus* L. J. Jpn Soc. Hortic. Sci.76:149-156.

Serrano M, Pretel MT, Martínez-Madrid MC, Romojaro F, Riquelme F (1998). CO_2 treatment of zucchini squash reduces chilling-induced physiological changes. J. Agric. Food Chem. 46:2465-2468.

Silva EAA, Pinho EVRV, Vieira MGGC, Carvalho MLM, Machado JC (2000). Alterações dos padrões de isoenzimas em sementes de milho infectadas por fungos. Pesquisa Agropecuária Brasileira. 35(9):1725-1732.

Silveira V, Floh EIS, Handro W, Guerra MP (2004). Effect of plant growth regulators on the cellular growth and levels of intracellular protein, starch and polyamines in embryogenic suspension cultures of *Pinustaeda*. Plant tissues organelles culture. 69:233-249.

Silveira V, Santa-Catarina C, Tun NN, Scherer GFE, Handro W, Guerra MP, Floh EIS (2006). Polyamine effects on the endogenous polyamine contents, nitric oxide release, growth and differentiation of embryogenic suspension cultures of *Araucáriaangustifólia* (Bert) O. Ktze. Plant Sci. 171:91-98.

Sood S, Vyas D, Nagar PK (2006). Physiological and biochemical studies during flower development in two rose species. Sci. Hortic. 108:390-396.

Souza AS (2008). Longevidade de Heliconia psittacorum x *H. spathocircinata* 'Golden Torch' e H. bilhai em resposta ao uso de reguladores de crescimento.158 f. Tese (Doutorado em Agronomia)- Universidade Federal de Viçosa.

Tabor CW, Tabor H (1984). Polyamines. Ann. Rev. Biochem. 53:749-790.

Thomashow MF (2001). So what's new in the field of plant cold acclimation. Plant Physiol. 125:89-93.

Vaslier N, Van Doorn WG (2003). Xylem occlusion in bouvardia flowers: evidence for a role of peroxidase and cathecol oxidase. Postharvest B and Technology. 28:231-237.

Vieira MRS, Teixeira da Silva JA, Lima GPP, Vianello F. (2010). Changes in polyamine, total protein and total carbohydrate content and peroxidase activity during the lifetime of chrysanthemum 'Faroe'. Floricult. Ornamental Biotech. 4(1):48-52.

Wang CY, Ji ZL (1989). Effect of low-oxygen storage on chilling injury and polyamines in zucchini squash. Sci Horticul. 39(1):17-19.

Woodson WR, Handa AK (1987). Changes in protein patterns and *in vivo* proteins synthesis during presenescence and senescence of hibiscus petals. J. Plant Physiol. 128:67-75.

Yang SF, Hoffman NE (1984).Ethylene biosynthesis and its regulation in higher plants. Ann. Rev. Plant Physiol. 35: 155-189.

Zauberman GY, Fuchs S, Akerman M (1985). Peroxidase activity in avocado fruit stored at chilling temperatures. Sci. Hortic. 26:261-265.

Phytochemical screening and antimicrobial studies on the methanolic bulb extract of *Allium sativum* L.

Ameh, G. I.[1]*, Eze, S. C.[2] and Omeje, F. U.[1]

[1]Department of Applied Biology and Biotechnology, Enugu State University of Science and Technology, Enugu, Enugu State, Nigeria.
[2]Department of Crop Science, University of Nigeria, Nsukka, Enugu State, Nigeria.

Methanolic extract of the bulb of *Allium sativum* was screened for phytochemical and antimicrobial properties. The crude extract of *A. sativum* bulb showed positive results for carbohydrates, glycosides and proteins in high concentrations; alkaloids, saponins, reducing sugars, oils and steroids in medium concentrations, while flavonoids and acidic compounds were present in low amounts. Tannins, resins and terpenoids were however absent in the extract. All the six tested microorganisms exhibited varied susceptibility to the plant extract. The minimum inhibitory concentration (MIC) of the organisms were, 50 mg/ml for *Salmonella paratyphi*, 25 mg/ml for *Bacillus subtilis*, 100 mg/ml for *Klebsiella pneumoniae*, 12.5 mg/ml for *Candida albicans*, 3.12 mg/ml for *Candida paralopsis* and 6.25 mg/ml for *Candida tropicalis*. *A. sativum* showed both antifungal and antibacterial properties.

Key words: Phytochemical screening, antimicrobial activity, *Allium sativum,* bulb extract.

INTRODUCTION

Plants have long been serving mankind as sources of useful drugs, food, additives, flavouring agents, colourants, binders and lubricants (Falodun et al., 2006). Medicinal plants are sources of important drugs used in the treatment of diseases either alone or in combination with other plants (Awonubi, 1988). Chemical substances found in plants include alkaloids, glycosides, essential oil, saponins, tannins, steroids, terpenoids, resins, flavonoids, proteins and others. These substances are potent bioactive compounds found in medicinal plant parts that can be used for therapeutic purposes (Soforowa, 1993; Nwadiaro and Nwachukwu, 2007). These inherent bioactive principles differ from plant to plant as a result of their biodiversity and they produce a definite physiological effect on human body. Several authors have screened different medicinal plants for the presence of these active principles. The knowledge of medicinal plants are important in pharmaceutical industry. Banson and Olutimayin (2001) have shown that Plants are composed of a wide variety of active principles. These established a good support to the use of these plants in herbal medicine and as base for the development of new drugs and phytomedicine. It has been demonstrated by several workers that some of these bioactive principles of plant origin possess antimicrobial properties. The increased reliance on the use of medicinal plants in industrialized countries had been traced to the extraction and development of several drugs and chemotherapeutics from these plants as well as from traditionally used rural herbal medicine (UNESCO, 1998).

Allium sativum L., commonly called garlic is a widely cultivated herbal plant, belonging to the family Alliaceae. Garlic is a plant with long, flat grass like leaves and a papery hood around the flowers (Milner, 2005). The greenish white or pink flowers are grouped together at the end of a long stalk. The stalk rises directly from the flower bulb, which is the part of the plant used as food and medicine (Awonubi, 1988). The most active components of fresh garlic are alliin and an enzyme called allinase. When garlic is chewed, chopped, bruised or cut, these compounds mix to form allicin, which is responsible for garlic's strong smell. It's medicinal claims have included cures for toothaches, cold, coughs and

*Corresponding author. E-mail: g_ameh@yahoo.com.

Table 1. Phytochemical analysis of garlic bulb extract.

Active Ingredient	Extract
Carbohydrates	+++
Reducing sugars	++
Alkaloids	++
Glycosides	+++
Saponins	++
Tannins	–
Flavonoids	+
Resins	–
Proteins	+++
Oils	++
Steroids	++
Terpenoids	–
Acidic compounds	+

- , Absent; +, low concentration; ++, medium concentration; +++ high concentration.

other viral infections, open wounds and evil demons (Fluck, 1973). The aim of this study was to evaluate the antimicrobial activities of active principles from *A. sativum* bulb extract.

MATERIALS AND METHODS

Collection and identification of plant material

Fresh bulbs of garlic were collected from Nsukka Central Market in Enugu State and authenticated by Mr. A. Ozioko, a Taxonomist of Bioresource Development and Conservation Programme (BDCP), Nsukka.

Processing of plant samples

The scales around the fresh bulbs of garlic were removed while the bulbs were washed and rinsed properly in tap and sterile distilled water, respectively. The bulbs were macerated, dried in an electric oven and milled mechanically before being stored in a plastic container.

Extraction of plant material

The methanol extract of the garlic bulb was prepared using the procedure described by Harbone (1994). Twenty five gram (25 g) of the powdered sample was soaked in a mixture of methanol and distilled water in the ratio of 3:2 for four days and later filtered to obtain the methanol extracts. The mixture was first concentrated by evaporation using water bath at 100°C for 1 h.

Test microorganisms

Clinical isolates of *Salmonella paratyphi, Bacillus subtilis, Klebsiella pneumoniae, Candida albicans, Candida paralopsis* and *Candida tropicalis* used in this study were obtained from Pharmaceutical Microbiology Laboratory of the University of Nigeria, Nsukka.

Sensitivity tests

The antimicrobial test was carried out using the agar diffusion method described by Opara and Anasa (1993). Nutrient agar and Sabouraund dextrose agar (SDA) were prepared for the sensitivity tests. The test organisms were inoculated on nutrient agar plates for bacteria and SDA for fungi, and spread uniformly using a sterile glass spreader. Cavities of 1 cm diameter were made on the nutrient agar using a sterile cork borer. The cut agar disks were removed with forceps sterilized by flaming. To each of the cavities was introduced 0.1 ml of the plant extract. One of the cavities was filled with 0.1 ml of absolute ethanol to serve as control. The plates were allowed to stand for 1 h on the bench for diffusion to occur before the growth of the organisms commenced and incubated at 37°C for 24 to 48 h. The zones of inhibition were then observed and recorded.

Determination of minimum inhibitory concentration (MIC) of the bulb extract on the test organisms

The minimum inhibitory concentration of the extract was determined according to the technique of Baron and Finegold (1990). Standardized suspensions of the test organisms were inoculated into a series of sterile Petri dishes of nutrient broth containing dilutions of the bulb extract (100, 50, 25, 12.5, 6.25, 3.125 and 1.563 mg/ml) and incubated at 37°C for 24 h. The MICs were read as the least concentration that inhibited the growth of the test organisms.

Phytochemical screening

The phytochemical screening of the garlic bulb extract was carried out according to the procedures described by Trease and Evans (1989).

RESULTS

The crude extract of *A. sativum* bulb showed positive results for alkaloids, glycosides, saponins, flavonoids, steroids, proteins, carbohydrates, oils, reducing sugars and acidic compounds (Table 1). The extract of the plant material was, however devoid of tannins, resins and terpenoids. Carbohydrates, glycosides and proteins occurred in high concentrations. Alkaloids, saponins, steroids, reducing sugars and oils were present in medium concentrations, while flavonoids and acidic compounds had low concentrations (Table 1).

The results of the sensitivity tests showed that all the six micro-organisms exposed to the methanolic extract of garlic bulb were sensitive to the plant extract. The micro-organisms studied had obvious differences in their susceptibility to garlic bulb extract. *C. tropicalis* showed the highest inhibition zone diameter (30 mm), followed by *C. albicans* (29 mm), while *S. paratyphi* showed the lowest (15 mm) inhibition zone diameter (Table 2).

Table 3 shows the MIC of the garlic bulb extract on the test isolates. The MIC results for the test organisms were *S. paratyphi* (50 mg/ml), *B. subtilis* (25 mg/ml), *K. pneumoniae* (100 mg/ml), *C. albicans* (12.5 mg/ml), *C.*

Table 2. Inhibition zone diameter produced by garlic bulb extract against the test organisms.

Test organism	Inhibition zone diameter (IZD) (mm)
Salmonella paratyphi	15
Bacillus subtilis	16
Klebsilla pneumoniae	25
Candida albicans	29
Candida paralopsis	21.8
Candida tropicalis	30

Table 3. Minimum inhibitory concentration (MIC) exhibited by the bulb extract against the test isolates.

Test organism	Zone of inhibition (mm)						
	100 mg/ml	50 mg/ml	25 mg/ml	12.5 mg/ml	6.25 mg/ml	3.125 mg/ml	1.563 mg/ml
S. paratyphi	+	10	+	+	+	+	+
B. subtilis	+	13.5	10	+	+	+	+
K. pneumoniae	25	+	+	+	+	+	+
C. albicans	+	19	16	14	+	+	+
C. paralopsis	+	25	23	21	13	11	+
C. tropicalis	+	20	16	13	11	+	+

+ = No inhibition.

paralopsis (3.125 mg/ml) and *C. tropicalis* (6.25 mg/ml).

DISCUSSION

Phytochemical screening of the garlic bulb extract revealed the presence of alkaloids, glycosides, saponins, flavonoids, steroids, proteins, carbohydrates, oils, reducing sugars and acidic compounds. These bioactive principles are believed to be responsible for the observed antimicrobial effect of the plant extract. Several workers have attributed the antimicrobial effect of plant extract to the presence of these secondary plant metabolites (Nweze et al., 2004; Gandhiraja et al., 2009). Phytochemical analysis of leaf extracts of *Ocimum gratissmum* revealed the presence of alkaloids, cardiac glycosides, flavonoids, glycosides, resins, steroidal terpenes and tannins (Mbata and Saika, 2008). The presence of active principles in the bulb extract of *A. sativum* could be used to establish a good support for the use of the plant in herbal medicine.

In this study, the six microorganisms studied were sensitive to the plant extract. The plant extract showed varying degrees of antimicrobial activity on the micro organisms. This is in agreement with the work of Rojas et al. (2006) on ten medicinal plants. The inhibitory activities of the plant extract agree with the report of Leven et al. (1979). Nwadiaro and Nwachukwu (2007) linked the antimicrobial activities of plants to the presence of tannins, alkaloids, flavoinoids and saponins. It has been reported that various plants that are rich in alkaloids, tannins and glycosides possess antimicrobial activity against a number of micro organisms (Adebajo et al., 1983). This suggests that the plant extract used in this study has a general antimicrobial activity. The antifungal potential of *A. sativum* bulb extract is more pronounced than its antibacterial properties. This is in consonance with the report of Nantitanon et al. (2007) on *Hyptis sauveolens*.

The various sizes of inhibition zones obtained in this study is an indication of the potency of the bioactive principles contained in the plant extract. The varied MICs of the plant extract on the test organisms correlates with the report indicating that micro organisms exhibits varied level of susceptibility to plant extracts. The MIC result in this study agrees with Banso and Adeyemo (2007) showing that antimicrobial agent with low activity against an organism has a high MIC, while a highly active antimicrobial agent gives a low MIC. The present study therefore shows that *A. sativum* bulb extract has useful antimicrobial properties. Further work is needed to isolate the active principles from the plant in order to test the specific antimicrobial activity of the respective phytochemical components.

REFERENCES

Adebajo, AO Adewumi, CO and Essien, EE (1983). Antiinfective agent of higher plants. International Symposium of Medium Plants. 5th ed. University of Ife, Nigeria. Pp. 152-158.

Awonubi MO (1988). Use of Local Herbs for Curing Diseases. Pharma. Herbal Med. 4(2):26 -27.

Banso A, Adeyemo SO (2007). Phytochemical and Antimicrobial

Evaluation of Ethanolic extract of *Dracaena mannii* Bark. Nig. J. Biotech. 18(1-2):27-32.

Banso A, Olutimayin T (2001). Phytochemical and Antimicrobial Evaluation of Aqueous extracts of *Daniella oliveri* and *Nauclea latifolia.* Nig. J. Biotechnol. 12(1):114-118.

Baron JE, Finegold SM (1990). Method for Testing Antimicrobial Effectiveness. In: Bailey Scotts Diagnostic Microbiol. Mosby, C.V. (ed.), Missouri, pp. 171-194.

Falodun A, Okunrobo LO, Uzoamaka N (2006). Phytochemical Screening and Anti-Inflamatory Evaluation of Methanolic and Aqueous extracts of *Euphorbia heterophylla* Lim. Afr. J. Biotechnol. 5(6):259-531.

Fluck H (1973). Medicinal Plants and their uses. W. feulsham and Coy. Ltd. New York, pp. 7-15.

Gandhiraja N, Sriram S Meenaa V, Srilakshini JK, Sasikumar C, Rajeswari R (2009). Phytochemical screening and antimicrobial activity of the plant extracts of *Mimosa pudica L.,* against selected microbes. Ethnobot. Leaflets 13:618-624.

Harbone NV (1994). Phytochemical methods. A guide to Modern Techniques of Plant analysis. 2nd Ed. Chapman and Hall, London, p. 425.

Leven M, Vanden-Berghe DA, Marten I, Vilientnick A, Lomweas EC (1979). Screening of higher plants for biological activity. Planta Medica 36:311-312.

Mbata TI, Saika A (2008). Antimicrobial activity and Phytochemical Screening of Crude ethanolic extract of leaves of *Ocimum gratissimum L.* on *Listeria monocytogenes.* Internet J. Microbiol. 4(2):1-13.

Milner JA (2005). Garlic *(Allium sativum). In:* Encyclopedia of Dietary Supplements. Black, M. and Cragg, G. (Eds.). Marcel Dekker, New York, pp. 229-240.

Nantitanon W, Chowwanapoonpohn S, Okonogi S (2007). Antioxidant and Antimicrobial activities of *Hyptis sauveolens* essential oil. Sci. Pharma. 75:35-46.

Nwadiaro PO, Nwachukwu I (2007). Inhibition of Pathogenic Fungi by ethnobotanical extracts of *Cymbopogon citrates, Ceiba pentandra* and *Loranthus bengwelensis.* Nig. J. Biotechnol. 18(1-2):61-65.

Nweze EI, Okafor JI, Njoku O (2004). Antimicrobial Activities of Methanolic Extracts of *Trema guineensis* and *Morinda lucida* used in Nigeria Herbal Medicinal Practice. J. Biol. Res. Biotechnol. 2(1):39-46.

Opara AA, Ansa MA (1993). The Antimicrobial activity of tea and Coffee on selected Organisms. J. Med. Lab. Sci. 3:45-48.

Rojas JJ, Ochoa VJ, Ocampo SA, Munoz JF (2006). Screening for Antimicrobial activity of ten Medicinal Plants in Colombian folkloric Medicines: A possible alternative in the treatment of non-nosocomial infections. Comp. Altern. Med. 6:2-12.

Soforowa A (1993). Medicinal Plants and Traditional Medicine in Africa. Chichester John, Willey and Sons, New York, p. 256.

Trease GE, Evans WC (1989). Pharmacognosy. 13th Edition. Bailliere Tindall Ltd., London, pp. 350-389.

UNESCO (1998). FIT/504 – RAF - 48. Terminal Report: Promotion of ethnobotany and the sustainable use of Plant Resources in Africa. Paris, p. 60.

Evaluation of growth performance and nutritional quality of diets using digestive enzyme markers and *in vitro* digestibility in Siamese fighting fish (*Betta splendens* Regan, 1910)

Karun Thongprajukaew[1,2,3], Uthaiwan Kovitvadhi[2,4]*, Satit Kovitvadhi[5], Arunee Engkagul[2,6] and Krisna Rungruangsak-Torrissen[2,7]*

[1]Interdisciplinary Graduate Program in Bioscience, Faculty of Science, Kasetsart University, Bangkok 10900, Thailand.
[2]Biochemical Research Unit for Feed Utilization Assessment, Faculty of Science, Kasetsart University, Bangkok 10900, Thailand.
[3]Department of Applied Science, Faculty of Science, Prince of Songkla University, Songkhla 90112, Thailand.
[4]Department of Zoology, Faculty of Science, Kasetsart University, Bangkok 10900, Thailand.
[5]Department of Agriculture, Faculty of Science and Technology, Bansomdejchaopraya Rajabhat University, Bangkok 10600, Thailand.
[6]Department of Biochemistry, Faculty of Science, Kasetsart University, Bangkok 10900, Thailand.
[7]Institute of Marine Research, Ecosystem Processes Research Group, Matre Research Station, N-5984 Matredal, Norway.

Digestive enzymes and their effects on *in vitro* digestibility of feeds and feedstuffs, as well as on growth performance quality were studied in Siamese fighting fish (*Betta splendens* Regan, 1910). The specific activities of total protease, amylase, trypsin and chymotrypsin increased during development and were higher in females than in males at maturation ($P < 0.05$). The activity ratio of trypsin to chymotrypsin corresponded to fish growth, and showed lower values in females than in males. White muscle levels of RNA decreased during development, with females having higher levels than males, while the protein levels increased with no difference between sexes. In the oocytes, trypsin-like and chymotrypsin-like specific activities were very low, and the concentrations of RNA, protein and protein/lipid ratio were higher than in the muscle. For *in vitro* digestibility, wheat gluten, soybean meal and fish meal were among good protein sources while the meals from peanut, fish and soybean were good carbohydrate sources. Golden apple snail meat was a good source for both protein and carbohydrate. The crude enzyme extracts from different growth stages and sexes had different abilities to digest the same feeds and feedstuffs. This will make it possible to preliminarily study the authenticated nutritional quality of raw materials for future feed formulations for *B. splendens*.

Key words: *Betta splendens*, digestive enzymes, feedstuff, *in vitro* digestibility, muscle quality, oocyte quality, Siamese fighting fish.

INTRODUCTION

The Siamese fighting fish (*Betta splendens* Regan, 1910), widely distributed throughout Southeast Asia, is one of the most popular species for freshwater aquarium. There are two domesticated forms; ornamental fish with long fins, and sport fish with short and rounded fins for improved fighting ability (Meejui et al., 2005). Long-finned males are very important economically, providing the highest income among the exported ornamental fishes of Thailand (Wiwatchaisaet, 2000). The development of

*Corresponding author. E-mail: fsciutk@ku.ac.th, Krisnart@imr.

economical and palatable dry feeds with optimized nutrient content is essential. However, there is no report on the nutrient requirements and their digestibilities for this species.

Studies on the main digestive enzymes have demonstrated changes in isoforms and activity profiles at different developmental stages of Siamese fighting fish, based on normal feeding habits (Thongprajukaew 2010). The nutritional requirements of the fish should be closely linked to the digestive physiology and growth during development, and protein digestion is the key factor for food utilization and growth (Rungruangsak-Torrissen et al., 2006). Trypsin specific activity and the protease activity ratio of trypsin to chymotrypsin (T/C ratio) have been used as enzymatic markers for growth rate and feed efficiency in association with amino acid absorption and transport (Sunde et al., 2004). Therefore, digestion of protein should rely primarily on the presence of pancreatic proteases such as trypsin (associated with growth) and chymotrypsin (associated with limited or reduced growth), as illustrated by Rungruangsak-Torrissen et al. (2006). This knowledge of the role of the key enzyme trypsin has made it possible to compare the nutritional qualities of feeds and feed raw materials for different growth stages, using *in vitro* protein digestibility technique standardized by trypsin activity which has been studied in relation to growth (*in vivo* trials) in different fish species (Rungruangsak-Torrissen, 2007). For carbohydrate, it has been investigated in herbivore (freshwater mussel) and shown that among different nutrient digestibilities; carbohydrate is the secondary factor after protein for nutritional quality of feeds (Areekijseree et al., 2006). *In vitro* lipid digestibility has not been shown to be associated with feed efficiency and the quality of the lipid itself (Areekijseree et al., 2006). Growth is not only associated with protein digestive efficiency, but also with muscle (filet) and gamete qualities. Changes in protein metabolism in the muscle and oocytes have been investigated in different species, and it was found that the levels of RNA and protein in these tissues are affected by different feedings. In addition, the level of trypsin-like specific activity in oocytes is associated with maturation rate; whereas females with higher maturation rates showed lower oocyte trypsin-like specific activity.

The objectives of this study were to examine whether changes in the levels of the main digestive enzymes (total protease, amylase, trypsin, and chymotrypsin) could affect the digestive efficiency and growth performance quality at different developmental stages and of different sexes of Siamese fighting fish feeding on natural foods. Analysis was conducted using: (1) the T/C ratio for indication of digestive efficiency and growth; (2) the activity ratio of amylase to trypsin (A/T ratio) as an indicator for carnivorous feeding habits (Hofer and Schiemer, 1981); (3) the levels of RNA, RNA/protein ratio, and protein/lipid ratio in the muscle and oocytes, as well as oocyte trypsin-like specific activity, for indications of a quality of growth performance; and (4) *in vitro* digestibility of feeds and feed raw materials for evaluation of protein and carbohydrate utilizations, a method which is simple, economical, and less time-consuming than *in vitro* growth trials. The results from *in vitro* digestibility will provide a basic knowledge for selecting appropriate feed raw materials for future feed formulations, whereas their effects on growth performance quality (including digestive enzyme expressions and qualities of muscle and oocytes) of Siamese fighting fish can be further studied.

MATERIALS AND METHODS

Fish rearing and preparation

Betta splendens were randomly collected at three different stages: 10 days old (completely developed digestive tract and having predatory behavior); 1.5 months old (sexually identified for individual rearing); and 3 months old (maturation stage). The experiment was performed in a completely randomized design (CRD) at a Thai fish farm in Nakorn Pathom province, which is a main production area in Thailand of fish for export. The experiment began in three cement tanks (75 cm diameter × 35 cm height). In each tank, 3 days old newly hatched larvae were pooled from five females (~400 larvae per female) and reared in 25 L of water. The water volume was increased every few days, up to 150 L within one week when the fish were 10 days old after hatching. The fish were reared at the same water volume until they were 1.5 months old. They were then reared individually in 200 mL glass bottles until maturation at 3 months old. The fish were cultured under natural light regime at an ambient temperature range of 27 to 29°C, pH 7.0 to 7.2, and 5.0 to 5.3 mg/L of dissolved oxygen. During the first week of feeding, the larvae were fed with cooked chicken egg yolk dissolved in water, followed by paramecium and then water flea (*Moina* sp.), twice daily. During 10 days old to 1.5 months old, the fish were fed with water flea; after that they were fed with mosquito larvae until maturation at 3 months old. They were starved for 2 h prior to sampling. Both males and females were studied at 1.5 and 3 months old. Females were identified by visual observation of an ovipositor, appearing as a small white spot in the anal region. At each stage and for each sex, 30 individuals were measured for average body weight and total length; five pooled samples (four fish per sample) were used for digestive enzymes study; and 20 pooled samples were used for white muscle and oocyte studies. In addition, three pooled samples (100 fish per sample) from each sex at 1.5 and 3 months old were used for determining *in vitro* digestibility.

Animal ethics

The use of Siamese fighting fish as an animal model was performed in accordance with the "Ethical Principles and Guidelines for the Use of Animals for Scientific Purposes", National Research Council, Thailand. The fish were sacrificed by chilling in ice before the white muscle, oocytes and digestive tracts were carefully collected.

Digestive enzyme study

Enzyme extraction

Enzyme extracts from *B. splendens* were prepared from the whole body of 10 days old juveniles, from the digestive area of 1.5 months old fish, and from the digestive tracts of 3 months old adults. The procedure was performed according to Rungruangsak-Torrissen (2007). Briefly, the samples were homogenized on ice in 50 mM Tris-HCl buffer pH 8 containing 200 mM NaCl (1:1 w/v). The homogenate was centrifuged at 4°C at 10000 × g for 20 min; then the supernatant was collected and kept at -80°C for later determination. Total protein content of each crude enzyme extract was determined according to Lowry et al. (1951).

Protease specific activity

Total protease activity was assayed by measuring the increase in cleaved short-chain polypeptides using azocasein as substrate, according to the study of Areekijseree et al. (2004). The 100 mM phosphate buffer pH 8 and temperature 50°C were chosen as the most suitable condition for total protease activity in Siamese fighting fish (Thongprajukaew, 2010). Total protease specific activity was expressed as the increase in absorbance at 440 nm h^{-1} mg $protein^{-1}$.

Amylase specific activity

Amylase activity was determined by measuring the increase in reducing sugar from starch solution using 3,5-dinitrosalicylic acid (DNS), according to the study of Areekijseree et al. (2004). The 100 mM phosphate buffer pH 8 and temperature 50°C were chosen as the most suitable condition for amylase activity in Siamese fighting fish (Thongprajukaew, 2010). The amylase specific activity was expressed as μmol maltose produced h^{-1} mg $protein^{-1}$.

Trypsin and chymotrypsin specific activities

Trypsin and chymotrypsin activities were determined by initial reaction rates at optimal temperatures according to the method described by Rungruangsak-Torrissen (2007), using BAPNA (*N*-*α*-Benzoyl-*DL*-arginine-*p*-nitroanilide HCl) and SAPNA (*N*-succinyl-*L*-ala-*L*-ala-*L*-pro-*L*-phe-*p*-nitroanilide) as specific substrates, respectively. The 100 mM phosphate buffer pH 8 and temperature 50°C were chosen as the most suitable condition for trypsin and chymotrypsin activities in Siamese fighting fish (Thongprajukaew, 2010). Both trypsin and chymotrypsin specific activities were expressed as μmol *p*-nitroanilide produced h^{-1} mg $protein^{-1}$.

Muscle and oocyte qualities

RNA and protein concentrations in white muscle and oocyte samples were determined as described in Rungruangsak-Torrissen (2007). The samples were extracted with monophasic solution of phenol and guanidine isothiocyanate (TRIzol® reagent; Invitrogen, Carlsbad CA, USA). The extinction coefficients for calculating RNA and protein concentrations are $E260$ = 40 μg RNA $ml^{-1}$and $E280$ = 2.1 mg protein ml^{-1}. For lipid determination, the samples were dried, ground and then extracted using ethyl acetate as described by Rungruangsak-Torrissen et al. (2009). Protein synthesis capacity and protein growth were expressed as RNA/protein ratio and protein/lipid (P/L) ratio, respectively. All values were expressed on wet weight basis.

Oocyte maturation

Enzyme extracts from the oocytes were obtained as described in enzyme extraction. Trypsin-like and chymotrypsin-like activities were determined by initial reaction rates at optimal temperatures using BAPNA and SAPNA as specific substrates, respectively according to the method described by Rungruangsak-Torrissen (2007). The pH and temperature profiles were studied to determine optimal conditions for further investigation. The pH profiles were performed at ambient temperature using 100 mM buffers of various pHs (6 to 11). The buffers used were phosphate buffer for the pH range 6 to 8, $NaHCO_3$-Na_2CO_3 buffer for the pH range 9 to 10, and Na_2HPO_4-NaOH for pH 11. For the temperature profile study, the reaction mixture was performed at optimal pH and at various temperatures (20 to 80°C). Both trypsin-like and chymotrypsin-like specific activities were expressed as μmol *p*-nitroanilide produced h^{-1} mg $protein^{-1}$.

Biochemical composition of diets

Live diets, commercial feeds with specific formulations for Siamese fighting fish at juveniles and mature stages, and feedstuffs were used. They were dried for 24 h at 105°C before analyses of other biochemical compositions, such as protein, lipid, fiber, and ash, as described by the AOAC (2005). The values were expressed on dry matter basis. Carbohydrate values or nitrogen free extract (NFE) were calculated by the difference.

In vitro digestibility

Protein and carbohydrate digestibilities of different feeds and feedstuffs using fish crude enzyme extracts standardized by enzyme activity were determined using the method described by Areekijseree et al. (2006). The feeds and feedstuffs were gently dried at 60°C for 48 h to control moisture level, and then ground. Blanks without crude enzyme extracts were used as controls to measure the levels of amino acids and sugar liberated from the feed ingredients. Protein digestibility (based on trypsin activity) was determined by measuring the increase in liberated reactive amino groups of cleaved peptides. *In vitro* protein digestibility was expressed as μmol *DL*-alanine equivalent g dried $feed^{-1}$ trypsin $activity^{-1}$. Carbohydrate digestibility (based on amylase activity) was determined by measuring the increase in reducing sugar. *In vitro* carbohydrate digestibility was expressed as μmol maltose g dried $feed^{-1}$ amylase $activity^{-1}$.

Statistical analysis

All analyses were performed using SPSS version 14 software (SPSS Inc., Chicago, USA). Mean and standard error of mean were calculated throughout. Statistical analysis at 95% significance level was determined using one-way ANOVA, and multiple comparisons were analyzed using Duncan's multiple range test (DMRT).

RESULTS

Fish growth

Growths of *B. splendens* were higher during the first 1.5 months (6 folds on average) compared to the last 1.5 months (2 folds on average), regardless of sex (Table 1). Males grew faster than females, indicating by higher total length and body weight at the same stage. During the last 1.5 months, both total length and body weight of males increased 1.8 and 2.9 folds, while in females they increased 1.3 and 2 folds, respectively (Table 1). Almost entirely female samples (> 90%) at 3 months old were mature.

Development of digestive enzymes

Specific activities of total protease, amylase, trypsin and chymotrypsin increased with age, and showed significant differences between sexes ($P < 0.05$) only at maturation (Table 1). Regression analyses between these enzyme specific activities and growth parameters (total length and body weight) indicated significantly positive relationships during development in both males and females ($P < 0.05$) (Table 2). Regardless of sex, these significant relationships were also observed, except for between total protease specific activity and total length and between chymotrypsin specific activity and all growth parameters (Table 2). According to the correlation coefficient values, females seemed to show more variations in relation to total length and less variation in relation to body weight, compared to males (Table 2). Interestingly, while the enzyme specific activities in females were similar to or higher than in males, the protease activity ratios of trypsin to chymotrypsin (T/C ratio) in females were lower than in males, and became significant at maturation in accordance with fish growth parameters (Table 1). The correlations between T/C ratio and growth parameters were negative and significant relationships were only observed in females (Table 2). The feeding marker of the activity ratio of amylase to trypsin (A/T ratio) fluctuated during development, and was not statistically different ($P > 0.05$) between sexes (Table 1). It also showed negative relationships with growth parameters during development, albeit insignificant (Table 2).

Muscle quality during development

The levels of RNA and RNA/protein ratio in the white muscle decreased with age, and females had significantly higher levels than males ($P < 0.05$) at all studied stages (Table 1). On the other hand, protein and lipid levels in the white muscle increased with age, and the levels were similar between sexes, except at maturation when significantly higher lipid contents ($P < 0.05$) were observed in females (Table 1). This resulted in significantly lower ratio of protein to lipid (P/L ratio) in the white muscle of females at maturation (Table 1). RNA concentrations were negatively correlated with trypsin expression ($r = -0.79$, $P < 0.01$), total length ($r = -0.83$, $P < 0.01$) and body weight ($r = -0.81$, $P < 0.01$), regardless of sex. A positive relationship between RNA concentration and T/C ratio was observed in females ($r = 0.73$, $P < 0.05$). Protein contents were positively correlated with trypsin expression ($r = 0.80$, $P < 0.01$), total length ($r = 0.82$, $P < 0.01$), and body weight ($r = 0.82$, $P < 0.01$), regardless of sex. In contrast, a negative relationship between protein content and T/C ratio was observed in females ($r = -0.73$, $P < 0.01$). Similar to RNA, RNA/ protein ratios were negatively correlated with trypsin expression ($r = -0.79$, $P < 0.01$), total length ($r = -0.82$, $P < 0.01$) and body weight ($r = -0.72$, $P < 0.01$). Interestingly, only the T/C ratio showed positive correlation with the muscle RNA/protein ratio ($r = 0.63$, $P < 0.01$), regardless of sex.

Oocyte quality

Characterization of trypsin-like and chymotrypsin-like enzymes in the oocytes

Oocyte trypsin-like specific activity showed optimum pH 10 (Figure 1A) where at least two optimal activities at temperature ranges of 20 to 30°C and 50 to 70°C were observed (Figure 1B). Similarly, chymotrypsin-like specific activity exhibited optimum pH 10 (Figure 2A) with at least two activity peaks at temperature ranges of 20 to 30°C and 70 to 80°C (Figure 2B). According to the characteristics observed, the optimum pH 10 and temperature 20°C were chosen for studying trypsin-like and chymotrypsin-like activities in the oocytes of Siamese fighting fish.

Oocyte quality at maturation

The levels of trypsin-like and chymotrypsin-like specific activities in the oocytes were rather low (Table 1). The concentrations of RNA (2.3 folds), protein (3.1 folds) and lipid (1.3 folds) were higher in the oocytes than in the white muscle (Table 1). These resulted in higher P/L ratio (2.3 folds) and lower RNA/protein ratio (0.8 fold) in the oocytes, compared to the white muscle (Table 1).

In vitro digestibility during development

The feeds and feedstuffs chosen for the *in vitro* digestibility study and their biochemical compositions are shown in Table 3. The enzyme extracts obtained from the different life stages of Siamese fighting fish showed different abilities to digest the same materials, and the digestion ability decreased with age for both protein (Figure 3) and carbohydrate (Figure 4). Moreover, differences in digestion between males and females were observed (Figures 3 and 4), especially in carbohydrate digestion (Figure 4).

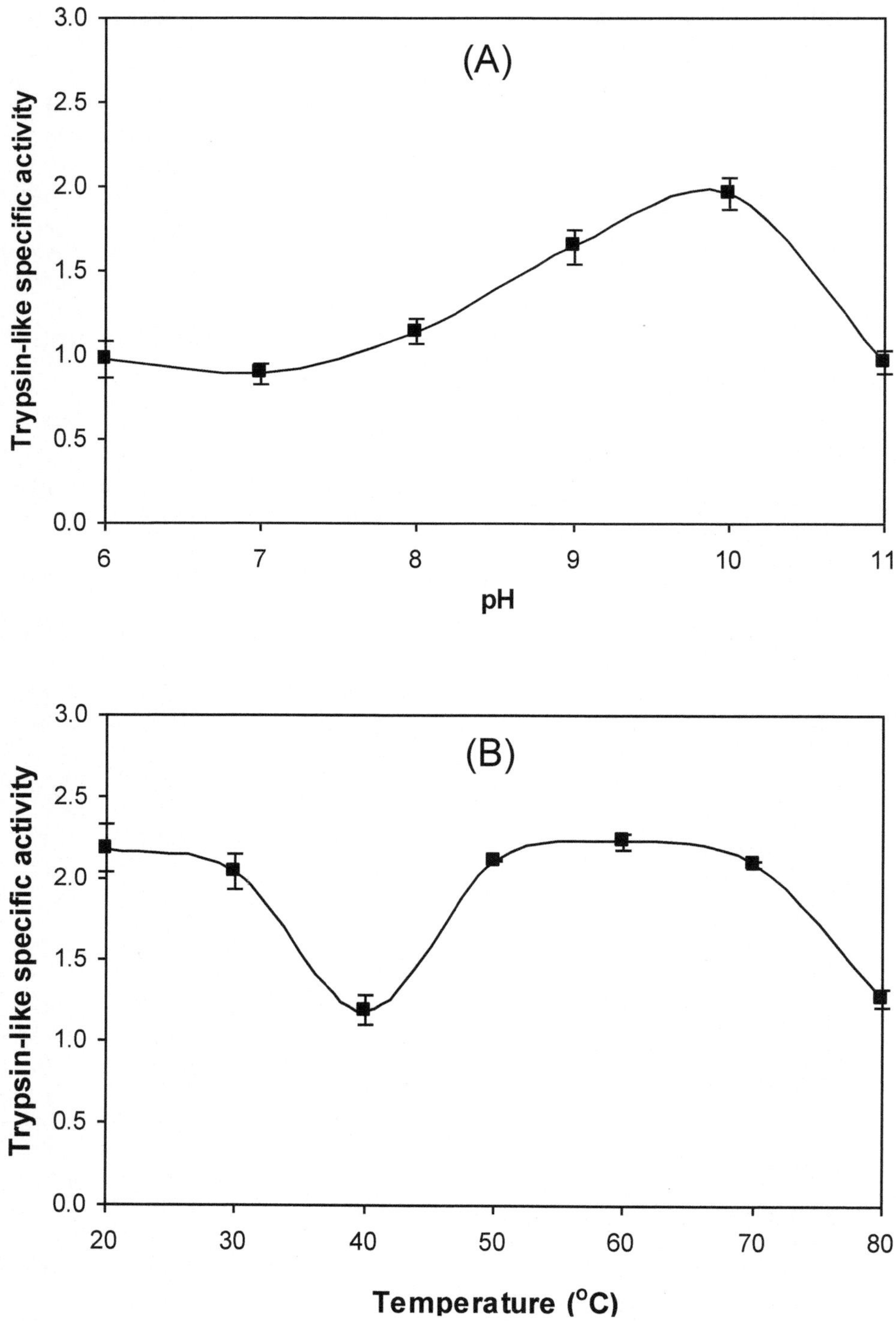

Figure 1. Trypsin-like specific activity (μmol *p*-nitroanilide produced h^{-1} mg $protein^{-1}$) in the oocytes of 3 months old females ($n = 3$), pH profile at ambient temperature (A) and temperature profile at pH 10 (B).

In vitro protein digestibility

Among the selected feeds and feed raw materials, wheat gluten had the highest protein digestibility at all three representative stages ($P < 0.05$) (Figure 3). Interestingly, golden apple snail (*Pomacea canaliculata*) meat could be an alternative source of protein for 10 days old juveniles, as it had a relatively high protein digestibility (Figure 3A). In addition, ranking from high to low digestibility, the meals from soybean, fish, meat and bone, peanut, and corn gluten, which are common ingredients in feed formulations, also showed high protein digestibility (Figure

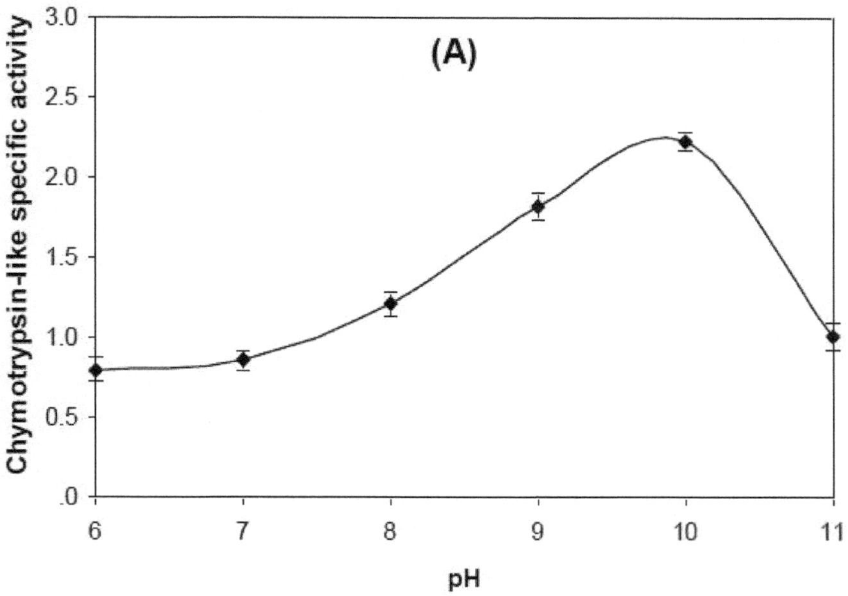

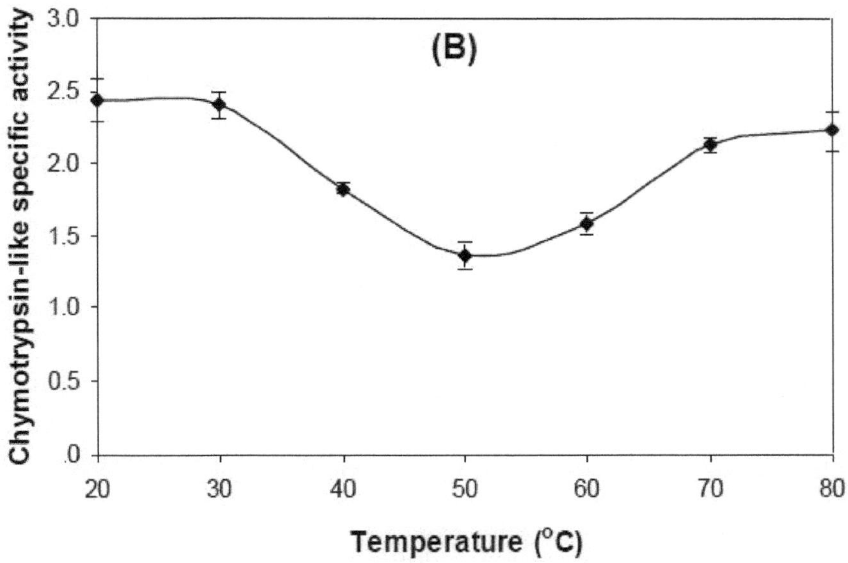

Figure 2. Chymotrypsin-like specific activity (μmol *p*-nitroanilide produced h^{-1} mg $protein^{-1}$) in the oocytes of 3 months old females (n = 3), pH profile at ambient temperature (A) and temperature profile at pH 10 (B).

3A). The protein sources suitable for 1.5 months old fish (Figure 3B) were similar to 10 days old juveniles (Figure 3A), regardless of sex. Females showed significantly higher protein digestion than males ($P < 0.05$) for *Spirulina* sp. meal, commercial feed formulation 2, blood meal, and white sorghum (Figure 3B). Similar suitable protein sources were also observed for 3 months old adults, but with no difference in protein digestion between sexes (Figure 3C). Blood meal could also be an alternative protein source at maturation stage (Figure 3C). Surprisingly, natural diets (water flea, mosquito larva, and chicken and duck egg yolks) had lower protein digestibility than golden apple snail meat, the main feedstuffs, and the commercial feeds (Figure 3).

In vitro carbohydrate digestibility

The digestibility of carbohydrate using the enzyme extracts exhibited low levels of digested products from

Table 1. Biological parameters at different stages of Siamese fighting fish. The values with asterisk (*) indicated significant differences between males and females in each stage (P < 0.05).

Parameter	10 days old	1.5 months old		3 months old		Pooled SEM
		Male	Female	Male	Female	
Growth parameters (n = 30)						
Length (mm)	6.63	30.31	27.33	53.79*	34.49*	0.20
Weight (mg)	48.75	413.33*	329.00*	1,190.48*	648.57*	8.26
Enzyme parameters (n = 5)						
[a]Total protease specific activity	0.30	3.19	3.18	5.42*	13.01*	0.29
[b]Amylase specific activity [A]	18.52	67.39	72.93	136.41*	169.76*	0.99
[c]Trypsin specific activity [T]	2.10	11.29	10.72	14.42*	22.40*	0.54
[c]Chymotrypsin specific activity [C]	1.79	13.89	14.67	14.41*	34.48*	0.29
T/C ratio	1.17	0.81	0.73	0.96*	0.69*	0.01
A/T ratio	9.58	6.03	6.81	9.11	7.41	0.26
White muscle parameters (n = 20)						
RNA (μg g^{-1})	6,198	3,607*	4,407*	1,830*	2,223*	33
Protein (mg g^{-1})	33.22	70.66	67.66	149.97	134.09	1.68
Lipid (mg g^{-1})	16.19	13.42	14.89	32.12*	42.58*	0.32
RNA/protein ratio (μg g^{-1})	186.74	48.52*	67.21*	12.35*	16.67*	0.75
Protein/lipid ratio (mg mg^{-1})	2.07	5.29	4.41	4.60*	3.33*	0.07
Oocyte parameters (n = 20)						
[c]Trypsin-like specific activity	-	-	-	-	2.18	0.06
[c]Chymotrypsin-like specific activity	-	-	-	-	2.44	0.06
RNA (μg g^{-1})	-	-	-	-	5,145	172
Protein (mg g^{-1})	-	-	-	-	412	23.07
Lipid (mg g^{-1})	-	-	-	-	55.64	1.58
RNA/protein ratio (μg g^{-1})	-	-	-	-	13.20	0.63
Protein/lipid ratio (mg mg^{-1})	-	-	-	-	7.65	0.40

[a] Expressed as increase in absorbance at 440 nm h^{-1} mg $protein^{-1}$, [b] Expressed as μmol maltose produced h^{-1} mg $protein^{-1}$, [c] Expressed as μmol *p*-nitroanilide produced h^{-1} mg $protein^{-1}$.

the selected feeds and feed raw materials (Figure 4). This could be affected by sex and developmental stages (Figure 4). The highest digestibility of carbohydrate was found in peanut meal for 10 days old juveniles (Figure 4A). In addition, white sorghum, water flea, corn, tapioca chip, wheat bran, and mosquito larva (ranked from higher to lower digestibility) were among the highly digestible carbohydrate sources for this early stage (Figure 4A). Males at 1.5 months old showed high carbohydrate digestibility for commercial feed formulation 5 > peanut meal > commercial feed formulation 6 > fish meal > soybean meal, with higher digestibility than females of the same age (Figure 4B). An indication that mosquito larvae are the ideal carbohydrate source for females during growth was observed, as they showed the highest carbohydrate digestibility by enzyme extracts from females with significantly higher (P < 0.05) than male enzyme extracts (Figure 4B). For fish at maturation, males showed high carbohydrate digestibility for peanut meal > golden apple snail meat > commercial feed formulation 2 > commercial feed formulation 5 > fish meal 58 > commercial feed formulation 8 > shrimp shell (Figure 4C). Females at maturation showed high carbohydrate digestibility for peanut meal > fish meal 58 > commercial feed formulation 8 > commercial feed formulation 5 > soybean meal > commercial feed formulation 2 (Figure 4C). At maturation, females exhibited higher carbohydrate digestibility than males for most of the feeds and feedstuffs selected (Figure 4C).

DISCUSSION

Specific activities of total protease, trypsin and chymotrypsin increased with age in Siamese fighting fish, and they were significantly different between sexes at maturation (P < 0.05) (Table 1). Proteolytic activity has been early reported to play an important role in carnivorous and omnivorous species (Hofer and Schiemer, 1981). Among acidic and alkaline proteases,

Table 2. Regression analysis (n = 15) between growth parameters (total length and body weight) and expression of each enzyme parameter in Siamese fighting fish during development of males, females, and regardless of sex. Significant correlation coefficients (r) are shown by the P values < 0.05 (**bold values**).

Enzymes	Sex	Total length		Body weight	
		r	P	*r*	P
Total protease	Regardless of sex	0.44	0.054	0.49	**0.034**
	Male	0.96	**0.000**	0.92	**0.000**
	Female	0.83	**0.002**	0.98	**0.000**
Amylase (A)	Regardless of sex	0.77	**0.000**	0.78	**0.000**
	Male	1.00	**0.000**	0.99	**0.000**
	Female	0.92	**0.000**	0.99	**0.000**
Trypsin (T)	Regardless of sex	0.60	**0.007**	0.59	**0.008**
	Male	0.91	**0.000**	0.92	**0.001**
	Female	0.90	**0.000**	0.99	**0.000**
Chymotrypsin (C)	Regardless of sex	0.46	0.050	0.42	0.076
	Male	0.83	**0.001**	0.86	**0.006**
	Female	0.93	**0.000**	1.00	**0.000**
T/C ratio	Regardless of sex	-0.30	0.214	-0.16	0.501
	Male	-0.42	0.170	-0.54	0.349
	Female	-0.94	**0.000**	-0.90	**0.003**
A/T ratio	Regardless of sex	-0.04	0.879	-0.32	0.679
	Male	-0.17	0.595	-0.28	0.808
	Female	-0.60	0.067	-0.68	0.156

only the alkaline protease trypsin is the key for feed efficiency and growth (Rungruangsak-Torrissen et al., 2006), while the alkaline protease chymotrypsin is the key for limited or reduced growth (Rungruangsak-Torrissen et al., 2006; Chan et al., 2008). This has made the protease activity ratio of trypsin to chymotrypsin (T/C ratio) the key factor for digestive efficiency and growth performance in aquatic animals (Sunde et al., 2004; Rungruangsak-Torrissen, 2007; Rungruangsak-Torrissen et al., 2006, 2009).

In Siamese fighting fish, acidic proteases are dominant in early life stage while alkaline proteases are dominant in later stages (Thongprajukaew, 2010). Trypsin contributes approximately 40 to 50% of overall protein digestion activity in carnivorous species (Eshel et al., 1993), and plays a crucial role in regulating the activity of pancreatic proteases (Stryer, 1988). Amylase specific activity also increased with age in Siamese fighting fish, similar to the observations in other species by Lazo et al. (2007), and the expressions were also significantly different between sexes at maturation (P < 0.05) (Table 1). In spite of significantly higher female digestive enzymes (either for protein or carbohydrate digestions) at maturation, the T/C ratio was significantly lower in females in association with slower growth, compared to the males (P < 0.05) (Table 1). Both T/C and A/T ratios showed negative correlations with total length and body weight due to reduction in growth rate at older stage, but significant relationships were only observed in the female T/C ratios (Table 2). On the other hand, all studied digestive enzymes showed positive relationships with total length or body weight or both growth parameters, and could be affected by sexes (Table 2). The results indicated that the T/C ratio was the most reliable key parameter for growth efficiency, independent of the digestive enzyme specific activity levels, as described by Rungruangsak-Torrissen et al. (2009). The A/T ratio was not significantly correlated to growth parameters (P > 0.05) (Table 2) and not associated with the levels of carbohydrate or protein in the diets (Tables 1 and 2), similar to the observation by Gamboa-Delgado et al. (2003) but in contrast to the suggestions by Hofer and Schiemer (1981) and Hidalgo et al. (1999). Higher digestive enzyme expressions could be associated with higher consumption or higher utilization or both, but the indication of food utilization for somatic growth efficiency (regardless of protein or lipid growth) is the T/C ratio, designated as digestive efficiency (Rungruangsak-Torrissen et al., 2006, 2009).

During development, the levels of white muscle RNA

Table 3. Biochemical compositions of feeds and feedstuffs used for *in vitro* digestibility studies. The commercial feeds are specific feed formulations for Siamese fighting fish, collected from a Thai aquatic animal market. The formulations 3 and 4 are used for rearing juveniles, and the formulations 1 and 2 and 5 to 8 are used for rearing maturing fish. The analyses were performed as described by the AOAC

Feed and feedstuff	Moisture $g\ kg^{-1}$	$g\ kg^{-1}$ on dry matter basis				
		Protein	NFE	Lipid	Fiber	Ash
Natural feeds						
Pomacea canaliculata meat	808.2	660.1	177.8	26.5	2.1	133.5
Moina sp.	856.4	557.1	279.9	62.0	47.4	53.6
Mosquito larva	734.2	526.3	242.0	127.5	27.8	76.4
Chicken egg yolk	480.6	302.7	86.3	566.0	nd	45.0
Duck egg yolk	417.4	283.2	24.0	654.1	nd	38.7
Commercial feeds						
Formulation 1	81.3	618.5	110.3	131.4	13.9	125.9
Formulation 2	45.8	533.5	216.7	134.4	5.4	110.0
Formulation 3	55.9	518.1	203.4	151.1	4.4	123.0
Formulation 4	66.1	517.0	273.8	95.0	9.6	104.6
Formulation 5	71.4	512.0	333.3	44.3	10.9	99.5
Formulation 6	81.3	390.9	477.9	50.4	6.5	75.0
Formulation 7	74.8	337.4	540.8	37.9	9.1	74.8
Formulation 8	67.8	338.2	526.2	37.9	22.2	75.5
Feedstuffs						
Animal sources						
Blood meal	59.8	903.9	71.4	4.3	3.6	16.8
Fish meal 59	89.4	592.5	64.0	93.5	7.5	242.5
Fish meal 58	73.8	585.4	71.0	80.2	7.9	255.5
Meat and bone meal	46.0	400.6	163.6	85.3	2.1	348.4
Shrimp shell	732.6	491.4	49.0	23.9	161.2	274.5
Plant and algal sources						
Wheat gluten	82.7	836.6	150.0	6.5	nd	6.9
Corn gluten meal	75.1	520.7	422.3	30.8	5.2	21.0
Spirulina sp. meal	87.4	461.4	390.5	2.3	4.2	141.6
Soybean meal	99.6	395.2	455.9	34.7	41.5	72.7
Peanut meal	68.2	385.3	424.2	51.8	67.9	70.8
Coconut meal	63.0	200.9	536.7	100.3	89.3	72.8
Wheat bran	100.5	140.2	709.5	33.0	71.2	46.1
Rice bran	98.3	115.9	596.7	150.7	57.3	79.4
White sorghum	120.9	115.3	811.7	23.0	22.2	27.8
Wheat flour	120.6	111.4	874.2	3.8	3.4	7.2
Corn	109.1	74.2	841.9	45.3	25.9	12.7
Steam broken rice	104.4	68.1	903.4	13.2	1.7	13.6
Tapioca chip	92.4	27.5	904.7	5.0	15.8	47.0

NFE = nitrogen free extract, nd = not detected.

decreased while protein increased, resulting in decreasing RNA/protein ratio (Table 1), similar to the observations in other species by Mathers et al. (1993) and Peragon et al. (2001). Increased white muscle protein concentrations during successive growth have also been observed in rainbow trout, *Oncorhynchus mykiss* (Rungruangsak-Torrissen et al., 2009). This indicates that more protein is retained at older age, and *B. splendens* females had higher protein synthesis and turnover rate (RNA concentration and RNA/protein ratio)

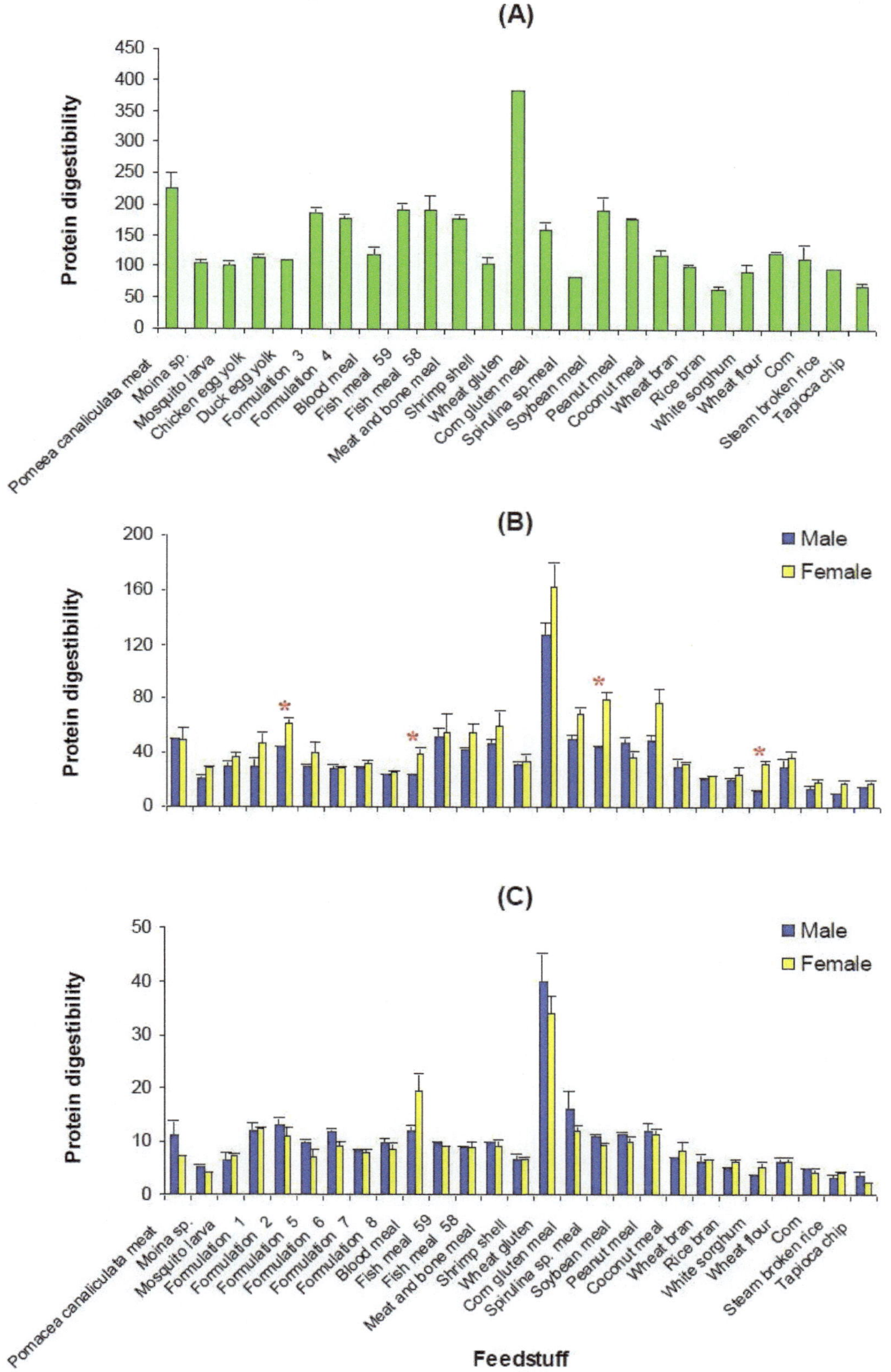

Figure 3. *In vitro* protein digestibility (μmol *DL*-alanine equivalent g dried feed^{-1} trypsin activity^{-1}) of feeds and feedstuffs using dialyzed crude enzyme extract from 10-day-old juveniles (A), males and females of 1.5-month-old (B), and males and females of 3-month-old (C). The bars with asterisks indicate significant differences between males and females (P < 0.05).

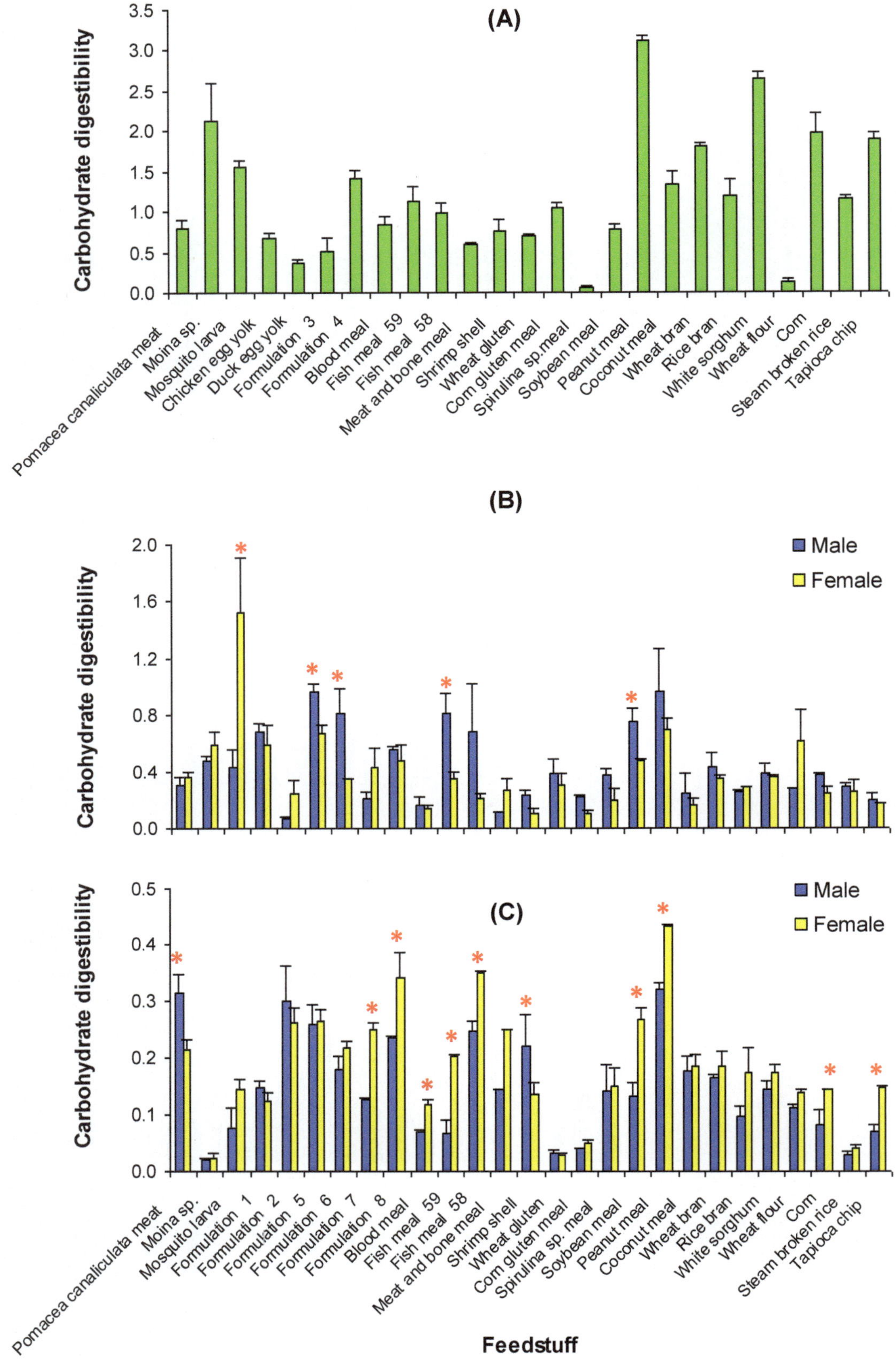

Figure 4. *In vitro* carbohydrate digestibility (μmol maltose g dried feed^{-1} amylase activity^{-1}) of feeds and feedstuffs using dialyzed crude enzyme extract from 10 days old juveniles (A), males and females of 1.5 months old (B), and males and females of 3 months old (C). The bars with asterisks indicate significant differences between males and females (P < 0.05).

than males (Table 1). White muscle protein and lipid concentrations were similar between sexes during development, except at maturation when lipid concentrations were significantly higher in females, resulting in significantly lower P/L ratio in female white muscle ($P < 0.05$) (Table 1). This indicates a higher energy requirement during maturation in females than males. Male Siamese fighting fish exhibit more aggressive and predatory behaviors than females (Jaroensutasinee and Jaroensutasinee, 2001); this may also result in a higher protein content and P/L ratio in male white muscle. At maturation, females showed a higher level of protein synthesis and turnover rate in the oocytes than in the white muscle (Table 1), as indicated by the levels of RNA or RNA/protein ratio or both. The specific activities of trypsin-like and chymotrypsin-like in the oocytes were at very low levels, similar to the levels of digestive trypsin and chymotrypsin in 10 days old juveniles (Table 1). However, these protease specific activity levels in the oocytes cannot be compared directly to those in the larval digestive tract which would increase after feeding. Sveinsdottir et al. (2006) reported increases of both protease expressions in Atlantic cod (*Gadus morhua*) embryos during segmentation (organogenesis) period, followed by a decrease to very low levels at hatching and early larval period (about 0.1 fold for trypsin-like and 0.5 fold for chymotrypsin-like specific activities, compared to non-fertilized eggs). This means that the levels of trypsin-like and chymotrypsin-like expressions in Siamese fighting fish embryos prior to hatching should be lower than the levels observed in the oocytes in Table 1. The expression levels observed in the 10 days old juveniles in Table 1 should have resulted from the new development of trypsin and chymotrypsin in the digestive system at first feeding period. The low protease specific activity of trypsin-like enzyme (probably also chymotrypsin-like enzyme) in the oocytes at maturation stage in Siamese fighting fish should be a common.

Both protein and lipid are important for egg development and newly hatched larvae. Lipid in the eggs appears to be a main metabolic fuel during egg stage and early yolk-sac stage in European sea bass (*Dicentrarchus labrax*), while protein seems to be mobilized for energy at late yolk-sac stage (Rønnestad et al., 1998). Both protein and lipid levels in Siamese fighting fish decreased after hatching, as the levels were 12.4 and 3.4 folds in the oocytes, respectively compared to 10 days old juveniles (Table 1). This indicates that the level of protein in the eggs should be relatively more important than the lipid level, resulting in significantly higher P/L ratio in the oocytes than in the 10 days old juveniles ($P < 0.05$) (Table 1). Serine proteases play an important role in yolk formation and degradation during embryogenesis (Hiramatsu et al., 2002; Sveinsdottir et al., 2006). Trypsin and chymotrypsin from the Siamese fighting fish digestive system, assayed at an optimal pH 10, showed optimal activities at temperature 30 to 35°C and 40°C, respectively (Thongprajukaew, 2010). On the other hand, both trypsin-like and chymotrypsin-like activities in the oocytes showed optimal conditions at pH 10 and 20 to 30°C, with respectively low activities at 40 and 50°C (Figures 1 and 2). The differences in temperature profiles of the two serine proteases between the oocytes (current work) and the digestive system (Thongprajukaew, 2010) in Siamese fighting fish indicate different forms of the serine proteases between these tissues. It is interes-ting to note that both trypsin and chymotrypsin in the digestive tract with an optimal pH 10 showed higher specific activities at 20 to 30°C in females at maturation, compared to immature females and males at all stages (Thongprajukaew, 2010). This probably indicates that the trypsin and chymotrypsin in the digestive system with activity characteristics at pH 10 and low tempe-ratures of 20 to 30°C (Thongprajukaew, 2010), similar to trypsin-like and chymotrypsin-like in the oocytes (current work), may influence maturation in female Siamese fighting fish. Surprisingly, the relatively high activities of the oocyte enzymes were observed at high temperatures of 70 to 80°C (Figures 1B and 2B). This phenomenon has also been observed in cod oocytes (unpublished data), and in the hepatopancreas of cuttlefish (*Sepia officinalis*), showing trypsin with optimal activity but less stability at 70°C (Balti et al., 2009).

The developmental variations in trypsin and amylase expressions has made it possible to use them to standardize *in vitro* digestibility values for protein and carbohydrate, respectively in feedstuffs for suitable future feed formulations for specific stages and sexes (Figures 3 and 4). The most appropriate protein source for all stages was wheat gluten (Figure 3) with protein content of 837 g kg^{-1} on dry matter basis (Table 3). Another interesting source for 10 days old juveniles was golden apple snail, a pest against plants in agricultural eco-systems and an intermediate host of aquatic parasites. Golden apple snail meat had highly digestible protein value (Figure 3) with protein content of 660 g kg^{-1} on dry matter basis (Table 3), and with relatively high essential amino acid index and essential fatty acid profile (Bombeo-Tuburan et al., 1995). It could be a good replacer for other animal protein sources in formulated diets because of low price, and its gonad and egg contain high content of the main carotenoids, astaxanthin and carotenoprotein (Dreon et al., 2007), which may improve survival rate of juveniles and coloration in Siamese fighting fish. Soybean meal and fish meal were the other good feed raw materials for all stages. Soybean meal is widely used as the most effective alternative for fish meal in artificial feeds, because of its well-balanced amino acid profile and reasonable price. Chou et al. (2004) reported that fish meal could be replaced up to 40% by soybean meal without causing reduction in growth and protein utilization in cobia (*Rachycentron canadum*). Sales (2008) demonstrated similarity in apparent protein digestibility values between fish meal (56 to 99%) and soybean meal

(50 to 99%).

The other suitable protein sources for the older stages were corn gluten, *Spirulina* sp. meal, and blood meal. Corn gluten meal is expensive, and caused low fiber and yellow color in salmonids (Hardy, 1996), which may also limit its use in feeds for Siamese fighting fish. *Spirulina* sp. contains high levels of essential amino acids, vitamins, and color enhancing pigments (Peirettie and Meineri, 2008), and has 461 g kg^{-1} in protein content on dry matter basis (Table 3) with moderate levels of digestible products (Figure 3B and C). These microalgae have no cell wall, resulting in improved digestion and absorption (Nandeesha et al., 1998), and *Spirulina* sp. meal has been identified as a potential protein source (Nandeesha et al., 2001). Blood meal is not an ideal main protein source because of very low isoleucine and glycine (Khawaja et al., 2007), although its digestibility values (55 to 99%) is relatively high (Sales, 2008). Moreover, during on-growing (1.5 months old), females showed significantly higher protein digestibility for *Spirulina* sp. and white sorghum than males (Figure 3B), which may indicate the presence of certain amino acids and other nutrients suitable for females in these feedstuffs.

Carbohydrate digestibility is low in carnivorous species whereas amylase expression is important, but it plays as secondary role in selecting appropriate feed ingredients having equally digestible protein. Areekijseree et al. (2006) reported protein digestibility to be the key factor for determining feed quality in an aquatic herbivore (freshwater mussel), followed by carbohydrate digestibility as secondary factor. A good alternative carbohydrate source for Siamese fighting fish rearing at all stages was peanut meal (Figure 4). White sorghum and water flea were the other alternative feed raw materials for 10 days old juveniles (Figure 4). The other suitable alternatives were fish meal and soybean meal for 1.5 months old males and adult females, mosquito larva and probably wheat flour for 1.5 months old females, and golden apple snail meat for adult males (Figure 4). Despite the lower carbohydrate content in feed raw materials from animal sources compared to plant sources (Table 2), animal feedstuffs demonstrated high *in vitro* carbohydrate digestibility (Figure 4). Using these animal protein sources will also provide good carbohydrate sources.

Siamese fighting fish at various growth stages and of different sexes had different abilities to digest the same raw materials and commercial feeds (Figures 3 and 4). Atlantic salmon with genetically different trypsin phenotypes have different abilities to digest the same fish meals (Bassompierre et al., 1998). Therefore, utilizing *in vitro* digestibility techniques using fish crude enzyme extracts from specific growth stages and sexes, and standardized by trypsin or amylase activity, is important in screening feedstuffs for developing formulated feeds with high nutritional qualities for optimizing growth performance quality. The work on feed formulations for *in vitro* and *in vivo* studies to optimize coloration and growth performance quality in Siamese fighting fish is currently under way.

ACKNOWLEDGEMENTS

We are grateful to Jarinporn Farm, Nakorn Phathom province, for kindly providing *B. splendens* samples. We would like to thank the Office of the Higher Education Commission (OHEC), Thailand, for research funding under the program Strategic Scholarships for Frontier Research Network for the Joint Ph.D. Program Thai Doctoral Degree. This work was also financially supported in part by the Thesis and Dissertation Research Fund, Graduate School, Kasetsart University, Thailand.

REFERENCES

AOAC (2005). Official Methods of Analysis of AOAC International. 18th eds. Association of Official Analytical Chemists, Maryland.

Areekijseree M, Engkagul A, Kovitvadhi S, Kovitvadhi U, Thongpan A, Rungruangsak-Torrissen K (2006). Development of digestive enzymes and *in vitro* digestibility of different species of phytoplankton for culture of early juveniles of the freshwater pearl mussel, *Hyriposis (Hyriopsis) bialatus* Simpson, 1900. Invert. Reprod. Dev. 49:255-262.

Areekijseree M, Engkagul A, Kovitvadhi U, Thongpan A, Mingmuang M, Pakkong P, Rungruangsak-Torrissen K (2004). Temperature and pH characteristics of amylase and proteinase of adult freshwater pearl mussel, *Hyriopsis (Hyriopsis) bialatus* Simpson 1900. Aquaculture 234:575-587.

Balti R, Barkia A, Bougatef A, Ktari N, Nasri M (2009). A heat-stable trypsin from the hepatopancreas of the cuttlefish (*Sepia officinalis*): Purification and characterization. Food Chem. 113:146-154.

Bassompierre M, Ostenfeld TH, McLean E, Rungruangsak-Torrissen K (1998). *In vitro* protein digestion and growth of Atlantic salmon with different trypsin isozymes. Aquacult. Int. 6:47-56.

Bombeo-Tuburan I, Fukumoto S, Rodriguez EM (1995). Use of the golden apple snail, cassava, and maize as feeds for the tiger shrimp, *Panaeus monodon*, in the ponds. Aquaculture 131:91-100.

Chan CR, Lee DL, Cheng YH, Hsieh DJY, Weng CF (2008). Feed deprivation and re-feeding on alterations of proteases in tilapia *Oreochromis mossambicus*. Zool. Stud. 47:207-214.

Chou RL, Her BY, Su MS, Hwang G, Wu YH, Chen HY (2004). Substituting fish meal with soybean meal in diets of juvenile cobia *Rachycentron canadum*. Aquaculture 229:325-333.

Dreon MS, Ceolin M, Heras H (2007). Astaxanthin binding and structural stability of the apple snail carotenoprotein ovorubin. Arch. Biochem. Biophy. 460:107-112.

Eshel A, Lindner P, Smirnoff P, Newton S, Harpaz S (1993). Comparative study of proteolytic enzymes in the digestive tracts of the European sea bass and hybrid striped bass reared in freshwater. Comp. Biochem. Physiol. 106:627-634.

Gamboa-Delgado J, Molina-Poveda C, Cahu C (2003). Digestive enzyme activity and food ingesta in juvenile shrimp *Litopenaeus vannamei* (Boone, 1931) as a function of body weight. Aquacult. Res. 34:1403-1411.

Hardy RW (1996). Alternate protein sources for salmon and trout diets. Anim. Feed Sci. Technol. 59:71-80.

Hidalgo MC, Urea E, Sanz A (1999). Comparative study of digestive enzymes in fish with different nutritional habits. Proteolytic and amylase activities. Aquaculture 170:267-283.

Hiramatsu N, Ichikawa N, Fukada H, Fujita T, Sullivan CV, Hara A (2002). Identification and characterization of proteases involved in specific proteolysis of vitellogenin and yolk proteins in salmonids. J. Exp. Zool. 292:11-25.

Hofer R, Schiemer F (1981). Proteolytic activity in the digestive tract of several species of fish with different feeding habits. Oecologica 48:342-345.

Jaroensutasinee M, Jaroensutasinee J (2001). Bubble nest habitat characteristics of wild Siamese fighting fish. J. Fish Biol. 58:1311-319.

Khawaja T, Khan SH, Ansari NN (2007). Effect of different levels of blood meal on broiler performance during two phase of growth. Int. J. Poult. Sci. 6:860-865.

Lazo JP, Mendoza R, Holt GJ, Aguilera C, Arnold CR (2007). Characterization of digestive enzymes during larval development of red drum (*Sciaenops ocellatus*). Aquaculture 265:194-205.

Lowry OH, Rosenbrough NJ, Farr AL, Randall RJ (1951). Protein measurement with the Folin phenol reagent. J. Biol. Chem. 193:265-275.

Mathers EM, Houlihan DF, McCarthy ID, Buren LJ (1993). Rates of growth and protein synthesis correlated with nucleic acid content in fry of rainbow trout, *Oncorhynchus mykiss*: effects of age and temperature. J. Fish. Biol. 43:245-263.

Meejui O, Sukmanomon S, Na-Nakorn U (2005). Allozyme revealed substantial genetic diversity between hatchery stocks of Siamese fighting fish, *Betta splendens*, in the province of Nakornpathom, Thailand. Aquaculture 250:110-119.

Nandeesha MC, Gangadhara B, Manissery JK, Venkataraman LV (2001). Growth performance of two Indian major carps, catla (*Catla catla*) and rohu (*Labeo rohita*) fed diets containing different levels of *Spirulina platensis*. Bioresour. Technol. 80:117-120.

Nandeesha MC, Gangadhara B, Varghese TJ, Keshavanath P (1998). Effect of feeding *Spirulina platensis* on the growth, proximate composition and organoleptic quality of common carp, *Cyprinus carpio* L. Aquacult. Res. 29:305-312.

Peirettie PG, Meineri G (2008). Effects of diets with increasing levels of *Spirulina platensis* on the performance and apparent digestibility in growing rabbits. Livest. Sci. 118:173-177.

Peragon J, Barroso JB, Garcia-Salguero L, de la Higuera M, Lupianez JB (2001). Growth, protein-turnover rates and nucleic-acid concentrations in the white muscle of rainbow trout during development. Int. J. Biochem. Cell Biol. 33:1227-1238.

Rungruangsak-Torrissen K (2007). Digestive efficiency, growth and qualities of muscle and oocyte in Atlantic salmon (*Salmo salar* L.) fed on diets with krill meal as an alternative protein source. J. Food Biochem. 31:509-540.

Rungruangsak-Torrissen K, Stien LH, Daae BS, Vågseth T, Thorsheim GB, Tobin D, Ritola O (2009). Different dietary levels of protein to lipid ratio affected digestive efficiency, skeletal growth, and muscle protein in rainbow trout families. Scholarly Research Exchange, vol. 2009, Article ID 709529, doi:10.3814/2009/709529.

Rungruangsak-Torrissen K, Moss R, Andresen LH, Berg A, Waagbo R (2006). Different expressions of trypsin and chymotrypsin in relation to growth in Atlantic salmon (*Salmo salar* L.). Fish Physiol. Biochem. 32:7-23.

Rønnestad I, Koven W, Tandler A, Harel M, Fyhn HJ (1998). Utilisation of yolk fuels in developing eggs and larvae of European sea bass (*Dicentrarchus labrax*). Aquaculture 162:157-170.

Sales J (2008). The use of linear regression to predict digestible protein and available amino acid contents of feed ingredients and diets for fish. Aquaculture 278:128-142.

Stryer L (1988). Biochemistry, 3rd ed. WH Freeman, New York.

Sunde J, Eiane SA, Rustad A, Jensen HB, Opstvedt J, Nygard E, Venturini G, Rungruangsak-Torrissen K (2004). Effect of fish feed processing conditions on digestive protease activities, free amino acid pools, feed conversion efficiency and growth in Atlantic salmon (*Salmo salar* L.). Aquacult. Nutr. 10:261-277.

Sveinsdottir H, Thorarensen H, Gudmundsdottir A (2006). Involvement of trypsin and chymotrypsin activities in Atlantic cod (*Gadus morhua*) embryogenesis. Aquaculture 260:307-314.

Thongprajukaew K (2011). Feed development using digestive enzyme technology for successive growth in Siamese fighting fish (*Betta splendens* Regan, 1910). Ph.D. thesis, Inter Departmental Multidisciplinary Graduate Program in Bioscience, Kasetsart University, Bangkok, Thailand.

Wiwatchaisaet Y (2000). Improvement of Siamese fighting fish for export. Thai Fish. Gaz. 53:169-179.

4

Biodeterioration of binding media in tempera paintings by *Streptomyces* isolated from some ancient Egyptian paintings

Akmal A. Sakr[1,3], M. F. Ghaly[1], M. F. Ali[2] and M. E. F. Abdel-Haliem[1]*

[1]Botany Department, Faculty of Science, Zagazig University, Zagazig, Egypt.
[2]Conservation Department, Faculty of Archaeology, Cairo University, Cairo, Egypt.
[3]Conservation Department, Civilization Museum, Cairo, Egypt.

Eight out of 46 *Streptomyces* isolates representing highest deterioration symptoms of disfiguration and scaling were isolated from mural paintings within tombs at Tell Basta and Tanis, Lower Egypt. These isolates were identified using traditional and 16S rDNA sequencing methods as *Streptomyces albidofuscus*, *Streptomyces ambofaciens*, *Streptomyces canarius*, *Streptomyces chibaensis*, *Streptomyces coelicolor*, *Streptomyces corchorusii*, *Streptomyces nigrifaciens* and *Streptomyces parvullus*. These identified *Streptomyces* isolates decomposed animal glue into amino acids with a high percentage of L-glutamic acid and ammonia as end product, depolymerized arabic gum into free mono-sugars, decomposed egg yolk into amino acids with low percentage of L-glutamic acid and ammonia and decomposed bees wax enzymatically into short chains of stearic acid. Finally, these isolates caused flaking of paintings layers due to penetration of filamentous mycelium within painting layers with the assistance of enzymatic hydrolysis.

Key words: Animal glue, arabic gum, bees wax, biodeterioration, egg yolk, *Streptomyces*, Tell Basta, Tanis, Egyptian tombs.

INTRODUCTION

Paintings in ancient Egyptian tombs were carried out using tempera technique, by mixing earthy pigments such as hematite, limonite, azurite blue, lamp black and green malachite or synhesis pigments such as Egyptian blue with different binding media such as arabic gum, animal glue, egg yolk and bees wax (Newman and Serpico, 2000).

The ability of *Streptomyces* to deteriorate mural paintings was attriubted to utilizing a wide range of carbon sources, possessing different adaptation methods against adverse environmental conditions such as biocides and heavy metals (McCarthy and Williams, 1992) and producing a wide range of specific enzymes that could decompose binding media, so every medium has its own deterioration mechanism and products (Salvadori et al., 2003).

Animal glue is composed mainly of amino acids and is exposed to biodeterioration by collagenase enzyme produced by *Streptomyces* that causes cross linking of amid bonds in collagen of animal glue into amino acids (Chadefaux et al., 2009), so these paintings lose its cohesiveness detected from plaster layers and stone support (Goshev et al., 2005). Biodeterioration of animal glue in mural paintings was seen, whereas *Streptomyces* colonized fresco paintings in Cathedral of Nativity of the Virgin in the Pafnutii-Borovski monastery, Russia, decomposed animal glue used as binding medium and restoration material (Karpovich-Tate and Rebrikova, 1991), paintings of Boyana church, Bulgaria, also *Streptomyces griseus*, *Streptomyces albus*, *Streptomyces flavus* and *Streptomyces fradiae* caused flaking and powdering of these paintings (Hadjivulcheva and Gesheva,1982) and gesso layers on Virgin Mary wooden statue in Church of

*Corresponding author. E-mail: farrag1999@yahoo.com.

Gora, Croatia, and *Streptomyces* sp. vigorously attacked animal glue in these layers (Briški et al., 2001).

Furthermore, arabic gum was the most common binding medium in ancient Egyptian paintings. This gum is a complex one and composed mainly of arabinogalactan oligosaccharides, and polysaccherides (Onyari et al., 2008). These polysaccharides are subjected to enzymatic depolymerization by *Streptomyces* into monmeric units of free mono sugars that modified red color estimated spectrophotometrically at 420 nm (Colombini et al., 2002).

The third binding medium was egg yolk that was used in many ancient Egypt paintings, or as a varnish layer to protect the paintings from adverse environmental conditions (Derric et al., 1999). Egg yolk contains 20 to 40% lipid, two thirds of these lipids from triglycerides that composed mainly of phosphatidylcholines lecithin (Odlyha et al., 2000), whereas lethcianease enzyme could decompose phospholipids into ammonia as end product (Arthur and Henerici, 1970).

Bees wax was used as protective layer and latter was used as binding medium. Bees wax subjected to biodeterioration by *Streptomyces* that could decompose hydrocarbons by producing biosurfactants that emulsify hydrophobic compounds and cause ring fission and carbon assimilation (Roper, 2004) or by producing esterase enzyme, since *Geodermatophilus* isolated from limestone blocks from the pyramid of the King Chevern and tombs of workers at Deir el Madeina, Upper Egypt, produced esterase enzyme (Gtari et al., 2012).

Finally, binding media were subjected to biomechanical deterioration by *Streptomyces* that could penetrate its hyphae within painting layers and cause dislodging of various layers, so these painting layers were detached from the plaster layers and stone supports (Altenburger et al., 1996).

Due to the lack of literature about the role of *Streptomyces* in biodeterioration of binding media in ancient Egyptian paintings and deterioration mechanisms and products, and since it was suggested that different binding media are susceptible to deterioration by *Streptomyces*, this study was conducted.

Streptomyces isolated from mural paintings were identified morphologically, biochemically and using 16S rDNA sequence method. Biodeterioration of binding media used in ancient Egyptian mural paintings such as animal glue, arabic gum, egg yolk and bees wax by *Streptomyces* were completely analyzed.

MATERIALS AND METHODS

Isolation locations

Forty six (46) *Streptomyces* isolates were collected from deterioration symptoms of scaling and disfiguration from mural paintings in tombs of Ankh m b3st, Ist, Ankh h.f and Ihi at Tell Basta and tomb of The King Oserkon II at Tanis (Figure 1), Lower Egypt, using sterile cotton swab method. These bacterial isolates were isolated and purified onto starch-nitrate-agar (SNA) plates (agar 20; starch 20; KH_2PO_4 1; $MgSO_4$ 0.5; NaCl 0.5; KNO_3 2 and $CaCO_3$ 3 g / l distilled water (Kuster and Williams, 1964) supplemented with antifungal (Dermatine 10 to 50 µg / l) to inhibit the growth of competitive fungi. The plates were incubated for 45 days at 30°C.

Identification of bacterial isolates

The most biodeterioration active bacterial isolates (eight isolates) were identified morphologically and biochemically according to the identification keys presented by Kämpfer (2006) and confirmed using 16S rDNA sequence method.

16S rDNA sequencing

Total DNA was extracted from eight isolated colonies of actinobacteria according to the method presented by Sambrook and Russel (2001). The gene coding for 16S rRNA was amplified from each isolate by polymerase chain reaction (PCR) with universal primers (forward primer [F27] 5'-AGAGTTTGATCCTGGCTCAG-3' (Chénbey et al., 2000) and reverse primer [R1492] 5'-GGTTACCTTGTTACGACTT-3') (Turner et al., 1999). These primers bind to universally conserved regions and permit the amplification of an approximately 1500 bp fragment. The PCR amplification was carried out in a Gene-Amp PCR system 9600 thermocycler (Perkin Elmer).

The amplification conditions were as follows: 94°C for 10 min and 35 cycles of denaturation at 95°C for 30 s, annealing-extension at 56°C for 1 min, 72°C for 1 min and an extension at 72°C for 10 min. Presence and yield of specific PCR products (16S rRNA gene) were monitored by running 1% agarose gels. Then, PCR product was cleaned up by using GeneJET™ PCR purification kit (Fermentas).

Amplified DNA fragments were partially sequenced at GATC Biotech AG (Konstanz, Germany) using ABI 3730xl DNA sequencer using forward primer (F27). The 16S rDNA sequences which have been determined in the present study were deposited at NCBI web server (www.ncbi.nlm.nih.gov; Table 1). Sequence analysis and comparison with published sequences was made using the Basic Local Alignment SearchTool (BLAST) program (http://www.ncbi.nlm.nih.gov/blast) (Altschul et al., 1997).

Deterioration of animal glue

Streptomyces isolates were cultured on starch-nitrate-agar free from carbon source whereas 2% animal glue was used as sole carbon source in 250 ml flasks (Petrova et al., 2006). Each flask contained 100 ml of broth meduim (pH was adjused to 7), inoculated with 10% spore suspension (1×10^6 spores / ml) and incubated at 30°C for two weeks. After the end of incubation period, the biomass was filtered off and the filterate was cleared by centrifugation at 3000 rpm for 15 min.

One milliliter of the filterate was analysed by high performance liquid chromatography (HPLC) amino acid analyzer LC300 Eppendorf Germany (National Research Centre, Dokky, Giza) for quantitive anaylsis of amino acids in the deteriorated samples.

Biodeterioration of arbic gum

Biodeterioration of arabic gum by *Streotpmyces* isolates was determined by estimating the functional groups in both deteriorated and control arbic samples by infra-red (IR) spectroscopy (JASCO FT-IR 61000, National Research Centre, Dokky, Giza).

Biodeterioration of egg-yolk

Biodeterioration of egg-yolk was investigated whereas *Streptomyces*

Figure 1. Location of sampling: **(a)** Azurite blue in tomb of Oserkon II with white crust, Tanis. **(b)** Red ochre, tomb Ankh m b3st. **(c)** Azurite blue, tomb of Ihi, Tell Basta. **(d)** Red and yellow paintings, tomb Ankh h3 f, Tell Basta. **(e)** Black color from tomb of Ankh h3 f, Tell Basta. **(f)** White crust of soduim chloride, tomb of Ist, Tell Basta.

isolates were cultured on liquid starch-nitrate broth (SNB) medium free carbon and nitrogen source, whereas 2% egg yolk was used as sole carbon and nitrogen source and incubated at 30°C for two weeks. Released ammonia was detected in the obtained filterate using Fourier transform infrared (FT-IR) spectroscopy (JASCO FT-IR 61000, National Research Centre, Dokky, Giza).

Table 1. Phylogenetic affiliation of inoculated strains (Homology of 16S r DNA and similarity in comparison with NCBI Data).

Location	Homology approximately	Similarity enter genes 16S r DNA (%)	Authors' accession number	G + C content (%)
Azurite blue, tomb of Ihi, Tell Basta.	*S. albidofuscus*	99	Later name is *S. pyridomyceticus* Bankit1507621 JQ625331	58
Yellow color of Southern wall of Ankh h3ftomb.	*S. ambofaciens*	99	Bankit1507642 JQ625332	60.6
Blue color, ceiling burial tomb of Oserkon II, Tanis.	*S. canaries*	99	Bankit1507650 JQ625337	58.9
Black color, tomb Ankh h.f, Tell Basta.	*S. chibaensis*	100	Bankit1507649 JQ625336	58.9
Red color, tomb Ankh m b3st, Tell Basta.	*S. coelicolor*	99	Bankit1507648 JQ625335	59.2
Limestone, tomb of Oserkon II, Tanis.	*S. corchorusii*	98	Bankit1507647 JQ625334	58.9
North wall, tomb of Ist, Tell Basta.	*S. nigrifaciens*	98	Later name is *S. flavovirens* Bankit1507149 JQ625330	~56
Yellow color of Southern wall of Ankh h3ftomb.	*S. parvullus*	99	Bankit1507645 JQ625333	56.3

Deterioration of bees wax

Degradation of bees wax by *Streptomyces* isolates was detected by inoculating *Streptomyces* isolates onto SNA plates supplemented with 2% bees wax as sole carbon source emulsified with 80 0. 1 ml/L Tween. Incubation period was two weeks at 30°C. Inoculated and non-inoculated samples were analyzed by mass spectra (JOLE, JMS-AX500 Mass spectra, National Research Centre, Dokky, Cairo).

Mechanical deterioration

The mechanical deterioration of binding media by *Streptomyces* was determined by covering glass slides with animal glue and inoculated with *Streptomyces* isolates, incubated at 30°C for two weeks then examined with scanning electron microscope (JOLE, SEM 6300, National Research Centre, Dokky, Cairo). The penetration of *Streptomyces* mycelium within the painting layers was observed according to colonizied method of Milanesi et al. (2006).

RESULTS AND DISCUSSION

Eight selected and identified isolates from different paintings (Figure 1) were attributed to genus *Streptomyces* (*Streptomyces albidofuscus*, *Streptomyces ambofaciens*, *Streptomyces canarius*, *Streptomyces chibaensis*, *Streptomyces coelicolor*, *Streptomyces corchorusii*, *Streptomyces nigrifaciens* and *Streptomyces parvullus*) with similiarity more than 98% and the authors' accession numbers in the International GenBank were illustrated in Table 1, since *Streptomyces* spp. are considered the typical and first colonizers of deteriorated cultural heritage objects (Lanoot, 2005) and involved significantly in biodeterioration of binding media due to their ability to decompose relatively complex organic substances such as cellulose, pectin, chitin and proteins (Borrego et al., 2012).

Our results indicate that ainmal glue enhanced growth and sporulation of *S. ambofaciens* more than other *Streptomyces* isolates. Furthermore, HPLC amino acid analyser data revealed that *S. ambofaciens* and *S. nigrifaciens* decomposed animal glue into amino acids, and ratios of amino acids of argnine and histidine were lower than that obtained for alanine and glycine. Also, ammonia was revealed (Figures 2 and 3), and its ratio reached 1024.76 and 1244.84 µg/ml with *S. ambofaciens* and *S. nigrifaciens,* respectively (Tables 2 and 3). On the other hand, FT-IR patterns of deteriorated animal glue revealed presence of L-glutamic acid at 1454 cm^{-1} and ammonia group at 3426 cm^{-1} (Figures 4 and 5). L-glutamic acid was the most present amino acid in the deteriorated samples, with ratio 824.32 µg / ml and 794.961 µg / ml for *S. ambofaciens* and *S. nigrifaciens,* respectively (Tables 2 and 3).

On the other hand, *Streptomyces* isolates increased pH of medium supplemented with animal glue as carbon source, whereas *S. ambofaciens* reduced pH of medium from 7 to 6.1 after two weeks, then increased final pH into 8.67, while *S. nigrifaciens* increased pH from 7 to 9 after three weeks of incubation (Table 4). This may be explained by the fact that animal glue was decomposed enzymatically by *Streptomyces* through two steps due to collagenolytic activity of isolated *Streptomyces* (Górny, 2004). The first step was decomposition of animal glue into free amino acids and the second step was decomposition of amino acids into ammonia as end product that increased pH of media, making favorable setting for the growth of *Streptomyces* and providing nutrition for other associated mciroorganisms in the ecosystems (Tomek and Pechova, 1992).

Streptomyces involved in the biodeterioration of arabic gum varied in their ability to depolymrise arabic gum,

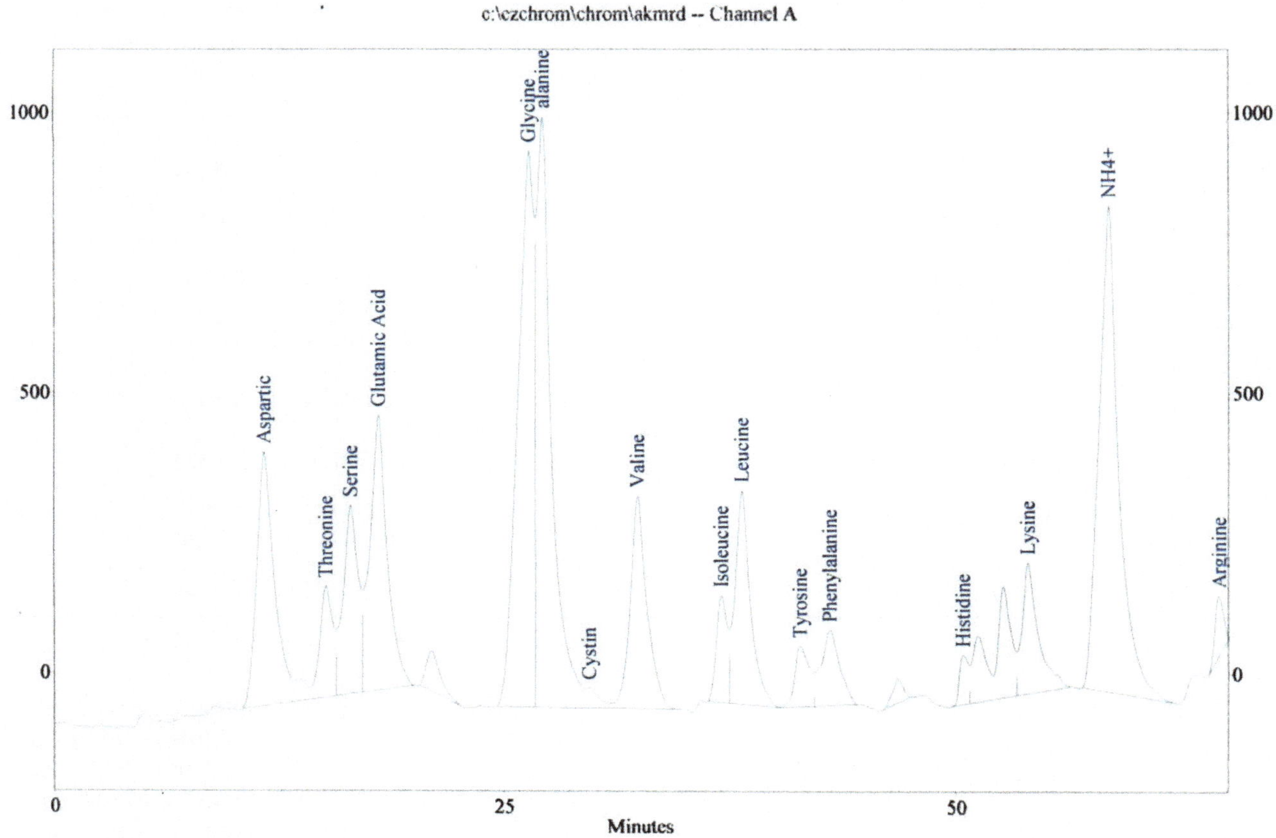

Figure 2. Quantitative analysis of amino acid from decomposition of animal glue by *S. ambofaciens*.

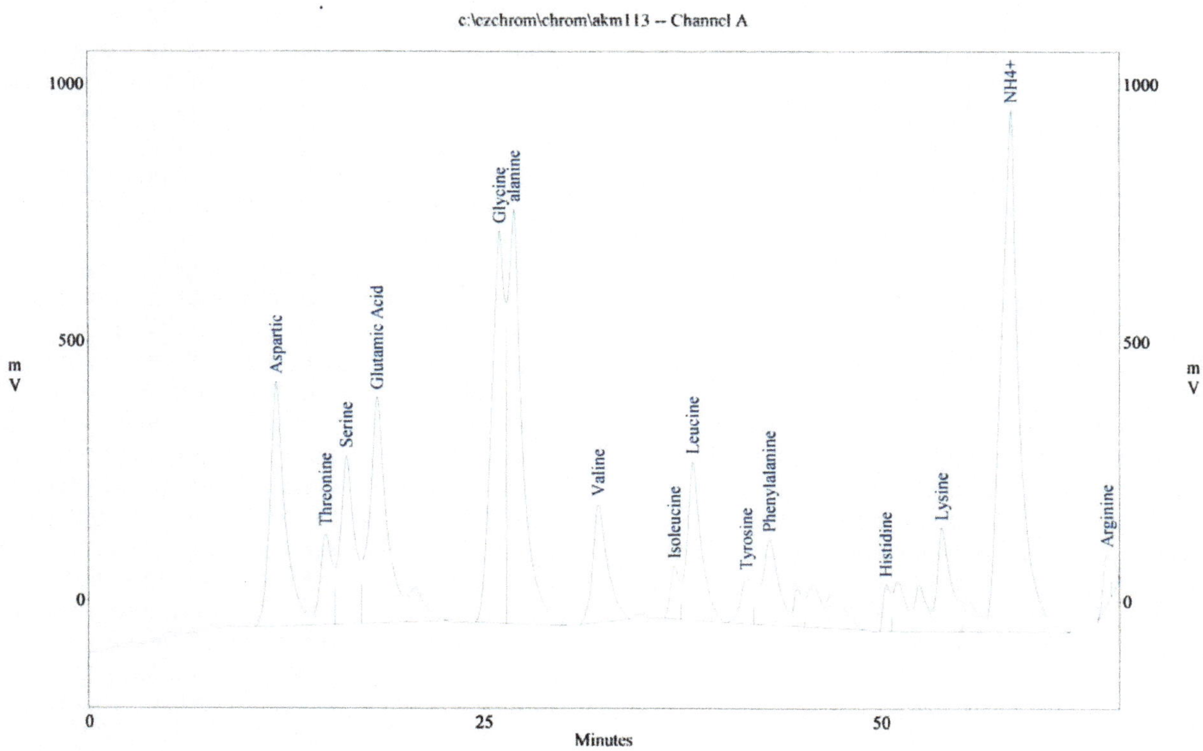

Figure 3. Quantitative analysis of amino acid from decomposition of animal glue by *S. nigrifaciens*.

Table 2. Quantitative analysis of amino acid resulting from decomposition of animal glue by *S. ambofaciens*.

Peak number	Name	Time (min)	Area	Concentration (µg/ml)
1	Aspartic	11. 63	30502852	605.28
3	Threonine	15.07	101135930	170.4
4	Serine	16.45	19306890	268.28
5	Glutamic acid	18.02	33690412	824.32
7	Glycine	26.33	60421520	400.88
8	Alanine	27.13	628234552	795.48
9	Cystin	29. 68	1764352	79.48
10	Valine	32.33	24626244	350
11	Isoleucine	36,97	7747364	168.84
12	Leucince	38.08	22831084	517.92
13	Tyrosine	41.33	6081601	202.8
14	Phenylalnine	42.98	9526860	285.88
16	Histidine	50.38	2989894	112.36
19	Lysine	53.95	13983382	356.28
21	NH4	58.40	6663752	1024.76
22	Arginine	64.47	3569638	155. 56

Table 3. Quantative analysis of amino acids resulting from decomposition of animal glue by *S. nigrifaciens*.

Peak number	Name	Time (min)	Area	Concentration (µg/ml)
1	Aspartic	11.90	313052	621.16
2	Threonine	14.98	9149939	135.8
3	Serine	16.33	20009218	278.04
4	Glutamic acid	18.27	32490380	794.961
6	Glycine	25.98	34583220	289.16
7	Alanine	26.90	51494672	625
8	Valine	32.20	14926400	212.12
9	Isoleucine	36.98	3947304	86.04
10	Leucince	38.15	18896858	428. 68
11	Tyrosine	41. 52	4588704	153.04
12	Phenylalnine	42.98	1240370	372.44
16	Histidine	50.30	2931907	110.2
19	Lysine	53.82	12272261	312.68
21	NH_4	58.13	80950976	1244.84
22	Arginine	64.20	2955407	128.8

whereas *S. ambofaciens*, *S. parvullus*, *S. nigrifaciens* and *S. canarius* had high depolymerisation ability of polysaccherides, *S. corchorusii* had a good ability, whereas *S. albidofuscus* had a moderate one, but *S. chibaensis* and *S. coelicolor* had a low growth rate (Figure 6). Furthermore, spectrophotometric data indicated that free mono-sugars obtained from depolmerization of arabic gum enzymatically by *Streptomyces* and the produced monosugar are not obtained from decompose phenolic compounds in the arabic gum, whereas FT-IR pattern gave 3897 cm^{-1} intense band characterizing for phenolic groups in both standard and deteriorated arabic gum samples (Figures 7 and 8). These results are similar to that of Onyari et al. (2008). Phenolic compounds in the arabic gum could not be decomposed by *Streptomyces* isolates, whereas FT-IR pattern gave an intense band at 3897 cm^{-1} characterized for phenolic groups (Derric et al., 1999), that appeared in both standard and deteriorated samples, since the presence of phenol groups in both deteriorated and the standard samples was attributed to toxicity of these groups for *Streptomyces* isolates (Derric et al., 1999).

Our finding indicates that growth rate of *Streptomyces* cultured onto charcoal with arabic gum as binding meduim was increased, this may be due to adsorbing phenolic compounds from arabic gum by charcoal and

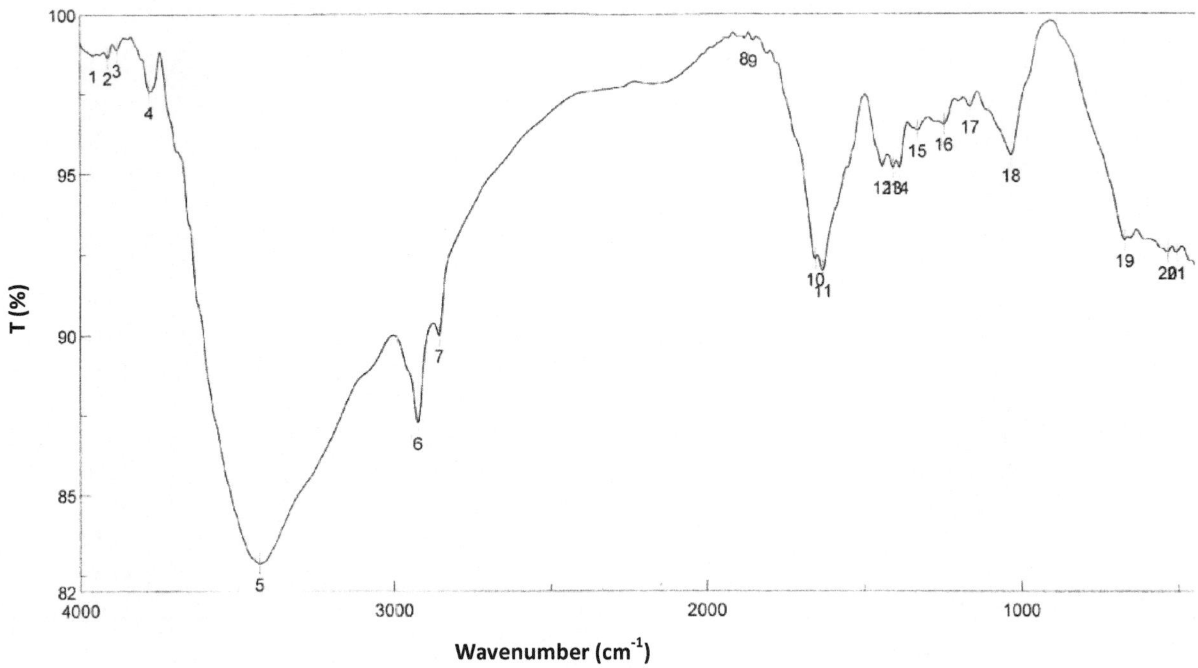

Figure 4. IR spectrum for standard animal glue.

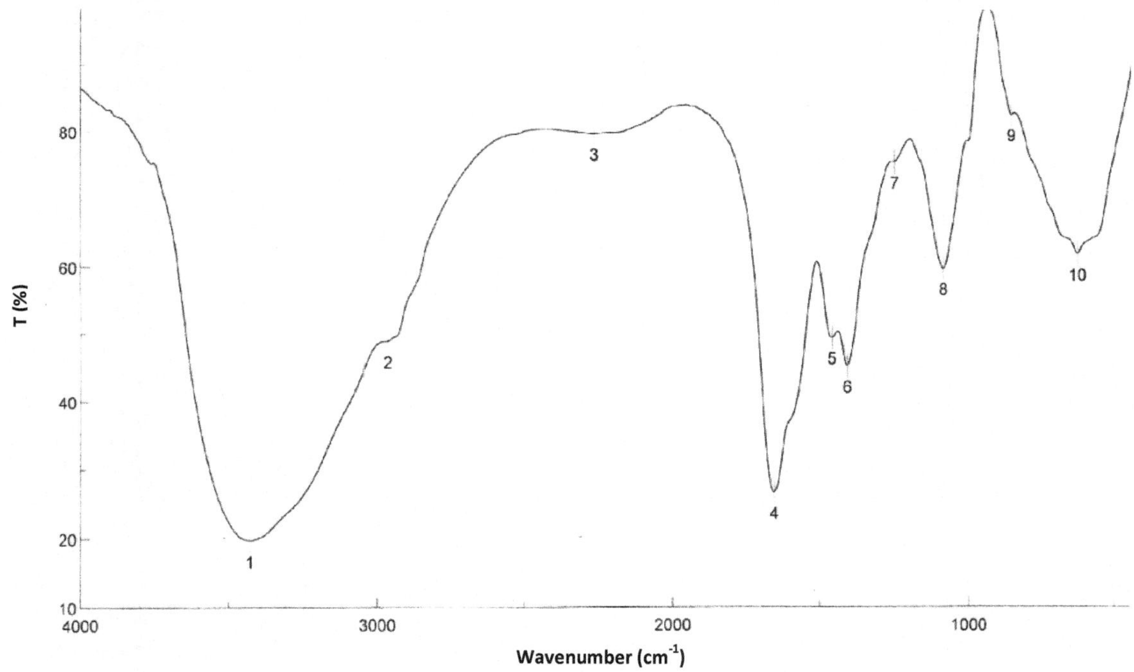

Figure 5. IR spectrum for deteriorated animal glue.

prevented them from access to *Streptomyces* mycelium (Normand, 2006).

Furthermore, FT-IR patterns of deteriorated egg yolk samples gave an intense band at 1500 cm^{-1} of (C-N-H) group orginating from amino acids and the band which appeared at 3426 cm^{-1} was characterized for ammonia group (NH_3), whereas egg yolk was decomposed by leucinase enzyme produced by *Streptomyces* isolates into amino acids and ammonia as end product (Peris-Vicente et al., 2006).

Table 4. Final pH after 2 to 3 weeks of incubating isolates *S. ambofaciens* and *S. albidofuscus* on starch-nitrate broth medium where animal glue was used as carbon and nitrogen source.

Isolate	Initial pH	Final pH after different periods	
		2 weeks	3 weeks
S. albidofuscus	7.0	8.1	8.1
S. nigrifacien	7.0	8.7	9
S. ambofaciens	7.0	6.1	8.67

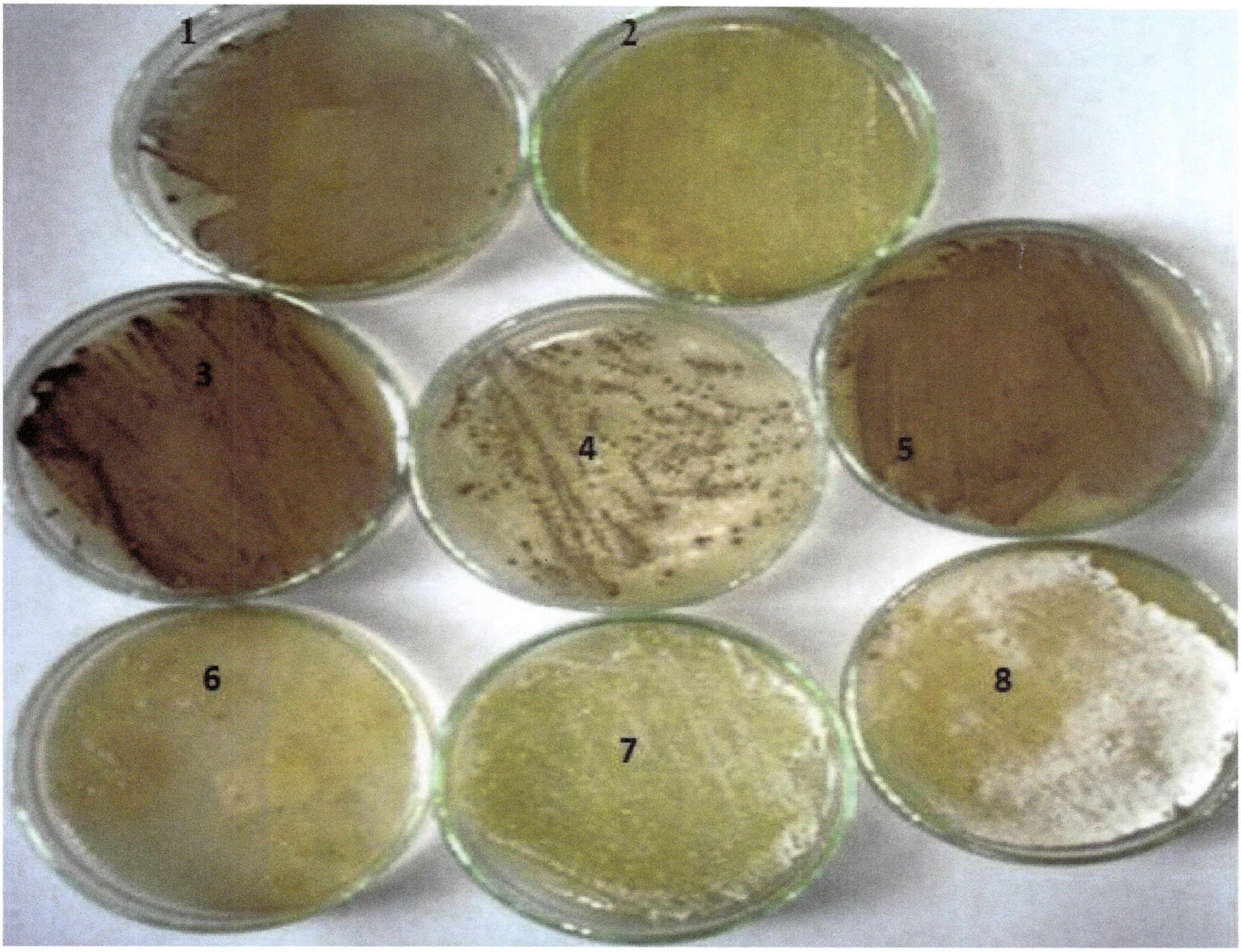

Figure 6. Growth of isolated *Streptomyces* on arabic gum as carbon source in starch-nitrate-agar media. **(1)** *S. corchorusii.* **(2)** *S. coelicolor.* **(3)** *S. parvullus.* **(4)** *S. ambofaciens.* **(5)** *S. nigrifaciens.* **(6)** *S. chibaensis.* **(7)** *S. albidofuscus.* **(8)** *S. canarius.*

Table 5 shows that *Streptomyces* isolates varied in their biodegradability of bees wax, whereas *S. ambofaciens*, *S. nigrifaciens*, *S. coelicolor* had high growth rate, whereas *S. corchorusii*, *S. chibaensis*, *S. albidofuscus* had middle growth rate and *S. canarius* had low rate (Figure 9). These isolates decomposed bees wax, where *S. corchorusii* and *S. ambofaciens* decomposed long chains of bees wax into short chains ($C_{18}H_{36}$) (Figure 10). These results are confirmed by the previous results of Regert et al. (2005). Massspectra data indicated that bees wax was decomposed by *Streptomyces* into short chains of hydrocarbon of stearic acid, since it was found that the number of carbon atoms gave the characteristic number of stearic acid (C_{18}:n) the main component of bees wax (Regert et al.,

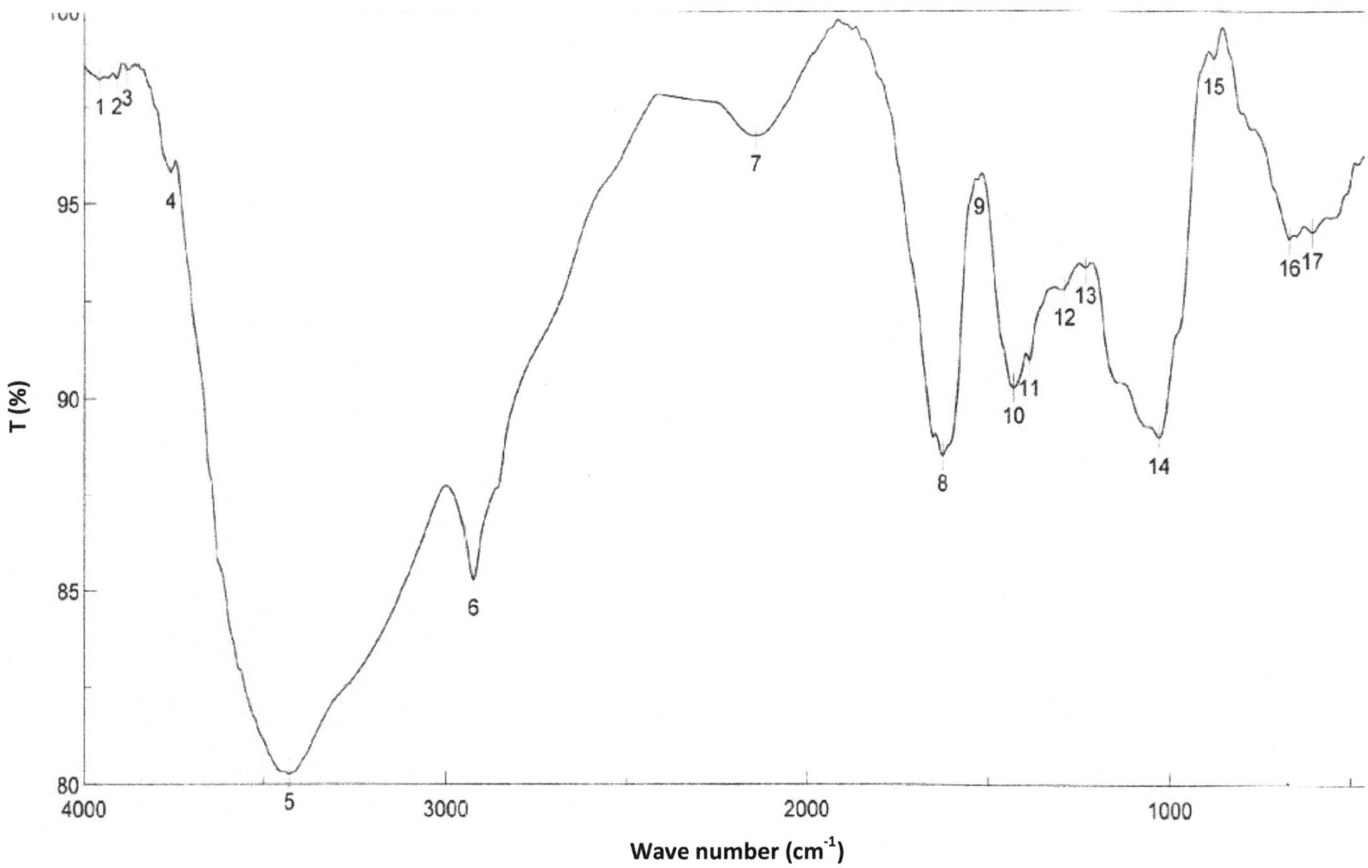

Figure 7. IR spectra of standard arabic gum.

Table 5. Growth of *Streptomyces* strains on bees wax.

***Streptomyces* isolate**	**Growth rate**
S. albidofuscus	++
S. ambofaciens	+++
S. canarius	++
S. chibaensis	++
S. coelicolor	+++
S. corchorusii	++
S. nigrifaciens	+++
S. parvullus	+++

+++, Good growth; ++, moderate growth.

2001). Decompostion of bees wax resulted in enzymatic action of esterase enzyme, whereas *Geodermatophilus* isolated from paintings in workers' tombs, Luxor, Upper Egypt, could decompose bees wax into short chains of stearic acid due to production of esterase enzyme (Essoussi et al., 2010), or this decomposition may have occurred as a result of biosurfactants produced by *Streptomyces* that emulsify hydrophobic compounds and cause ring fission and carbon assimilation (Pizzul et al., 2006). Due to decomposition of bees wax by *Streptomyces*, varnish layers lost its hydrophobicity and plasticity because of decrease of hydrocarbons in the ancient bees wax (Roper, 2004).

Finally, SEM micrographs of inoculated and non inoculated glass slides covered with film of animal glue revealed that *S. ambofaciens* made holes in this film (Figures 11a and b). On the other hand, mycelium of *S. coelicolor* penetrated within painting layers of red hematite and azuirte blue mixed with arabic gum as binding meduim, whereas the spores connected with mycelium were observed (Figure 11c) and caused flaking of these layers deteched from plaster and stone support (Pepe et al., 2009).

Moreover, mycelium of *S. coelicolor* and *S. canarius* penetrated into painting layers of red hematite and azuircte blue, since the biodeteriorative role of *Streptomyces* includes both chemical and mechanical destruction due to swelling and shrinking of mycelium within painting layers (Altenburger et al., 1996).

Conclusions

Streptomyces colonizing mural paintings and deteriorating

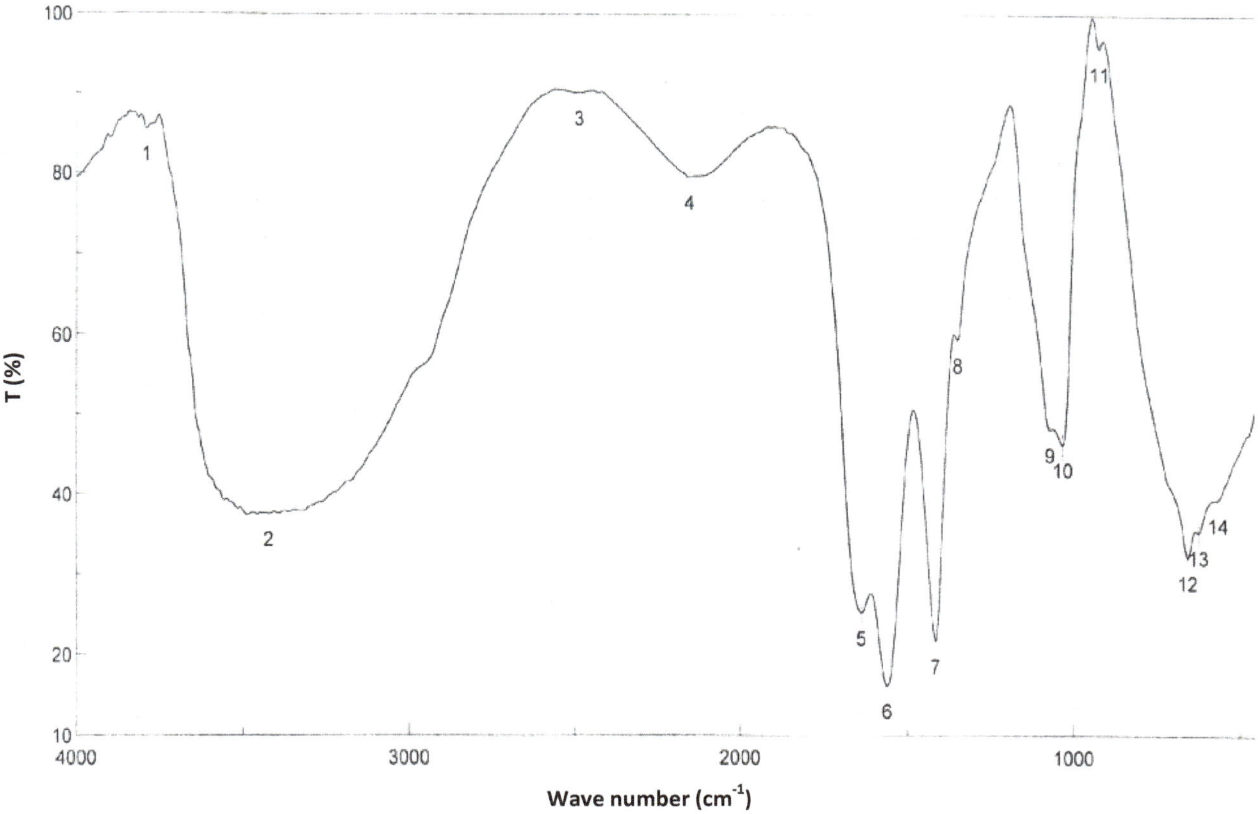

Figure 8. IR spectra of deteriorated arabic gum.

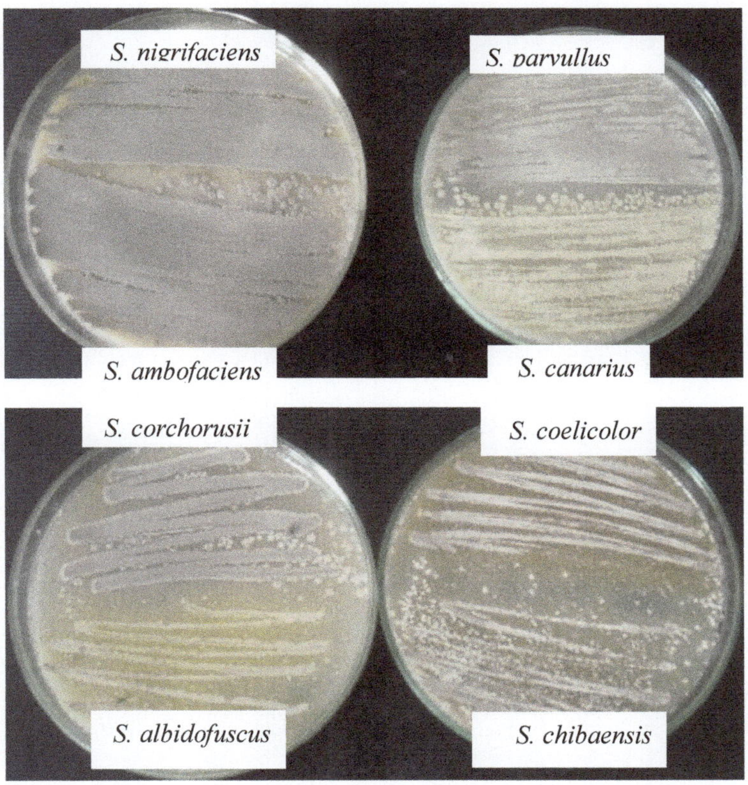

Figure 9. Effect of *Streptomyces* isolates on bees wax.

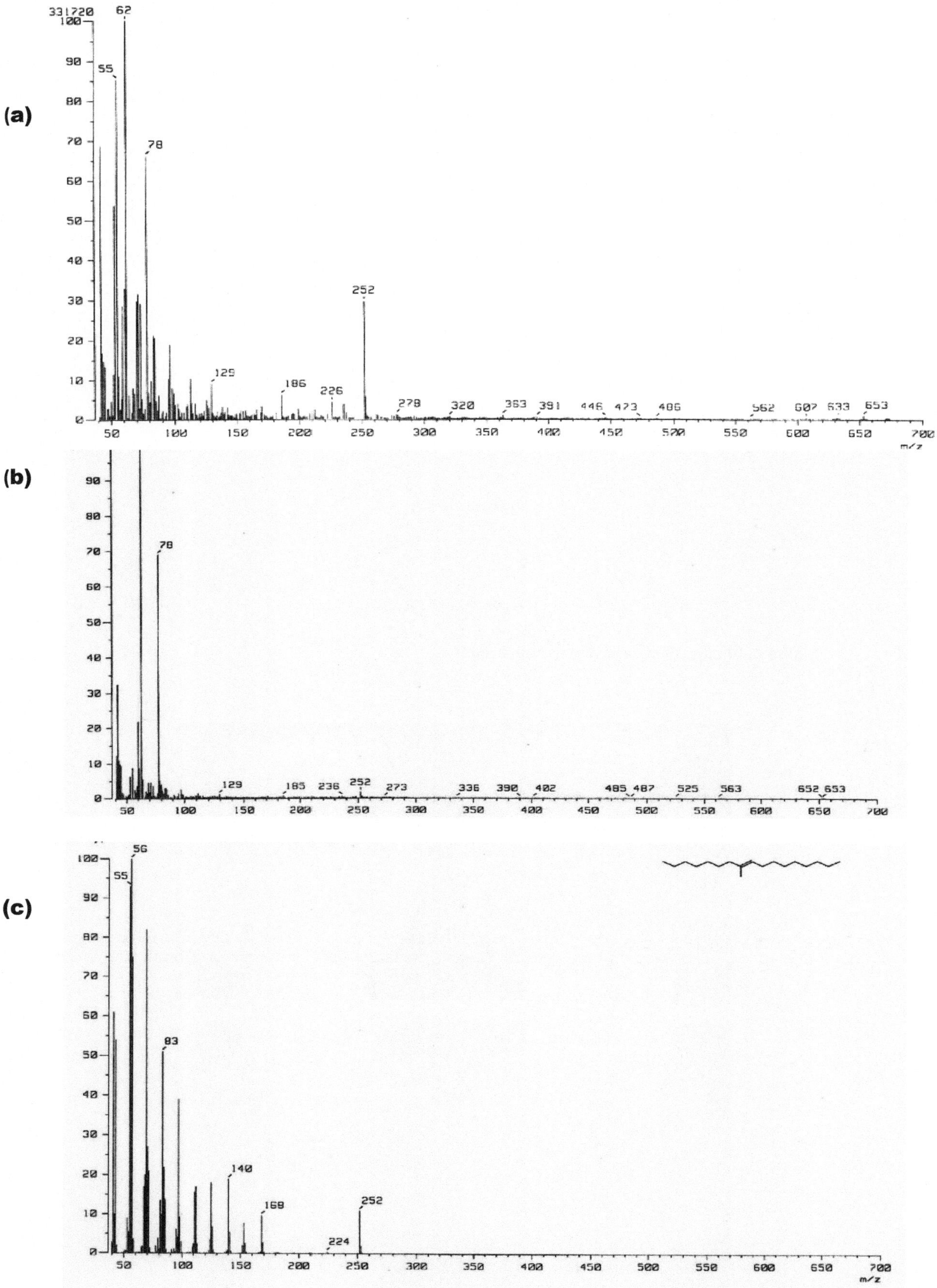

Figure 10. Mass spectra of bees wax inoculated: (a) control, (b) *Streptomyces ambofaciens*, (c) *Streptomyces nigrifaciens*.

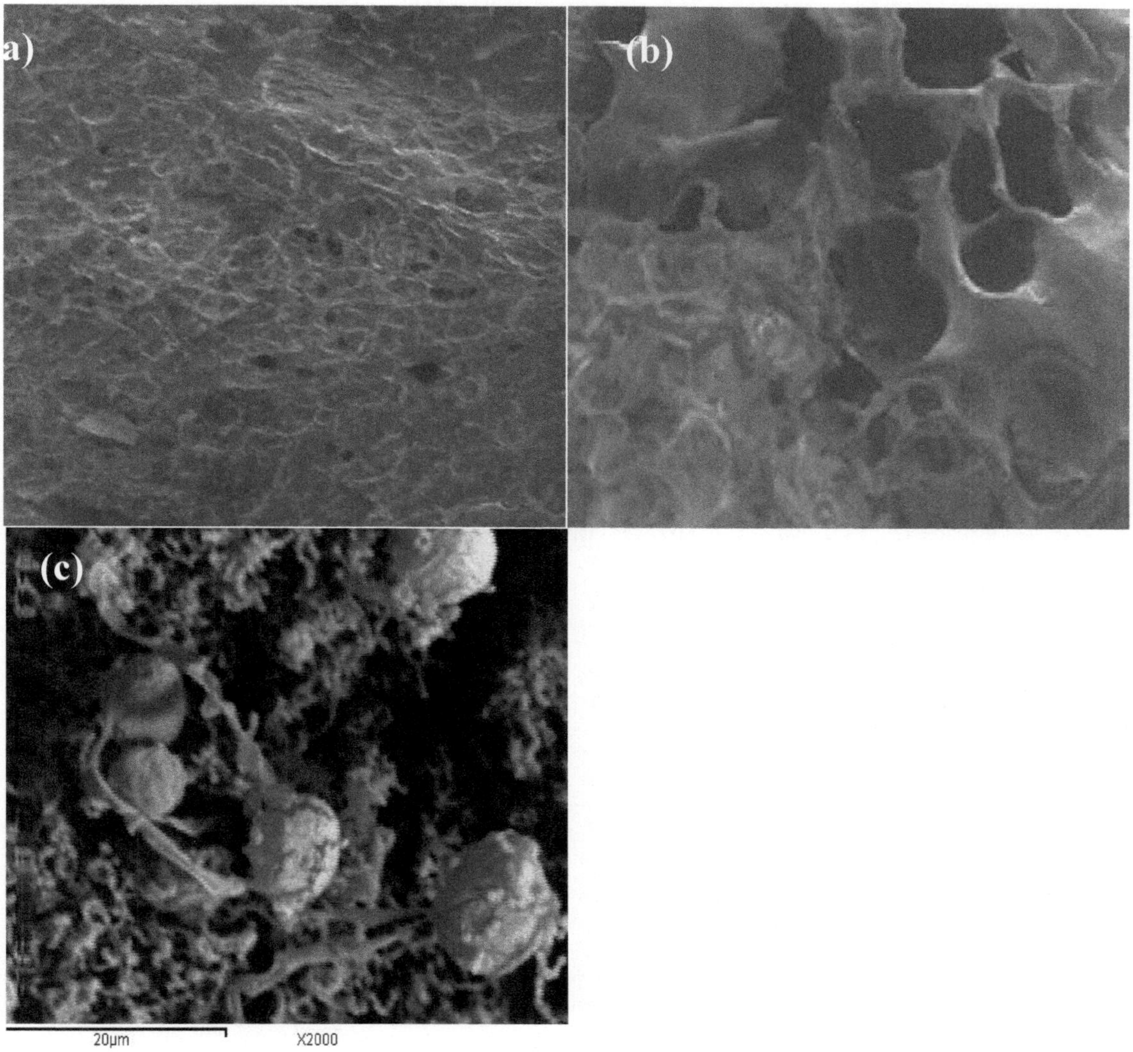

Figure 11. SEM micrograph of animal glue; (a) control, (b) inoculated by *S. ambofaciens*, 1200x, (c) pentration of hyphae of *S. coelicolor* within binding meduim of arabic gum causing flaking of painting layers, 2000x.

binding media were characterized. Scientific analyses bring new information to the light on mechanisms of deterioration of binding media by *Streptomyces* and the end products of their biodeterioration.

REFERENCES

AltenburgerP, Kämpfer P, Makristathis A, LubitzW, Brusse H (1996). Classification of bacteria isolated from a medival wall paintings.J. Biotechnol.47: 39-52.

Altschul SF,Madden TL,Schäffer AA,Zhang J,Zhang Z,MillerW,Lipman D J(1997). Gapped BLAST and PSI-BLAST: a new generation of protein database search programs. Nucleic Acids Research17: 389-402.

Arthur T, HenericiMD(1970). Molds, Yeasts and Actonomycetes, London.

BorregoS, Lavin P, Perdomo I, Gómez de Saravia S, Guiamet P (2012). Determination of Indoor Air Quality in Archives and Biodeterioration of the Documentary Heritage. Intern. Scholarly Research Network, doi:10.5402/2012/680598 (in press).

Briški F, Krstić D, Jagić R(2001). Microbial species on polychromee sculpture from a ruined church, evaluation of the microbicide PBK aginst further biodeterioration.Stud.Conserv.46: 14-22.

Chadefaux A, Le Ho A, Bellot-Gurlet B, Reiche I (2009). Curve –Fitting Micro-ATR-FTIR studies of the amide 1 and 11 bands of type1 collagen in archaeological bone materials, E-Preserv. 6: 129- 173.

Chénbey D, Philippot L, Hartmann A, Hénalut C, Germon JC(2000). 16S rDNA analysis for characterization of denitrifying bacterial isolated from three agricultural soils. FEMS Microb. Ecol. 24:121-128.

Colombini M, Ceccarini A, Carmignani A(2002). Ion chromatography characterization of polysacchairdes in ancient wall paintings.J. Chromat. A968: 79-88.

Derric MR, Stulik D, Landry J (1999). Infrared Spectroscopy in Conservation Science, in The Getty Conservation Institute, Ball T, Tidwell S, (eds.):183.

Essoussi I, Ghodhbane-Gatri F, Amiri H, BrusettiL, Gtari M(2010). Esterase as an enzymatic signature of *Geodematophilaceae* ability to Sahara desert stones and mounments.J. Appl. Microbiol.108: 1723–732.

Górny R(2004). Filamentous Microorganisms and their fragments in indoor air.Ann. Agric. Environ. Med.11:185-197.

Goshev I, Gousterova A, Vasileva-Tonkova E, Nedkov P (2005).

Characterization of the enzyme complexes produced by two newley isolated thermophylic actinomycete strains during growth on collagen-rich materials.Process Biochem. 40: 1627-1631.

Gtari M, Essoussi I, Maaoui R, Sghaier H, Boujmil R, Gury J, Pujic P, Brusetti L, Chouaia B, Crotti E, Daffonchio D, Boudabous A, Normand P (2012). Contrasted resistance of stone dwelling Geodermatophilaceae species to stresses known to give rise to reactive oxygen.FEMS Microbiol. Ecol. 80: 566-77.

Hadjivulcheva EN, GeshevaR(1982). Actinomycetes isolated from theBoyana Church mural paintings.Comptes Rendus de 1'Academie Bulgare des Sci. 35: 71-74.

Kämpfer P(2006). The Prokaryotes. A Hand book on the biology of bacteria. Vol. 3, 3rd ed., Dworkin M, Falkow S, Rosenberg E, Schleifer K-H, Stackebrandt E (Editors). Springer. The family *Streptomycetaceae*, Part I, Taxonomy:538-604.

Karpovich-Tate N, Rebrikova NL (1991). Microbial communities on damaged frescoes and building materials in the Cathedral of Nativity of the Virgin in the Pafnutii-Borovski monastery. Russia. Inter. Biodet. 27: 281-296.

Kuster E,Williams ST(1964). Selection of media for isolation of *Streptomyces*. Nature 22: 928-929.

Lanoot B (2005). Improved taxonomy of the genus *Streptomyces*, Ph D Thesis, Gent Universiteit, FaculteitWetenschappenVakgroepBiochemie, physiologie&MicrobiologieLaboratoriumvoorMicrobiologie: 36.

Mc Carthy AJ, Williams ST(1992). Actinomycetes as agent of biodeterioration in the environment – a review.Gene155: 189-192.

Milanesi C, Baldi F, Borin S, Vignani R, Ciampolini F, Faleri C, Cresti M(2006). Biodeterioration of a fresco by biofilm forming bacteria, Inter. Biodet.Biodeg.57: 168–173.

Newman R, Serpico M(2000). Adhesive and binders. In: Ancient Egyptian Materials and Technology, P.T. Nicholson and I. Shaw (eds.), Cambridge University Press, Cambridge:475-494.

Normand P(2006). The families *Frankiaceae*, *Geodermatophilaceae*, *Acidothermaceae* and *Sporichthyaceae*.Prokaryotes3: 669-681.

Odlyha M, Cohen SN, Foster GM(2000). Dosimetry of paintings: determination of the degree of chemical change in museum exposed test paintings (smalt tempera) by thermal analysis.Thermochimica Acta. 365: 35 – 44.

Onyari JM, Mulaa F, MuiaJ, Shiundu P(2008). Biodegrability of poly (lactic acid), preparation and characterization of PLA/ Gum Arabic blends. J. Polym. Environ.16: 205-212.

Pepe O, Sannino L, Palomba S, Anastasio M, Blaiotta G (2009). Heterotrophic microorganisms in deteriorated medival wall paintings in southern italian churches.Microbiol.Resear. 33: 1-10.

Peris-Vicente J, Adelantado JV, Domenech Carbo MT, Castro RM, Bosch Reig F(2006). Characterization of proteinaceous glues in old paintings by separation of o–phatalaldehyde derivatives of their amino acides by liquid chromatography with fluorescence detection. Talanta 68:1648-1654.

Petrova D, Derekova A, Vlahon S(2006). Purification and properities of individual collagenases from *Streptomyces* sp. strain 3B.Folia Microbiol. 51:93-98.

Pizzul L, Castillo MP, Stenstrom J(2006). Characterization of selected actinomycetes degrading polyaromatic hydrocarbons in liquid culture and spiked soil. World J. Microbiol. Biotechnol. 22: 745-752.

Regert M, Colinart S, Degrand L, Decavallas O,(2001). Chemical alteration and use of beeswax through time: Accelerated ageing tests and analysis of archaeological samples from various environmental contextes. Archaeometry 43: 549 – 569.

RegertM. Langlois J, Colinart S(2005). Characterization of wax works of art by gas chromarographic procedures. J. Chromatography A. 1091: 124-136.

Roper MM(2004). The isolation and characterisation of bacteria with the potential to degrade waxes that cause water repellency in sandy soils.Australian J. Soil Res. 42:427 – 434.

Salvadori B, Errico V, Mauro M, Melinek E, Dei L(2003). Evalulation of gypsum and oxalates in deterioration of mural paintings by quantative FTIR spectroscopy. Spectroscop. Lett. 36: 501-513.

Sambrook J,Russel D(2001). Molecular cloning: A laboratory manual, 3rd ed. Cold Springs Harbour Press.

Tomek J, Pechova P(1992). A Note on the thin layer chromatography of media in paintings. Stud. Conserv. 37:39-41.

Turner S, Pryer KM, Miao VPW, Palmer DJ (1999). Investigation deep phylogenatic relationships among cyanobacteria and Plastids by small subunit rRNA sequence analysis. J. Eukaryotic Microbiol. 46:327-338.

A low-cost *Lactobacillus salivarius* L29 growth medium containing molasses and corn steep liquor allows the attainment of high levels of cell mass and lactic acid production

KiBeom Lee[1]*, Sang-Kee Kang[2] and Yun Jaie Choi[2]*

[1]Biotechnology and Business Center, Incheon Technopark, 7-50 Songdo-Dong, Yeonsu-Gu, Incheon 406-840, South Korea.
[2]School of Agricultural Biotechnology, Seoul National University, Seoul 151-742, South Korea.

The aim of the present work was to formulate a *Lactobacillus salivarius* L29 industrial fermentation medium. High cell numbers and good levels of lactic acid by a *L. salivarius* L29 were obtained after shake flask fermentation using molasses as the sole carbon source and corn steep liquor (CSL (industrial grade); an organic source of N) as the principal nitrogen source. The optimum concentrations of molasses and CSL facilitating good cell growth and high-level lactic acid production were found to be 6 and 6% (both v/v), respectively. The maximum cell yield was 2.02 × 10^9 CFU/mL, thus about 15% lower than that obtained when MRS broth was employed for 5-L fermenters culture. Lactic acid production upon growth in industrial broth was 105 g/L; the total sugar content of the medium was 118 g/L (sucrose: glucose: fructose 68:14:18; w/w/w). Upon growth in De Man, Rogosa and Sharpe (MRS) broth (the total sugar content of which was 127 g/L, all of which was glucose), the lactic acid yield was 120 g/L. The optimized industrial growth medium was significantly more economical than were conventional broths.

Key words: *Lactobacillus salivarius* L29, molasses, corn steep liquor, culture medium optimization, lactic acid.

INTRODUCTION

Probiotics are defined as viable microorganisms that when consumed in adequate amounts, confer health benefits by improving the properties of indigenous microflora (Guillot, 1998). Antimicrobial polypeptides synthesized by bacteria can be admixed with live probiotic bacteria, thus increasing antimicrobial activity against pathogens (Dover et al., 2007). Probiotics have been used as an alternative to antibiotics in animals and humans. The bacteria most commonly associated with probiotic activity are lactobacilli and bifidobacteria (Gionchetti et al., 2003). The development of live microbial feed supplements currently attracts much research attention. Any probiotic benefiting human or animal health must fulfill several criteria (Gaurav et al., 2012). First, the probiotic must be amenable to large scale production. Second, the probiotic must be stable in food, in adverse environmental conditions associated with mixing, and in the gastrointestinal tract of the host.

Lactobacillus strains are the most widely used probiotic bacteria. Lactobacilli produce lactic acid as the sole or principal end-product of carbohydrate metabolism. However, *Lactobacillus* strains are nutritionally fastidious (Holt et al., 1994) and culture can be difficult, because such microbes have limited biosynthetic capacities and it is necessary that expensive nutrients (carbohydrates, amino acids, peptides, nucleic acid derivatives, and vitamins) be added to culture media. De Man, Rogosa and Sharpe (MRS) broth is the medium most commonly employed for the growth of lactobacilli.

However, the medium is complex and is too costly for industrial use. Therefore, it is important to reduce the cost of growth medium; low-cost (but effective) medium components must be identified. Molasses is the most important world-wide raw material for lactic acid production (Rashid and Altaf 2008; Ortiz et al., 2012). It provides a cheap source of fermentable sugars and contains some fermentive nutrients as well (Aslam, 1999). Corn steep liquor (CSL), a major by-product of the corn wet-milling industry, is also an inexpensive source of nutrients available on a large scale (Parekh et al., 1999; Altaf et al., 2007; Salgado et al., 2009).

L. salivarius L29, isolated from pig feces, is a probiotic bacterium that resides in the gastrointestinal tracts of animals (Yun et al., 2009). Earlier studies showed that this microorganism exhibited some valuable characteristics (Yun et al., 2009). It inhibited the growth of various pathogens including *Enterococcus faecalis, E. faecium, Escherichia coli, Staphylococcus aureus,* and *S.agalactiae*. The low pH created upon lactobacterial growth, and secretion by lactobacilli of a bacteriocin termed salivaricin (Ocana et al., 1999; Vera Pingitore et al., 2007; Juarez 2003; O' Shea et al., 2012) was considered to explain the observed beneficial effects. Thus, *L. salivarius* L29 is potentially a health-promoting probiotic.

For this reason, industry interest in production of *L. salivarius* L29 and products thereof has grown. The objective of the present work was to develop a low-cost culture medium that nonetheless facilitated efficient production of both *L. salivarius* L29 biomass, and lactic acid.

MATERIALS AND METHODS

Bacterial strain and preparation of inocula

A strain of *L. salivarius* L29 isolated from a pig farm in Korea (Yun et al., 2009) was used in the present study. A stock culture of *L. salivarius* L29 was maintained at -80°C in MRS broth containing 40% (v/v) glycerol. Each inoculum was prepared as follows. A 1 mL amount of frozen stock was inoculated into 50 mL MRS broth and grown at 37°C for 16 h. Aliquots (2.5 mL) of such cultures were transferred to 50 mL amounts of broth in 250 mL Erlenmeyer flasks and incubated at 37°C for 16 h on a rotary shaker operating at 150 rpm. The broth composition varied with the experimental design.

Culture media and growth conditions

Molasses-plus-CSL-based media supplemented with yeast extract (3.1 g/L), $MnSO_4$ (0.04 g/L), sodium citrate (0.01 g/L), K_2HPO_4 (4 g/L), and Tween-80 (0.2 g/L), containing different amounts with specified increments in concentration of both molasses and CSL, were prepared. Medium (50 mL aliquots) in 250 mL Erlenmeyer flasks was autoclaved at 121°C for 15 min and next inoculated to 5% (v/v) using a 16 h seed culture, prepared as described above. Incubation at 37°C followed. Sixteen hours later, the viable cell count and the lactate level of each culture were measured. Viability was determined by counting of colony-forming units (CFUs).

Optimization of industrial medium

The levels of molasses (the carbon source) and CSL (an organic nitrogen source) in the fermentation medium varied in the following manner. Initially, the molasses levels ranged from 2 to 10% (v/v) increments in the medium described above. Upon optimization of the molasses level, the CSL level varied between 2 to 10% (v/v) increments, again in the medium described above, but now containing the optimum concentration of molasses. The experiments were repeated in the reverse order (that is, vary CSL levels first; optimize; and next test molasses levels).

Bioreactor fermentation

Scale-up employed 5-L fermenters (Fermentec Inc. Co. Daejeon, Korea) containing 3-L working volumes of optimized medium. The control medium was MRS medium. In parallel, the experimental medium was: 6% (v/v) molasses, 6% (v/v) CSL, yeast extract (3.1 g/L), $MnSO_4$ (0.04 g/L), sodium citrate (0.01 g/L), K_2HPO_4 (4 g/L), and Tween-80 (0.2 g/L). Fermentation media were inoculated with flask cultures to 5% (v/v). Fermentation was conducted at 37±1°C with agitation of 150 rpm/min. The pH was maintained at pH 5.5 by automatic addition of 10 M NaOH. The total fermentation time was approximately 24 h. All experiments were performed at least in triplicate. Culture broth samples were taken periodically and analyzed in terms of lactate and sugar content, pH, and viable cell numbers.

Analytical methods

Cell growth was measured spectrophotometrically employing a UV-visible spectrometer (Beckman-Coulter Inc. Co. Fullerton, USA) operating at 600 nm. Sugar and lactic acid concentrations were determined via high- performance liquid chromatography (HPLC). An Agilent Technologies 1200 series column equipped with an RI detector was used (Agilent Technologies Inc. Co. Santa Clara, USA). The ion exclusion column (Aminex HPX-87H, Biorad, Hercules, CA, USA) was operated at a temperature of 40°C; the mobile phase was 0.06 N H_2SO_4; and the flow rate was 0.6 mL/min. The sample injection volume was 20 μL.

RESULTS AND DISCUSSION

Flask fermentation

MRS broth is commonly used for small-scale cultivation of *L.salivarius* L29. However, if industrial-scale growth of *L. salivarius* L29 in bioreactors is contemplated, it is necessary to develop a new low-cost medium. We sought

Table 1. Optimization of molasses concentration for the growth of *L. salivarius* L29.

Molasses concentration (%)	Maximum number of viable cells (CFU/mL)
2.0	1.25 ± 0.00 ×10^{6}
4.0	1.95 ± 0.01 × 10^{6}
6.0	3.70 ± 0.05 × 10^{6}
8.0	4.20 ± 0.01 × 10^{6}
10.0	4.00 ± 0.01 × 10^{6}

Shake-flask cultivation, 37°C, 150 rpm, 5% (v/v) inoculums, 16 h. Each number represents the mean ± SD of three replicates.

Table 2. Optimization of CSL concentration for the growth of *L. salivarius* L29 in medium containing 8.0% (v/v) molasses.

CSL concentration (%)	Maximum number of viable cells (CFU/mL)
2.0	0.57 ± 0.01 × 10^{9}
4.0	0.81 ± 0.02 × 10^{9}
6.0	1.26 ± 0.01 × 10^{9}
8.0	1.09 ± 0.00 × 10^{9}
10.0	0.75 ± 0.00 × 10^{9}

to optimize bacterial yield and lactic acid production and reduce production cost, by evaluating the utility of various media upon shake flask culture (250 mL flasks). We tested inexpensive growth media containing molasses and CSL. Refined sugars (glucose or sucrose) are commonly used as a carbon source for the growth of probiotics, but these are economically not feasible due to high cost of pure sugars whereas the product (lactic acid) is relatively cheap (Oh et al., 2005). Molasses, a food industrial waste, high in moisture and rich in carbon source have been considered as attractive nutrient source for industrial lactic acid production. The large amount of inorganic salts in molasses, exerting high osmotic pressure and the toxicity to cells are important factor during fermentation (Maiorella et al., 1984).

First, we explored the effects of various concentrations of molasses (2 to 10%, v/v) on the maximum yield of viable *L. salivarius* L29 cells. Cell levels were highest (4.20 × 10^{6} CFU/mL) when 8% (v/v) molasses was employed (Table 1). The low number of viable cells yielded upon culture in medium containing 2% molasses was presumably attributable to a paucity of nutrients. Higher levels of molasses (ca. 10%, v/v) appeared to inhibit cell growth, possibly because of the presence of high levels of salt (Mendes et al., 2011). Yeast extract is commonly used as a nitrogen source for the growth of probiotics, but CSL, an agricultural waste product, may be more cost-effective when industrial fermentation is contemplated. Although CSL is a good source of nutrients and elements, certain required growth components are lacking, and yeast extract (3.1 g/L), $MnSO_4$ (0.04 g/L), sodium citrate (0.01 g/L), K_2HPO_4 (4 g/L), and Tween-80 (0.2 g/L) must be added as supplements. We also measured the numbers of viable cells in cultures grown with different concentrations of CSL in the presence of 8% (v/v) molasses. The highest numbers of viable cells (ca. 1.26 x 10^{9} CFU/mL) were obtained in cultures grown with 6% (v/v) CSL (Table 2). Viable cell numbers did not rise when the CSL level was further increased. Higher CSL levels may be inhibitory because of distortion of the C: N ratio and impurities remaining in CSL during lactic acid fermentation (Underwood et al., 2004).

To better understand the effects of 8% (v/v) molasses and 6% (v/v) CSL on growth and lactic acid production by *L. salivarius* L29, CSL was added at 6% (v/v) to cultures growing in the presence of molasses at 2 to 10% (v/v); growth proceeded for 16 h. The optimum concentration of molasses was 6% (v/v); viable cell numbers and lactic acid yield were 2.0 × 10^{9}CFU/ mL and 20 g/L, respectively (Table 3). These findings indicate that when molasses is used as sole (main) carbon source, *L. salivarius* L29 may require relatively high levels of the molasses, possibly because of a shortage of other nutrients. However, CSL is also a carbon source; hence, the optimal concentration of molasses upon the addition of CSL appeared to be 6% (v/v). The adverse effects of high molasses concentrations (over 6% [v/v]) may be attributable to osmotic effects and the toxicity. Beyond a critical substrate concentration, a reduction in water activity combined with plasmolysis can lower the fermentation rate. To reconfirm the optimal concentrations identified, we measured the effects of different concentrations of CSL on growth and lactic acid production by *L. salivarius* L29 in the media containing 6% (v/v) molasses. In all instances, both cell growth and lactic acid yield were maximal at 6% (v/v) CSL (Table 4). Thus, we found that optimal conditions for the industrial growth of *L. salivarius*, in a medium containing inexpensive nutrients, are 6% (v/v) molasses and 6% (v/v) CSL.

Cultivation of *L. salivarius* L29 in an optimal medium in 5-L fermenters

To explore the ability of *L. salivarius* L29 to utilize sugars contained in molasses, batch fermentation was performed in 5-L fermenters, accompanied by pH control. Cell growth was dependent on the culture medium employed (data not shown). The cell number attained 9.57[OD_{600nm}] in MRS broth; the figure upon growth in industrial broth was 7.71[OD_{600nm}]. In both broths, *L. salivarius* L29 grew exponentially for the first 10 h and then entered a typical stationary phase. Growth between 10 and 24 h was probably limited by the presence of inhibitory compounds including organic acids, aldehydes,

Table 3. Optimization of molasses concentration for the growth of *L. salivarius* L 29 in medium containing 6% (v/v) CSL.

Molasses concentration (%)	Maximum number of viable cells (CFU/mL)	Lactic acid yield (g/L)
2.0	$1.02 \pm 0.00 \times 10^9$	16.1 ± 0.08
4.0	$1.29 \pm 0.04 \times 10^9$	18.6 ± 0.04
6.0	$2.00 \pm 0.02 \times 10^9$	20.0 ± 0.06
8.0	$1.20 \pm 0.04 \times 10^9$	17.83 ± 0.15
10.0	$1.12 \pm 0.04 \times 10^9$	16.88 ± 0.03

Table 4. Optimization of CSL level for the growth of *L. salivarius* L29 in medium containing 6% (v/v) molasses.

CSL concentration (%)	Maximum number of viable cells (CFU/mL)	Lactic acid yield (g/L)
2.0	$1.20 \pm 0.02 \times 10^9$	15.52 ± 0.14
4.0	$1.26 \pm 0.04 \times 10^9$	16.93 ± 0.17
6.0	$1.97 \pm 0.01 \times 10^9$	19.76 ± 0.19
8.0	$1.29 \pm 0.04 \times 10^9$	17.19 ± 0.08
10.0	$1.29 \pm 0.03 \times 10^9$	17.19 ± 0.07

phenols, and heavy metals (Roukas, 1998). We also measured medium pH values during growth (the initial pH was 6.0). Similar decreases in pH, from pH 6.0 to 5.5, were seen in the earlier stages of growth in either broth. This may be attributable to formation of organic acids via both deamination of medium amino acids and production of organic acids during growth. Later, the pH was held relatively constant at 5.5 via pH control.

The nutritional requirements of lactic acid bacteria are complex (Dalié et al., 2010), and only some available peptides are metabolized. To improve cell growth and lactic acid production, basic culture media, usually high in protein, are commonly supplemented with yeast extract, CSL, molasses, minerals and inorganic supplements (Amrane and Prigent, 1998; Amrane, 2000; Ha et al., 2003; Liu et al., 2004; Panesar et al., 2007). The sugar utilization and lactic acid production profiles are shown in Figure 1. Sugars in both MRS and industrial broth were directly fermented to lactic acid by *L. salivarius* L29. However, a lag in lactic acid production was evident between 12 to 24 h of fermentation. This may be attributable to temporary inactivation of *L. salivarius* L29 because the undissociated form of lactic acid can inhibit cell membrane functions (Ariyapitipun et al., 1999), thus compromising both nutrient transport and growth. The levels of lactic acid produced from MRS broth (127 g total sugar/L) were 120 g/L whereas that produced when the industrial medium (118 g total sugar/L) was employed was 105 g/L.

The industrial broth contained 80 g sucrose/L, 21.8 g glucose/L, and 16.3 g fructose/L. *L. salivarius* L29 can use both the di- and mono-saccharides of molasses as carbon sources for lactic acid fermentation. *L. salivarius* L29 is homofermentative and efficiently produces lactic acid from sucrose. Table 5 summarizes our data on fermentation by *L. salivarius* L29 of MRS and industrial broth. The maximum viable cell counts obtained were 2.38×10^9 and 2.02×10^9/mL in MRS and industrial broth, respectively. The lactic acid yield was considerably higher (by 94%) when MRS broth was employed. This is attributable to the high nutritive value of MRS broth, which contains glucose, proteins, amino acids, and vitamins. The lower lactic acid yield observed when industrial broth was used may be attributable to the presence of inhibitory compounds such as inorganic salts. The lactic acid production rate was as 5 g/L/h when MRS broth was used, and 4.38 g/L/h when the industrial inhibitory compounds. Further, when molasses and CSL are used together in fermentation medium, serious problems in product purification may become evident (Timbuntam et al., 2006). These problems can be medium was employed. The lower production evident in industrial medium may again reflect the presence of overcome by the use of ultra filtration and electrodialysis.

In conclusion, we assessed the ability of *L. salivarius* L29 (in terms of both bacterial growth and lactic acid production) to grow in an inexpensive industrial broth containing molasses as the carbon source and CSL as the principal nitrogen source; minor nutrients were also added. The optimal medium composition featured 6% (v/v) molasses and 6% (v/v) CSL. Yeast extract (3.1 g/L), $MnSO_4$ (0.04 g/L), sodium citrate (0.01 g/L), K_2HPO_4 (4 g/L), and Tween-80 (0.2 g/L) were also required. We have thus shown that high numbers of viable cells and good levels of lactic acid may be achieved via fermentation in an industrial broth. This medium is simpler and less expensive than MRS medium. We have thus laid the foundation for the development of cost-

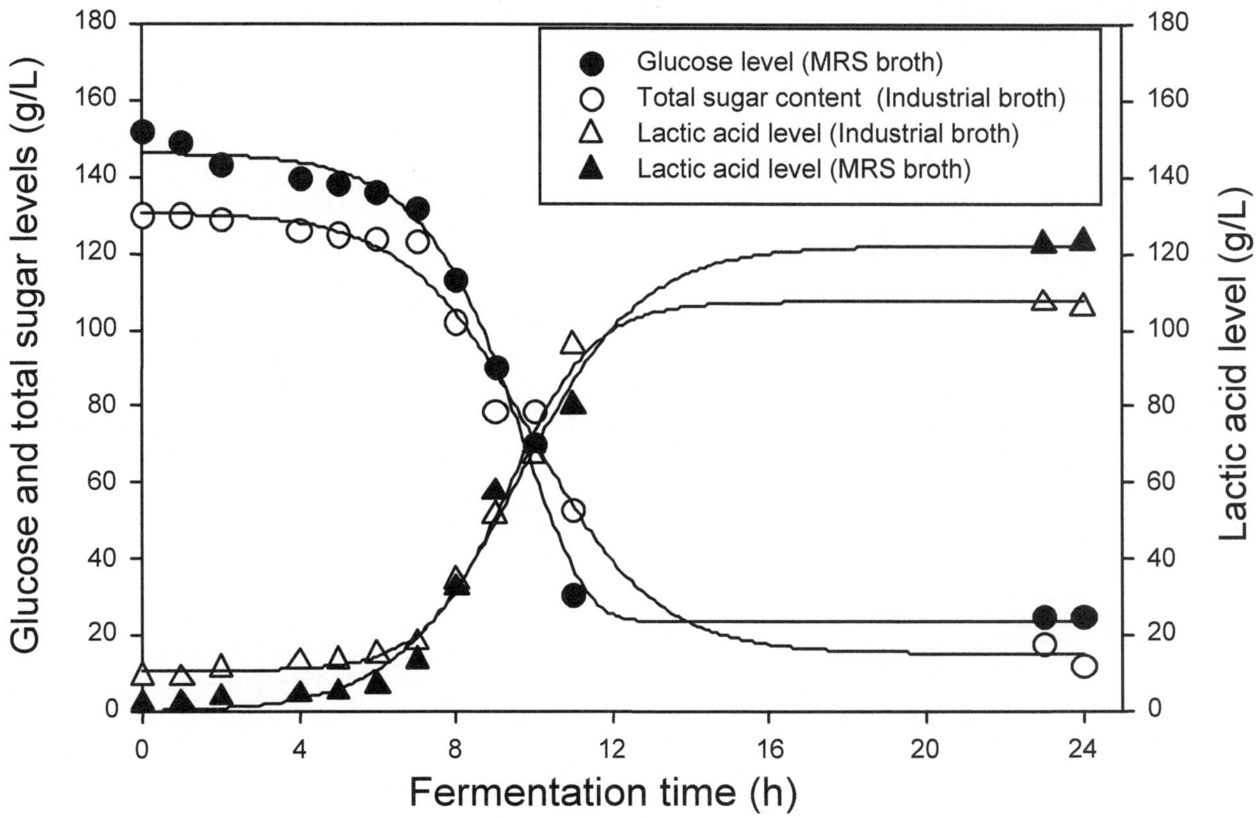

Figure 1. Glucose and total sugar consumption levels, and lactic acid yield, by *L. salivarius* L29 grown in batch fermentations using MRS and industrial broth. Each point represents the mean value of data obtained in three independent experiments.

Table 5. Summary of fermentation parameters derived from experiments using MRS and industrial medium for the growth of *L. salivarius* L29.

Parameter	MRS broth	Industrial broth
Maximum CFU (× 10^9)	2.38 ± 0.01	2.02 ± 0.02
Initial sugar level (g/L)	152	130
Sucrose	0	88
Glucose	152	18
Fructose	0	24
Residual sugar level (g/L)	25	12
Sucrose	0	6.1
Glucose	25	2.9
Fructose	0	2.9
Sugar consumption (g/L)	127	118
Total lactic acid yield (g/L)	120	105
% Yield $_{p/s}$	94	89
Productivity after 24 h (g/L/h)	5	4.38

The fermentation time was 24 h.

effective lactic acid fermentation.

ACKNOWLEDGEMENT

This research was supported by Technology Development Program for Agriculture and Forestry, Ministry for Food, Agriculture, Forestry and Fisheries, Republic of Korea (Project No. 610005031SB130).

REFERENCES

Altaf M, Venkateshwar M, Srijana M, Reddy G (2007). An economic approach for L-(+) lactic acid fermentation by *Lactobacillus amylophilus* GV6 using inexpensive carbon and nitrogen sources. J. Appl. Microbiol. 103:372-380.

Amrane A (2000). Effect of inorganic phosphate on lactate production by *Lactobacillus helveticus* grown on supplemented whey permeate. J. Chem. Technol. Biotechnol. 75:223-228.

Amrane A, Prigent Y (1998). Influence of yeast extract concentration on batch cultures of *Lactobacillus heveticus*: growth and production coupling. World J. Microbiol. Biotechnol. 14:529-534.

Ariyapitipun T, Mustapha A, Clarke AD (1999). Microbial shelf life determination of vacuum-packaged fresh beef treated with polylactic acid, lactic acid, and nisin solutions. J. Food Prot. 62(8):913-920.

Aslam M (1999). Ethanol production for motor vehicles and chemical feedback. Pakistan Academy of Science, Islamabad, pp. 3-37.

Dalié DKD, Deschamps AM, Richard-Forget F (2010). Lactic acid bacteria-Potential for control of mould growth and mycotoxins: A review. Food Control 21(4):370-380.

Dover SE, Aroutcheva AA, Faro S, Chikindas ML (2007). Safety study of an antimicrobial peptide lactocin 160, produced by the vaginal *Lactobacillus rhamnosus*. Infect. Dis. Obstet. Gynecol. 2007:78248.

Gaurav T, Ruchi T, Saurabh P, Preeti P (2012). Promising future of probiotics for human health: Current scenario. Chron. Young Sci. 3:17-28.

Gionchetti P, Rizzello F, Helwig U, Venturi A, Lammers KM, Brigidi P, Vitali B, Poggioli G, Miglioli M, Campieri M (2003). Prophylaxis of pouchitis onset with probiotic therapy: a double-blind, placebo-controlled trial. Gastroenterology 124(5):1202-1209.

Guillot JF (1998). Les probiotiques en alimentation animale. Cahiers Agricultures 7:49-54.

Ha MY, Kim SW, Lee YW, Kim MJ, Kim SJ (2003). Kinetics analysis of growth and lactic acid production in pH-controlled batch cultures of *Lactobacillus casei* KH-1 using yeast extract/corn steep liquor/glucose medium. J. Biosci. Bioeng. 96:134-140.

Holt JG, Krieg NR, Sneath PHA, Staley JT, Williams ST (1994). Group 19. Regular, nonsporing Gram-positive rods. In: Holt JG, Krieg NR, Sneath PHA, Staley JT, Williams ST (Eds.), Bergey's manual of determinative bacteriology, Williams & Wilkins Co., Baltimore, pp. 565-570.

Juarez Tomas MS, Ocana V, Wiese B, Nader-Macias ME (2003). Growth and lactic acid production by vaginal *Lactobacillus adidophilus* CRL 1259. Inhibition of uropathogenic *Escherichia coli*. J. Med. Microbiol. 52:1117-1124.

Liu C, Liu Y, Liao W, Wen Z, Chen S (2004). Simultaneous production of nisin and lactic acid from cheese whey: optimization of fermentation conditions through statistically based experimental designs. Appl. Biochem. Biotechnol. 114:627-638.

Maiorella BL, Blanch HW, Wilke CR (1984). Feed component inhibition in ethanolic fermentation by *Saccharomyces cerevisiae*. Biotechnol. Bioeng. 24:1155-1166.

Mendes FS, Pajuelo MG, Cordier H, Francois JM, Vasconcelos I (2011). 1,3-Propanediol production in a two-step process fermentation from renewable feedstock. Appl. Microbiol. Biotechnol. 92(3):519-527.

Ocana VS, De Ruiz Holgado AP, Nader-Macias ME (1999). Characterization of a bacteriocin-like substance produced by a vaginal *Lactobacillus salivarius* strain. Appl. Environ. Microbiol. 65:5631-5635.

Oh HR, Wee YJ, Yun JS, Han SH, Jung SW, Ryu HW (2005). Lactic acid production from agaricultural resources as cheap raw materials. Biores. Technol. 96:1492-1498.

Ortiz ME, Fornaguera MJ, Raya RR, Mozzi F (2012). Lactobacillus reuteri CRL 1101 highly produces mannitol from sugarcane molasses as carbon source. Appl. Microbiol.Biotechnol. 95:991-999.

O' Shea EF, O' Connor PM, Raftis EJ, O' Toole PW, Stanton C, Cotter PD, Ross RP, Hill C (2012). Subspecies diversity in bacteriocin production by intestinal *Lactobacillus salivarius* strains. Gut Microbes. 3(5):468-473.

Panesar PS, Kennedy JF, Gandhi DN, Bunko K (2007). Bioutilisation of whey for lactic acid production. Food Chem. 105:1-14.

Parekh M, Formanek J, Blaschek HP (1999). Pilot-scale production of butanol by *Clostridium beijerinckii* BA101 using a low-cost fermentation medium based on corn steep water. Appl. Microbiol. Biotechnol. 51:152-157.

Rashid T, Altaf Z (2008). Potential and environmental concerns of ethanol production from sugarcane molasses in Pakistan. Nature Precedings. hdl.10101/npre. 1499.1.

Roukas T (1998). Pretreatment of beet molasses to increase pullulan production. Proc. Biochem. 33:805-810.

Salgado JM, Rodriguez N, Cortés S, Dominguez JM (2009). Development of cost-effective media to increase the economic potential for larger-scale bioproduction of natural food additives by *Lactobacillus rhamnousus*, *Debaryomyces hansenii*, and *Aspergillus niger*. J. Agric. Food Chem. 11:10414-10428.

Timbuntam W, Sriroth K, Tokiwa Y (2006). Lactic acid production from sugar-cane juice by a newly isolated *Lactobacillus* sp. Biotechnol. Lett. 28:811-814.

Underwood SA, Buszko ML, Shanmugam KT, Ingram LO (2004). Lack of protective osmolytes limits final cell density and volumetric productivity of ethanologenic, *Escherichia coli* k011 during xylose fermentation. Appl. Environ. Microbiol. 70:2734-2740.

Vera Pingitore E, Salvucci E, Sesma F, Nader-Macias ME (2007). Different strategies for purification of antimicrobial peptides from Lactic Acid Bacteria (LAB), in communicating current research and educational topics and trends in applied microbiology, In: Mendez-Vilas A (Ed), Formatex, Badajoz pp. 557-568.

Yun JH, Lee KB, Sung YK, Kim EB, Lee HG, Choi YJ (2009). Isolation and characterization of potential probiotic lactobacilli from pig feces. J. Basic. Microbiol. 49:220-226.

6

Overexpression of *BrSAC1* encoding a phosphoinositide phosphatase isolated from Chinese cabbage (*Brassica rapa* L) improved tolerance to cold, dehydration, and salt stresses in transgenic tobacco

Kyung Hee Han[1#], Yu Jin Jung[2,3#], Uuganchimeg Bayarsaikhan[1], In Hye Lee[1], Jang Sun Choi[6], Ill Sup Nou[4], Yong Gu Cho[5] and Kwon Kyoo Kang[1, 2,3*]

[1]Department of Horticulture, Hankyong National University, 67 Seokjeong-dong, Ansung 456-749, South Korea.
[2]Institute of Genetic Engineering, Hankyong National University, Ansung 456-749, Korea.
[3]Clone Company, 67 Seokjeong-dong, Ansung 456-749, South Korea.
[4]Department of Horticulture, Sunchon National University, Suncheon 540-742, Korea.
[5]Department of Crop Science, Chungbuk National University, Cheongju, Korea.
[6]National Institute of Horticultural and Herbal Science, Suwon, 440-746, Korea.

This study demonstrates the isolation and characterization of cDNA encoding a phosphoinositide phosphatase (PIP) from a stem cell cDNA library of Chinese cabbage *(Brassica rapa)* seedling. The full length gene (*BrSAC1*; GenBank accession no., GU434275) contained 1999 base pairs (bp), with an open reading frame of 1785 bp, encoding a polypeptide of 594 amino acids with a predicted molecular weight of 65 kDa, including a putative N-terminal signal peptide (the signal peptide counted within the 594 residues). Other regions found within the sequence include a conserved KXKXX COPI-binding motif and a consensus Cx5R (T/S) catalytic motif. BrSAC1 protein shares 92% identity with AtSac1B, and 86% identity with AtRHD4 at the amino acid level. Gene expression analyses revealed that *BrsSAC1* was constitutively expressed at high levels in the pistil, stamen and flower bud, whereas it was expressed at low levels in the leaf and stem. In addition to injury, *BrSAC1* expression was also induced in response to different types of stress condition, namely cold, desiccation, salt, submergence, abscisic acid and heavy metals. Overexpression of *BrSAC1* in transgenic tobacco conferred tolerance to cold, dehydration, and salt stress at the seed germination/seedling stage as reflected by the percentage of germination/green seedlings, the fresh weight of seedlings and their development pattern. Our data suggest that *BrSAC1* is an important stress response determinant in plants.

Key words: Abiotic stress, *Brassica rapa*, phosphoinositide phosphatase, transgenic plant.

INTRODUCTION

Plant growth is strongly affected by a variety of major obstacles affecting plant growth and crop environmental stresses. Cold, salt, and drought are the productivity. In response to these stresses, plants initiate signaling cascades leading to the development of defense responses. Phosphoinositides (PIs) are signaling molecules that regulate cellular events, including vesicle targeting and interactions between membrane and cytoskeleton. To date, various enzymes involved in PI

*Corresponding author. E-mail:kykang@hknu.ac.kr.

#These authors contributed equally to this work.

signaling and biosynthesis have been identified and characterized in animals (Martin, 1998; De Matteis et al., 2002). Compared to animals, in plants the role of PIs is still poorly understood. During the last decade, however, new information has emerged showing involvement of PIs in various plant processes, such as calcium mobilization, signal transduction, cytoskeletal rearrangements and membrane trafficking (Wang, 2004) in addition to the defense against attack by pathogens (Laxalt and Munnik, 2002). Recent research indicates that PIs play important roles in development, including stress signaling (Meijer et al., 2001; Xiong et al., 2001), growth (Williams et al., 2005), and vascular patterning (Carland and Nelson, 2004). PI phosphatases (PIP) are classified into three main groups, 3-, 4-, or 5-phosphatases on the basis of the position of the phosphate-inositol head group being hydrolyzed (Takenawa and Itoh, 2001). The group of phosphatases called suppressor of actin (SAC) domain phosphatases have been identified in yeast and animal; these phosphatases were shown to hydrolyze the phosphates on multiple positions on the inositol head group of PIs (Hughes et al., 2000). As in other eukaryotic systems, PI metabolism in higher plants is considered important in signal transduction. The involvement of phospholipid signaling in the cell responses to stress stimuli, including salt, osmotic, temperature, and pathogen stresses, has been shown, and multiple members of the PI pathway are thought to mediate stress responses through different mechanisms.

In this study, we isolated the *BrSAC1* gene encoding a PIP from the cDNA of *Brassica rapa* seedling and characterized its modulation of PI signals during abiotic stress response. Also, introgression of *BrSAC1* into tobacco led to stress tolerance when assessed at the seedling stage.

MATERIALS AND METHODS

Plant materials and stress treatments

Chinese cabbage (*B. rapa* var. OSOME) was aseptically grown on Murashige Skoog (MS) agar medium (Murashige and Skoog, 1962) in a culture room under a 16:8 h light:dark period at 25°C. Salt and hormone treatments were conducted, after nine days of growth, on seedlings that were transferred to fresh liquid half-strength MS medium without sucrose (MSH medium) containing either 250 mM NaCl or 100 µM abscisic acid (ABA) for 24 h. Induction of cold stress was examined on seedlings grown in the above conditions for 24 h after transferring to 4°C. Membrane fluidity treatments using chemical modulators were conducted on cut leaves of seedlings exposed to 5, 10, and 15 mM benzyl alcohol at 25°C for 3 h followed by growth at 8°C for 24 h. Cut leaves of seedlings were kept at 25°C for 6 h in the presence of 2, 4 and 6% dimethyl sulfoxide (DMSO), while the control leaves were kept in water at 25°C.

Isolation of full-length cDNA of *BrSAC1*

A partial *BrSAC1* sequence was identified in the dbEST division of GenBank and searches were conducted using basic local alignment search tool (BLAST) with a specific SAC domain motif as the query sequence. The full-length cDNA was isolated by RACE technology using the SMART RACE-PCR kit (Clontech, Palo Alto, CA, USA) (Lee et al., 2009).

RNA isolation and real-time PCR

Total RNA from leaf tissues was extracted using Trizol reagent (Invitrogen, USA) according to the manufacturer's protocol. The specific primers for the analysis of *BrSAC1* expression in germinating seeds were: *BrSAC1*F3 5′-CGGAAAGAGTTGCATTCAGC-3′ and *BrSAC1*R3 5′-ATGATGGCTGGAATGCCCT-3′. Real-time polymerase chain reaction (qRT-PCR) was performed using a Bio-RAD I Cycler IQ5 machine as previously described (Ali-Benali et al., 2005) using RT pre-mix (TOYOBO Co., Japan). The mean threshold cycle (Ct) values of triplicate PCR reactions were computed and relative quantification of expression levels was performed using the comparative Ct method (Livak and Schmittgen, 2001). The fold change in total RNA of the target gene relative to the reference gene (actin) was determined by the following formula: fold change = $2^{-\Delta\Delta Ct}$, where $\Delta\Delta Ct = (Ct_{target\ gene} - Ct_{actin\ gene})$ transgenic plants - $(Ct_{target\ gene} - Ct_{actin\ gene})$ wild-type plants.

First-strand cDNA was generated using Superscript™ III reverse transcriptase according to manufacturer's instructions (Invitrogen, USA). BrSAC1 was amplified using 1 ul cDNA template added with the primer pairs *BrSAC1*F2 5′-ATGGAGGCTCCTAAGCA-3′ and *BrSAC1*R2 5′-CGCAAAGAGTTGCATTCAGC-3′. The PCR amplification program consisted of an initial step at 94°C for 5 min followed by 30 cycles of 94°C for 1 min, 58°C for 1 min, 72°C for 2 min, and a final step at 72°C for 10 min.

Generation of *BrSAC1* transgenic tobacco plants

Full-length sequence of *BrSAC1* amplified from a pGEM-T easy vector with the primers *BrSAC1*F1 (5′-ATGGAGGCTCCTCCTAAGCACA-3′) and *BrSAC1*R1 (5′-TCAAGGTCTGGGCTTGTGCAG-3′) was confirmed by sequencing and inserted into the *SfiI* site of a pBig plant binary vector system for overexpression; the vector also contained a modified cauliflower mosaic virus (CaMV) 35S promoter. The pBig plant binary vector containing complete coding sequence of *BrSAC1* was introduced into tobacco seedlings via *Agrobacterium tumefaciens*-mediated transformation according to the procedure of Horsch et al. (1985).

Four-week-old T2 generation tobacco seedlings were analysed for the presence of the transgene. DNA was extracted from the leaves of the tobacco seedlings by using cetyltrimethyl ammonium bromide (CTAB) method (Rogers and Bendich, 1994). The primers used for PCR amplification were *BrSAC1*F2 and *BrSAC1*R2, with reaction conditions as described above.

Analysis of transgenic plants for abiotic stress tolerance

Four homozygous T2 generation transgenic tobacco (*Nicotiana tabacum* cv. Samson) plants were used in all subsequent abiotic stress assays. The seeds of transgenic and non-transgenic (wild type, WT) plants were germinated and grown on MS medium for 16 days in a culture room under a 16:8 h light:dark period at 24°C.

For salt, cold, and dehydration stress treatments, (Figure 4), the seedlings were germinated and grown according to the above procedure for 14 days and then transferred to ½ MS media containing 250 mM NaCl, 4°C cold, and 0.3 M and 0.4 M mannitol. After 14 days, the fresh weight (FW) and shoot and root lengths were measured. Statistical analyses were performed using a Student's *t* test in Microsoft Excel. P-values below 0.05 were considered significant. To determine the effects of salt, cold, and dehydration stress on transgenic tobacco plants (Figures 5 to 7), seeds of WT and transgenic plants were germinated and grown on ½ MS medium for various times (noted below) in a culture room according to the procedure described above. The effect of salt stress resulting in chlorosis was examined in WT and transgenic seedlings (21 days) that had been transferred to sterile tissue paper soaked with a 250 mM NaCl MSH solution for four days. After treatment, seedlings were washed briefly in sterile Milli-Q water and allowed to grow and recover for eight days on MSH in a culture room, and then the number of green seedlings was determined. Four leaf discs of transgenic (lines 2 and 4) and WT plants (21 days) were floated in 250 mM NaCl MSH solution treatment versus no salt control MSH solution for four days to examine status and chlorophyll content. The leaf disk chlorophyll content (mg/g FW) was measured using the method described by Aono et al. (1993) and was shown as mean ± standard deviation. To examine the effects of cold stress treatment, transgenic and WT plants (21 days) were transferred to a cold chamber (4°C) for 15 days. After treatment, the plants were transferred to culture room conditions and allowed to grow and recover for 15 days. Individual plant FW (mg) was measured and is shown as the mean ± standard deviation. To examine the effects of dehydration stress on germinating seeds of transgenic and WT plants, the seeds were germinated on 0.3 and 0.4 M mannitol and observations (percent germination, FW) of representative plants were recorded over a period of eight days according to the method of Mukhopadhyay et al. (2004).

Multiple sequence alignment and phylogenetic analysis

Multiple sequence alignment of the *BrSAC1* gene was performed using Clustal X version 1.83 (Thompson et al., 1997). A phylogenetic tree was constructed using DNA Data Bank of Japan (DDBJ; http://clustalw.ddbj.nig.ac.jp/top-e.html), and bootstrap analysis was performed with 1,000 samplings.

RESULTS

Cloning and sequence analysis of *BrSAC1*

A partial cDNA of *BrSAC1* was identified by a database search of the dbEST division of GenBank with a specific SAC domain motif as the query sequence. The full-length cDNA of BrSAC1 was isolated using rapid amplification of cDNA ends-polymerase chain reaction (RACE-PCR), revealing the gene to be 1,999 bp in length and consisting of a 114 bp 5′-untranslated region (UTR), a complete opening reading frame (ORF) of 1,785 bp encoding a polypeptide of 594 amino acids (65 kDa) and a 3′-UTR of 100 bp (data not shown).

Sequence BLAST analysis showed that the BrSAC1 protein exhibited identity with AtSac1B (*Arabidopsis thaliana*), NP201403 (92%); AtRHD4 (*A. thaliana*), NP190714 (86%); SAC1 (*Saccharomyces cerevisiae*), CAA82057 (38%); Sac1 (*Drosophila melanogaster*), NM138243 (37%); Rat Sac1 (*Rattus norvegicus*), NP446250 (35%); mouse Sac1 (*Mus musculus*), NM030692 (35%); human Sac1 (*Homo sapiens*), NM014016 (35%); rice Sac1 (*Oryza sativa*), NP001067755 (54%); and maize Sac1 (*Zea mays*), BT086272 (48%). A comparative analysis of the *BrSAC1* gene sequence with known SAC domain containing sequences revealed the presence of a KXKXX conserved COPI binding motif and the consensus CX5R(T/S) sequence for the Sac1-catalytic domain (Figure 1A).

Further, both BrSAC1 and AtSac1B contain N-terminal KXKXX motifs that serve as binding sites for the coatomer (COPI) complex. This COPI interaction motif is considered a unique feature of mammalian Sac1s, as it is not conserved in other plants, yeast, or Drosophila Sac1 proteins (Figure 1B). The phylogenetic relationship of the deduced amino acid sequence of *BrSAC1* to the SAC domains from other organisms was determined using Clustal W within the BioEdit program (Figure 1C).

Expression of the *BrSAC1* gene in various organs and during seed germination

Expression analysis of the *BrSAC1* gene transcript was performed in various vegetative tissues (leaf, stem and root), reproductive tissues (stamen, pistil, petal, flower bud, calyx of flower, and flower stalk), and during seed germination. Semi-quantitative reverse transcription (RT)-PCR analysis showed varying expression of *BrSAC1* gene across tissues, highest in the stamen among reproductive tissues and in leaf among vegetative tissues (Figure 2A). An examination of *BrSAC1* expression during the first 72 h of seed germination revealed an increase in mRNA levels of *BrSAC1* that peaked at 36 h, about 20 fold higher than that at 0 h, with a fast decrease at subsequent times examined (Figure 2B).

Expression pattern of *BrSAC1* under different abiotic stresses

The results of qRT-PCR analysis showed that *BrSAC1* was induced under various abiotic stresses. *BrSAC1* transcripts showed rapid induction (9.0 fold) peaking at 1 to 2 h after exposure to cold (4°C) conditions, followed by a gradual decrease in expression levels to pre-treatment levels over the next several hours of exposure. In the presence of salt stress (250 mM NaCl), transcript levels were rapidly induced several fold by 1 h, maintaining a high level of induction after 3 to 8 h, then returning to pre-treatment levels (Figure 3). Expression of *BrSAC1* in response to the stress hormone ABA was also determined and exogenous ABA treatment (100 μM) resulted in strong *BrSAC1* induction within 1 h, gradually decreasing to pre-treatment levels by 8 h. We also wanted to examine the role of membrane fluidity and the cold induction of *BrSAC1* transcript. Membrane rigidification at low temperatures can be prevented by benzyl alcohol (BA) that acts as a membrane fluidizer. Similar examinations can also be simulated at room temperature by treatment with a membrane rigidifier DMSO (Mukhopadhyay et al., 2004). In order to examine

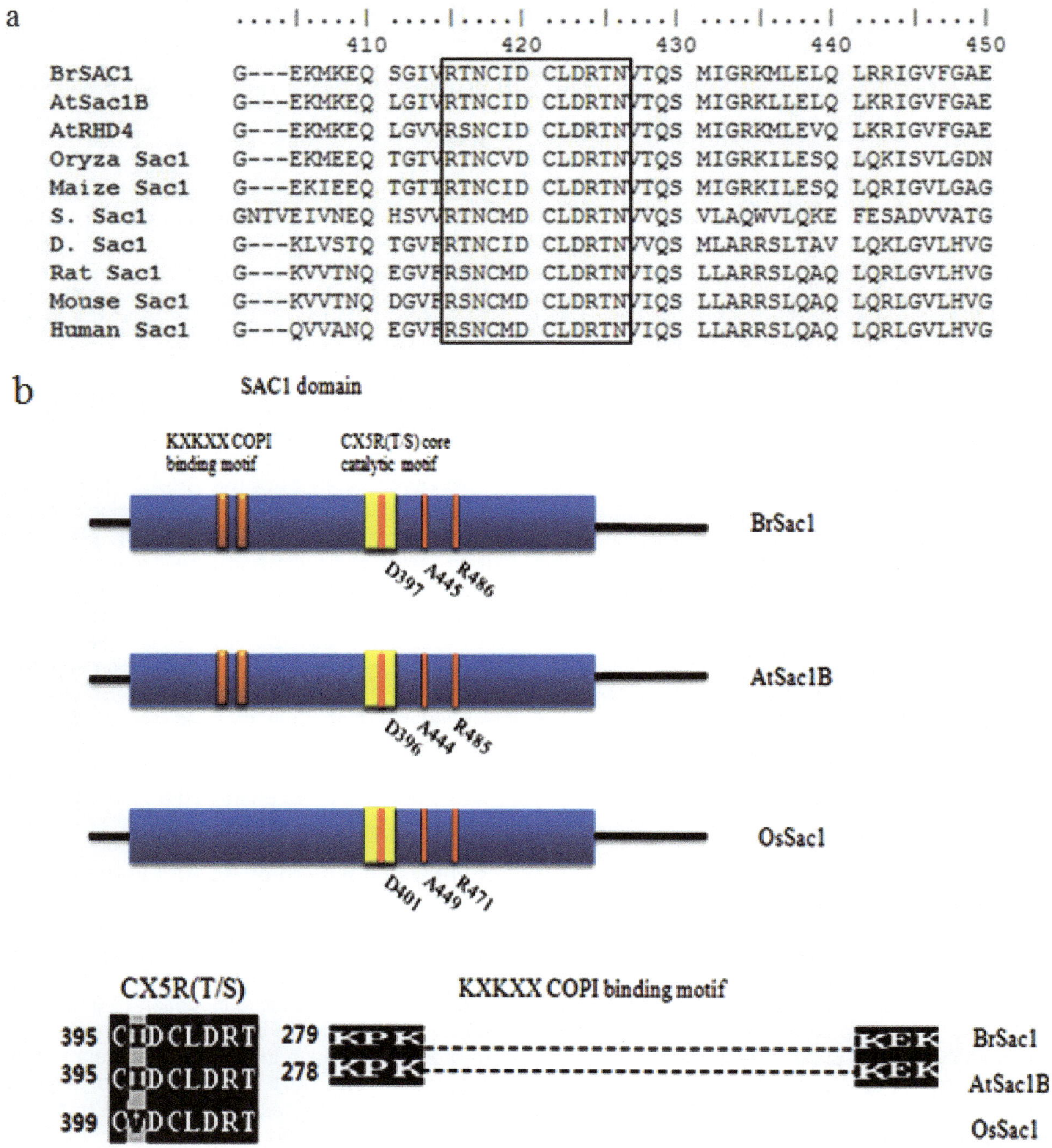

Figure 1. Sequence alignment, domain structure and phylogenetic relationship of *BrSAC1*. **A.** Comparison and deduced amino acid sequence of *BrSAC1* and other Sac1 proteins, AtSac1B (*Arabidopsis thaliana*), NP201403; AtRHD4 (*A. thaliana*), NP190714; SAC1 (*Saccharomyces cerevisiae*), CAA82057; Sac1 (*Drosophila melanogaster*), NM138243; rat Sac1 (*Rattus norvegicus*), NP446250; mouse Sac1 (*Mus musculus*), NM030692; human Sac1 (*Homo sapiens*), NM014016; rice Sac1 (*Oryza sativa*), NP001067755; and Maize Sac1 (*Zea mays*), BT086272. Sequence alignment of the core catalytic domains and the KXKXX COPI binding motifs of the SAC1 protein family. The CX_5R(T/S) core catalytic domain of Sac1 is highly conserved in *yeast*, higher plants, *Drosophila*, and *mammals.* Identical and conserved residues in all aligned sequences are indicated by the rectangle box. **B.** (a) Diagrams of the domain organization of Sac1 proteins. Domains and motifs of interest are highlighted by colored boxes. Blue boxes show the Sac1-catalytic domain, yellow boxes show the CX5R(T/S) core catalytic motif, and orange boxes show the KXKXX COPI binding motif. Three key amino acids within the Sac1 domain known to affect catalytic activity (Rivas et al., 1999; Liu et al., 2008) are indicated in red. (b) Sequence alignment of the core catalytic domains and the KXKXX COPI binding motifs of the Sac1 protein families. Identical amino acids are highlighted by black boxes. **C.** The phylogenetic tree of the Sac1 protein family. *BrSAC1* isolated in this study. A phylogenetic tree was constructed using CLC Main workbench 5 with CLUSTAL W alignments as input using the neighbor-joining method (Saitou and Nei 1987) to construct the tree.

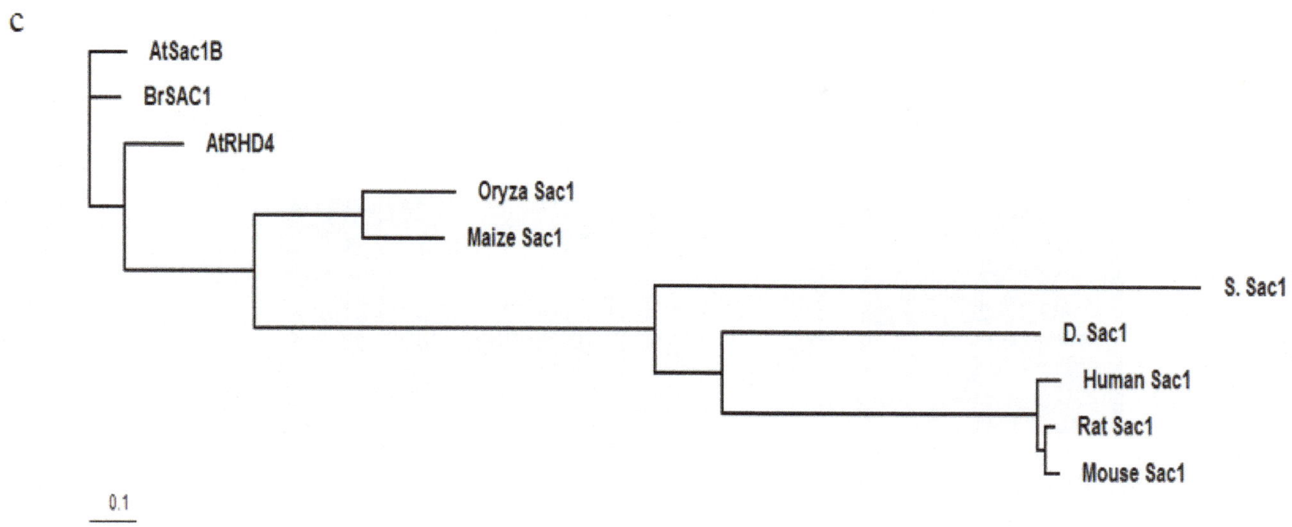

Figure 1. Contd.

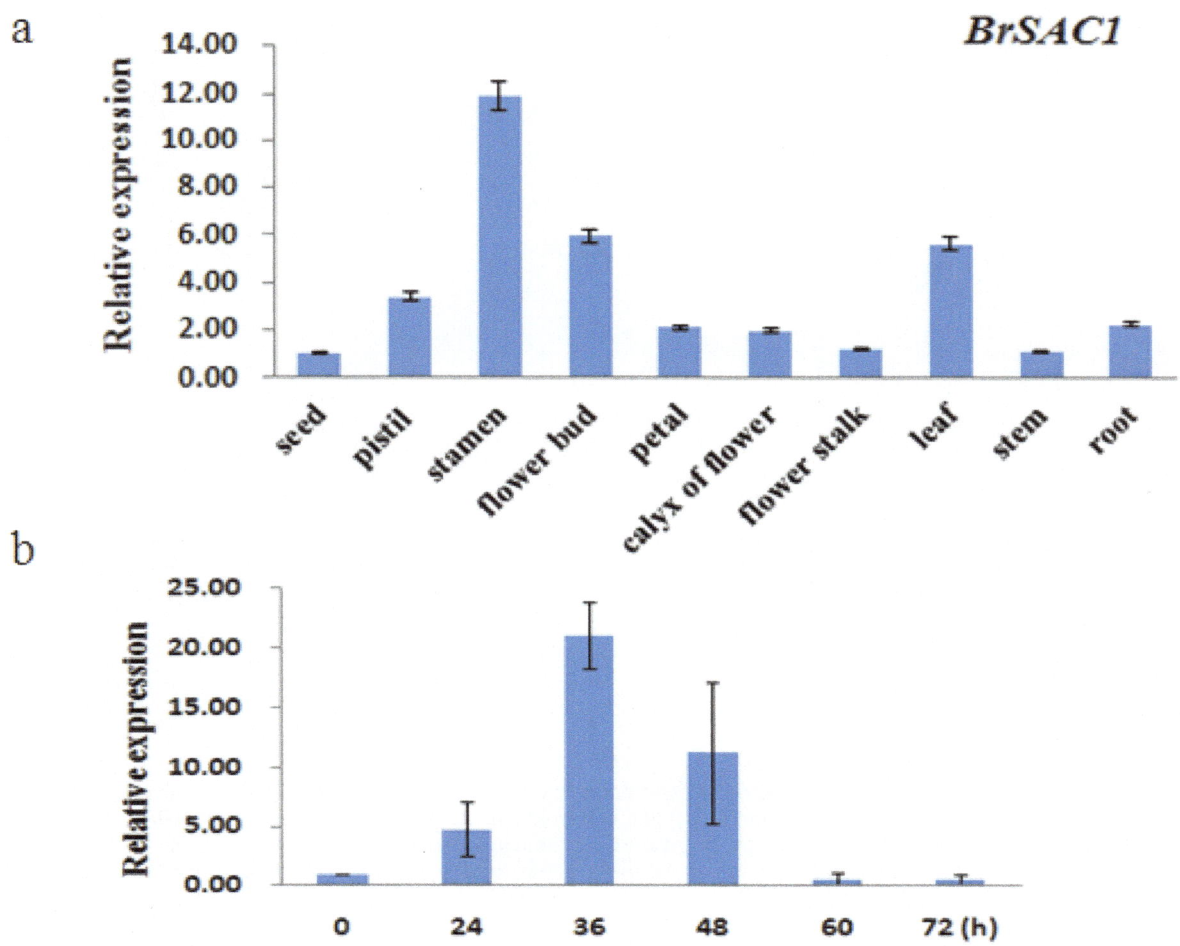

Figure 2. *BrSAC1* showing the broad expression across plant tissues using real-time PCR analysis. **A.** Expression levels of *BrSAC1* were measured in various tissues collected from *Brassica rapa.* **B.** Real-time PCR analysis of *BrSAC1* in germinating seedlings (0-72 h after germination). Amounts of template were calibrated using the mRNA as a reference. Data are presented as means ± standard deviation of three independently replicated experiments.

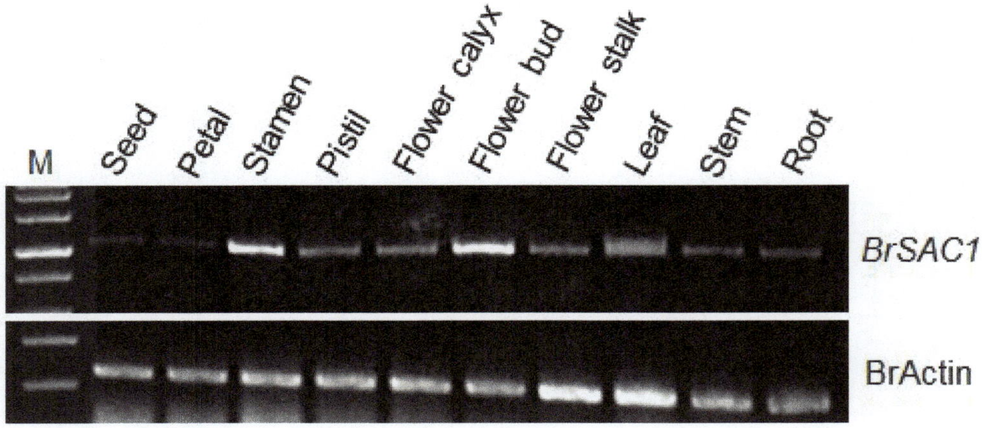

Supplemental Figure 1 for Figure 2. Han et al. *BrSAC1* showing broad expression across plant tissues. Semi-quantitative reverse transcription-polymerase chain reaction (RT-PCR) analysis of *BrRZFP1* was conducted in various tissues collected from *Brassica rapa.* Actin was used as a positive control and was detected in all tissues shown as the set of lower bands.

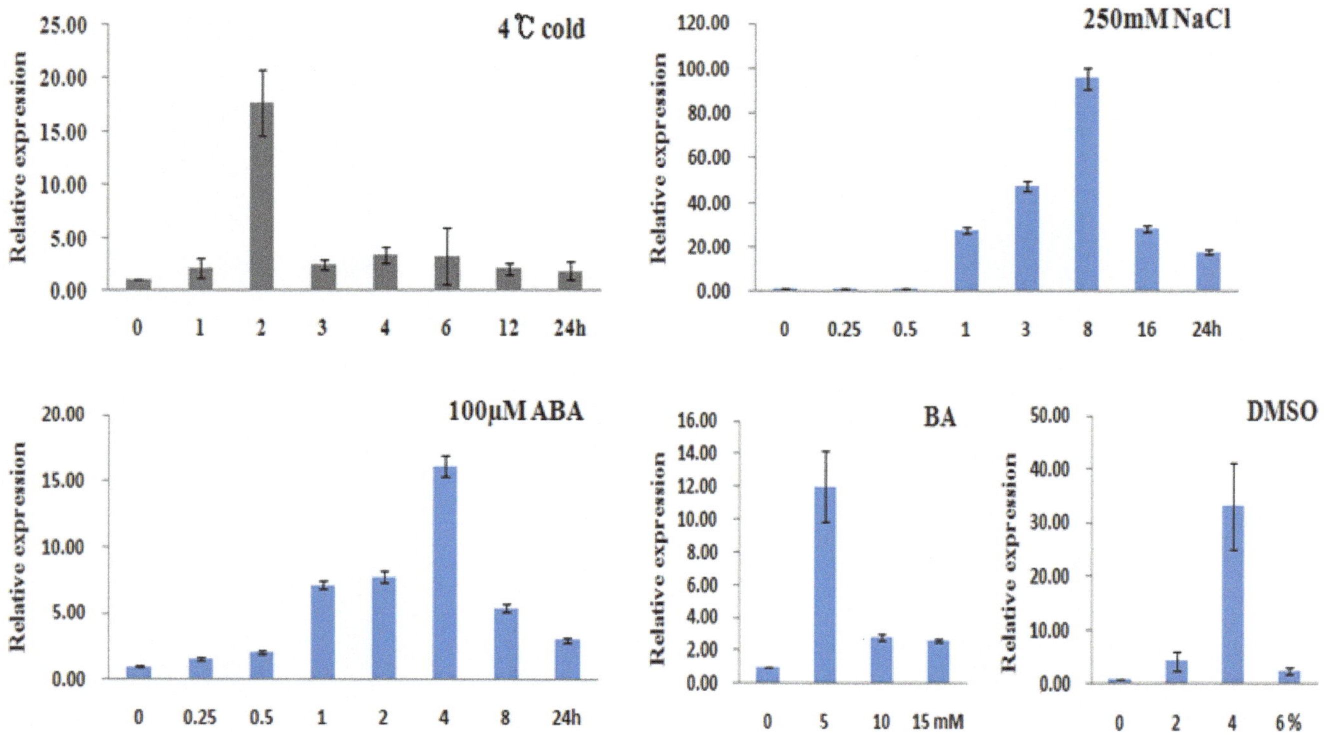

Figure 3. *BrSAC1* expression is induced by abiotic stress treatments in *Brassica* seedlings. Expression levels were examined by RT-PCR at various times after treatment with cold (4°C), salt (250 mM NaCl), or 100 μM abscisic acid (ABA). Expression profiles were also examined in the presence of various concentrations of chemical modulators of membrane fluidity benzyl alcohol (BA) at 4°C and DMSO at room temperature. Actin was used as a positive control and to determine relative expression for each treatment.

the role of membrane fluidity in the cold induction of the *BrSAC1* gene, plants were examined either after cold treatment for 3 h in the presence of BA or at room temperature for 6 h in the presence of DMSO. The cold-induced accumulation of the *BrSAC1* gene was reduced by the inhibition of membrane rigidification by increasing concentrations of BA. Additionally, *BrSAC1* expression declined after treatment of seedlings with elevated concen-

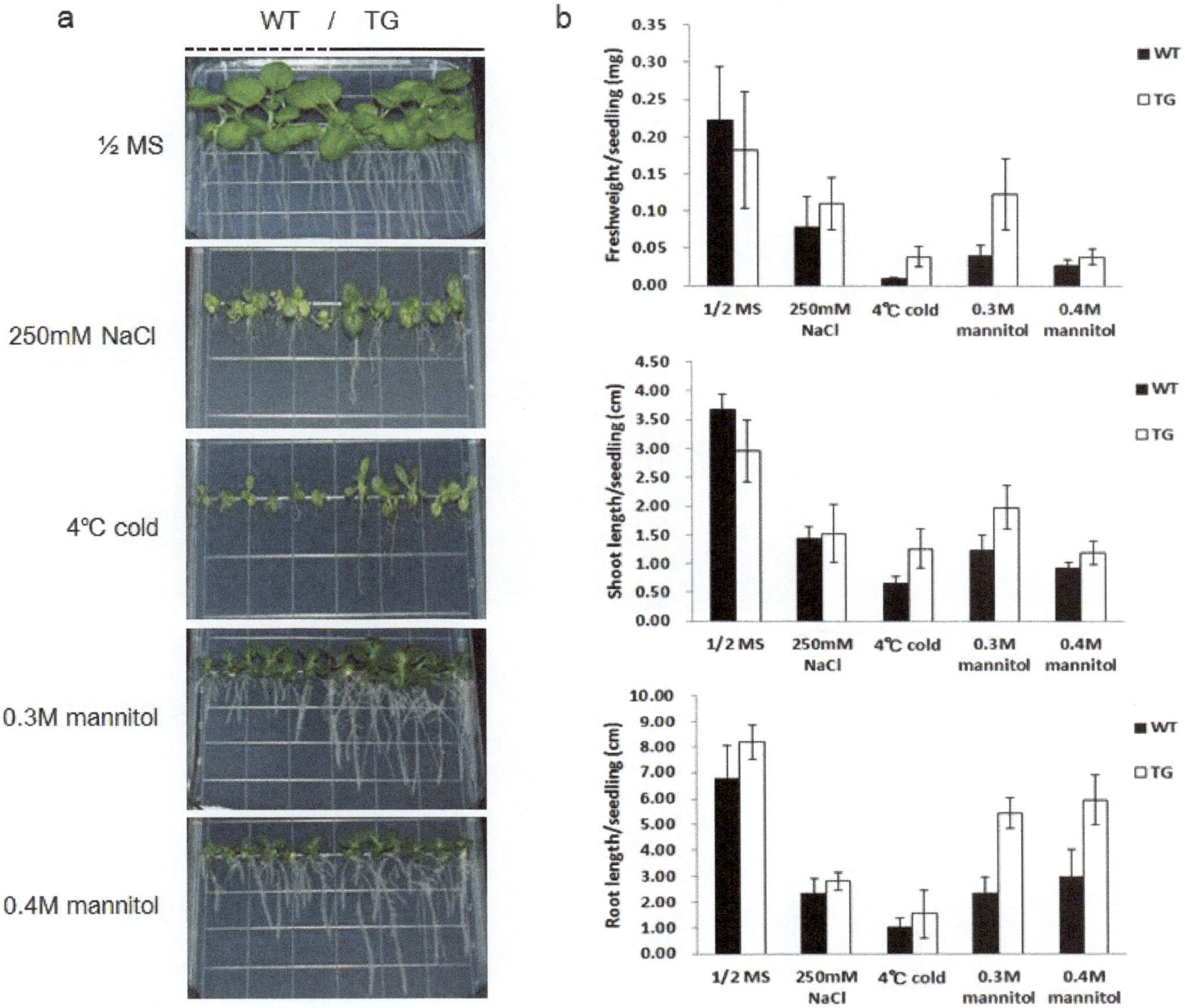

Figure 4. Plants overexpressing *BrSAC1* showing enhanced tolerance to stress. Tobacco plants overexpressing *BrSAC1* (transgenic lines: TG) and WT plants were examined under various stress conditions. The 10-day-old seedlings grown on ½ MS medium were transferred to new media containing 250 mM NaCl for salt stress, 0.3 M or 0.4 M mannitol for dehydration stress, or placed at 4℃ for cold stress. After 18 days of stress treatment, plants were photographed (A) and examined for relative fresh weight and shoot and root lengths. Data are represented as average ± standard deviation for 5 plants from 1 representative experiment (B). In this analysis, significant *P* values of 0.015 to 0.02 (FW), 0.015 to 0.02 (shoot lengths), and 0.02 to 0.025 (root lengths) were obtained between WT and transgenic lines (TG1 and TG2).

trations of the membrane rigidifier, DMSO, greater than 4% (Figure 3). The decline was thought to be caused by the toxic influence of this compound.

BrSAC1 overexpression increases stress tolerance

To study the physiological functions of *BrSAC1*, we generated several transgenic tobacco plant lines that overexpressed *BrSAC1* by fusing its full-length cDNA sequence to the CaMV 35S promoter in the pBig vector (data not shown). Four positive transgenic plant lines were confirmed by genomic PCR, Southern blot, and RT-PCR (data not shown). No significant morphological or growth differences were observed between the transgenic and non-transgenic plants under standard growth conditions. Each of the transgenic lines was examined for responses to salt stress in comparison to WT. Seedlings were stressed for 18 days with salt (250 mM NaCl), cold (4°C), or dehydration (0.3 M and 0.4 M mannitol) stresses, and FW, and shoot and root lengths were examined (Figure 4). After salt stress, all seedlings (WT and TG) showed significant reductions in each of the physiological categories measured. Overexpressing

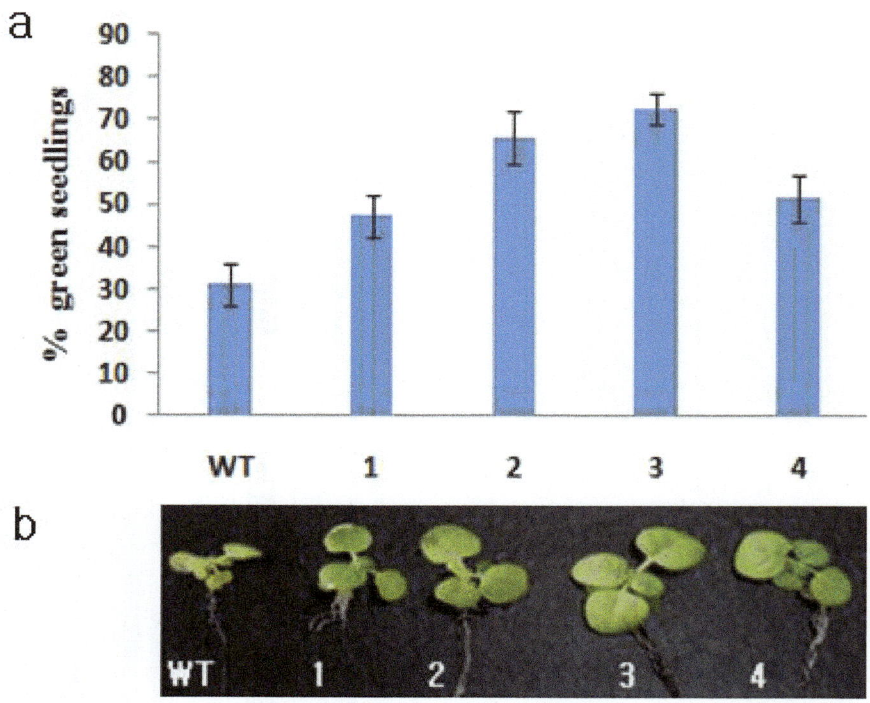

Figure 5. Plants overexpressing *BrSAC1* showed increased salt tolerance. Seedlings of tobacco seedlings from transgenic lines (1 to 4) overexpressing *BrSAC1* were examined for effects of salt treatment. **A.** The number of non-chlorotic (Green) 21 day old seedlings (WT and transgenic lines) were counted following a 4 day 250 mM NaCl salt-stress treatment and eight day recovery period. **B.** Pictures of representative seedlings from (A) are shown at the end of the recovery period.

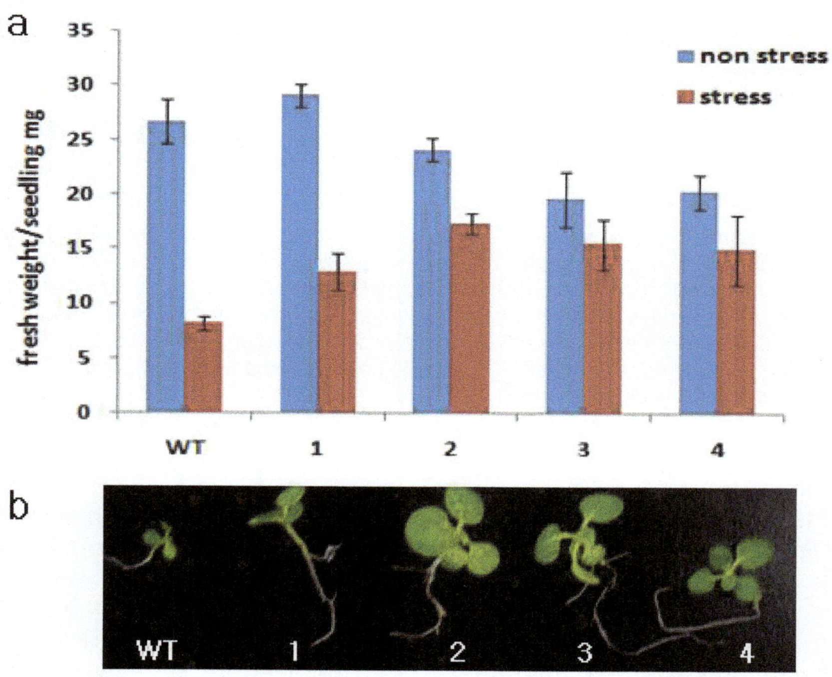

Figure 6. Plants overexpressing *BrSAC1* showing increased cold tolerance. Tobacco seedlings from transgenic lines (1 to 4) overexpressing *BrSAC1* were examined for effects of cold stress: **A.** Fresh weight of 21-day-old seedlings (WT and transgenic lines) that were cold-stressed (4°C) for 15 days, followed by a 15 day recovery, was measured for comparison with that of unstressed controls. **B.** Pictures of representative seedlings from (A) at the end of the recovery period.

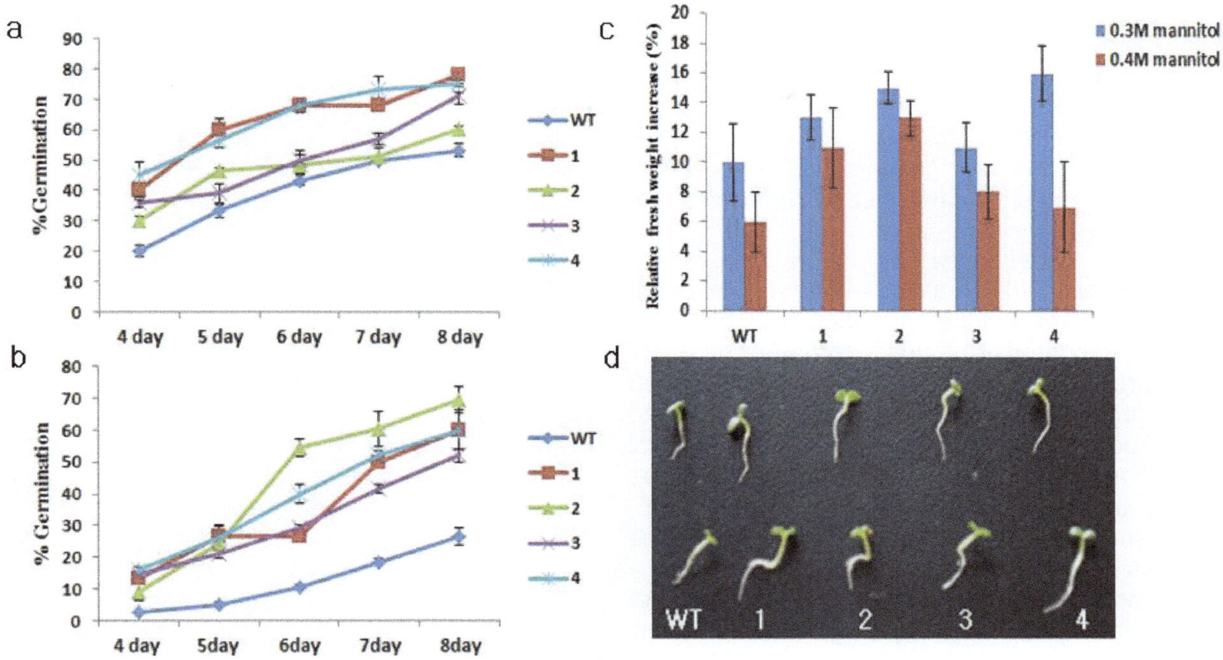

Figure 7. Plants overexpressing *BrSAC1* showing increased dehydration tolerance. Tobacco seedlings overexpressing *BrSAC1* (transgenic lines 1 to 4) were examined for effects of dehydration stress. Percent germination of seeds (WT and transgenic lines) sown on 0.3 M (A) and 0.4 M (B) mannitol is shown for four to eight days after the start of germination. (C) Percent increase in relative fresh weight (FW) of 8-day-old seedlings germinated on 0.3 M or 0.4 M mannitol is shown relative to the FW of unstressed seedlings. (D) Pictures of representative seedlings of WT and transgenic lines taken 8 days after germination on 0.3 M (upper) and 0.4 M (lower) mannitol.

BrSAC1-carrying lines showed greater tolerance to salt stress than WT having significant increases in FW and shoot and root length of almost twice those of WT-levels. Plants were also examined after salt stress followed by a recovery period. Seedlings were subjected to salt stress in 250 mM NaCl for 4 days and then returned to non-stressed conditions for eight days after which the number of plants without any chlorosis (green seedlings) was counted (Figure 5) while about 30% of WT seedlings remained green; at the end of this period several overexpressing lines showed significant increases in the number of green seedlings, some with more than 70% (Figure 5A). Representative examples of surviving plants from this experiment are shown (Figure 5B) (P values of 0.015 and 0.02, respectively, by student's t-test).

Overexpressing lines were also examined to determine the role of *BrSAC1* during cold stress by growing seedlings at 4°C for 15 days followed by a recovery period of the same time (Figure 6). An increase in FW was noted in some transgenic lines with advantage levels accounted to 50 to 100% of WT FW (Figure 6A) ($P < 0.02$, by student's t-test). In addition, transgenic seedlings appeared to be larger and healthier than WT seedlings after cold stress treatment showing recovery marked by emergence and expansion of third and fourth leaves compared to WT seedlings that only produced undeveloped first two leaves (Figure 6B).

In order to examine the role of *BrSAC1* in dehydration stress tolerance, WT and transgenic lines were grown and analyzed in the presence of stressing levels of mannitol over an eight-day period (Figure 7). Germination was dramatically effected in WT plants, with levels reduced to less than 20% by day 4. At the same time point, transgenic *BrSAC1* lines showed almost three fold increases in percent germination in both the lower (0.3 M) and higher (0.4 M) concentrations of mannitol examined (Figures 7A and B). Similar increased levels of percent germination were observed in almost every *BrSAC1* transgenic line over the period examined, often at levels two-fold or greater compared to WT, suggesting an important role for this gene in dehydration stress tolerance. In addition, the FW of transgenic lines stressed by mannitol was also increased in several cases to two-fold above that of WT plants (Figures 7A and B) (P values of 0.01 and 0.015, respectively, by student's t test).

Together these results suggest that *BrSAC1* plays an important role in several stress related processes, and overexpression of *BrSAC1* may lead to transgenic plants becoming tolerant to salt, cold, and dehydration stresses.

DISCUSSION

The SAC domain was first identified in the yeast (*S.*

cerevisiae) Sac1p PIP protein and subsequently in a number of proteins from yeast and animals. It is approximately 400 amino acids in length and characterized by seven conserved motifs. The Sac1 phosphatases represent a major class of PIP phosphatases and have a signature feature of being integral membrane proteins (Cleves et al., 1989; Whitters et al., 1993).

PIPs are phosphorylated derivatives of phosphatidylinositol (Ptdlns) that serve a variety of primary intracellular roles in the specification of dedicated membrane microdomains that organize signal transduction processes, as co-factors for the regulated activities of proteins, and as precursors for second messengers such as diacylglycerol and soluble inositol phosphates (Fruman et al., 1998; Di Paolo and De Camilli, 2006). In plant cells, all PI except PI (3, 4 and 5) P3 have been identified. Several lines of evidence suggest that as in yeast, PIs in plants may regulate many cellular activities such as vesicle trafficking (Matsuoka et al., 1995; Kim et al., 2001), pollen tube growth (Kost et al., 1999), and responses to stress and hormonal treatments (Meijer et al., 1999, 2001; DeWald et al., 2001). A number of kinases and phospholipase Cs involved in the metabolism of PIs have been characterized in plants (Cleves et al., 1991; Stevenson et al., 2000).

Homologs of the Sac1 PIP phosphatase are disseminated throughout the eukaryotic kingdom as evidenced by inspection of insect, plant, and mammalian genomes (Figure 1). The intracellular distribution of Sac1 PIP phosphatases in the endoplasmic reticulum (ER) and golgi of mammalian cells recapitulates aspects of what is observed in yeast (Xie et al., 1998; Nemoto et al., 2000), but with some unanticipated twists. Of interest, hSac1 also contains a C-terminal KXKXX motif that serves as binding site for the coatomer (COPI) complex. This COPI interaction motif is a unique feature of mammalian Sac1s as it is not conserved in yeast, plant or Drosophila Sac1 proteins (Figure 1).

Our examinations reveal that *BrSAC1* plays a role in salt, cold and drought responses, as each of these processes regulated *BrSAC1* expression. Rapid induction of *BrSAC1* protein was seen within 1 h of both salt (NaCl) and cold (4°C) treatments. Although *BrSAC1* is rapidly induced by both of these stresses, the pattern of transcript induction over 24 h between them is quite different. *BrSAC1* induction by cold was rapid and peaked at 17-fold in the first 2 h, then declined rapidly until it was close to pre-induced levels by 24 h (Figure 3). The difference in transcript induction pattern may suggest different modes of action for *BrSAC1* in response to stress. Stress responses, such as salt and cold, have been proposed to be activated through different signaling pathways: ABA dependent or ABA independent (Xiong et al., 2002; Shinozaki et al., 2003; Bartels and Sunkar, 2005; Tuteja, 2007). The influence of *BrSAC1* on responses to abiotic stresses of salt, cold and dehydration were further examined through the creation of several overexpressing transgenic lines in tobacco.

In conclusion, we cloned, characterized, and investigated the function of Sac1; PIP phosphatase encoded by the *BrSAC1* gene of *B. rapa* L. Overexpression of the *BrSAC1* gene in tobacco plants enhanced their tolerance to salt, cold and dehydration stresses. These results suggest that *BrSAC1* may be important to the understanding, and potential breeding, of plants towards improved abiotic stress response.

ACKNOWLEDGEMENTS

We thank Dr. Philip Becraft for the critical reviews and comments. This work was supported by a grant from the Next-Generation BioGreen 21 Program (Plant Molecular Breeding Center No. PJ008129), Rural Development Administration and Technology Development Program (No. 111098-03-1-HD110) for Agriculture and Forestry, Ministry for Food, Agriculture, Forestry and Fisheries, Republic of Korea and 2012 Overseas Benchmarking Program supported by Planning Office of Hankyong National University.

REFERENCES

Ali-Benali MA, Alary R, Joudrier P, Gautier MF (2005). Comparative expression of five Lea Genes during wheat seed development and in response to abiotic stresses by real-time quantitative RT-PCR. Biochim. Biophys. Acta 1730:56-65.

Aono M, Kubo A, Saji H, Tanaka K, Kondo N (1993). Enhanced tolerance to photooxidative stress of transgenic *Nicotiana tabacum* with high chloroplastic glutathione reductase activity. Plant Cell Physiol. 34:129-135.

Bartels D, Sunkar R (2005). Drought and salt tolerance in plants. Critical Reviews in Plant Sciences 24:23-58.

Carland FM, Nelson T (2004). Cotyledon vascular pattern 2-mediated inositol (1,4,5) triphosphate signal transduction is essential for closed venation patterns of *Arabidopsis* foliar organs. Plant Cell 16:1263-75.

Cleves AE, McGee TP, Bankaitis VA (1991). Phospholipid transfer proteins: a biological debut. Trends Cell Biol. 1:30-34.

Cleves AE, Novick PJ, Bankaitis VA (1989). Mutations in the *SAC1* gene suppresses defects in yeast Golgi and yeast actin function. J Cell Biol 109:2939-2950.

De Matteis M, Godi A, Corda D (2002). Phosphoinositides and the Golgi complex. Curr Opin Cell Biol 14: 434-447.

DeWald DB, Torabinejad J, Jones CA, Shope JC, Cangelosi AR, Thompson JE, Prestwich GD, Hama H (2001). Rapid accumulation of phosphatidylinositol 4,5-bisphosphate and inositol 1,4,5-triphosphate correlates with calcium mobilization in salt-stressed *Arabidopsis*. Plant Physiol 126:759-769.

Di Paolo G, De Camilli P (2006). Phosphoinositides in cell regulation and membrane dynamics. Nature 443:651-657.

Fruman DA, Meyers RE, Cantley LC (1998). Phosphoinositide kinases. Annu. Rev. Biochem. 67:481-507.

Horsch RB, Fraley RT, Rogers SG, Sanders PR, Lloyd A (1985). A simple and general method for transferring genes into plants. Science 227:1229-1231.

Hughes WE, Cooke FT, Parker PJ (2000). Sac phosphatase domain proteins. Biochem J 350:337-52

Kim DH, Eu YJ, Yoo CM, Kim YW, Pih KT, Jin JB, Kim SJ, Stenmark H, Hwang I (2001). Trafficking of phosphatidylinositol 3-phosphate from the trans-Golgi network to the lumen of the central vacuole in plant cells. Plant Cell 13:287-301

Kost B, Lemichez E, Spielhofer P, Hong Y, Tolias K, Carpenter C, Chua NH (1999). Rac homologues and compartmentalized phosphatidylinositol 4,5-bisphosphate act in a common pathway to regulate polar pollen tube growth. J. Cell Biol. 145:317-330

Laxalt AM, Munnik T (2002). Phospholipid signaling in plant defense. Curr Opin Plant Biol 5:332-338.

Lee SY, Jung YJ, Kang KK (2009). Isolation and characterization of *BrMDR1* a novel MDR-type ATP-binding cassette (ABC) transporter in *Brassica rapa* L. Korean J. Plant Res 22:273-280.

Livak KJ, Schmittgen TD (2001). Analysis of relative gene expression data using real time quantitative PCR and the $2^{-\Delta\Delta}$Ct method. Methods 25:402-408.

Martin TF (1998). Phosphoinositide lipids as signaling molecules: common themes for signal transduction, cytoskeletal regulation, and membrane trafficking. Annu. Rev. Cell Dev. Biol. 14:231-264.

Matsuoka K, Bassham DC, Raikhel NV, Nakamura K (1995). Different sensitivity to wortmannin of two vacuolar sorting signals indicates the presence of distinct sorting machineries in tobacco cells. J. Cell Biol. 130:1307-1318.

Meijer HJ, Berrie CP, Iurisci C, Divecha N, Musgrave A, Munnik T (2001). Identification of a new polyphosphoinositide in plants phosphatidylinositol 5-monophosphate (PtdIns5P) and its accumulation upon osmotic stress. Biochem. J. 360:491-498.

Meijer HJG, Divecha N, van den Ende H, Musgrave A, Munnik T (1999). Hyperosmotic stress induces rapid synthesis of phosphatidyl-d-inositol 3,5-bisphosphate in plant cells. Planta 208:294-8.

Mukhopadhyay A, Vij S, Tyagi AK (2004). Overexpression of a zinc-finger protein gene from rice confers tolerance to cold, dehydration, and salt stress in transgenic tobacco. Proceedings of the National Academy of Sciences 101:6309-6314.

Murashige T, Skoog F (1962). A revised medium for rapid growth and bio assays with tobacco tissue cultures. Physiologia Plantarum 15:473-497.

Nemoto Y, Kearns BG, Wenk MR, Chen H, Mori K, Alb Jr JG, et al (2000). Functional characterization of a mammalian Sac1 and mutants exhibiting substrate-specific defects in phosphoinositide phosphatase activity. J. Biol. Chem. 275:34293-34305.

Rogers SR, Bendich AJ (1994). Extraction of total cellular DNA from plants, algae and fungi; In: Gelvin SB and Schilperoort RA (Eds) Plant Molecular Biology Manual (Dordrecht, The Netherlands: Kluwer Academic Publishers) 2nd edition:1-8.

Shinozaki K, Yamaguchi-Shinozaki K, Seki M (2003). Regulatory network of gene expression in the drought and cold stress responses. Curr. Opin. Plant Biol. 6:410-417.

Stevenson JM, Perera IY, Heilmann I, Persson S, Boss WF (2000). Inositol signaling and plant growth. Trends Plant Sci. 5:252-258.

Takenawa T, Itoh T (2001). Phosphoinositides, key molecules for regulation of actin cytoskeletal organization and membrane traffic from the plasma membrane. Biochim. Biophys. Acta 1533:190-206.

Thompson JD, Gibson TJ, Plewniak F, Jeanmougin F, Higgins DG (1997). The CLUSTAL_X windows interface: flexible strategies for multiple sequence alignment aided by quality analysis tools. Nucleic Acids Research 25: 4876-4882.

Tuteja N (2007). Abscisic acid and abiotic stress signaling. Plant Signal Behav. 2:135-138.

Wang X (2004). Lipid signaling. Curr. Opin. Plant Biol. 7:329-336.

Whitters EA, Cleves AE, McGee TP, Skinner HB, Bankaitis VA (1993). SAC1p is an integral membrane protein that influences the cellular requirement for phospholipid transfer protein function and inositol in yeast. J. Cell Biol. 122:79-94.

Williams ME, Torabinejad J, Cohick E, Parker K, Drake EJ, Thompson, JE, Hortter M, Dewald DB (2005). Mutations in the *Arabidopsis* phosphoinositide phosphatase gene SAC9 lead to over accumulation of PtdIns(4,5)P2 and constitutive expression of the stress-response pathway. Plant Physiol. 138:686-700.

Xie Z, Fang M, Rivas MP, Faulkner AJ, Sternweis PC, Engebrecht JA, Bankaitis VA (1998). Phospholipase D activity is required for suppression of yeast phosphatidylinositol transfer protein defects. Proc Natl Acad Sci USA95: 12346-51.

Xiong L, Lee Bh, Ishitani M, Lee H, Zhang C, Zhu JK (2001). FIERY1 encoding an inositol polyphosphate 1-phosphatase is a negative regulator of abscisic acid and stress signaling in *Arabidopsis*. Genes Dev. 15:1971-1984.

Xiong L, Schumaker KS, Zhu JK (2002). Cell signaling during cold, drought and salt stress. Plant Cell 14:165-183.

Effects of natural long storage duration on seed germination characteristics of *Periploca angustifolia* Labill.

Raoudha Abdellaoui[1]*, Aymen Souid[1,2], Dhikra Zayoud[1,2] and Mohamed Neffati[1]

[1]Laboratoire d'Ecologie Pastorale, Institut des Régions Arides, 4119 Médenine, Tunisie.
[2]Département de Biologie, Faculté des Sciences de Tunis, 1060 Tunis, Tunisie.

This study was carried out to evaluate the effect of long-term natural aging on germination ability and several biochemical characteristics regarding soluble sugars and polyphenol matter contents and radical scavenging activity of *Periploca angustifolia* Labill. (Asclepiadaceae) stored seeds for 1, 3, 7, 10, 11 and 15-years, dry storage compared to freshly collected seeds. The long-term aging caused an important decrease of germination percentage, seed vigor index, seeds' viability, moisture content and seed vigor except for seeds stored for seven-years. The latter showed the highest percentages of germination and viability, seed vigor index and seed vigor under a 7.3% moisture content when compared to the oldest seeds (15-years old) which presents the lowest moisture content. In our study, aged seeds showed the lowest radical scavenging percentage activity and amounts of polyphenol, keeping free radicals and peroxides at high levels causing thereby seeds deterioration. *P. angustifolia* seeds undergo a process of after-ripening under the storage conditions, possibly depending on the low, but steady water loss down to an optimal storage water content of approx. 7.3%, thereafter undergoing some deterioration as indicated by reduced amounts of soluble sugars by polyphenol contents and experimentally tested antioxidant activities, which is in line with increased membrane leakages as indicated by increased electrical conductivities of solution from experimentally soaked seeds.

Key words: *Periploca angustifolia* Labill, polyphenol, germination.

INTRODUCTION

The necessity of seed storage was firstly recognised when humans began to domesticate plants thousands of years ago. Viable seeds had, and still have, to be maintained from one growing season to the next and the necessity of seed conservation (*in-situ* or *ex-situ*) is one of the best strategies for the protection of plant diversity (Rajjou and Debeaujon, 2008). The goal of gene banks is to maintain seed viability for indefinite periods of long term storage, typically for 10 to 100-years or more. According to Rajjou and Debeaujon (2008), seed longevity varies among, and within species due to the differences in genotype, provenance and the cumulative effects of the environment.

Understanding the germination process will help in the conservation strategies. Seed germination considered as a critical phase in the reproductive cycle of great importance for species fitness, and variation in germination percentage has been interpreted as an adaptation to ecological conditions (Navarro and Guitián, 2003; Rajjou and Debeaujon, 2008). In the context of climate change, plant genetic composition may change in response to the selection pressure and some plant communities or

*Corresponding author. E-mail: raoudha.abdellaoui@ira.agrinet.tn.

species associations may be lost as species move and adapt at different rates (Donohue et al., 2005). Seeds of many plant species are extremely tolerant to harsh environmental conditions provided they are in a state of desiccation. In this dry state, their metabolic activity is drastically reduced to a very low level (quiescence) while retaining their ability to germinate for considerable periods (Buitink and Leprince, 2008).

Active oxygen species (AOS) are involved in various aspects of seed physiology (Quan et al., 2008). Their generation, which occurs during seed desiccation, germination and ageing, may lead to oxidative stress and cellular damage, resulting in seed deterioration (Møller et al., 2007). However, cells are endowed with detoxifying enzymes and antioxidant compounds that scavenge AOS and participate in seed survival (Bailly et al., 2008). The detoxifying mechanisms play a key role in acquisition of desiccation tolerance of developing seeds, completion of seed germination and seed storability (Garnczarska et al., 2009). Accumulation of antioxidant components in dry seeds during the tardy maturation step on the mother plant contributes to control their storability potential (Bailly, 2004).

The protective role of antioxidant secondary metabolites such as total phenols and soluble sugars during aging or oxidative stress is well documented. Knowing and understanding the complex features that govern seed longevity are therefore of major ecological, agronomical and economical importance.

Periploca angustifolia Labill. (Asclepiadaceae) is a perennial shrub widely distributed in southern Spain, Silice and North Africa. It is a Saharan-Mediterranean species common in semi-arid and arid bioclimatic zones of Tunisia (Le Floc'h et al., 1989). This species is an evergreen shrub with a branching bush-type that reaches 3.5 to 4 m and dominant in active sand dunes and stabilized sand fields in southern Tunisia (Neffati, 1994). It is browsed by livestock and its leaves are valuable animal forage and fodder under open grazing conditions (Neffati, 1994). Le Floc'h (1983) reported that a decoction of leaves is as sweet as tea and that the roots are used as a source of steroids. The roots are used for medicinal purposes as treatment against hypertension Ferchichi (1990) and as antidiabetic (Askri, 1988). These characteristics gave to *P. angustifolia* a valuable multi-purpose shrub for semi-arid to arid ecological areas.

The present study was undertaken to provide basic information on life span of *P. angustifolia* seeds, to verify the possibility of their conservation and to recognize possible relationships between seed viability, germination, enzymatic and non-enzymatic antioxidants. The polyphenol and soluble sugar contents and radical scavenging activity were determined over the storage period of the seeds. Also, the focus of the work is to study the after-ripening phenomenon under long-term seed storage. Information from this study will be used in establishing programs for reintroduction of this shrub in arid and semi-arid regions of Tunisia as a measure to rehabilitate degraded lands.

MATERIALS AND METHODS

Seed collection

Seeds were collected from 12 plants in early June at the experimental site of the Arid Land Institute-Medenine (El Fjé 65°25'E, 37°07'N; South-East Tunisia). This area is arid to semi-arid with a typical Mediterranean climate, characterized by irregular rainfall events and a harsh dry summer and cold winter periods. Annual rainfall is around 144 mm and annual mean evapotranspiration 1096 mm. Mean annual temperature is 20.5°C with a minimum temperature 6.2°C in January and 36.8°C maximum in August (Gorai and Neffati, 2007). Seeds were stored in air-dry storage in the seed bank of the Range Ecology Laboratory in which relative humidity was set at 30% and temperature was maintained at 20°C. The seeds used dated from 1994 to 2009. Seeds collected in June, 2009 were considered as controls compared to other dates of collect (1994, 1998, 1999, 2002, 2006 and 2008 that correspond to 15, 11, 10, 7, 3 and 1-year aged seeds, respectively).

Germination experiments

Seeds were surface sterilized with sodium hypochlorite prior to any experimental usage to avoid fungal attack. Petri dishes (90 mm) containing two disks of Whatman filter papers soaked in distilled water. During ten days, germination experiments were conducted in an incubator at 25°C in complete darkness (Lumінincube II, analis, MCAS-350, Belgium). A completely randomized design in the germination tests (five replicates of 20 seeds) was used. Petri dishes were inspected daily until the first seed had germinated to identify the onset of germination and distilled water was added, if necessary, to keep the paper moist. Seeds were scored as germinated when at least 2 mm of the radicle was visible.

Seed viability test

Seed viability was evaluated by measuring the respiratory capability of isolated embryos according to Tommasi et al. (2006) slightly modified method. The isolated embryos (25) were incubated for 1 h in 5 mL of a solution of 1% (w/v) 2,3,5 triphenyl tetrazolium choride (TTC) in phosphate buffer 50 mM, pH 7.3 for 3 to 4 h at 25°C in darkness. Seeds of which the embryo exhibited no overall carmine red staining were scored as non-viable. Viability percentage (VP) was calculated as the number of viable embryos/total number of embryos × 100.

Determination of seed vigor (SV)

The electrical conductivity (EC) of the soak water of stored seeds was measured according to Goel et al. (2003) with slight modification. Three replicates, each one was carried out with 50 seeds which were weighted and then soaked in 5 mL of deionised water at 20°C for 24 h and the EC of the seed soak water determined using a conductivity meter (INOLAB). The EC was expressed as $\mu S\ m^{-1}\ seed^{-1}$.

Moisture content (MC)

Moisture content (MC) was estimated by drying 100 mg of seeds for

24 h at 105°C. The MC (mL g^{-1} FW) = (FW-DW/FW) × 100 was determined on a fresh weight basis; where, FW is the fresh weight and DW is the dry weight. All measurements were done in triplicates.

Methods of germination expression

From the germination data collected, the following variables were determined according to these formulas:
1. Germination (%) = (number of germinating seeds/total number of seeds) × 100.
2. Germination index = Σ *G*/*t*; where, *G* is the relative germination percentage at two-day intervals and *t* is the total germination period.
3. Vigor index (VI) = $G \times S_x$; where, *G* is the germination index and S_x is the radical mean length 10 days after germination (Yi et al., 2008).

Total phenol assay

The total phenols in seeds were determined by the Folin-Ciocalteu method as described by Li et al. (2007) with some modifications. A lot of 200 mg of seeds were homogenized with 5 mL of methanol. A volume of 1 mL of Folin-Ciocalteu's (10 X) reagent was then added to the 200 µL of the extract and stirred vigorously by vortex and left to stand for 5 min. Finally, 800 µL of saturated sodium carbonate solution (5%) was added, stirred vigorously and left to stand in darkness at room temperature for 2 h. Absorbance was measured at 765 nm using a spectrophotometer (Jenway). Determination of total phenols was carried out in triplicates, and the results are mean values, calculated on the starting material weight basis and given as µg of gallic acid equivalent (GAE) per gram of seed FW from the calibration curve of acid gallic standard solutions (0 to 25 µg mL^{-1}).

2,2-Diphenyl-1-pycrilhydrazil hydrate (DPPH) radical scavenging activity assay

DPPH (2,2-diphenyl-1-pycrilhydrazil hydrate) radical scavenging activity was determined according to the method of Amarowicz et al. (2004) that is slightly modified by Yesil-Celiktas et al. (2007). A lot of 100 mg of seed FW was homogenized with 2 mL of methanol and centrifuged at 9,000 g for 15 min. To 100 µL of extract, 2.9 mL of 0.1 mM methanolic solution of DPPH was added. The contents were stirred vigorously and then left to stand at room temperature for 30 min in dark. Decrease in colorization was measured spectrophotometrically at 517 nm. The radical scavenging activity (RSA) was calculated using the equation: RSA (%) = 100 x (1 − AE/AD); where, AE represent the absorbance of the solution containing antioxidant extract and AD is the absorbance of the DPPH solution. All measurements were done in triplicates.

Determination of soluble sugars

Soluble sugar was determined following the phenolsulphuric acid method according to (Robyt and White, 1987). 100 mg of seeds were grinded in fine powder and moistened in 5 mL of methanol (80%) then incubated at 70°C for 30 min. The extract was centrifuged at 5,000 rpm for 10 min. To 1 mL of supernatant, 1 mL of 5% phenol and 5 mL of concentrated sulphuric acid, were added.

Optic density was read by a spectrophotometer at 490 nm. Soluble sugar concentrations were expressed as mmol g^{-1} seed FW using D-glucose as standard. All measurements were done in triplicates.

Statistical analysis

All data analyses were performed with the software SPSS 13.0 (SPSS Inc, Chicago, Illinois, USA). Analysis of variance was used to assess the effect of storage period on the studied parameters. Statistical comparisons between means were performed with Student-Newman-Keuls' test. Differences were considered significant at $P<0.05$. The Pearson Correlation was used to establish relationships between the measured parameters.

RESULTS

Germination percentages

As shown in Figure 1a, the seeds collected in June 2009, considered as controls in entire study, reached a germination percentage of 56%. When compared to control and other storage period, the highest germination percentage (93%) was attained for seven-year old seeds. After one to three-year storage period, germination percentages were lower (72 and 81%, respectively) than those after seven-years and did not show any significant differences, whereas they became significantly higher than control. The lowest germination percentage was obtained after eleven-year storage period (31%). However, no germination was recorded after 15-years of storage. Also, all seeds of the different storage periods showed the same delay of germination (two days). However, 11-years aged seeds germinated after one day. The curve of survival presented in Figure 1b that represent the mean final germination to the period of storage, showed that seeds longevity decreased after seven-years of storage. Before this period, final germination percentage increased significantly to reach 93%.

Seed viability

Due to the storage period, the overall seed viability decreased except those recorded after seven years (Table 1). Before the storage, seed viability was 80%, while it decreased significantly by increasing the storage time (78.67, 73.33, 45.33, 41.33 and 24% at 1, 3, 10, 11 and 15-years, respectively). Compared to other periods, after seven-year storage period, the viability reached a higher value of 94.67%. The correlation coefficient (R^2) between germination percentage and seed viability suggested significant relationship (R^2 = 0.89) when seeds were stored for different periods (Table 2).

Seed moisture content

MC recorded for fresh collected seeds (June 2009) was 9.33% (Table 1). This value decreased significantly with storage period to reach 2.83% after 15-years of storage. This parameter had a significant correlation with germination percentage (R^2 = 0.68) and seed viability (R^2 = 0.84)

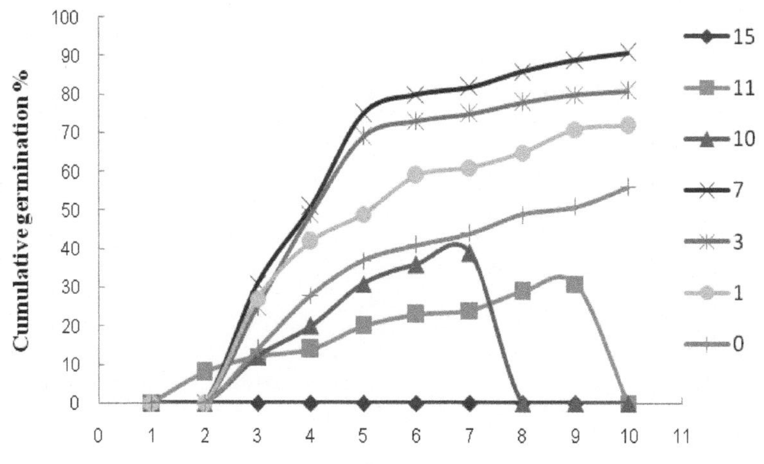

(a)

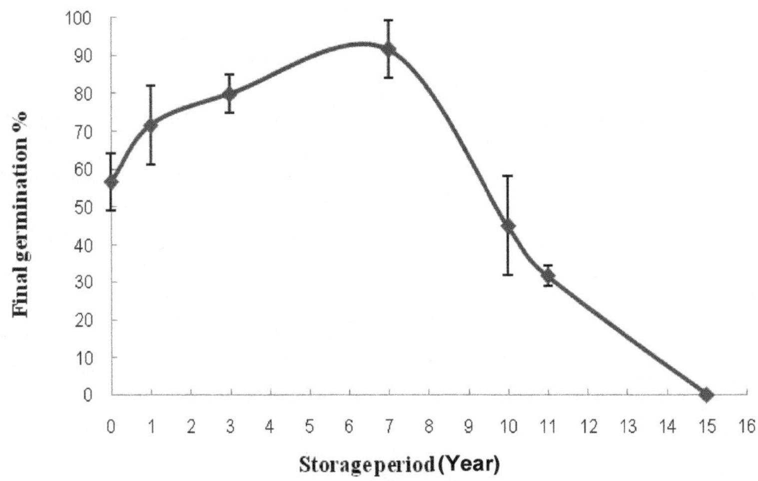

(b)

Figure 1. (a) Cumulative germination percentage of *Periploca angustifolia* Labill. seeds stored for different periods. After 15-years of storage seeds germination was completely inhibited (n = 5). (b) Survival curve, representing mean final germination, of *Periploca angustifolia* Labill. seeds under long term storage. Values of the final germination percentages (mean, 95% confidence limits, n=5), having the same letter are not significant different (P>0.05) (Student-Newman-Keuls' test).

Table 1. Vigor index (VI), viability percentage (V) and moisture content (MC) of *Periploca angustifolia* seeds over the storage period.

Storage period (year)	VI	V	MC
0	$42.48^{c} \pm 0.11$	$80.00^{b} \pm 4.00$	$9.33^{a} \pm 1.53$
1	$52.87^{b} \pm 2.12$	$78.67^{b} \pm 10.07$	$9.00^{a} \pm 1.00$
3	$50^{b} \pm 1.65$	$73.33^{b} \pm 6.11$	$7.5^{ab} \pm 0.50$
7	64.26 ± 4.04^{a}	$94.67^{a} \pm 6.11$	$7.33^{ab} \pm 2.08$
10	17.13 ± 0.02^{d}	$45.33^{c} \pm 6.11$	$5.00^{bc} \pm 1.00$
11	8.64 ± 0.92^{e}	$41.33^{c} \pm 6.11$	$4.83^{bc} \pm 0.76$
15	0.00^{f}	$24^{d} \pm 10.58$	$2.83^{c} \pm 1.04$

Means (n = 5 for VI, n = 3 for V and MC±SD) within a column followed by the same letter are not significantly different at the 0.05 probability level according to Student-Newman-Keuls test.

Table 2. Correlation coefficients (R^2) between pairs of the studied parameters of *Periploca angustifolia* seeds over the storage period.

Parameter	DPPH	EC	SS	V	MC	PP	GP
EC	*-0.33*						
SS	0.50	-0.94					
V	*0.23*	-0.95	0.83				
MC	0.58	-0.80	0.79	0.84			
PP	0.60	-0.79	0.85	0.75	0.80		
GP	*0.17*	-0.90	0.81	0.89	0.68	0.59	
VI	*0.22*	-0.93	0.81	0.95	0.78	0.68	0.94

GP, Germination percentage; V, viability percentage; DPPH, radical scavenging activity (%RSA); MC, moisture content; PP, polyphenol content; SS, soluble sugar; EC, electric conductivity; VI, vigor index. Italic values are not significant at $P < 0.05$.

(Table 2). The highest germination and viability percentages were recorded for seeds stored for seven-years, having a moisture content of 7.33%. When MC exceeded this value, as recorded for freshly collected seeds or after one year of storage, the germination percentage decreased suggesting that storability could be improved with MC around 7%.

Seed vigor

EC increased significantly after 10, 11 and 15-years of storage (302, 336 and 404 μSm^{-1} seed^{-1}, respectively) showing that seed vigor was drastically affected with storage period (Figure 2a). The lowest EC (197 μSm^{-1} seed^{-1}) was recorded for seven-year aged seeds. For the other periods (control, one and three years of storage), the EC were 226.33, 242 and 263.33 μSm^{-1} seed^{-1}, respectively. The EC and MC were inversely proportional by over the storage period (Table 1). This indicates that the integrity of the membrane system depends on MC. Within certain limits, MC has negative effects on seed vigor of stored *Periploca* seeds. The correlation coefficients (Table 2) identified between EC and viability, germination percentage and MC suggested significant relationships with R^2 = -0.95, -0.90 and -0.80, respectively.

Vigor index

Seven-year aged seeds had significantly higher seed vigor index (64.26) compared with the control and other storage periods (Table 1). The lowest VI was registered for 11-year aged seeds; however, no germination occurred for 15-year old seeds. This parameter was highly correlated with MC, soluble sugars and viability (R^2 = 0.78, 0.81 and 0.95) (Table 2). It was negatively correlated with EC (R^2 =-0.93). At low MC (<7%), the VI was significantly affected reaching 0 for seeds having 2.83% of MC with no germination. In addition with MC >7%, the VI decreased significantly reaching 42.48 when MC was 9.3% for freshly collected seeds.

Soluble sugars content

As shown in Figure 2b, seeds stored for 15-years attained lower soluble sugars concentration (6.27 mmol g^{-1} of seed) compared with control and other storage periods. For fresh seeds and those stored for one and seven years, soluble sugars concentration (10.63, 10.54 and 10.34 mmol g^{-1} of seed, respectively) did not show any significant difference. When seeds were stored for three or 10 years, the soluble sugars concentration reached similar values ranging between 9.22 and 9.13 mmol g^{-1} of seed and no significant difference was marked. Correlation analyses revealed positive correlations between soluble sugars concentration and viability (R^2 = 0.83), germination percentage (R^2 = 0.81) or MC (R^2 = 0.79) (Table 2). However, the electric conductivity and the former variable was negatively correlated and suggested significant relationships with (R^2 = -0.94).

DPPH and total phenol contents

Seven-year aged seeds showed the lowest antioxidant activity (74.5%) followed by 15 and 11-year old seeds (76.0 and 76.6%, respectively) (Figure 3a). The highest antioxidant activities were recorded for fresh collected seeds and one-year aged seeds (83.8 and 83.2%, respectively). The total phenols varied from 67.3 to 244.7 mg GAE g^{-1} of seed (Figure 3b). The lowest content was registered for aged seeds (15-years) and the highest content was obtained for fresh collected seeds. Regression models were tested to establish relationships between total polyphenol contents and antioxidant activity and the linear regression test show a positive relationship with antioxidant activity and polyphenol contents with R^2 = 0.60.

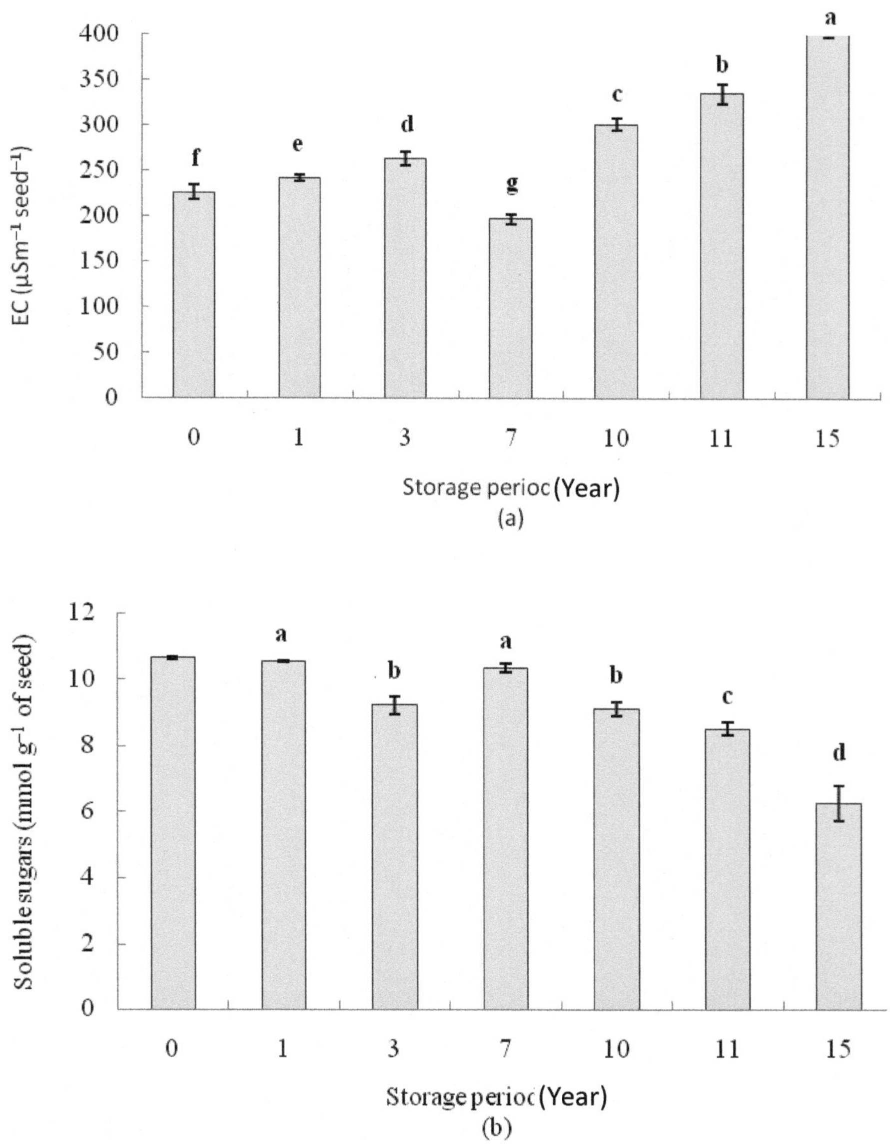

Figure 2. (a) Changes in electric conductivity (μSm^{-1} $seed^{-1}$) and (b) soluble sugar concentrations (mM. g^{-1} seed FW) of *P. angustifolia* seeds over the storage period. Data represent mean ± SE, *n* = 3.

DISCUSSION

Our results corroborate several other studies and reveal that seed germination characteristics were affected by over storage period (McDonald, 1999; Hsu et al., 2003). Freshly collected seeds presented low percentage of germination however, stored for seven-years they reached its maximum and beyond this period, germination decreased drastically. As indicated by McDonald (1999) and (Hsu et al., 2003), seeds that deteriorated rapidly by increasing storage duration generally showed a marked decline in their ability to germinate. Our data shows that seed viability decreased with storage period except that registered for seven-year old seeds suggesting high correlation with germination percentage. In aged seeds, because of the destruction of the membrane system, many electrolytes flowed out of cells, so the vigor of seed reduced (Wang et al., 1999). Seed deterioration could be explained by the findings that the ageing of seeds leads to lipid peroxidation that subsequently causes membrane perturbation (Goel and Sheoran, 2003). The change of electric conductivity during seed soaking is commonly used as an indicator for testing the integrity of plasma membrane (Bewley and Black, 1994; Wang et al., 2003). The electric conductivity was therefore evaluated showing significant increase of solute leakage with the storage period that reflects thereby the seed vigor. The lowest electric conductivity was registered for seven-year

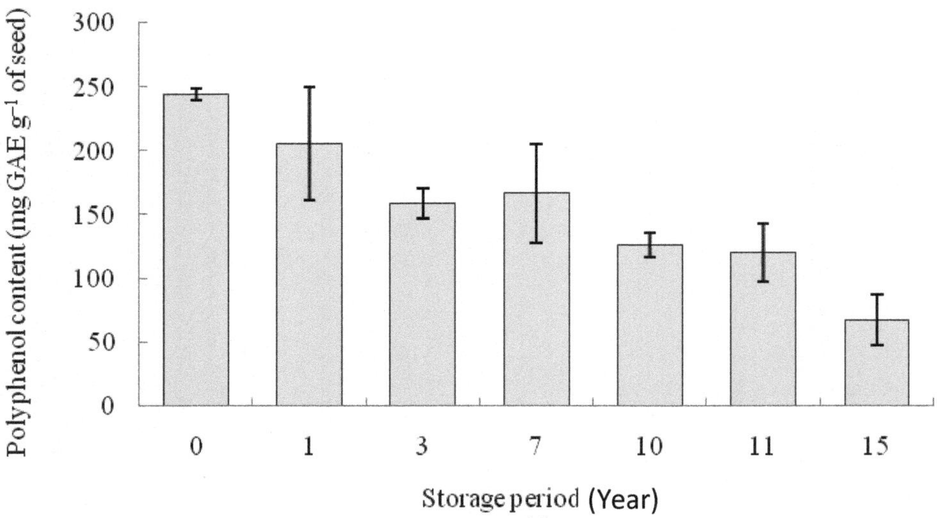

(a)

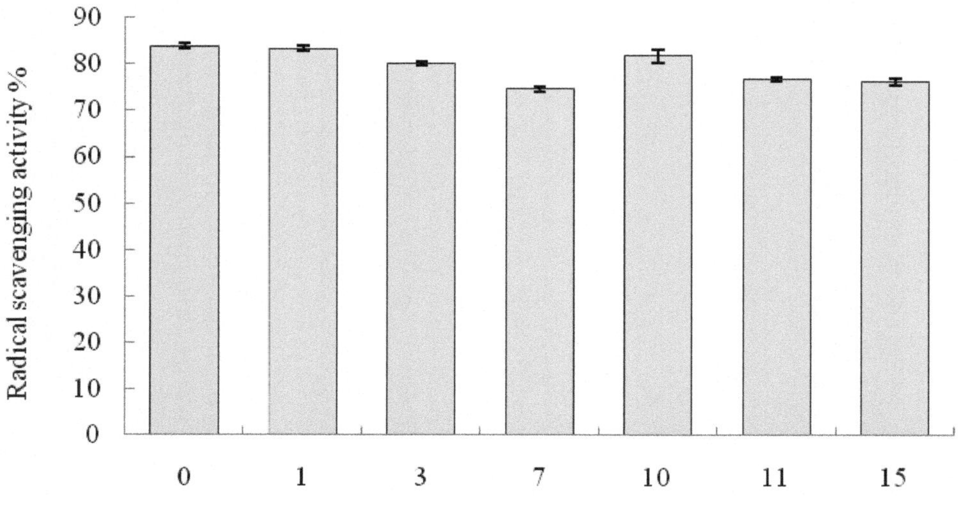

(b)

Figure 3. (a) Changes in radical scavenging activity (%RSA) and (b) polyphenol amounts (µg GAE.$g^{-1}$of seed FW) of *Periploca angustifolia* seeds over the storage period. Data represents mean ± SE, *n* = 3.

aged seeds. The data analysis reveals positive correlations between seed vigor, viability and germination percentage. These results prove that germination process is associated with the damage occurring at the membrane as shown by Fujikura and Karssen (1995). McDonald (1999) proved that accelerated aging of seeds is recognized as an accurate indicator of seed vigor and storability. The highest germination percentage, viability and vigor index were registered for seven-year aged seeds having a MC of 7.33%. As proved by Walter et al. (2005), the rate of aging is strongly influenced by environmental and genetic factors such as storage temperature, seed moisture content and seed quality. Rajjou and Debeaujon (2008) showed that when seeds deteriorate during storage, they lose vigor, become more sensitive to stress during germination and ultimately become unable to germinate. *Periploca* seeds are orthodox type achieving their maturity at low moisture content as many other seeds from the same class (De Tullio and Arrigoni, 2003). They are usually described as dry, although they contain some water, typically in the range of 5 to 10%, depending on the species. Results show that *Periploca* seeds with MC of 7.33% presented the highest germination percentage, viability and seed vigor index. Our approach provides optimum MC of 7.33% for *Periploca* seeds stored at 20°C. When MC was ≤7%, seed vigor dropped. Analy-zing *Zygophyllum xanthoxylon* (Bge.) Maxim seeds, Yi et al. (2008) proved that the best MC maintaining the mem-brane system was 3.81%. These authors showed that as seed MC declined, the vigor was reduced and the inte-grity of membrane decreased suggesting that seeds with low MC have shortened longevity. Consequently, low MC storage not only can be used to maintain the quality of seeds but also improve their storability as shown by Wang et al. (2003).

During storage, a large number of reactive oxygen species (ROS) are generated in the seed during ageing which causes lipid peroxidation (McDonald, 1999) and may lead to oxidative stress and cellular damage, resulting in seed deterioration (Rajjou and Debeaujon, 2008). However, cells are endowed with detoxifying enzymes and antioxidant compounds that scavenge ROS and play a key role in seed survival, completion of seed germination and seed storability. Substances that could play a role in maintaining seed viability such as phenolic compounds and total soluble sugars were estimated in freshly collected seeds of *Periploca* and those stored from one to 15 years. Seed vigor index showed significant relationships with total soluble sugars (R^2 = 0.81) and total phenolic contents (R^2 = 0.677). Our results are similar to those reported by Pukacka and Ratajczak (2007) on *Fagus sylvatica* L. seeds stored from two to 10 years showing that germination capacity was strongly and positively correlated with amounts of total phenolic compounds.

During maturity, seeds characteristically accumulate soluble sugars (Amuti and Pollard, 1977). These solutes are known to contribute to the development of tolerance to desiccation and to the longevity itself during storage (Bernal-Lugo and Leopold, 1995; Obendorf, 1997). The protective effect of these solutes is thought to occur through maintaining the structural integrity of membranes and providing stability for macro molecules such as proteins (Lee and Timasheff, 1981; Crowe and Crowe, 1986). Our study shows a high negative correlation between electric conductivity and soluble sugars emphasizing on their role in protecting membranes. Many researchers showed that the presence of these soluble sugars have been found to correlate with longevity (Horbowicz and Obendorf, 1994; Steadman et al., 1996). This fact was also proved in our study, as seed viability was highly correlated with soluble sugars amounts.

In terms of seed germinability, capacity to scavenge ROS seems to be of a particular interest. In our study, fresh seeds showed the highest DPPH scavenging activity percentage, and amounts of polyphenol keeping free radicals and peroxides were at low levels. These findings are aligned with those found by Chang and Sung (1998) and McDonald (1999) suggesting that free radicals and peroxides in freshly collected seeds are generally kept at low levels by cooperative reactions of enzymatic and non-enzymatic anti-oxidative systems. Though aged seeds of *Periploca* (15-year old seeds) showed a seed viability of 24%, they did not germinate. During hydration of the aged seed early in germination, catahydroperoxide lyase becomes active and breaks down oxygenated fatty acids accumulated during storage (Wilson and McDonald, 1986). This reaction may increase the level of free radicals to a point in which the detoxifying capacity of the aged seed has been oversaturated (Gidrol et al., 1994). Therefore, it could well be that seed ageing resulted in an impairment of the enzymatic systems involved in the elimination of the toxic $^{\bullet}O_2$ intermediate species generated either during dry storage or during germination (Sung and Chiu, 1995; Bailly et al., 1996). This mechanism could explain why intact ungerminated seeds are often metabolically alive, yet the seed dies a day or two after hydration.

P. angustifolia seeds undergo a process of after-ripening under the storage conditions, possibly depending on the low, but steady water loss down to an optimal storage water content of 7.3%, thereafter undergoing some deterioration as indicated by reduced amounts of soluble sugars. When seeds are stored as genetic resources, viability must be maintained for several decades or even centuries. The stringency of the conditions for seed storage increases as the required longevity increases.

REFERENCES

Amarowicz R, Pegg RB, Moghaddam PR, Barl B, Weil JA (2004). Free-radical scavenging capacity and antioxidant activity of selected plant species from Canadian prairies. Food Chem. 84:551-62.

Amuti KS, Pollard CJ (1977). Soluble carbohydrates of dry and developing seeds. Photochemistry. 16:529-32.

Askri M (1988). Contribution à l'étude de *Periploca angustifolia*. Thèse de spécialité. Faculté des Sciences de Monastir, 189p.

Bailly C, Benamar A, Corbineau F, Come D (1996). Changes in malondialdehyde content and in superoxide dismutase, catalase and glutathione reductase activities in sunflower seeds as related to deterioration during accelerated ageing. Physiol Plant. 97:104-10.

Bailly C (2004). Active oxygen species and antioxidants in seed biology. Seed Sci. Res. 14:93-107.

Bailly C, El-Maarouf-Bouteau H, Corbineau F (2008). From intracellular signaling networks to cell death: the dual role of reactive oxygen species in seed physiology. C. R. Biol. 331:806-814.

Bernal-Lugo I, Leopold AC (1995). Seed stability during storage: raffinose and seed glassy state. Seed Sci. Res. 5:75-80.

Bewley JD, Black M (1994). Seeds Physiology of development and germination. 2nd Ed. New York: Plenum Press.

Buitink J, Leprince O (2008). Intracellular glasses and seed survival in the dry state. Comptes Rendues Biologies. 331:788-795.

Chang SM, Sung JM (1998). Deteriorative changes in primed shrunken-2 sweet corn seeds during storage. Seed Sci. Technol. 26:613-626.

Crowe JH, Crowe LM (1986). Stabilization of membranes in anhydrobiotic organisms. Membranes, meta-bolism, and dry organisms. Ithaca, NY: Cornell University Press,188-209.

De Tullio MC, Arrigoni O (2003). The ascorbic acid system in seeds: to protect and to serve. Seed Sci. Res. 13:249-60.

Donohue K, Dorn L, Griffith C, Kim E, Aguilera A, Polisetty CR, Schmitt J (2005). Niche construction through germination cueing: Life-history responses to timing of germination in *Arabidopsis thaliana*. Evolution. 59:771-785.

Ferchichi A (1990). *Periploca laevigata* (Ait.) Ecologie et aptitude germinative. DEA Eco. Végét. Univ. Tunis II, p 108.
Fujikura Y, Karssen CM (1995). Molecular studies on osmoprimed seeds of cauliflower: a partial amino acid sequence of a vigour-related protein and osmopriming-enhanced expression of putative aspartic protease. Seed Sci. Res. 5:177-181.
Garnczarska M, Bednarski W, Jancelewicz M (2009). Ability of lupine seeds to germinate and to tolerate desiccation as related to changes in free radical level and antioxidants in freshly harvested seeds. Plant Physiol. Biochem. 47:56-62.
Gidrol X, Lin WS, Degousee N, Yip SF, Kush A (1994). Accumulation of reactive oxygen and oxydation of cytokinin in germinating soybean seeds. Eur. J Biochem. 224:21-28.
Goel A, Goel AK, Sheoran IS (2003). Changes in oxidative stress enzymes during artificial ageing in cotton (*Gossypium hirsutum* L.) seeds. J. Plant Physiol. 160:1093-1100.
Goel A, Sheoran IS (2003). Lipid peroxidation and peroxide scavenging enzyme in cotton seeds under natural ageing. Biol. Plant. 46:429-434.
Gorai M, Neffati M (2007). Germination responses of Reaumuria vermiculata to salinity and temperature. Ann. Appl. Biol. 151:53-59.
Horbowicz M, Obendorf RL (1994). Seed desiccation tolerance and storability: dependence on flatulence-producing oligosaccharides and cyclitols review and survey. Seed Sci. Res. 4:385-405.
Hsu CC, Chen CL, Chen JJ, Sung JM (2003). Accelerated aging-enhanced lipid peroxidation in bitter gourd seeds and effects of priming and hot water soaking treatments. Sci Hortic. 98:201-212
Lee JC, Timasheff SN (1981). The stabilization of proteins by sucrose. J. Biol. Chem. 256:7193-7201.
Le-Floc'h E (1983). Contribution à une étude ethnobotanique de la flore de la Tunisie.
Le Floc'h E, Nabli A, Valdeyron G (1989). Biologie et écologie des principaux taxons.Essai de synthèse sur la végétation et la phyto-écologie Tunisienne.Element de botanique et de phyto- écologie coordonné par Nabli M.A. 51-193.
Li HB, Cheng KW, Wong CC, Fan KW, Chen F, Jiang Y (2007). Evaluation of antioxidant capacity and total phenolic content of different fractions of selected microalgae. Food chem. 102:771-776.
Li J, Zhang Y, Yu Z, Wang Y, Yang Y, Liu Z, Jiang J, Song M, Wu Y (2007). Superior storage stability in low lipoxygenase maize varieties. J Stored Prod Res. 43: 530-534.
McDonald MB (1999). Seed deterioration: physiology, repair and assessment. Seed Sci. Technol 27:177-237.
Møller IM, Jensen PE, Hansson A (2007). Oxidative modifications to cellular components in plants. Ann. Rev. Plant Biol. 58:459-481.
Navarro L, Guitian J (2003). Seed germination and seedling survival of two threatened endemic species of the northwest Iberian Peninsula. Biol Conserv. 109:313-320.
Neffati M (1994). Caractérisation morpho-biologique de certaines espèces végétales nord-africaines, implication pour l'amélioration pastorale. Thèse Doc. Univ. Gent, 264 p + annexe.
Obendorf RL (1997). Oligosaccharides and galactosyl cyclitols in seed desiccation tolerance (Review Update). Seed Sci. Res. 7:63-74.
Pukacka S, Ratajczak E (2005). Production and scavenging of reactive oxygen species in *Fagus sylvatica* seeds during storage at varied temperature and humidity. J. Plant Physiol. 162:873-885.
Quan LJ, Zhang B, Shi WW, Li HY (2008). Hydrogen peroxide in plants: a versatile molecule of reactive oxygen species network. J. Integr. Plant Biol. 50:2-18.
Rajjou L, Debeaujon I (2008). Seed longevity: Survival and maintenance of high germination ability of dry seeds. C. R. Biol. 331:796-805.
Robyt JF, White BJ (1987). Biochemical Techniques- Theory and Practice. Books/ Cole Publishing Company, Monterey, CA, USA, 267-275.
SPSS (2002) SPSS 11.5 for Windows Update. SPSS Inc., USA.
Steadman KJ, Pritchard HW, Dey PM (1996). Tissue-specific soluble sugars in seeds as indicators of storage category. Ann. Bot. 77:667-674.
Sung JM, Chiu CC (1995). Lipid peroxidation and peroxide-scavenging enzymes of naturally aged soybean seed. Plant Sci. 110:45-52.
Tommasi F, Paciolla C, De Pinto MC, De Gara L (2006). Effects of storage temperature on viability, germination and antioxidant metabolism in *Ginkgo biloba* L. seeds. Plant Physiol. Biochem. 44:359-368.
Walter LM, Wheeler J, Grotenhuis M (2005). Longevity of seeds stored in a genebank: species characteristics. Seed Sci. Res. 15:1-20.
Wang XF, Zheng GH, Yang SJ (1999). Effect of ultradry storage on fluidity of plasma membrane of seeds. Chin Sci Bull. 44:1008-1013.
Wang XF, Jing XM, Lin J (2003). Studies on membrane function and sugar components of ultra dried seeds. Acta Bot Sin. 45:23-31.
Wilson JDO, McDonald JMB (1986). The lipid peroxidation model of seed ageing. Seed Sci. Technol. 14:269-300.
Yesil-Celiktas O, Girgin G, Orhan H, Wichers HJ, Bedir E, Vardar-Sukan F (2007). Screening of free radical scavenging capacity and antioxidant activities of Rosmarinus officinalis extracts with focus on location and harvesting times. Eur. Food Res. Technol. 224:443-51.
Yi L, Jian JQ, Zhi BD, Tao W, Li ZA (2008). Storage behavior of Zygophyllum xanthoxylon (Bge.) Maxim seeds at low moisture contents. Acta Physiol. Plant. 30:651-656.

8

In vitro cell culture of *Charybdis congesta* for enhanced production of secondary metabolites: Proscillaridin A, Scillaren A and Scilliroside

A. Shiva Reddy[1], P. Sita Devi[1] and S. Ravi Kiran[2]

[1]Analytical Chemistry Division, Indian Institute of Chemical Technology, Hyderabad – 500007, India.
[2]Department of Biochemistry and Biotechnology, Aurora's Degree & PG College, Chikkadpally, Hyderabad – 500020, India.

Callus cultures of *Charybdis congesta* were initiated *in vitro* and the effect of growth regulators was tested on callus growth and secondary metabolite production. Among several standard media formulated for use in the present study, MS and B5 were found to be potentially active and facilitated the calculation of callus induction frequency (CIF). The CIF was higher in both MS (70%) and B5 (63%) media supplemented with 1-naphthalene acetic acid (NAA) (9.0 µM) and benzyl amino purine (BAP) (0.9 µM). However, with indole-3-butyric acid (IBA) (9.0 µM) and BAP (0.8 µM), less amount (22.6%) of CIF was observed in MS medium but no callus formation was noticed in B5 medium. Rapid high performance thin layer chromatography (HPTLC) screening of callus extracts revealed that the callus established in MS medium supplemented with 4.5 µM NAA and 0.46 µM BAP produced the highest yield of Proscillaridin A (4.51 mg/g DW), Scilliroside (3.3 mg/g DW), Scillaren A (2.35 mg/g DW) and desacetylscilliroside (8.62 mg/g DW), which was higher than from the intact plants. The results obtained indicate that the *in vitro* cultures of *C. congesta* might be an excellent source of secondary metabolites and further metabolic profiling may provide insights into up scaling of the compounds which lead to greater commercial interest and continuous supply of cultures.

Key words: Squill, *Charybdis congesta*, bufadienolides, callus cultures, reflectance scanning densitometry.

INTRODUCTION

Charybdis congesta, formerly known as *Urginea congesta*, belonging to the family Hyacinthaceae is a perennial bulbous species found in dry sandy places and is commonly known as Indian Squill. The bulbs of Squill contain various steroidal glycosides known as bufadienolides (0.15 to 2.4%) which include Scillarin A (SA), Scilliroside (SC), Scilliglauside (SG), Proscillaridin A (PA), Desacetyl-scillirosidine (DAS), and Scillirosidine (SDN) (Kopp et al., 1996), possessing interesting medicinal, rodenticidal and insect repellent properties (Verbiscar et al., 1986; FAIR961436, 1999; Pascual-Villalobos, 2002). At present, the most exploited source of these bufadienolides is bulbs of *C. congesta* and often requires large amounts of plant material which makes the plant endangered because of over-harvesting (Babu et al., 2001; Namdeo, 2007). Further, the conventional methods of propagation suffer from various limitations such as non-availability of plants on time due to geographical and unforeseen environmental conditions. Therefore, there is a need for effective and alternative sources for the propagation and production of

Figure 1. *C. congesta* bulb collected from Osmania University, Hyderabad, India.

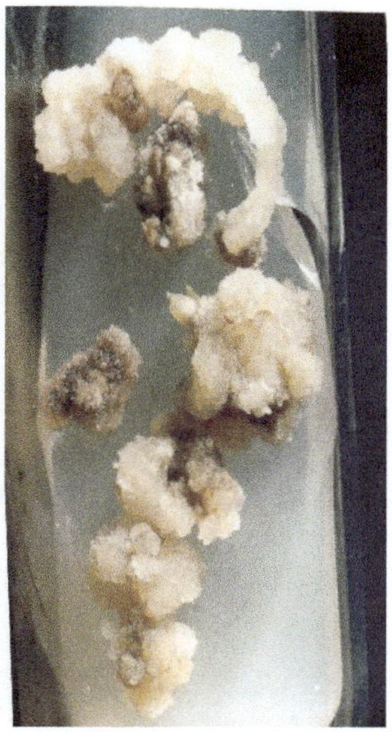

Figure 2. Callus obtained from bulbs as explants cultured on MS medium.

secondary metabolites. In this connection, biotechnological approaches, specifically, plant tissue culture becomes a powerful tool for mass multiplication and is found to have right potential for the production of many secondary metabolites from elite and threatened species (Dicosmo and Misawa, 1995; Thanonkeo and Panichajakul, 2005; Liu et al., 2006, Roostika et al., 2007). Cell cultures often have a higher rate of metabolism than intact plants due to fast proliferation and condensed biosynthetic cycle which makes them model systems for the study of secondary metabolite production as well as micropropagation. The tissue culture technique has been developed for many species of the family (Jha et al., 1984, 1991; Garari and Backhauss, 1987; Wawrosch et al., 2005) but *C. congesta* which is a South Indian variety, has not been so far studied and to the best of our knowledge, there are no published reports. Therefore, the aim of the present study was to develop a methodology for *in vitro* culture of *C. congesta*, extract bufadienolides from callus and evaluate the results to see how quantitative the present studies are with respect to tissue culture programme and the potential utility of the same. Here in, the investigative study carried out is described and the results are discussed.

MATERIALS AND METHODS

Bulbs of *C. congesta* (Figure 1) were procured from Osmania University, Hyderabad, India. A voucher specimen has been deposited at the Department of Botany, Osmania University, Hyderabad, India (No. OU BOT 4964). Reference Standard of Proscillaridin A, Scilliroside, desacetylscilliroside, Scillaren A, solvents and reagents were purchased from Merck chemicals, India.

Callus induction

To establish callus cultures, the bulbs cut into small explants consisting of two scale segments attached to a portion of the basal plate were used. Explants were inoculated individually into various media supplemented with 3% (w/v) sucrose, 0.8% (w/v) agar and various combinations of growth regulators, auxin (1.5, 3.0, 4.5, 6.0 and 9.0 μM 2,4-D, NAA and IAA) and cytokinins (0.25, 0.45, 0.75 and 0.95 μM BAP and Kinetin). The cultures were kept under a 16 h photoperiod of 40 μmol $m^{-2}s^{-1}$ (Osram Biolux tubes) at 25 ± 1°C. Subcultures were routinely carried out every two weeks and all experiments were performed in quintuplicate. The suitable phytohormone combination was selected based on the more CIF obtained.

Measurement of callus growth

The cultures were scored for their ability to produce callus in terms of callus induction frequency (CIF) at the end of four to six weeks. The frequency was calculated as the ratio of number of explants responding to the total number of explants inoculated and expressed in percentage. Subsequently, growth measurement of callus was measured as dry weight (DW) by harvesting the biomass and drying at 60°C in a hot air oven to a constant weight.

Statistical analysis

All the experiments were performed in quintuplicate and the significance of the results obtained was verified by using Statistical Analysis Software (SAS®, 2000) and the data were presented as mean ± standard error (SE).

Extraction of PA, SC and SA from bulbs and callus cultures

Callus (Figure 2) obtained at the end of 7th week in MS (Murashige and Skoog 1962) and B5 (Gamborg et al., 1968) media supplemented with NAA and BAP was selected for extraction and quantification of active compounds. The callus was powdered and extracted each time with methanol, chloroform: methanol (8:2), chloroform: ethanol (8:2) and chloroform. The solution was subsequently

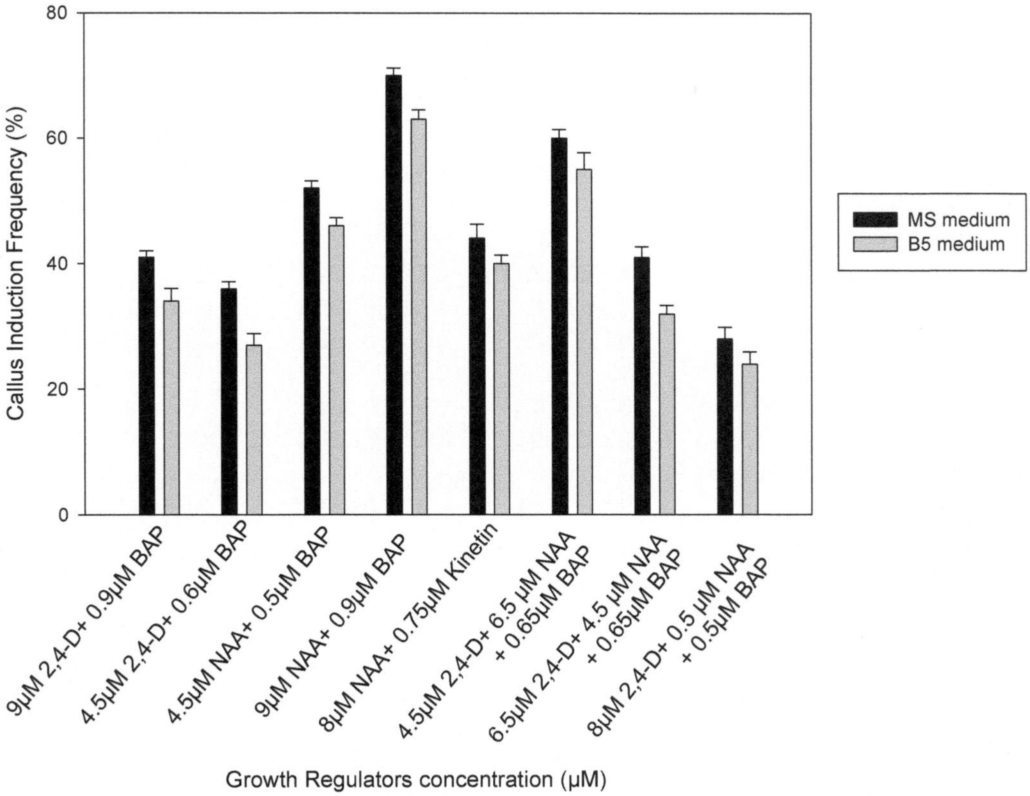

Figure 3. Callus induction frequency (%) of bulb explants cultured on MS and B5 media at different concentrations of growth regulators.

filtered and excess solvent was evaporated under vacuum in a rotary evaporator. Similarly, the crude extracts were prepared using bulbs as such by the same procedure described. The samples were assayed by high performance thin layer chromatography for their identity and quantification of active compounds.

High performance thin layer chromatography (HPTLC)

HPTLC is a major instrumental innovation in the field of separation science and thin layer chromatography is sensitive and robust both qualitatively and quantitatively. It consists of a semi-automatic applicator Linomat IV equipped with 100 μl syringe (Camag, Switzerland). The sample was applied in the form of bands of 8 mm width on aluminium plates precoated with silica gel $60F_{254}$ (20 x 10 cm with 0.25 mm thickness, Merck, Germany) used as stationary phase. The plates were developed in a twin trough glass chambers (Camag, Switzerland) saturated with the mobile phase chloroform : ethanol (9:1 v/v) and a scanner3 with optional deuterium, tungsten and mercury emitting a continuous spectrum of 190 to 800 nm as a source for reflectance scanning densitometry and evaluation of the compound at its λ_{max} 298 nm.

Micro preparative thin layer chromatography

Sample solution of 1 mg/ml concentration was applied as a single band of 170 mm width with a constant flow rate of 6 μl/s on 60 F_{254} silica gel plates. The twin trough chambers were saturated with the mobile phase, chloroform: ethanol in the ratio of 9:1 (v/v) and linear ascending development was carried up to 70 cm from the application level. The developed plate was visualized in a UV cabinet for the detection of the separated compounds. The desired compounds appearing as broad bands at different R_f values were scrapped carefully. The obtained compounds were eluted in micro columns with same mobile phase, chloroform: ethanol (9:1 v/v) to isolate the desired compound from silica gel. The collected fractions were concentrated by evaporating the solvent in a rotavapor at a temperature of 35 to 40°C and analyzed by H^1 NMR and MASS spectroscopy to distinguish and characterize the separated compounds.

RESULTS AND DISCUSSION

Scope of the work and method optimization

As a first step, two sets of media (MS and B5) have been optimized, relating to growth of biomass and yield of secondary metabolites. Both media were found to be potentially active and facilitated the calculation of CIF. The extent of callus proliferation was determined visually and the results are depicted in Figure 3. The CIF was more in both MS (70%) and B5 (63%) media supplemented with NAA (9.0 μM) and BAP (0.9 μM). With IBA (9.0 μM) and BAP (0.8 μM), lower CIF (22.6%) was observed in MS medium but no callus formation was noticed in B5 medium. The callus obtained was profuse, creamy yellow, friable and fast growing, although it was friable but compactwithothercombinationsofgrowthregulators. As evident from the data in Table 1, highest cell dry weight of 1.02 mg

Table 1. Effect of growth regulators on biomass of *C. congesta* bulb explants cultured on MS and B5 media.

Growth regulators concentration (µM)	MS		B5	
	FW (mg)	DW (mg)	FW (mg)	DW (mg)
2,4-D + BAP (9 + 0.9)	3.23 ± 0.201	0.31 ± 0.032[b]	2.10 ± 0.211	0.15±0.011[d]
2,4-D + BAP (4.5 + 0.6)	4.56 ± 0.334	0.42 ± 0.054	2.30 ± 0.256	0.17 ± 0.015
2,4-D + Kinetin (9 + 0.8)	-	-	-	-
2,4-D + Kinetin (9 + 0.5)	-	-	-	-
NAA + BAP (4.5 + 0.46)	5.45 ± 0.420	0.72 ± 0.071	2.11 ± 0.378	0.31 ± 0.025
NAA + BAP (9 + 0.9)	6.78± .511[a]	1.02 ± 0.138	3.53 ± 0.510[c]	0.40 ± 0.028
NAA + Kinetin (8 + 0.75)	3.66 ± 0.222	0.45 ± 0.066	2.00 ± 0.257	0.35 ± 0.030
IBA + BAP (8 + 0.75)	4.13 ± 0.318	0.56 ± 0.072	-	-
2,4D + NAA + BAP (4.5 + 6.5 + 0.65)	4.67 ± 0.287	0.50 ± 0.051	3.00± .612[c]	0.39 ± 0.033
2,4D + NAA + BAP (6.5 + 4.5 + 0.65)	6.23±0.509[a]	0.73 ± 0.091[b]	4.32 ± 0.615	0.51±0.045[d]
2,4D + NAA + Kinetin (2 + 1 + 0.25)	-	-	-	-
2,4D + BAP + Kinetin (8 + 0.5 + 0.5)	3.97 ± 0.330	0.41 ± 0.074	2.45 ± 0.292	0.51 ± 0.038

FW, Fresh weight; DW, dry weight. Numbers with the same letters in each column means differences are not significant (P > 0.05).

and secondary metabolite content was obtained in MS media supplemented with NAA (9.0 µM) and BAP (0.9 µM), followed by 2, 4-D, NAA and BAP (6.5 + 4.5 + 0.65 µM) whereas, the maximum dry weight of 0.51 mg was obtained in B5 medium supplemented with 2, 4-D, NAA and BAP (6.5 + 4.5 + 0.65 µM). Further increase in concentration of growth regulators did not in any way further enhance a quantifiable growth in biomass formation. In contrast, 2,4-D and kinetin combination resulted in no callus formation proving that NAA and BAP combination is crucial and effective in the induction of callus and enhanced production of bioactive compounds which is in agreement with those of published reports (Zhao et al., 2001; Dhar and Joshi, 2005; Liu et al., 2006).

Plants are rich source of bioactive constituents which include alkaloids, glycosides, flavonoids and volatile oils with wide range of phytochemical activity, and most of the secondary metabolites which are responsible for this activity can be alternatively produced by plant cell cultures for their enhanced production (Ramachandra and Ravishankar, 2002; Vanisree et al., 2004). It is also well known that the specified composition of nutrient medium with mutual ratio of auxin to cytokinin plays a major role not only in successful establishment of cell cultures but also in the production of secondary metabolites (Narayanaswamy, 1994). Therefore, it is necessary to find an optimal combination of various media and they have to be frequently tested for each and every new system. The present study facilitated an attractive offer in this direction and successful results were derived.

Extraction yield and time

Bulb explants cultured showed appreciable callus induction at the end of 40 to 45 days of culture. The dried callus was extracted and analyzed by HPTLC. The peaks obtained at R_f 0.12, 0.32 and 0.56 for standards and the same obtained for callus extract confirms the separated compound as SA, SC and PA respectively (Figure 4), which appeared as very bright fluorescent spots. The compounds isolated by micropreparative TLC were subjected to NMR and Mass spectroscopic studies and when compared with those of standard reference samples and that of published reports, it confirmed that the compounds under investigation were SA, SC and PA (Kopp et al., 1996). It was found that SA and PA content in the crude extract were found to be 1.34 and 3.94 mg/g DW whereas, in callus extract, it was 2.35 and 4.51 mg/g DW (Table 2) which indicates that tissue culture studies in our first attempt facilitated a reasonable enhancement in the formation of the main and most important secondary metabolite, SA and PA. The major toxic elements, SC and DAS that actively contribute to the highly toxic nature of *C. congesta* were 2.08 and 7.69 mg/g DW, respectively, which were found to be slightly enhanced in callus extracts, that is, 3.3 and 8.62 mg/g DW, respectively.

Proscillaridin A

^{13}CNMR, 400 MHz (ppm): 162.6, 149.3, 147.8, 147.2, 122.9, 121.4, 115.2, 109.6, 86.5, 77.7, 74.2, 73.7, 72.4, 72.4, 50.0, 48.9, 48.4, 42.9, 40.1, 38.0, 36.0, 33.6, 32.8, 29.2, 29.0, 27.5, 21.5, 19.3, 17.0, 16.8.

Scillaren A

^{13}CNMR, 400 MHz (ppm): 162.6, 149.3, 147.8, 147.2, 122.9, 121.4, 115.2, 110.4, 109.6, 86.6, 86.5, 81.5, 77.7, 76.8, 74.1, 73.8, 72.3, 71.8, 71.5, 62.2, 50.0, 48.9, 48.4, 42.5, 40.1, 38.0, 36.0, 33.6, 32.8, 29.2, 29.0, 27.5, 21.5, 19.3, 17.3, 16.8.

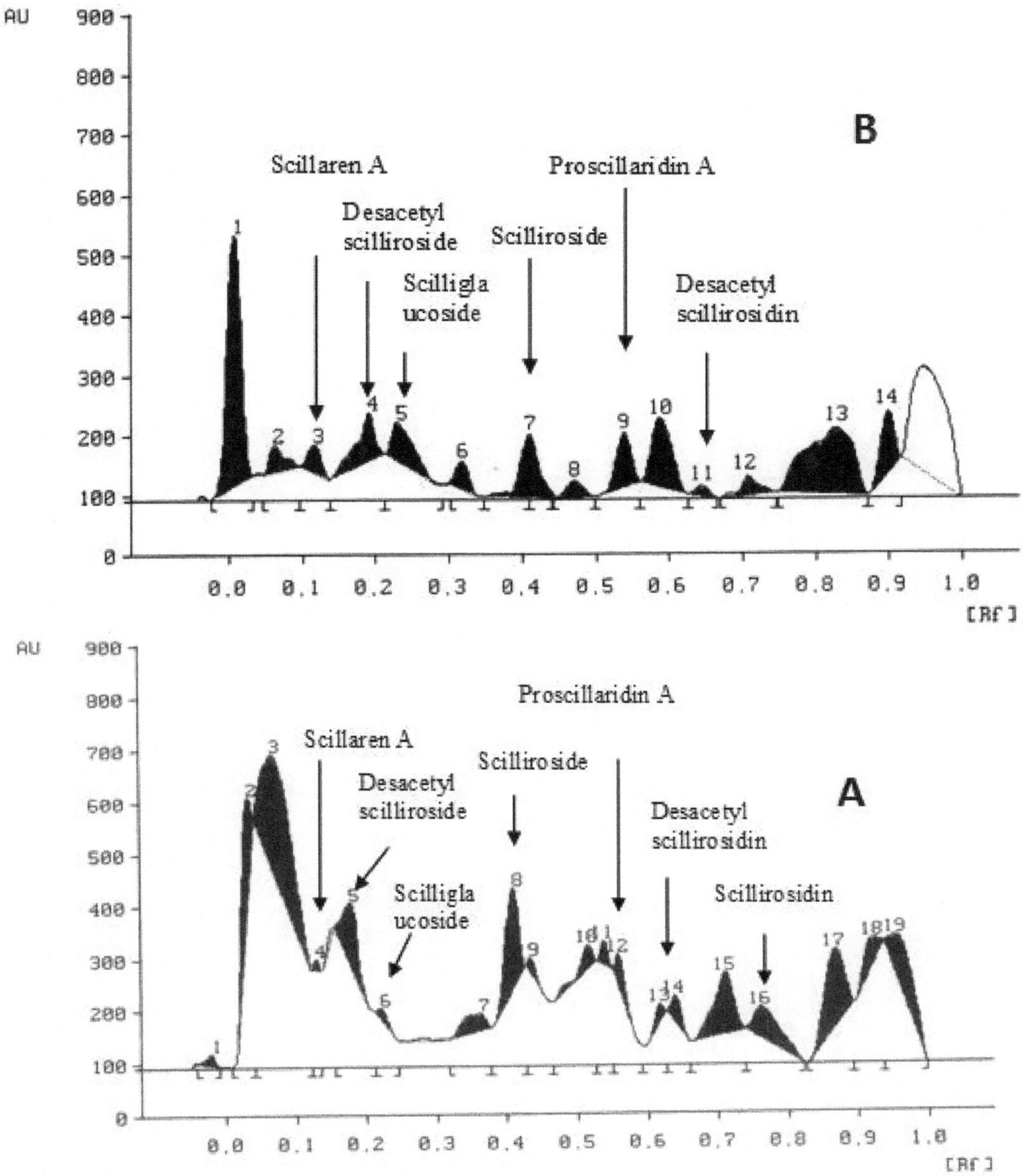

Figure 4. Densitograms showing separation of individual components in (A) Crude extract, (B) Callus extract by reflectance scanning densitometry on HPTLC plates.

Scilliroside

^{13}CNMR, 400 MHz (ppm): 170.2, 162.6, 149.3, 147.0, 142.0, 126.5, 122.9, 115.2, 109.9, 95.1, 81.5, 80.7, 76.8, 74.2, 71.5, 70.5, 62.2, 50.3, 48.5, 47.6, 40.4, 39.6, 36.3, 34.9, 32.1, 29.5, 29.3, 27.5, 21.9, 21.1, 19.6, 17.8, 17.1.

Conclusion

Adopting tissue culture for the enhancement of comercially important and potentially valuable secondary metabolites from a phytochemical source is a rational approach and a very practical strategy for micropropagation studies. The authenticity as has been observed from the information gathered is comprehensive and supports the need of the present study of the Indian herbal based formulation to evaluate allelochemical activity. Evaluation of the results led to the selection of optimal conditions for the induction of callus and production of proscillaridin A, scilliroside and desacetylscilliroside by *in vitro* culture of *C. congesta*. The results obtained are consistent, promising and indicate that the *in vitro* cultures of *C. congesta* might represent an excellent method to produce the bioactive compounds of commercial significance in great yields. Further investigations are underway to scale up the process in a bioreactor to achieve high growth rates and accumulation

Table 2. Composition of Secondary metabolites in crude and callus extracts of *C. congesta* obtained by calculating average peak areas from HPTLC chromatograms

Compound*	R_f	Composition in Crude extract (mg/g DW)	Composition in Callus extract (mg/g DW)
Scillaren A	0.12	1.38	2.35
Desacetylscilliroside	0.18	7.69	8.62
Scilliglaucoside	0.23	1.40	2.68
Scilliroside	0.32	2.08	3.30
Proscillaridin A	0.56	3.94	4.51
Desacetylscillirosidin	0.62	2.16	1.02
Scillirosidin	0.77	3.67	-

*All the compounds were identified using standard reference samples obtained from Merck Chemicals, India. Crude extract was prepared from bulbs using the same procedure like preparation of callus extract

of secondary metabolites.

ACKNOWLEDGEMENT

One of the authors, Ravi Kiran gratefully acknowledges Ms. Saroja Ranganath, Former Assistant Professor, Aurora's Degree and PG College, Hyderabad for the timely help in the preparation of the manuscript.

REFERENCES

Babu KN, Ravindran PN, Peter KV (2001). Spices. In: Parthasarthy VA, Bose TK, Das P (Ed) Biotechnology of Horticultural crops 3rd volume Nayaprakash, Calcutta, India pp. 329.

Dicosmo F, Misawa M (1995). Plant cell and tissue culture: alternatives for metabolite production. Biotechnol. Ads. 13:425-435.

FAIR961436 (1999). European Communities Research Project: Improvement of efficiency and reduction of application rates of preferably naturally grown biocides by complexation with gamma-cyclodextrin, 1997-1999.

Gamborg O, Miller R, Ojima K (1968). Nutrient requirement suspensions cultures of soybean root cells. Exp. Cell. Res. 50:151-158.

Garari RE, Backhaus RA (1987). In vitro propagation of red squill, *Urginea maritima* Baker. Plant cell Tiss. Org. cult. 10:65-71.

Jha S, Mitra GC, Sen S (1984). In vitro regeneration from bulb explants of Indian squill, *Urginea indica* Kunth. Plant cell Tiss. Org. cult. 3:91-100.

Jha S, Sahu NP, Mahato SB (1991). Callus induction, organogenesis and somatic embryogenesis in three chromosomal races of *Urginea indica* and production of bufadienolides. Plant cell Tiss. Org. cult. 25:85-90.

Kopp B, Krenn L, Draxler M, Hoyer A, Terkola R, Vallaster P, Robien W (1996). Bufadienolides from *Urginea maritima* from Egypt. Phytochemistry. 42:513-522.

Liu HJ, Xu Y, Liu YJ, Liu CZ (2006). Plant regeneration from leaf explants of *Rhodiola fastigiata*. *In vitro* Cell Dev. Biol. Plant. 42:345-347.

Murashige T, Skoog F (1962). A revised medium for rapid growth and bioassays with tobacco tissue cultures. Physiol. Plant. 15:473-497.

Namdeo AG (2007). Plant cell elicitation for production of secondary metabolites: A Review. Pharmacog. Rev 1:69-79.

Narayanaswamy S (1994). In: Plant Cell Tissue Culture. 72 Tata McGraw Hill Publishing Co. Ltd, New Delhi.

Pascual-Villalobos M.J (2002). Anti-insect activity of bufadienolides from *Urginea maritima*. In: J. Janick and A. Whipkey (ed) Trends in new crops and new uses. ASHS Press, Alexandria, VA . pp. 564-566.

Ramachandra Rao S, Ravishankar GA (2002). Plant cell cultures: Chemical factories of secondary metabolites. Biotechnol. Ads. 20:101-153.

Roostika I, Purnamaningsih R, Darwati I, Mariska I (2007). In vitro culture manipulation on *Pruatjan* for secondary metabolite production. J. Agro. Biogen. 3:55-59.

SAS (2000). proprietary Software Release 8.1 (TS1M0), SAS Institute Inc, Cary, NC, USA.

Thanonkeo S, Panichajakul S (2005). Production of isoflavones, daidzein and genistein in callus cultures of *Pueraria candollei* Wall. Ex Benth. Var. mirifica. Nutraceu. Funct. Food. 28:45-53.

Vanisree M, Lee CY, Lo SH, Nalawade SM, Lin CY, Tsay SH (2004). Studies on the production of some important secondary metabolites from medicinal plants by plant tissue cultures. Bot. Bull. Acad. Sin. 45:1-22.

Verbiscar AJ, Patel J, Banigan TF, Schatz RA (1986). Scilliroside and other scilla compounds in red squill. J. Agric. Food. Chem. 34:973-979.

Wawrosch Ch, Kongbangkerd A, Kopf A, Kopp B (2005). Shoot regeneration from nodules of *Charybdis sp.*: a comparison of semisolid, liquid and temporary immersion culture systems. Plant cell Tiss. Org. cult. 81:319-322.

9

The combined effects of starvation and pH on the virulence of *Shigella sonnei* ATCC25931

Ali Ellafi*, Besma Harbi, Rihab Lagha and Amina Bakhrouf

Laboratory of Analysis, Treatment and Valorization of Environment Polluants and Product, Faculty of Pharmacy, Monastir (Tunisia), Tunisia.

***Shigella sonnei* encounter numerous different stresses during their growth, survival and infection. In this study, the effect of stress response to pH and starvation was investigated. We studied the survival, adhesion and the morphology of *Shigella* after its incubation in several pH. Our results show that after 2 h of incubation, the rate of cell survival was proportional to the decrease of pH of the medium, but we also noted that the reduction of the cells viability is significant for the normal cells compared to starved cells. In addition to that, the results proved the evidence that the pH can influence significantly the hydrophobicity of *S. sonnei* (normal and starved) and their capacity to produce biofilm on Congo red agar and on polystyrene microplate wells. The atomic force micrographs showed a reduction of the cells size after stress.**

Key words: *Shigella sonnei*, starvation, pH, survival, adhesion, morphology.

INTRODUCTION

Shigella is a member of the Enterobacteriaceae family and causes shigellosis characterized by bacillary dysentery (mucoid bloody stool). *Shigella* is commonly found in water contaminated with human feces and fecal–oral route is the primary mode of transmission. It is responsible for a worldwide health problem; however, it is a serious concern in developing countries. It is estimated that *Shigella* spp. account for 1.1 million deaths and 165 million cases of dysentery annually worldwide. Shigellosis has a high morbidity and mortality rate in children less than 5 years of age. Malnourished children are highly susceptible and *Shigella* infection further promotes impaired nutrition, recurring infection and retarded growth. Antibiotic resistant strains are continuously emerging thus treatment regiments become very difficult against shigellosis.

The environmental adaptation plays an important role for pathogens to survive in natural environment or invade hosts. When microorganisms encounter nutrient deprivation or starvation conditions, they respond to them by arresting all metabolic activity and growth. They can then carry out starvation-induced activities such as production of enzymes. This includes many degradative enzymes such as proteases, lipases, and substrate capturing enzymes such as glutamine synthetase, and alkaline phosphatase (Kjelleberg et al., 1987; Matin et al., 1989). During starvation, many bacterial species show increased resistance to a number of environmental stresses such as low pH and heat and oxidative stress (Hartke et al., 1994; Watson et al., 1998).

The enteric bacteria exhibit a stress response to sublethal environmental stresses (Ryu and Buchat, 1998; Leenanon and Drake, 2001). These stress responses can result in resistance to a variety of environmental or processing parameters (Abee and Wouters, 1999; Law, 2000). The nature and the intensity of the stress response may vary between bacteria (Leenanon and Drake, 2001). So, cross-protection is referred to as the ability of one stress condition to provide protection against other stresses. Microbial adaptation responses to one stress can lead to enhanced resistance to a different stress (Cheville et al., 1996).

Starvation conferred less acid tolerance to *Escherichia coli* O157:H7 than did stationary phase (Cheville et al., 1996). However, starved or stationary phase cells showed significantly more acid tolerance than acid-adapted (growth

*Corresponding author. E-mail: ali_lafi160@yahoo.fr.

at pH 5) mid and late-log phase cells (Arnold and Kaspar, 1995; Cheville et al., 1996). The aim of this work was to study the effect of stress response to pH and cross-protection: pH and starvation. Slime production was realized on Congo red agar. *Shigella sonnei* capacity to biofilm formation was tested. The morphology of cells was examined by atomic force microscopy.

MATERIALS AND METHODS

Bacterial strains

Shigella strain: *S. sonnei* was obtained from Monastir hospital, Tunisia and maintained at -80°C in Luria-Bertani broth (LB) supplemented with glycerol (15%, vol/vol). For the starved *Shigella* was obtained after incubation in seawater during eight months. For the experiments, the cells were grown at 37°C in tryptic soy broth (TSB, (Difco)) for 24 h. TSB microcosmes (100 ml) was achieved to pH 3.5, 4.5, 5.5, 6.5 and 7.3 and autoclaved (115°C for 15 min) in 100 ml Erlenmeyer flasks. *Shigella* cells were washed three times by centrifugation (13000 rpm for 10 min at 20°C) with TSB and then suspended in 10 ml of autoclaved TSB. The microcosms (100 ml) were inoculated with these suspensions (approximately 10^9 CFU/ml) and then incubated at 37°C.

Enumeration techniques

Plate counts of culturable cells were determined by the drop plate method (Hoben and Somasegaran, 1982), using Tryptic soy agar (TSA; Difco). Microcosms were sampled daily during the 30, 60, 120 min. Time zero (inoculation time) and subsequent samples were taken for plate counts. The plates were incubated at 37°C, and the number of colonies was counted after 24 and 48 h.

Phenotypic characterization of slime-producing bacteria

Qualitative detection of biofilm formation by tested strains was studied by culturing the strains on Congo red agar (CRA) plates as described previously (Freeman et al., 1989). *Shigella* strains were inoculated onto the surface of CRA plates, made by mixing 0.8 g Congo red with 36 g saccharose (Sigma) in 1 L of brain heart infusion agar, and were incubated for 24 h at 30°C under aerobic conditions and followed overnight at room temperature (Ellafi et al., 2009). Slime producing bacteria appeared as black colonies, whereas non-slime producers remained non pigmented (Subashkumar et al., 2006).

Cell surface hydrophobicity

Hydrophobicity was measured by the hexadecane partitioning method of Loosdrec et al. (1987). Bacterial cells grown overnight in tryptic soy broth were washed with PBS and resuspended in 4 ml of PBS, and the absorbance (DO_{540}) was determined. One milliliter of hexadecane was added to each cell suspension and equilibrated for 10 min. Each suspension was reincubated at 37°C for 30 min. The aqueous layer was removed and aerated to remove all traces of hexadecane, and absorbance (DO_{540}) was measured against a hexadecane-extracted PBS blank. The hydrophobicity index was expressed as the ratio of absorbance of the hexadecane-extracted sample to absorbance of the sample before extraction.

Adherence assay to measure biofilm production by *Shigella sonnei* cells under stress

Biofilm production by *Shigella sonnei* strains grown in TSB with different conditions was determined using a semi-quantitative adherence assay on 96-well tissue culture plates (Nunc, Roskilde, Denmark) as described previously (Chaieb et al., 2007) with some modifications. An overnight culture grown in TSB (Biorad) at 37°C was diluted to 1:100 in TSB supplement with 2% (wt/vol) glucose and different pH as reported elsewhere (Rachid et al., 2000). A total of 200 µl of these cell suspensions was transferred in a U-bottomed 96-well microtiter plate. Each strain was tested in triplicate. Wells with sterile TSB alone was served as controls.

After their incubation for 24 h at 37°C, the culture was removed and the plates were washed three times with 200 µl of phosphate-buffered saline (7 mM Na_2HPO_4, 3 mM NaH_2PO_4 and 130 mM NaCl at pH 7.4) to remove non adherent cells and dried in an inverted position. Adherent cells were fixed with 95% ethanol and were stained with 100 µl of 1% (wt/vol) crystal violet (Merck) for 5 min. Then, unbound crystal violet was removed and the wells were washed three times with 300 µl of sterile distilled water. The water was then cleared and the microtiter plate was air dried for 3 with 4 h. The optical density (OD) of each well was measured at 570 nm using an automated Multiskan reader (GIO. De Vitae, Rome, Italy). Biofilm formation was categorized as highly positive ($OD_{570} \geq 1$), low-grade positive ($0.1 \leq OD_{570} < 1$), or negative ($OD_{570} < 0.1$) (Chaieb et al., 2007)

Determination of morphological changes by AFM

In order to visualize any morphological changes in the stressed cells, *Shigella* cells were examined, in triplicate, by atomic force microscopy (AFM) (Nanoscope IIIA, Digital Instrument, VEECO). For the experiments the cells were collected, washed three times with PBS, and centrifuged. The final pellet was resuspended in PBS, placed on a round microscope cover slide and was simply dried in air according to the method previously described (Braga and Ricci, 1998).

Molecular identification of stressed bacteria

Bacteria were cultured on TSA for 24 h at 37°C. One colony was cultured in TSB for 24 h at 37°C, and 1.5 ml was centrifuged. The deoxyribonucleic acid (DNA) was extracted by boiling for 5 min and centrifugation at 13,000 rpm for 8 min. The supernatant was used for amplification by PCR with *Shigella* primers of the target *IpaH* gene. Polymerase chain reaction (PCR) were performed in 25 µl containing 50 ng of extracted DNA, 5 µl green Go *Taq* buffer (5×), 0.25 µl dNTPs (10 mM), 0.5 µl $MgCl_2$ (50 mM), 1 µl of *IpaH* forward primer 5′-GTTCCTTGACCGCCTTTCCGATAC-3′ (25 pM) and *IpaH* reverse primer 5′-CATTTC CTTCACGGCAGTGGA-3′ (25 pM) (Hartman et al., 1990), 1 U of GO *Taq* DNA polymerase (Promega, USA).

Amplification was conducted in the Thermocycler PTC 100 (BioRad). The reaction mixtures were heated at 94°C for 5 min and were subjected to 35 cycles of denaturation at 94°C for 90 s, annealing at 57°C for 30 s, and elongation at 72°C for 90 s, followed by 10 min of final extension period at 72°C. PCR products (5 µl) were analysed on 1% agarose gels stained with ethidium bromide (0.5 mg/ml) at 90 V for 1 h and visualized under ultraviolet transillumination. The amplification products were photographed and their sizes determined with a 100-bp molecular size marker (Promega, France).

Statistical analysis

Statistical analysis was performed using the S.P.S.S. 13.0 statistics package for Windows. The differences in the degree of adhesion assay were examined by the Friedman test, followed by the Wilcoxon signed ranks test. P-values of < 0.05 were considered as significant.

RESULTS

Survival *Shigella* with several pHs

We investigated the viability of the normal and starved *S. sonnei* in TSB with several pH (Table 1). After 2 h of incubation in the different pH, we noted a significantly decrease of the cell numbers. The cell remained constant in TSB at pH 7.3. Populations of cells inoculated in to TSB acidified to pH 3.5 rapidly decreased after 2 h of incubation, while a more gradual decrease was observed at pH 4.5 and 5.5. Populations of cells remained nearly constant at pH 6.5. These plots reveal that survival was proportional increased with an increase in pH. We also noted that the reduction of the cells viability is significant for the normal cells compared to starved cells. Indeed, the starved cells in seawater exhibited higher resistance to acid stress.

Phenotypic determination of slime production of *Shigella* sonnei under several pHs

Production of slime by all investigated strains was assessed by culture on CRA plates. Normal *S. sonnei* cells formed colonies with a red centre and a lighter outer zone and the starved cells in seawater developed colonies with black center and red contour. After 24 h of their incubation in TSB at different pH, all the strains developed colonies with black center and red contour (Figure 1).

Effect of stress on the surface hydrophobicity

The decrease of pH had a significant effect on the surface hydrophobicity of *S. sonnei* (Table 2). *S. sonnei* showed a significant increase (P<0.05) in surface hydrophobicity when subjected to diminution of pH. There was a 15 to 52% increase in the hydrophobicity for all the strains. This increase is important for the starved cells in seawater compared with the normal cells.

Biofilm formation by *Shigella* under several pHs

Biofilm formation of *S. sonnei* strains was evaluated in 96 wells plate with TSB at different pH (3.5, 4.5, 5.5, 6.5 and 7.3). The results of the OD_{570} presented in the Table 2 showed that the strain did not show any biofilm formation in the normal condition ($OD_{570}<0.1$). After incubation in several pH some normal or starved strain are able to form biofilm: were considered as low-grade positive ($0.1 \leq OD_{570} < 1$).

Morphological changes of *Shigella* under several pHs

The cell morphology under several pHs was examined by AFM (Figure 2). The control *S. sonnei* cells, whose length is about 3.4 µm have a normal rod shape with a smooth surface (Figure 2a). Whereas the cells starved in seawater present two coccoid shape forms whose length is less than 1 and 0.4 µm. After 24 h of incubation in the different medium we have noted a decrease of the cell size at 5.5, 4.5 and 3.5 for the normal strain. The size of cells is at 5.5 and 4.5 (Figure 2b); while to 3.5 it is lower to 1 µm. (Figure 2c). For the starved cells, the size is between 1 and 0.4 µm at different medium (Figure 2c).

Molecular confirmation of stressed strains

We used the technique of PCR to identify the stressed strains. After amplification of ipaH gene by PCR, we confirmed the identity of the investigated *S. sonnei* (Figure 3).

DISCUSSION

Bacterial cells have the ability to sense and respond to changes in their external environment (Rosen et al., 2001). The ability of bacteria to sense and respond effectively to changes in the environment is crucial for their survival (Ben Abdallah et al., 2009). Enteric bacteria such as *Shigella* are found in many diverse and extreme environments, and as a result it has developed responses to combat these adverse conditions (Foster and Spector, 1995). The present study showed that the survival of normal or starved *Shigella sonnei* is greatly influenced by the pH. We also noted that the reduction of the cells viability is significant for the normal cells compared to starved cells. Indeed, the starved cells in seawater exhibited higher resistance to acid stress.

During starvation, many bacterial species show increased resistance to a number of environmental stresses such as low pH and heat and oxidative stress (Hartke et al., 1994; Watson et al., 1998). The significant decrease of *Shigella* cells numbers at the beginning of the experiment may be a result of the death bacterial cells caused by the decrease of pH. These results are in accordance with those reported by Zaika and Phillips (2005), who demonstrated that the survival of *Shigella* is affected by the diminution of pH.

The persistence of *S. sonnei* under starvation and/or pH stress induced modifications in cell properties. Indeed, before their treatment, the normal *S. sonnei* cells formed colonies with a red center and a lighter outer zone and the starved cells in seawater developed colonies with black center and red contour. After 24 h of their incubation in TSB at different pH, all the strains developed colonies with black center and red contour. This state is

Table 1. Effect at pH on the survival of *Shigella sonnei* incubated at 37°C.

Percent of salt	Time (h)	Strains: Log_{10} CFU/ml ± Standard deviation	
		S	Si
3.5	0	9 ± 0.35	9 ± 0.28
	1	5.12 ± 0.5	5.43 ± 0.21
	2	5.033 ± 0.62	5.32 ± 0.16
4.5	0	9 ± 0.11	9 ± 0.14
	1	6.13 ± 0.77	6.4 ± 0.23
	2	6.28 ± 0.82	6.3 ± 0.51
5.5	0	9 ± 0.25	9 ± 0.27
	1	7.22 ± 0.11	7.81 ± 0.70
	2	6.3 ± 0.35	6.41 ± 0.15
6.5	0	9 ± 0.8	9 ± 0.78
	1	7.6 ± 0.65	7.7 ± 0.71
	2	7.3 ± 0.33	7.9 ± 0.91
7.3	0	9 ± 0.65	9 ± 0.2
	1	8.51 ± 0.35	8.16 ± 0.9
	2	8.7 ± 0.71	8.5 ± 0.6

S: strain before incubation in seawater; Si: strain incubated during eight month in seawater microcosms at room temperature.

considered a variable phenotype (Ellafi et al., 2009). This main difference between the normal and stressed strains is that there can be a state of passage toward the biofilm formation in case of stress elongation. According to Adam et al. (2002) the biofilm is a protective system under environment stress conditions (physical, chemical treatment).

Bacterial colonization of solid surfaces has been described as a basic and natural bacterial strategy in a wide variety of environments (Kjelleberg et al., 1983; Jana et al., 2000). Our observation showed that the low pH can influence significantly the capacity of *S. sonnei* cells to adhere to polystyrene and their hydrophobicity. This increase is important for the starved cells compared with the normal cells. Starvation is known to alter bacterial surface characteristics such as hydrophobicity, charge and irreversible attachment, which are essential factors in biofilm formation (Brown et al., 1977). According to Allan et al. (2002), *Citrobacter* sp biofilm formation was affected signifcantly by nutrient limitation.

Various studies suggest that fimbrial expression is associated with biofilm formation; fimbriae are known to be involved in attachment (Low et al., 1996) and biofilm development (Pratt and Kolter, 1998) and are likely to be involved in the mechanism of biofilm formation in the present case. Previous studies have shown that regulation of fimbrial production is affected by carbon source, nitrogen source, pH and temperature (Xie et al., 1997). As a consequence, this modification caused a general increase in virulence and resistance against stress (Givskov et al., 1994). Both acid-shocked and -adapted cells showed cross-protection against heat or other stresses (Ryu and Beuchat, 1998).

The AFM revealed reduction in the *S. sonnei* cells size caused by pH stress. After 24 h of incubation in the different medium we noted a decrease of the cell size at 5.5, 4.5 and 3.5 for the normal strain. The size of cells is at 5.5 and 4.5 (Figure 2b); while to 3.5 it is lower to 1 µm (Figure 2c). For the starved cells, the size is between 1 and 0.4 µm at different medium (Figure 2c). The reduction of cell size and the increase of the adhesion property are probably linked. Indeed, Bower et al. (1996) observed that spores adhere more quickly than vegetative cells to food contact surfaces due to higher hydrophobicity of their cell surfaces. According to Morita (1993), several bacteria such as *Shigella*, can survive for a long period under stressing environmental conditions owing to gradual changes in cellular physiology and morphology.

The reduction of the bacteria size, such as *Shigella,* during the stress is a strategy of survival (Jiang and Chai, 1996). In addition, the signification difference between normal and starved cells is caused by cross-protection.

These results are in agreement with those reported by

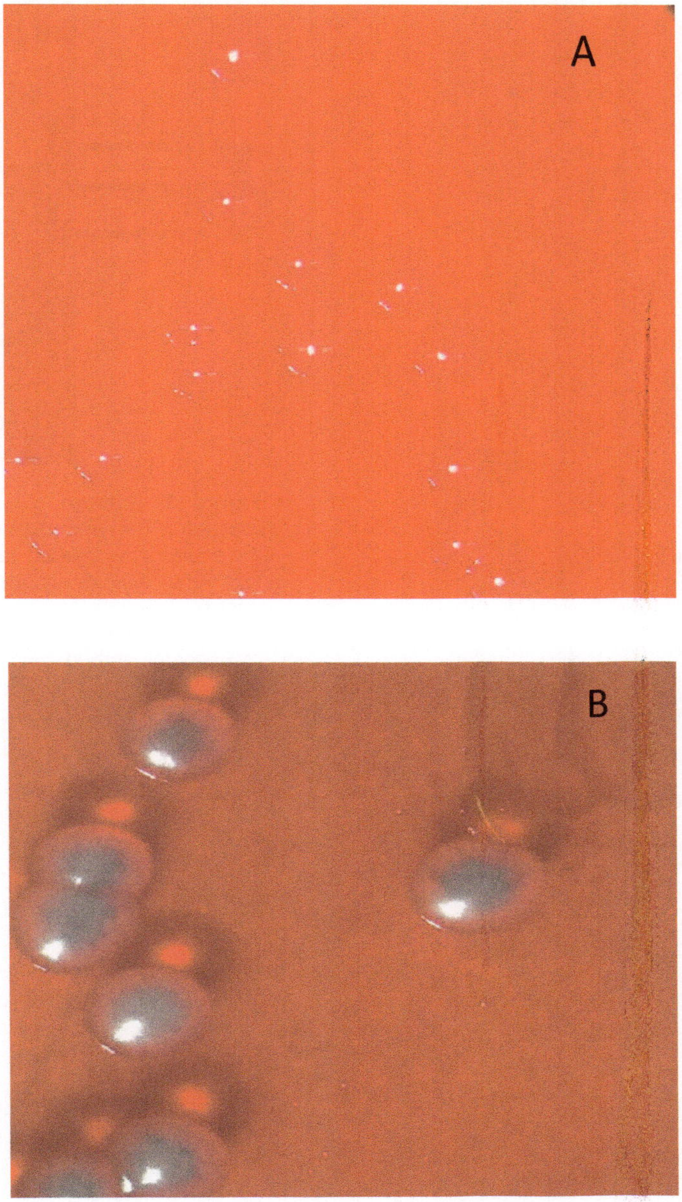

Figure 1. Detection of slime production by *S. sonnei* (a) before and (b) after incubation in several pH and starvation in seawater microcosms.

Table 2. Effect of pH on surface hydrophobocity and biofilm formation by *Shigella sonnei.*

Strain	pH									
	7.2		6.5		5.5		4.5		3.5	
	HI	Sd[1]	HI	Sd[1]	HI	Sd[1]	HI	Sd[1]	HI	Sd[1]
S1	15.2	0.53	16.83*	0.24	19.92*	0.2	39.25*	0.82	29.8*	0.66
S2	27.9	0.65	32.2*	0.13	50.9*	0.15	52.06*	0.15	23.89*	0.17
	DO_{570nm}	Sd[1]	DO_{570nm}	Sd[1]	DO_{570nm}	Sd[1]	DO_{570nm}	Sd[1]	DO_{570nm}	Sd[1]
S1	0.092	0.007	0.172*	0.022	0.195*	0.044	0.363*	0.22	0.225*	0.025
S2	0.189	0.033	0.214*	0.27	0.247*	0.37	0.308*	0.033	0.206*	0.17

[1]:Standard deviation. *:P<0.05.

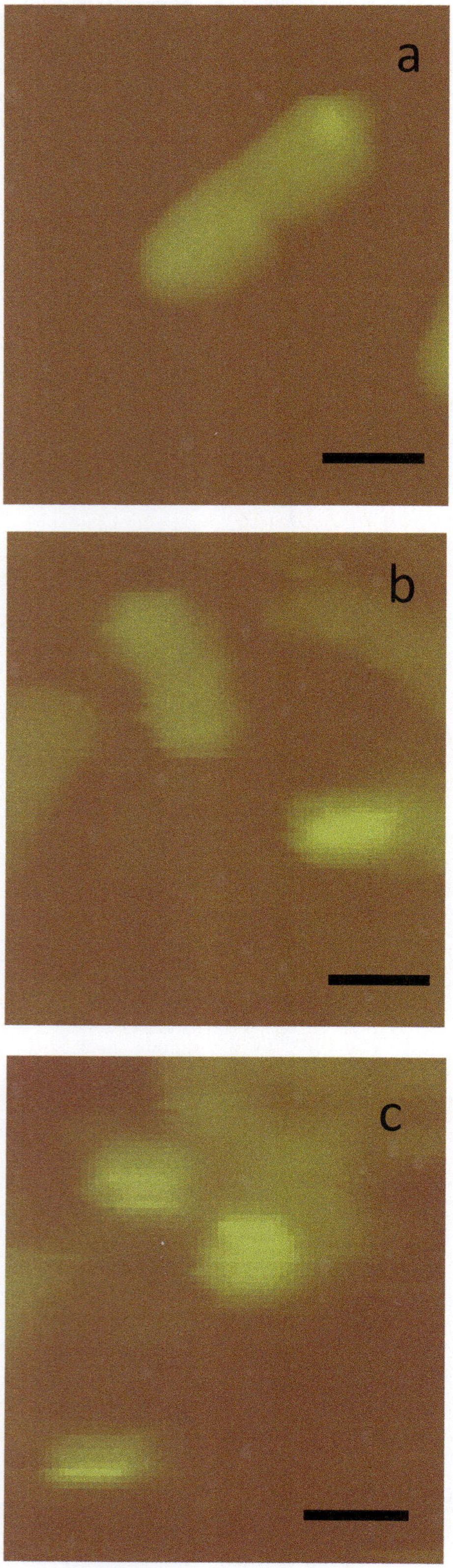

Figure 2. Atomic force micrographs of stressed S. sonnei cells. (a) control and cells stressed at pH: 6.5 ; (b) cells stressed at pH: 5.5 and 4.5; (c) cells starved and stressed in different pH and stressed at pH: 3.5. Bars: 1 µm.

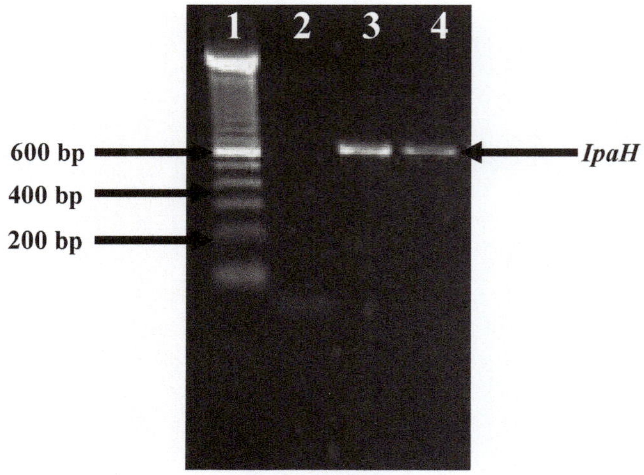

Figure 3. Agarose gel electrophoresis of polymerase chain reaction (PCR) amplification of *ipaH* gene. Lane 1, 100 bp DNA molecular size marker; lanes 2 to 4, PCR amplicons obtained with DNA amplification of *Shigella* strains. Lane 2, negative control; lane 3, S; lane 4, Si. **S:** strain before incubation in seawater; **Si:** strain incubated during eight month in seawater microcosms at room temperature.

Kalchayanand et al. (2004), which showed that hydrostatic pressure and bacteriocin mixture induce change in the morphology of *Escherichia coli* O157:H7 and *Salmonella typhimurium*. The stress response of *S. sonnei* is complex, robust, and versatile. Bacterial stress adaptation of *Shigella* and the potential for stress-associated enhanced virulence need to be addressed in more detail to prevent potential risk of disease. An increased understanding of mechanisms and regulation of the stress adaptation will provide information for pathogenic control, increase effective design of novel control methods.

REFERENCES

Abee T, Wouters JA (1999). Microbial stress response in minimal processing. Int. J. Food Microbiol. 50:65-91.

Adam B, Baillie GS, Douglas LJ (2002). Mixed species biofilms of *Candida albicans* and *Staphylococcus epidermidis*. J. Med. Microbiol. 51:344-349.

Allan VJ, Callow ME, Macaskie LE, Paterson-Beedle M (2002). Effect of nutrient limitation on biofilm formation and phosphatase activity of a *Citrobacter sp*. Microbiol. 148:277-288.

Arnold KW, Kaspar CW (1995). Starvation- and stationary–phase-induced acid tolerance in *Escherichia coli* O157:H7. Appl Environ Microbiol 61:2037-2039.

Ben Abdallah F, Bakhrouf A, Kallel H (2009). Enzymatic, outer membrane proteins and plasmid alterations of starved *Vibrio parahaemolyticus* and *Vibrio alginolyticus* cells in seawater. Arch. Microbiol. 191:493-500.

Bower C, McGuire J, Daeschel M (1996). The adhesion and detachment of bacteria and spores on food-contact surfaces. Trends Food Sci. Technol. 7:152-157.

Brown CM, Ellwood DC, Hunter JR (1977). Growth of bacteria at surface influence of nutrient limitation. FEMS Microbiol. Lett. 1:163-166.

Braga PC, Ricci D (1998). Atomic force microscopy:application to investigation of *Escherichia coli* morphology before and after exposure to cefodizime.Antimicrob. Agents Chemother 42:18-22.

Chaieb K, Chehab O, Zmantar T, Rouabhia M, Mahdouani K, Bakhrouf A (2007). *In vitro* effect of pH and ethanol on biofilm formation by clinical *ica* positive *Staphylococcus epidermidis* strains. Ann Microbiol. 57:431-437.

Cheville AM, Arnold KW, Buchrieser C, Cheng CM, Kaspar CW (1996). *rpoS* regulation of acid, heat, and salt tolerance in *Escherichia coli* O157:H7. Appl. Environ. Microbiol. 62:1822-1824.

Ellafi A, Denden I, Ben Abdallah F, Souissi I, Bakhrouf A (2009). Survival and adhesion ability of *Shigella* spp. strains after their incubation in seawater microcosms. World J. Microbiol. Biotechnol. 25:1161-1168.

Foster JW, Spector M (1995). How *Salmonella* survives against the odds. Annu. Rev. Microbiol. 49:145-174.

Freeman DJ, Falkiner FR, Keane CT (1989). New method for detecting slime production by coagulase-negative *staphylococci.* J. Clin. Pathol. 42:872-874.

Givskov M, Eberl L, Moller S, Poulsen, L, Molin S (1994). Responses to nutrient starvation in *Pseudomonasputida* KT2442:analysis of general cross-protection, cell shape, and macromolecular content. J. Bacteriol. 176:7-14.

Hartman AB, Venkatesan MM, Oaks EV, Buysse JM (1990). Sequence and molecular characterization of multicopy invasion plasmid antigen gene, ipaH, of *Shigella flexneri.* J. Bacteriol. 172(4):1905-1915.

Hoben HJ, Somasegaran P (1982). Comparison of the pour, spread, and drop plate methods for enumeration of *Rhizobium* spp. In inoculants made from presterilized peat. Appl. Environ. Microbiol. 44:1246-1247.

Jiang X, Chai TJ (1996). Survival of Vibrio parahaemolyticus at low temperatures under starvation conditions and subsequent resuscitation of viable nonculturable cells. Appl. Environ. Microbiol. 62 :1300-1305.

Kalchayanand N, Dunne P, Sikes A, Ray B (2004). Viability loss and morphological change of foodborne pathogens following exposure to hydrostatic pressures in the presence and absence of bacteriocins. Inter. J. Food Microbiol. 91:91-98.

Kjelleberg S, Hermansson M, Marden P (1987). The transient phase between growth and nongrowth of heterotrophic bacteria, with emphasis on the marine environment. Ann. Rev. Microbiol. 41:25-49.

Kjelleberg S, Humphrey BA, Marschall K (1983). Initial phases of starvation and activity of bacteria st surface. Appl. Environ. Microbiol. 46:978-984.

Law D (2000). Virulence factors of *Escherichia coli* O157:H7 and other Shiga toxin producing.

Leenanon B, Drake MA (2001). Acid stress, starvation, and cold stress affect post stress behavior of *Escherichia coli* O157:H7 and nonpathogenic *E. coli.* J. Food Prot. 64:970- 974.

Loosdrec MCM, Van Lyklema J, Norde W, Norde W (1987). The role of bacterial cell wall hydrophobicity in adhesion. Appl. Environ. Microbiol. 53:1893-1897.

Low D, Braaten B, van der Woude M (1996). Fimbriae. In:*Escherichia coli* and *Salmonella.* Neidhardt FC (eds). Washington, DC:American Society for Microbiol. pp. 146-157.

Matin A, Auger EA, Blum PH, Schultz JE (1989). Genetic basis of starvation survival in nondifferentiating bacteria. Annu. Rev. Microbiol. 43:293-316.

Morita RY (1993). Bioavailability of energy and the starvation state. In:Kjelleberg S (ed) Starvation in Bacteria. Plenum Press NewYork, pp. 1-23.

Pratt LA, Kolter R (1998). Genetic analysis of *Escherichia coli* biofilm formation, roles of Flagella, motility, chemotaxis and type I pili. Mol. Microbiol. 30:285-293.

Rachid S, Ohlsen K, Wallner U, Hacker J, Michael Hecker M, Ziebuhr W (2000). Alternative transcription factor sigma (B) is involved in regulation of biofilm expression in a *Staphylococcus aureus* mucosal isolate. J. Bacteriol. 182:6824-6826.

Ryu JH, Beuchat LR (1998). Influence of acid tolerance responses on survival, growth, and thermal cross-protection of *Escherichia coli* O157:H7 in acidified media and fruit juices. Int. J. Food Microbiol. 45:185-193.

Subashkumar R, Thayumanavan T, Vivekanandhan G, Perumalsamy L (2006). Occurrence of *Aeromonas hydrophila* in acute gasteroenteritis among children. Indian J. Med. Res. 123:61-66.

Watson SP, Antonio M, Foster SJ (1998). Isolation and characterization of *Staphylococcus aureus* starvation-induced, stationary-phase mutants defective in survival or recovery. Microbiol 144:3159-3169.

Xie H, Cai S, Lamont RJ (1997). Environmental regulation of Fimbrial gene expression in *Porphyromonas gingivalis.* Infect. Immun. 65:2265-2271.

Zaika LL, Phillips JG (2005). Model of the combined effects of temperature, pH and sodium chloride concentration on survival of *Shigella flexneri* strain 5348. Int. J. Food Microbial. 101:179-187.

Lead induced dyslipidemia: The comparative effects of ascorbate and chelation therapy

Regina Ngozi Ugbaja*, Beno Okechukwu Onunkwor and Demilade Akinbola Omoniyi

Department of Biochemistry, College of Natural Sciences, Federal University of Agriculture, Abeokuta, Nigeria.

To investigate the comparative effects of ascorbate and chelating agents on some markers of lipid metabolism in lead exposed rats, 35 male Wistar rats were used. They were grouped randomly into five (n=7); 28 of which were administered 75 mg/kg body weight lead acetate (PbAc) orally for 14 days after which their blood samples were assayed for lead. Three of the groups were further administered 30 mg/kg body weight D-penicillamine (D-pen), 30 mg/kg body weight succimer (DMSA) and 500 mg/kg body weight ascorbate (Asc) daily orally, respectively. The control group was however administered normal saline. The blood lipid profiles were determined spectrophotometrically. Lead exposure resulted in significant dyslipidemia ($p < 0.05$), characterized by 50% hypercholesterolemia and hypertriglyceridemia and 132% hyperphospholipidemia (plasma) while in the red blood cells, hypocholesterolemia and hypophospholipidemia were observed. During the therapeutic doses, the groups administered chelating agents and Asc showed a significant amelioration in the plasma and red blood cell levels of total cholesterol, triacylglycerols and phospholipids in the order, DMSA > Asc > D-pen. Decrease in blood lead levels after therapy indicated that the chelating agents have an advantage over Asc. The study indicates that administration of the antioxidant, Asc may not be more efficacious than the chelating agents but could be a cheaper and more convenient therapy for lead toxicity.

Key words: Ascorbate, chelating agents, dyslipidemia, lead exposure, plumbism.

INTRODUCTION

Metals are elements found in nature usually in the form of their respective compounds. They have been found to have much relevance in many industries; but as economic as most of them are, a couple of them also have adverse effects on man (Patil et al., 2006; Ponce-Canchihuaman et al., 2010; Rao et al., 2007). Most notably, are the heavy metals which have specific density greater than 5 g/cm^3 such as mercury (Hg), cadmium (Cd), arsenic (As) and lead (Pb). Most of these heavy metals persist in the environment and produce a variety of adverse effects because they are generally not biodegradable (Alissa and Ferns, 2011).

One of the leading metallic xenobiotic is lead (Pb). It is a stable heavy metal found in the air, water, soil and food as a contaminant (Allouche et al., 2011; Ait Hamadouche et al., 2009; Ibrahim et al., 2012). Occupational exposure has the highest account of lead poisoning. Environmental sources of lead include inhalation of automobile exhaust from gasoline containing alkyl lead additives, ingestion of dust contaminated with lead, lead-based paints and drinking water that had passed through lead piping (Verma and Dubey, 2003; Flora, 2009; Ait Hamadouche et al., 2009; Kumar et al., 2011; Alissa and Ferns, 2011). Other significant sources include cosmetic products, food-can soldering, toys, ceramic glazes and folk remedies (Mohammed et al., 2008; Sajitha et al., 2010).

Lead is relatively poorly absorbed into the body, but once absorbed it is slowly excreted and so accumulate in the body, especially in the bone, causing several tissue and organ damage (Alissa and Ferns, 2011; Zhang et al., 2012). The severity of cases of plumbism has been of serious concern to international health organizations such as the Centres for Disease Control and Prevention (CDC)

*Corresponding author. E-mail: regiugbaja@yahoo.com.

and World Health Organisation Occupational Safety and Health Administration (WHO-OSHA). The CDC redefined elevated blood lead levels (BLL) as that ≥10 μg/dl in children and 40 μg/dl in adults. Even at these safety levels, cases of lead poisoning have occurred (Mohammed et al., 2008; Flora, 2009; Olympio et al., 2009; Miranda et al., 2011).

Lead has been proven to produce a series of adverse effects on numerous organs and biochemical indices (Gurer and Ercal, 2000; Chen et al., 2003; Flora et al., 2004; Ademuyiwa et al., 2005; Dosumu et al., 2005; Choudhary et al., 2007). The adverse effects can be hematologic, reproductive, neurobehavioral, nephrotoxic, cytotoxic and cardiovascular (Patra et al., 2000; Moreira et al., 2001; Ademuyiwa et al., 2005; Diamond, 2005; Patil et al., 2006; Ait Hamadouche et al., 2009a; Olympio et al., 2009; Raafat et al., 2009; Ponce-Canchihuaman, 2010; Harishekar and Kiran, 2011; Mrugesh et al., 2011; Patra et al., 2011).

Lead poisoning has also been reported to cause oxidative damage to DNA by interfering with the incision step of DNA repair system, thus inducing carcinogenicity (Koedrith and Young, 2011). Also, a couple of proteins *in vivo* and the plasma membrane have lost their integrity and hence function as a result of plumbism (Okediran et al., 2009; Abam et al., 2008). Cases of lead induced dyslipidemia, hypertension and atherosclerosis have also been reported (Ademuyiwa et al., 2005; Heo et al., 2004). Chronic lead exposure has been demonstrated to alter fatty acid composition of erythrocyte membranes (Donaldson and Knowles, 1993). Acute lead exposure affects the cardiac function while chronic lead exposure affects the electrical and mechanical activities of the heart and alters the vascular smooth muscle function in experimental animals (Howard, 2001; Patrick, 2006; Mohammed et al., 2008; Alissa and Ferns, 2011).

Several therapies had been introduced by health professionals to treat acute and chronic lead poisoning. Noteworthy is the age-long use of chelating agents such as British Anti Lewisite (BAL), though its use has long been withdrawn. The use of chelating agents such as $CaNa_2EDTA$, meso-2,3-dimercaptosuccinic acid (DMSA or succimer), D-penicillamine is based on their ability to chelate heavy metals such as lead and consequently aid their excretion from the body (Flora et al., 2004; Kalia and Flora, 2005). They have always been used in severe cases of lead poisoning as the first line of treatment. The major problem about the efficacy and safety of these chelators is their non-specificity, which results in the mobilization of the heavy metals and also the essential elements like iron, zinc, calcium and a host of other divalent elements in the biological system. This is due to the rebound effects of the chelators (Gurer and Ercal, 2000). Aside this, the chelation therapy cannot be started where the subjects are still near or exposed to the source of pollution (Staudinger and Roth, 1998).

Recent findings on the use of antioxidants such as ascorbate (Asc), vitamin E and α-lipoic acid on lead poisoned subjects have been helpful. They have been found to be capable of restoring the levels of oxidative stress markers such as superoxide dismutase (SOD), catalase and reduced glutathione (Flora et al., 2008; Ait Hamadouche et al., 2009; Alissa and Ferns, 2011; Koedrith and Young, 2011). Precisely, the effectiveness of Asc, a water soluble vitamin, has been attributed to its ability to scavenge or quench free radicals or decrease the intestinal absorption of lead by reducing ferric iron to ferrous iron in the duodenum, Asc then increases the availability of iron which competes with lead for intestinal absorption (Garrow et al., 2000; Erdogan et al., 2005; Bashandy, 2006; Patrick, 2006; Baseem et al., 2009). Also, it forms a complex with lead through a covalent bonding with its hydroxyl groups and being water soluble, it could therefore be easily excreted through urine. An increasing number of scientists are now advocating for their use over conventional chelating agents as they generally do not have any side effects (Abam et al., 2008; Koedrith and Young, 2011).

Although there are indications that chronic lead exposure may affect systemic lipid metabolism, causing dyslipidemia, much investigation has not been done on the comparative efficiency of the chelating agents and Asc. Dyslipidemia is defined as a deviation from the normal lipid profile levels of a subject (Allouche et al., 2011). In Nigeria, where little attention is given to public health effects of environmental pollution like exposure to heavy metals such as lead, this study is with the view of evaluating the effect of sub-chronic lead exposure on plasma and erythrocyte lipid profiles. It is also aimed at comparing the efficacy of these chelating agents with Asc in reversing any observed effect of sub-chronic lead exposure in the rat.

MATERIALS AND METHODS

Chemicals

Lead acetate (ACS reagents grade >99% pure), diethyl ether, chloroform, isopropanol and all other chemicals used were of analytical grade.

Experimental design

Healthy 35 male Wistar rats purchased from the Department of Anatomy, Faculty of Veterinary Medicine, University of Ibadan, Ibadan, were used in this investigation. The animals were kept housed under standard conditions of temperature and natural light-dark cycle. All the animals had access to feed and clean water *ad libitum* and all conditions of animal experimentation conformed to the NIH guidelines as outlined in NIH publication 80-23 (revised, 1978). They were allowed two weeks for acclimation prior to experimental treatment.

The animals with body weight of 150-200 g were then randomly and evenly distributed into five groups (n=7) including the control group. Four groups were administered 75 mg/kg body weight lead acetate (PbAc) orally for 14 days and then followed by the various therapeutic interventions for another ten days (first five days treatment, followed by five days of rest to allow for redistribution of lead and then a second five days treatment). Two groups were administered chelating agents: D-Penicillamine (D-pen) and meso-2,3-dimercaptosuccinc acid (DMSA) (30 mg/kg body weight) respectively while the third group was administered an antioxidant (500 mg/kg body weight Asc). During the therapeutic period, the fourth group was administered normal saline (0.9% NaCl). The control group (not administered lead) was administered normal saline throughout the period of the study; and all administrations were done orally once daily.

At the end of the 14-day lead acetate (PbAc) administration, blood was collected from the four groups by tail incision blood lead level (BLL) before the commencement of therapeutic intervention.

After the 10 days of administering therapeutics, blood was collected into heparinised tubes via cardiac puncture under light ether anesthesia after an overnight fast. Aliquots of the blood samples were for BLL while the remaining blood samples were centrifuged to separate plasma from red blood cells.

Biochemical analyses

Blood lead analysis

For the estimation of BLL before and after therapeutic administration, 1 ml each of the whole blood samples were acid digested with concentrated nitric acid. BLL was then determined in duplicates using a Thermo Scientific S Series Atomic Absorption Spectrophotometer (Model Type S4 AA System).

Plasma lipid profiles

The plasma concentrations of total cholesterol and triacylglycerols were determined by spectrophotometric using Cypress diagnostic kits. HDL cholesterol and triacylglycerols were determined in plasma with the same diagnostic kits for total cholesterol and triacylglycerols after very low density lipoproteins and low density lipoproteins were precipitated using the method described by Gidez et al. (1982). Total phospholipids in plasma were extracted with chloroform-methanol mixture (2:1, v/v) as described by Folch et al. (1957). Phospholipid concentration was then assessed with ammonium ferrothiocyanate by the method of Stewart (1980). An aliquot of the extract (0.1 ml) was evaporated to dryness at 60°C. After cooling, 2 ml of chloroform was added to the dried extract, mixed and 2 ml of ammonium ferrothiocyanate was then added and then mixed for 1 min. The mixture was left for 10 min for separation to occur. The chloroform layer was then taken and the absorbance read at 488 nm. Phospholipid concentrations were determined using a phospholipid standard as reference.

Red blood cell lipid profile

Since the Folch et al. (1957) method of lipid extraction produced highly pigmented extracts, an improved procedure for red blood cell lipid extraction using chloroform - isopropanol (7:11, v/v) described by Rose and Oklander (1965) was used. For cholesterol determination, 0.1 ml of the extract was evaporated to dryness at 60°C and 20 µl of Triton X-100/chloroform mixture (1:1, v/v) was added to the dried extract for resolution. This was evaporated again and then 1 ml of the cholesterol kit reagent was added, mixed and incubated for 30 min before reading the absorbance spectrophotometrically. The triacylglycerol concentration was determined by evaporating to dryness 0.1 ml of the extract and adding 0.1 ml of 97% ethanol to re-suspend the dried lipid. To this, 1 ml of the triacylglycerol kit reagent was added, mixed and incubated for 30 min before the absorbance reading was taken. Determination of phospholipids in the red blood cells followed the same procedure as described for plasma.

Estimation of lipid peroxidation

Lipid peroxidation in plasma and red blood cells was estimated colorimetrically by thiobarbituric acid reactive substances (TBARS) method of Buege and Aust (1978). In brief, 0.1 ml of test sample (plasma and red blood cell) was treated with 2.0 ml of TBA-TCA-HCl, 1:1:1 reagent (thiobarbituric acid 0.37%, 0.25N HCl and 15% TCA) and incubated in a water bath at 95°C for 15 min. The tube was then placed on ice, centrifuged and the absorbance of clear supernatant was measured against blank at 535 nm. TBARS (malondialdehyde MDA) content was determined using the extinction coefficient of 155 $nM^{-1}cm^{-1}$.

Statistical analysis

The results obtained are expressed as mean ± standard deviation (SD). One-way analysis of variance (ANOVA) followed by Duncan's multiple range test (DMRT) was used to analyze the results. Values with $p<0.05$ were regarded as being significant, using the Statistical Package for Social Sciences (SPSS) version 16.0.

RESULTS

Figure 1 shows the blood lead levels of the animals after therapeutics while the concentration of lead in the control remained insignificantly different. Throughout the period of experiment, marked changes were recorded in the BLL of the other four groups. The three groups administered therapeutics showed a significant decrease ($p<0.05$) in the BLL at the end of the experiment. The decrease ranged from 20.8% in the lead + Asc group to 43.5% in the Lead + D-pen group. The lead only group showed no significant change in the level of concentration of lead in the blood.

Figures 2 and 3 depict the plasma and HDL lipid profiles of animals after therapeutics, respectively. The plasma concentrations of total cholesterol and triacylglycerol of the groups treated with lead only were significantly higher ($p<0.05$) than the control by two folds while for the phospholipids, it was by 2.4 fold. While in the HDL, the cholesterol and triacylglycerol values of the lead only group were lower and higher respectively, though not significantly different ($p>0.05$) from the other groups. The plasma concentrations of total cholesterol and triacylglycerol of the animals treated with DMSA and Asc were not significantly different ($p>0.05$) from the control and they showed a significant reduction ($p<0.05$) compared with the lead only group. However, there was still a significant difference ($p<0.05$) in the concentration of phospholipids found in the plasma of groups treated with D-pen and Asc compared with the control.

Figure 4 shows the red blood lipid profile of animals after therapeutics. The red blood cell concentrations of total cholesterol and phospholipids of the groups treated

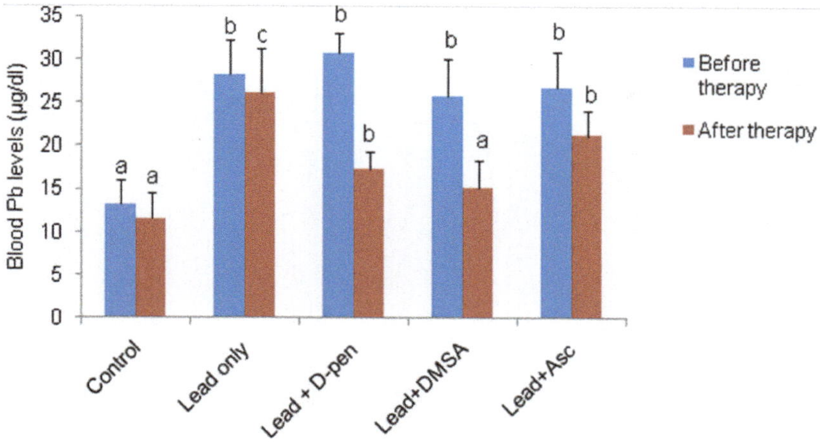

Figure 1. Blood lead levels in male Wistar rats exposed to lead (PbAc) after treatment with the chelating agents D-penicillamine (D-pen), meso-2,3-dimercaptosuccinic acid (DMSA or succimer) and the antioxidant ascorbate (Asc).

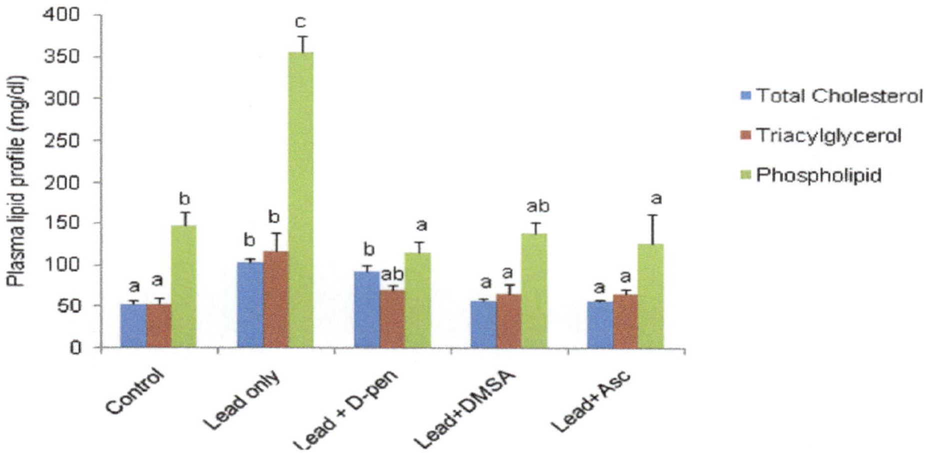

Figure 2. Plasma lipid profile in male Wistar rats exposed to lead (PbAc) after treatment with the chelating agents D-penicillamine (D-pen) meso-2,3-dimercaptosuccinic acid (DMSA or succimer), and the antioxidant ascorbate (Asc).

with lead only were significantly reduced ($p<0.05$) compared with the control. The triacylglycerol concentration of the lead only group was however not significantly different ($p>0.05$) from the other groups. The red blood cell concentrations of total cholesterol and phospholipids of the animals treated with the chelating agents (D-pen and DMSA) and Asc were not significantly different ($p>0.05$) from the control and they showed a significant increase ($p<0.05$) compared with the lead only group.

Figure 5 depicts the lipid peroxidation in the plasma and red blood cells. Lead exposure resulted in a significant increase ($p < 0.05$) of lipid peroxidation both in the plasma and red blood cells. While the increase in plasma lipid peroxidation was by 2.65 fold, it was by 1.23 fold in the red blood cells compared with the control. The therapeutics ameliorated the levels in the two compartments to near normal. Comparing the effects of the therapeutics, the DMSA was observed to be most effective in the plasma while Asc was most effective in the red blood cells.

DISCUSSION

Lead exposures in humans and animals have been reported to produce a series of adverse effects in the biological systems. One of these is the perturbations in lipid metabolism in different compartments of organisms (Ademuyiwa et al., 2008). These perturbations were reported as up-/down- regulation of the concentrations of the lipids. Lead exposure was able to cause these perturbations through the activities of some of the enzymes involved in the metabolism of these lipids being up-/down- regulated. The effects of a sub-chronic lead exposure on the lipid profiles of rats along with the

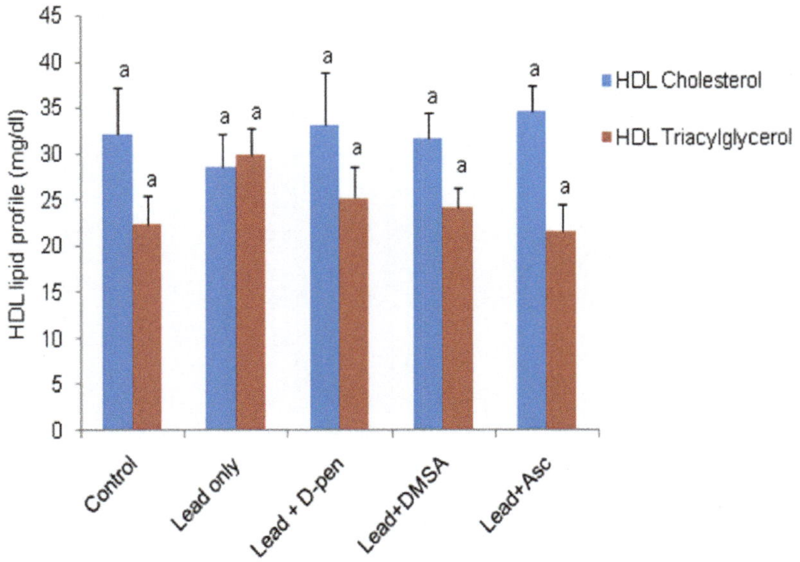

Figure 3. HDL lipid profile in male Wistar rats exposed to lead (PbAc) after treatment with the chelating agents D-penicillamine (D-pen), meso-2,3-dimercaptosuccinic acid (DMSA or succimer) and the antioxidant ascorbate (Asc).

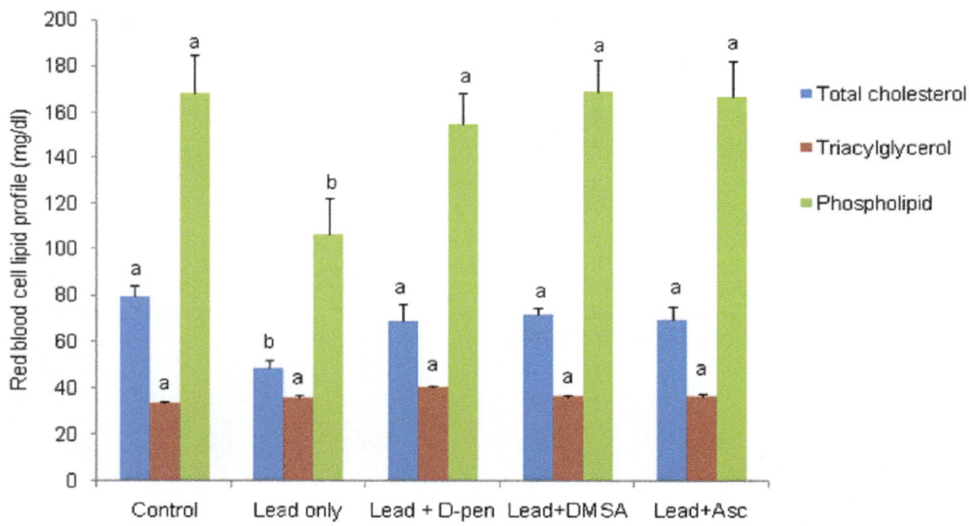

Figure 4. Red blood cells lipid profile in male Wistar rats exposed to lead (PbAc) after treatment with the chelating agents D-penicillamine (D-pen), meso-2,3-dimercaptosuccinic acid (DMSA or succimer) and the antioxidant ascorbate (Asc).

efficacy of chelating agent in comparison with an antioxidant vitamin (Asc) were examined in this study. The exposure of lead acetate was observed to induce plumbism (an elevation in blood lead levels) in the animals (Figure 1). This may be an indication of the extent of risk humans in an environment where such compound of lead exists. Lead still finds an active use in paints and as a fuel additive in developing countries like Nigeria. A higher risk of atherogenesis and cardiovascular diseases has been observed in those who are occupationally exposed to this, such as auto mechanics and painters to mention but a few (Ademuyiwa et al., 2005; Nriagu et al., 1997).

The control group also had minor quantities of lead in the blood. This may be as a result of contamination from the air they inhaled, or the feed and water ingested. This further affirms the ubiquity of lead poisoning observed in early reports (Zhang et al., 2012; Hassan and Jassim,

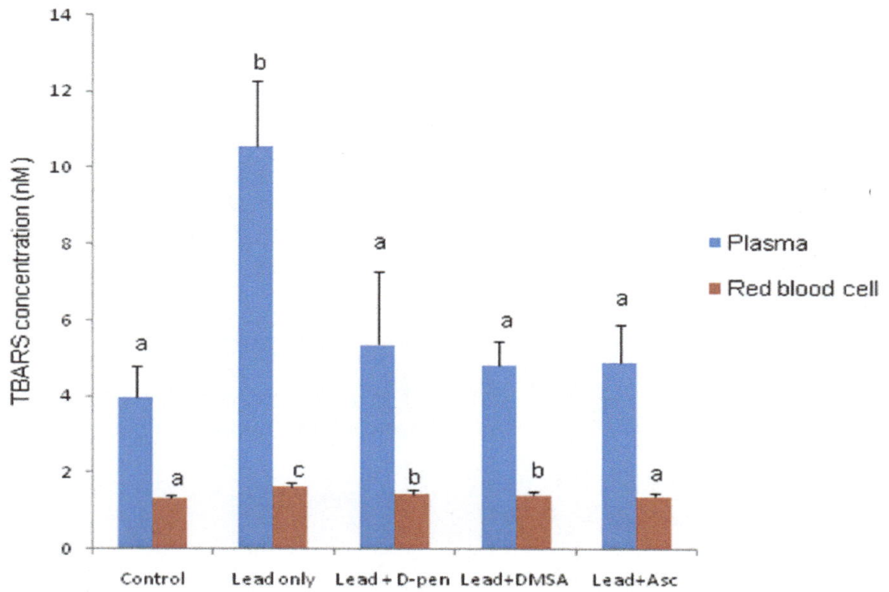

Figure 5. TBARS concentrations in the plasma and red blood cells of male Wistar rats exposed to lead (PbAc) after treatment with the chelating agents D-penicillamine (D-pen), meso-2,3-dimercaptosuccinic acid (DMSA or succimer), and the antioxidant ascorbate (Asc).

2010).

Lead was found to up-regulate the plasma concentrations of cholesterol, triacylglycerol and phospholipids and down-regulate their concentrations in the red blood cells significantly. These were in agreement with the findings of other researchers (Kristal-Boneh et al., 1999; Ait Hamadouche et al., 2009a). The disruption of the pro-oxidant/antioxidant balance that induces tissue injury via oxidative damage to biomolecules may be responsible (Erdogan et al., 2005; Bashandy, 2006; Flora, 2009; Sajitha et al., 2010). The high level of the lipids in the plasma could be attributed to changes in the rate of a number of processes. For cholesterol: a) altered distribution between the plasma and the tissues; b) enhanced absorption of exogenous cholesterol from diet; c) enhanced synthesis of endogenous cholesterol; d) decreased cholesterol excretion in the form of neutral sterols and e) decreased cholesterol transformation to bile acids (Kilic, 1993). It is also possible that the cholesterol synthesis and transport pathways may be adversely affected as observed in the significant decrease in the cholesterol: phospholipids ratio of the group administered lead only (data not shown). This may also imply an alteration in major cation fluxes of the cell membrane. It could also be as a result of the overproduction of very low density lipoprotein (VLDL), thus increasing the burden of triacylglycerol-rich lipoproteins on the common lipolytic pathway (Hassan and Jassim, 2010). Lead has also been known to bind directly to phosphatidylcholine in the red blood cell membrane, leading to a decrease in phospholipids levels (Ademuyiwa et al., 2005; Patrick, 2006; Abam et al., 2008). Lead exposure has been reported to cause a cytotoxic effect on the red blood cells due to its high potential in changing the osmotic fragility, interaction with membrane proteins and some essential trace metals in the blood (Gurer and Ercal, 2000; Mrugesh et al., 2011). From results, there were decreases in the red blood cell lipid concentration, indicating an efflux of the lipids from the intracellular to the extracellular fluid. This may be arising from the fact that the function of the plasma membrane may have been compromised.

At the end of the ten days of therapeutics, a marked change was observed in the blood lead levels. This confirms the findings of other reports which indicate the effectiveness of chelating agents and the recent use of antioxidant vitamins in aiding the excretion of absorbed lead (Gurer and Ercal, 2000; Onunkwor et al., 2004; Bashandy, 2006; Flora, 2009). This result however reveals that chelating agents remain a more rapid treatment for acute lead poisoning since they were more able to mobilize the removal of lead from the blood. Normally, after an ingestion or exposure to lead, it is primarily found in the red blood cells as it has a very high affinity for the cells. However, as blood flows through the soft tissues, it is deposited and bio accumulated (Babalola et al., 2010). During chelation therapy, lead in the blood is cleared giving rise to a redistribution of the lead from the tissues into the blood again for further clearance. The rate of mobilization from our results may be depicted as: DMSA > D-pen > Asc. The higher rate of reversal observed for DMSA may be as a result of its greater ability to mobilise lead even from soft tissues (Flora, 2009; Kalia and Flora, 2005). The lesser efficacy of Asc may be attributed to the fact that it is a water-soluble vitamin, only protective enough to intercept oxidants in the aqueous phase before they attack or cause any detectable damage to the lipids (Hassan and Jassim, 2010).

Also noteworthy is the result of the BLL obtained from the group administered lead only. There was no significant difference ($p<0.05$) in the BLL before and after, but there was a numerical decrease. It may be predicted that there is a gradual mobilisation of the accumulated lead from the blood into soft tissues such as the liver and kidney and hard tissues such as the bones (Flora, 2009).

The chelators and as well Asc were observed to decrease the accumulated lipids in the plasma (Figure 2). For total cholesterol concentration, the rate of reversal to normal level may be depicted as Asc > DMSA > D-pen while for phospholipids, it can be stated as DMSA > Asc > D-pen. For the rate of reversal of triacylglycerol concentration, however, it may be depicted as DMSA > Asc > D-pen. Virtually the reverse is observed for the red blood cells of the animals. The mechanisms by which Asc, a water soluble vitamin could be able to reduce the plasma concentration of the lipids especially cholesterol can be as follows: (i) it has been reported that the oxidation of cholesterol to bile acids is dependent on Asc status (Kilic, 1993), so having administered 500 mg/kg body weight of Asc to the animals in the first five days treatment, it could be said that the animals were already on an adequate intake of the vitamin. This adequacy is actually necessary for the transformation of cholesterol to bile acids at the rate limiting steps of bile acid biosynthesis; (ii) for the hydroxylation of carbon 7 (C7) of the cholesterol nucleus, Asc is important in the catalytic reaction by 7-alpha hydroxylase. In an Asc deficient condition, this reaction is inhibited, leading to a high plasma cholesterol concentration (Holloway and Rivers, 1984; Hemila, 1992). Although D-pen was able to reduce the BLL of subjects significantly, it was ineffective in restoring normal total cholesterol and triacylglycerol levels. This may be due to the fact that D-pen can penetrate the cell membranes and get metabolized easily before being able to accomplish the needed cholesterol and triacylglycerol reversal (Flora, 2009). In general, the chelating agents and Asc could have bound to the lead in the system thereby releasing the bound enzymes involved in the metabolism of these lipids so that their normal homeostasis can be maintained.

In conclusion, the results of this study have again indicated that lead poisoning is capable of inducing dyslipidemia, therefore capable of predisposing subjects to other risks such as atherosclerosis. The therapeutic interventions have also proved effective, although the DMSA has proved to be most effective of the three, in ameliorating the perturbations observed in lipid metabolism in this study. The chelating capacity of Asc is limited in sub-chronic dosage, so possibly a combination therapy could be more desirable. The therapeutic use of Asc would however be more advantageous than the chelating agents since the occupationally exposed subjects undergoing treatment would not need to be removed from their means of livelihood, it is very cheap and its use does not require the expertise of a physician. Further study may be carried out at lower doses of lead for a longer period as it is often the case practically, perhaps Asc may be more effective in chronic lead exposure due to the reasons above.

ACKNOWLEDGEMENT

The authors are grateful to Dr. T. O. Sunmonu for proof reading the paper and giving constructive criticism.

REFERENCES

Abam E, Okediran B, Odukoya O, Adamson I, Ademuyiwa O (2008). Reversal of ionoreulatory disruptions in occupational lead exposure by vitamin C. Environ. Toxicol. Pharmacol. 26(3):297-304.

Ademuyiwa O, Agarwal R, Chandra R, Behari JR (2008). Lead-induced phospholipidosis ans cholesterogenesis in rat tissues. Chem. Biol. Interact. Doi: 10.1016/j.cbi.2008.10.057 (pp. 7).

Ademuyiwa O, Ugbaja RN, Idumebor F, Adebawo, O (2005). Plasma lipid profiles and risk of cardiovascular disease in occupational lead exposure in Abeokuta, Nigeria. Lipids Health and Dis. 4(19).

Ait Hamadouche M, Slimani M, Aous AEK (2009). Biochemical parameters alterations induced by chronic oral administration of lead acetate in albino rats. Am. J Sci. Res. 4:5-16.

Ait Hamadouche M, Slimani M, Merad-Boudia B, Zaoui M (2009a). Reproductive toxicity of lead acetate in adult male rats. Am. J. Sci. Res. 3:38-50.

Alissa EM, Ferns GA (2011). Heavy metal poisoning and cardiovascular disease. J. Toxicol. Vol, 2011, Article ID 870125. 21 pp. Doi: 10.1155/2011/870125.

Allouche L, Ait Hamadouche M, Towabti A, Khennouf S (2011). Effect of ong-term exposure to low or moderate lead concentration on growth, lipid profile and liver function in albino rats. Adv. Biol. Res. 5(6):339-347.

Babalola OO, Okonji RE, Atoyebi JO, Sennuga TF, Raimi MM, Ejim-Eze EE, Adeniran OA, Akinsiku OT, Areola JO, John OO, Odebunmi SO (2010). Distribution of lead in selected organs and tissues of albino rats exposed to acute lead toxicity. Sci. Res. Essays 5(9):845-848.

Baseem MR, Medhat AR, Amal AE, Ahmad S (2009). Ameliorating effects of vitamin C against acute lead toxicity in albino rabbits. Aust. J. Basic Appl. Sci. 3(4):3597-3608.

Bashandy SAE (2006). Beneficial effect of combined administration of vitamin C and vitamin E in amelioration of chronic lead hepatotoxicity. Egypt. J. Hospit. Med. 23:371-384.

Buege JA, Aust SD (1978). Microsomal lipid peroxidation. Methods. Enzymol. 52:302-305.

Chen L, Yang X, Jiao H, Zhao B (2003). Tea catechins protect against lead-induced ROS formation, mitochondrial dysfunction, and calcium dysregulation in PC12 cells. Chem. res. Toxicol. 16:1155-1161.

Choudhary M, Jetley UK, Abagh Khan A, Zutshi S, Fatma T (2007). Effect of heavy metal stress on proline, malondialdehyde and superoxide dismutase activity in the cyanobacterium Spirulina platensis-S5. Ecotoxicol. and Environ. Saf. 66(2):204-209

Diamond GL (2005). Risk assessment of nephrotoxic metals. In: Tarloff J, Lash L (Eds), The toxicology of the kidney. CRC Press, London, pp. 1099-1132.

Donaldson WE, Knowles SO (1993). Is lead toxicosis a reflection of altered fatty acid composition of membranes? Comp. Biochem. Physiol. 104C:377-379.

Dosumu O, Onunkwor B, Odukoya O, Arowolo T, Ademuyiwa O (2005). Biomarkers of lead exposure in auto-mechanics in Abeokuta, Nigeria. Trace Elem. Electrolyte 22(3):185-191.

Erdogan Z, Erdogan S, Aksu T, Baytok E (2005). The Effects of dietary lead exposure and ascorbic acid on performance, lipid peroxidation status and biochemical parameters of broilers. Turk. J. Vet. Anim. Sci. 29:1053-1059.

Flora SJS (2009). Metal poisoning: threat and management. Al Ameen J. Med. Sci. 2(2): 4-26.

Flora SJS, Mittal M, Mehta A (2008). Heavy metal induced oxidative stress and its possible reversal by chelation therapy. Indian J. Med. Res. 128:501-523.

Flora SJS, Pande M, Kannan GM, Mehta A (2004). Lead ilnduced oxidative stress and its recovery following co-administration of melatonin or N-acetylcysteine during chelation with succimer in male rats. Cell. Mol. Biol. 50: 543-551.

Folch M, Lees M, Stanley GH (1957). A simple method for the isolation and purification of total lipids from animal tissues. J. Biol. Chem. 226:97-509.

Garrow JS, James WPT, Ralph A (2000). Human nutrition and dietetics. 10th ed. London Churchill Livingstone. pp137-148.

Gidez LL, Miller GJ, Burnstein M, Slagle S, Eder HA (1982). Separation and quantitation of subclass of human plasma high density lipoproteins by a simple precipitation procedure. J. Lipids Res. 23:1206-1223.

Gurer H, Ercal N (2000). Can antioxidants be beneficial in the treatment of lead poisoning? Free Rad. Biol.Med. 29(10):927-945.

Harishekar MB, Kiran B, (2011). Studies on lipid peroxidation induced by lead, alcohol on liver and amelioration by vitamin E. Asian J. Pharm. Clin. Res. 4(2):164-167.

Hassan AA, Jassim HM (2010). Effect of treating lactating rats with lead acetate and its interaction with vitamin E or C on neurobehaviour, development and some biochemical parameters in their pups. Iraqi J. Vet. Sci. 24(1):45-52.

Hemila H (1992). Vitamin C and plasma cholesterol. Crit. Rev. Food Sci. Nutr. 32:33-37.

Heo Y, Lee BK, Ahn KD, Lawrence DA (2004). Serum IgE elevation correlates with blood lead levels in battery manufacturing workers. Hum. Exp. Toxicol. 23:209-213.

Holloway DE, Rivers JM (1984). Long term effects of adequate and excessive dietary ascorbate on bile acid metabolism in the guinea pig. J. Nutr. 114:1370-1376.

Howard H (2001). Poorly controlled hypertension in a painter with chronic lead toxicity. Environ. Health Perspect. 109(11):95-99.

Ibrahim NM, Eweis EA, El-Bertagi HS, Abdel-Mobdy YE (2012). Effect of lead acetate toxicity on experimental male albino rat. Asian Pac. J. Trop. Biomed. 41-46.

Kalia K, Flora SJS (2005). Strategies for safe and effective therapeutic measures for chronic arsenic and lead poisoning. J. Occup Health. 47:1-21.

Kilic N (1993). The effect of ascorbic acid on liver and plasma cholesterol levels of male rats. J. Islam. Acad. Sci. 6(4):249-252.

Koedrith P, Young RS (2011). Advances in carcinogenic metal toxicity and potential molecular markers. Int. J. Mol. Sci. (12):9576-9595.

Kristal-Bonneh E, Coller D, Froom P, Harari G, Ribak J (1999). The association between occupational lead exposure and serum cholesterol and lipoprotein levels. Am. J. Public Health. 89 (7):1083-1087.

Kumar M, Kaur H, Hossain SKA, Sontakke UB, Tyagi A, Chandra G (2011). Role of copper and zinc supplementation on immunity of animals exposed to lead and cadmium: A Review. Wayamba J. Anim. Sci. 85-91.

Miranda ML, Anthopolos R, Douglas H (2011). A geospatial analysis of the effects of aviation gasoline on childhood blood lead levels. Environ. Health Perspect. 119:1513-1516.

Mohammed IK, Mahdi AA, Raviraja A, Najmul I, Iqbal A, Thuppil V (2008). Oxidative stress in painters exposed to low lead levels. Arg. Hig Rada Toksikol. 59:161-169.

Moreira EG, Rosa GJ, Baros SBM, Vassilieff VS, Vassilieff I (2001). Antioxidant defense in rat brain regions after developmental lead exposure. Toxicol. 169:145-151.

Mrugesh T, Dipa L, Manishika G (2011). Effects of lead on human erythrocytes. An *in vitro* study. Acta Poloniae Pharmacuetica - Drug Res. 68(5):653-656.

Nriagu J, Jinabhi CC, Naidoo R, Coutsoudis A (1997). Lead poisoning of children in Africa III. Kaduna, Nigeria. Sci. Total Environ. 197:13-19.

Okediran BS, Abam E, Odukoya OO, Adamson I, Ademuyiwa O (2009). Membrane, intracellular, plasma and urinary sodium and potassium in occupational lead exposure: effects of vitamin C supplementation. Trace elements Electrolyte. 26(2):49-59.

Olympio KPK, Gonçalves C, Günther WMR, Bechara EJH (2009). Neurotoxicity and aggresiveness triggered by low-level lead in children. Pan Am. J. Pub. Health 29(3):266-275.

Onunkwor B, Dosumu O, Odukoya OO, Arowolo T, Ademuyiwa O (2004). Biomarkers of lead exposure in petrol station attendants and auto-mechanics in Abeokuta, Nigeria: effects of 2-week ascorbic acid supplementation. Environ. Toxicol. Pharmacol. 17:169-176.

Patil AJ, Bhagwat VR, Patil JA, Dongre NN, Ambechar JG, Jailkhani R Das KK (2006). Effect of lead (Pb) exposure on the activity of superoxide dismutase and catalase in battery manufacturing workers (BMW) of Western Maharashtra (India) with Reference to heme biosynthesis. Int. J. Environ. Res. Pub. Health. 3(4):329-337.

Patra RC, Rautray AK, Swarup D (2011). Oxidative stress in lead and cadmium toxicity and its amelioration. Vet. Med. Internation. ID 457327.

Patra RC, Swarup D, Dwivedi SK (2000). Antioxidant Defense and Lipid Peroxide Level in Liver and Kidneys of Lead Exposed Rats. Asian-Aus. J. Anim. Sci. 13(10):1433-1439.

Patrick L (2006). Lead Toxicity II: The role of free radical damage and the use of antioxidants in the pathology and treatment of lead toxicity. Alternative Med. Rev. 11(2):114-127.

Ponce-Canchihuaman JC, Perez-Mendez O, Hernandez-Munoz R, Torres-Duran PV, Juarez-Oropeza MA (2010). Protective Effects of Spirulina maxima on hyperlipidemia and oxidative stress induced by lead acetate in the liver and kidney. Lipids Health Dis. 9(35).

Raafat BM, Shafaa MW, Rizk RA, Elgohary AA, Saleh A (2009). Ameliorating effects of vitamin C against acute lead toxicity in albino rabbits. Aust. J. Basic Appl. Sci. 3(4):3597-3608.

Rao GM, Shetty BV, Sudha A (2007). Evaluation of lead toxicity and antioxidants in battery workers. Biomed. Res. 19(1):1- 4.

Rose HG, Oklander M (1965). Improved procedure for the extraction of lipids from human erythrocytes. J. Lipids Res. 6:428-431.

Sajitha GR, Jose R, Andrews A, Ajantha KG, Augustine P, Augusti KT (2010). Garlic oil and vitamin E prevent the adverse effect of lead acetate and ethanol separately as well as in combination in the drinking water of rats. Ind. J. Clin. Biochem. 25(3):280-288.

Staudinger KC, Roth VS (1998). Occupational lead poisoning. Am. Fam. Phys. 57:719-726.

Stewart J (1980). Colorimetric determination of phospholipids with ammonium ferrothiocyanate. Anal Biochem. 104:10-14.

Verma S, Dubey RS (2003). Lead toxicity induces lipid peroxidation and alters the activities of antioxidant enzymes in growing rice plants. Plant Sci. 164:645-655.

Zhang A, Hu H, Sanchez BN, Ettinger AS, Park SK, Cantonwine D, Schnaa SL, Wright RO, Lamadrid-Figueroa H, Tellez-Rojo MM (2012). Association between prenatal lead exposure and blood pressure in children. Environ. Health Perspect. 120(3):445-450.

11

Insecticidal properties of materials used by resource-limited farmers to control fleas in free-range chickens in the Eastern Cape Province, South Africa

S. Moyo[1] and P. J. Masika[2*]

[1]Department of Livestock and Pasture Science, Faculty of Science and Agriculture, University of Fort Hare, P/Bag X 1314, Alice 5700, South Africa.

[2]Agricultural and Rural Development Research Institute (ARDRI), Faculty of Science and Agriculture, University of Fort Hare, P/Bag X 1314, Alice 5700, South Africa.

Fleas are obligate blood feeders that infest free-range chickens, thereby impeding their productivity. Commercial insecticides used in controlling fleas are expensive and inaccessible, hence making farmers to resort to low cost and easily available alternatives. The study was conducted to assess the insecticidal properties of ethno-veterinary medicine used in controlling fleas in free-range chickens in the Eastern Cape. *In vitro* repellency and contact bio-assay models were used to assess the insecticidal properties of *Tagetes minuta, Calpurnia aurea, Clutia pulchella*, used engine oil, paraffin, Jeyes fluid (carbolic acid 13%) on fleas. Distilled water was used as the negative control. Positive controls for the repellency and contact bio-assays, were Tabard (35% diethyltoluamide) and Karbadust (carbaryl 5%) respectively. *Tagetes minuta* was the most effective ($P<0.05$), demonstrating a repellency level of 75 at 100% concentration. *Clutia pulchella, C. aurea*, used engine oil and paraffin showed insignificant repellency ($P>0.05$). The repellency of *T. minuta,* Jeyes fluid and Tabard lasted for 4, 5 and 5 h, respectively. For the contact bio-assay, 100% concentrations of *C. pulchella* and *C. aurea* produced flea mortality of 82.5 and 75%. Paraffin, used engine oil, and Jeyes fluid (19.2%) caused flea mortality, which was not significantly different from Karbadust. The materials assessed showed various degrees of insecticidal properties. This justifies their use by resource-limited farmers in South Africa.

Key words: Fleas, medicinal plants, mortality, repellency.

INTRODUCTION

Fleas are one of the most important external parasites with more than 2000 species and sub-species affecting birds (Boughton et al., 2006). The flea (*Echidnophaga gallinacean*) is a major insect of domestic chickens (*Gallus gallus*) that can cause severe pathology or death if untreated (Boughton et al., 2006). These are blood feeding external parasites responsible for causing ulcerations on soft body parts like the comb and wattles, areas where the fleas commonly feed (Agboola et al., 2007). The ulcerations are caused by both lacerations from the flea mouthparts as well as resultant infections due to subcutaneously laid eggs. The resulting infection and blood loss, can compromise growth and even kill young chickens (Boughton et al., 2006; Agboola et al., 2007). Additionally, flea infestation has been found to co-occur with fowl pox in chicken, a viral diseases spread by the flea bite (Boughton et al., 2006; Gyimesi et al., 2007). Despite the prevailing negative impact of fleas, which include reduced growth, production and death, most of the resource-limited farmers carry out control measures irregularly. This could probably be due to financial const-

*Corresponding author. E-mail: pmasika@ufh.ac.za.

raints or ignorance (Luseba and Van der Merwe, 2006).

Currently, most flea control efforts by farmers worldwide are based on convenient animal spot-on treatments or oral medications (Power and Sudakin, 2007). The most widely used products are generally applied as spot-on on a monthly schedule (Boughton et al., 2006). Despite the variety of available products and different application methods, fleas remain a perpetual problem for many farmers and have developed resistance to a number of insecticides. As a result, management strategies to reduce the development of resistance are needed (Snyder et al., 2007). This could be addressed by searching for alternative flea control methods that are effective, economical and socially acceptable. Plants produce a diverse array of secondary metabolites with functions, such as defense against diseases and parasites (Pieters and Vlietinck, 2005). As a result natural products, including plant extracts, might be suitable for such alternative approaches (Luseba and Van der Merwe, 2006).

Basically, fleas are controlled by the use of commercial drugs, however use of commercial drugs have limitations such as being harmful to other non-target species and leave persistent residues (Snyder et al., 2007; Power and Sudakin, 2007). This has resulted in resource-limited farmers to use low cost alternative remedies for animal health care, which are cheap, locally available and culturally accepted in their communities. Plant derived products are increasingly being used to combat insects because they are natural and often assumed to be safe to the environment (Pieters and Vlietinck, 2005). Plant derived products may possess complex chemical structures that are not available in synthetic compounds, hence they are potentially useful biologically active compounds (Pieters and Vlietinck, 2005; Isman, 2008). Plants are used as singles or in combinations by resource-limited farmers as reported by Masika and Afolayan (2003) that more than one plant species is combined for the treating livestock in the Eastern Cape. On the other hand, Luseba and Van der Merwe (2006) found that most remedies were used in the form of a single plant. The combinations are intended to attain increased potency due to synergistic interactions between the individual components, such that the combination is superior to individual herbal treatment alone (Adams et al., 2006).

A previous study by Moyo and Masika (2009) revealed that farmers used various materials like Jeyes fluid, and used engine oil to control cattle ticks. Used engine oil is any oil that has served its service properties in a vehicle withdrawn from the meant area of application and considered not fit for its initial purpose (Delistraty and Stone, 2007). However, there is dearth of information on insecticidal properties of materials used in the ethno-veterinary control of fleas. Therefore, the aim of the study was to assess the insecticidal properties of selected remedies used in ethno-veterinary control of fleas in free-range chickens by resource-limited farmers in Eastern Cape, South Africa.

MATERIALS AND METHODS

Plant material collection

The fresh leaves of *Calpurnia aurea, Cultia pulchella* and *Tagetes minuta* were collected at Umdeni village in Amatola Basin (32°40'38"S, 26°59'79"E) in the Amathole District, Eastern Cape Province, South Africa. They were authenticated by Tony Dold at the Albany herbarium, Rhodes University. The leaves of *C. aurea* voucher no. SM01-010/2008, *C. pulchella* voucher no. SM01-011/2008 and *T. minuta* voucher no. SM01-012/2008 specimens were deposited at the Giffen herbarium of the University of Fort Hare.

Preparation of the materials

Aqueous extraction of fresh plant materials

The fresh leaves of *T. minuta*, *C. pulchella* and *C. aurea* were crushed separately using a meat mincer (No 10, Shandong Branch, China). The concentrations of the extracts were determined on a weight per volume basis Yin and Kwok (2005) to obtain 10, 25, 50 and 75% (w/v). The crushed test materials were squeezed to obtain 100% concentration extracts. Plant materials of *C. pulchella* and *C. aurea* were mixed in different proportions of 1:1, 6:4, 7:3, 8:2 and 9:1 to determine the most effective ratio. The mixture was stored at room temperature overnight and later strained using a muslin cloth and filtered using a Whatman No. 1 filter paper, respectively. The filtrates were stored in capped labeled bottles and kept in the refrigerator at 6°C until use.

Preparation of non-plant materials

Jeyes fluid was diluted using distilled water to make the following concentrations: 0.6, 1.2, 4.8, 9.6, 19.2, 38.4, 76.8, and 100%. This was done to determine the effective ratio. Paraffin and used engine oil were used undiluted. The majority of engine oils are derived from petroleum. It mainly consists of hydrocarbons; organic compounds entirely of hydrogen and carbon polycyclic aromatic hydrocarbons are the major constituents of used engine oil (Dominguez et al., 2004). Paraffin also known as kerosene is a hydrocarbon that is highly toxic with a flash point of 43°C (Muller et al., 2007). A commercial insecticide Kabadust (carbaryl 5%) (Agro-serve, Bryanston, South Africa) registered for the control of fleas was used as the positive control and distilled water as negative control.

Fleas

Fleas, *C. felisfelis* at pupae stage were obtained from Clinvet International, Bloemfontein, South Africa. The fleas were held in well ventilated small plastic vial, which was incubated at a temperature of 27 ± 3°C with a relative humidity of 80±10% (Mustapha et al., 2006). After five days, the adult emerged and these 5-day old fleas were used in the experiment.

In vitro repellency bioassay

Test tubes with diameters of 4 cm and lengths of 29 cm were used for this bioassay. The cat fleas, *C. felisfelis* were used because they can also attack chickens and they are the preferred model fleas for repellency testing (Mustapha et al., 2006). Filter papers (Whatman No. 1) with a diameter of lightly less than 4 cm were placed at the bottom of the test tubes. Aliquots of 1 ml of the test materials were applied separately on filter paper as uniformly as possible using a

Table 1. Least square means showing flea repelling activity of *T. minuta* and Jeyes fluid at different concentrations.

Material	Flea repellency (%)								
	20 min	40 min	1 h	2 h	3 h	4 h	5 h	6 h	7 h
T. minuta (100%)	100	100	100	100	100	76	56	36	30
T. minuta (75%)	100	100	100	100	90	66	46	33	26
T. minuta (50%)	100	100	100	100	83	60	36	26	16
Jeyes fluid (100%)	100	100	100	100	100	83	63	50	30
Jeyes fluid (76.8%)	100	100	100	100	86	46	33	20	16
Jeyes fluid (38.4%)	100	100	100	73	43	26	16	13	6
Jeyes fluid (19.2%)	100	100	83	50	30	13	10	0	0
Tabard	100	100	100	100	100	90	70	30	20
Distilled water	0	0	0	0	0	0	0	0	0

micropipette. Distilled water and Tabard (35% diethyltoluamide Acorn (Pty) Ltd, Strubens Valley, South Africa) were used as negative and positive control, respectively. Ten (5 day old fleas) of either sex, anaesthetized using carbon dioxide (CO_2) were placed 8 cm away from test material. The test tubes were closed with fine gauze secured with rubber band and held horizontally for 5 min to allow fleas to gain consciousness. The bottom end of the test tubes was put in a temperature regulated water bath at 41.8°C simulating chicken body temperature so as to attract fleas. Treatments were replicated 3 times. Observations were recorded after 20 and 40 min then hourly for 7 h. Any flea found 4 cm away from the test material upwards was considered having been repelled. Repellency was expressed as number of fleas moving away from the stimuli to the total number of fleas at each occasion;

$$\text{Repellency (\%)} = 100 - \frac{[T \times 100]}{N}$$

Where, T is the number of fleas found in the treated area and N is the total number of fleas used (Usavadee et al., 2007). The average repellency was calculated from the values obtained in three replicates.

Contact bioassay on fleas

The test to evaluate the efficacy of insecticidal activity followed the World Health Organization standard protocols (WHO, 1977) with slight modifications. For the experiment, filter papers were impregnated with test materials at concentrations used in the repellency assay. Distilled water and Karbadust® (carbaryl 5%) were used as negative and positive control, respectively. Ten fleas were anesthetized with CO_2 and introduced into the test tube 8 cm away from the test material with the aid of a suction tube. Test tubes were closed with fine gauze supported by a rubber band and were held horizontally until the fleas just gained consciousness. Then, the test tubes were held vertically so that the fleas would be in contact with the test material. Test tubes were then incubated for 24 h at a temperature of 27 ± 3°C with a relative humidity of 80± 10% (Mustapha et al., 2006). Mortality observations were done after 24 h. The mortality rate was calculated according to Abbot (1925) as follows:

$$\text{Flea mortality} = \frac{(1-T)}{Co} \times 100$$

Where, T is the number of live fleas after treatment with test material; Co is the number of live fleas after treatment with control. Dead adult fleas were identified when they failed to move after probing with a needle (Maheswaran et al., 2008).

Statistical analysis

In vitro repellency bioassay

The percentage flea repellency was calculated according to Usavadee et al. (2007). Mean repellency for all materials used was calculated. Data were analysed using Proc mixed for repeated measures (SAS 2002) to compute statistical differences of least square means.

Contact bio-assay

Mortality counts were corrected by using the formula of Abbot (1925). Data obtained were tested for normality before analyzed using SAS (2002). The General Linear Models (GLM) procedure was used to analyse the effect of treatment on fleas. Comparison of means was performed using the Tukey's student range test.

RESULTS

In vitro repellency bioassay

Tagetes minuta demonstrated a significant repellency ($P<0.05$) at a concentration of 75% at a concentration of 100% for 4 h (Table 1). Extracts of *C. aurea, C. pulchella* (10, 25, 50, 75 and 100%) and their proportions of 1:1, 6:4, 7:3, 8:2 and 9:1, used engine oil and paraffin did not show ($P>0.05$) repellency. Jeyes fluid at concentrations of 38.4, 76.8 and 100% lasted for 2, 3 and 5 h, respectively, while the standard Tabard was effective for 5 h after application (Table 1). The repellency efficacy of test materials decreased with time. Differences of least square means with time are reflected on Table 2. Extracts of *T. minuta* at 50, 75, and 100% concentrations showed significant difference at 3 to 4 h range. Jeyes fluid at 76.8% compared well with Tarbard which showed the significant difference at 4 to 5 h range.

Table 2. Differences of LSM with treatment and time.

Material	Time frame range (h)							
	0.33 - 0.67	0.67 - 1	1 - 2	2 - 3	3 - 4	4 - 5	5 - 6	6 - 7
T. minuta (100%)	-124E-16	2.31E-14	1.78E-14	-551E-16	23.3**	20**	20**	16.7
T. minuta (75%)	-115-16	0	2.75E-14	10	23.3**	20**	13	6.6
T. minuta (50%)	-191E-16	1.11E-14	1.95E-14	16.7	23.3**	23**	10	10
Jeyes fluid (100%)	-213-16	0	2.31E-14	-249E-16	16.7	20**	13.3	20
Jeyes fluid (76.8%)	-147E-16	-107E-16	3.73E-14	13	40**	13	3.3	13
Jeyes Fluid (38.4%)	-213-16	5.33-15	26.7**	30**	16.7	10	3.3	6.67
Jeyes fluid (19.2%)	-213-16	16.7	33.3**	20**	16.7	3.3	10	-155E-16
Tarbard	3.13E-13	-256E-15	7.11E-4	-128E-15	10	20**	40	10

**Differences of LSMs in each row with the two stars are significantly different at (P< 0.001). *Standard error (±4.4)

Table 3. Mean percentage mortality (±S.E) after 24 hour exposure of fleas to the plant material.

Material	Concentration (%) (w/v)				
	10	25	50	75	100
T. minuta	2.5^{q}	7.5^{q}	20^{opq}	32.5^{mno}	42.5^{ljk}
C. aurea	30^{nop}	37.5^{klm}	50^{ghl}	67.5^{def}	75^{bc}
C. pulchella	15^{pq}	30^{nop}	62.5^{def}	72.5^{cd}	82.5^{ab}
C. aurea + *C. pulchella*1:1	20^{opq}	42.5^{ljk}	56.2^{efg}	60^{efg}	72.5^{cd}
C. aurea + *C. pulchella*8:2	20^{opq}	45^{hlj}	50^{ghl}	65^{def}	70^{cd}
C. aurea + *C. pulchella*9:1	36.3^{klm}	51.3^{ghl}	56.3^{efg}	67.5^{def}	72.5^{cd}
C. pulchella + *C aurea*8:2	30^{nop}	45^{hlj}	55^{efg}	65^{def}	70^{de}
C. pulchella + *C aurea*9:1	35^{lmn}	50^{ghl}	65^{def}	70^{de}	72.5^{cd}
Karbadust	-	-	-	-	97.5^{a}
Control	0^{q}	0^{q}	0^{q}	0^{q}	0^{q}

Mortality means for each test material with the same superscripts are not significantly different at P<0.001 level. *Standard error (± 3.60).

Contact bioassay

The plant extracts of *C. pulchella*, C. *aurea* and *T. minuta* at concentration of 100% had flea mortality of 83.5, 73.3 and 42.5%, respectively, while the combination of *C. pulchella* and *C. aurea* in proportions of 1:1, 8:2 and 9:1 had mean mortality of more than 70% at higher concentrations (Table 3). Other proportions 6:4 and 7:3 of both plants showed insignificant flea mortality compared to single plant extracts. Paraffin and used engine oil exhibited flea mortality of 92.5 and 95%, respectively, and were not significantly different (P<0.05) to the positive control Karbadust (carbaryl 5%) that exerted a mean mortality of 97.5%. Jeyes fluid at 19.2, 38.4 and 76.8% concentration had a flea mortality of 82.5, 95 and 100%, respectively. Lower concentrations exhibited insignificant flea mortality efficacy. Although farmers use Jeyes fluid to control fleas, however, in this study the concentration level of 1.2% which is used for general disinfection did not exert significant flea mortality. Other proportions 6.4 and 7.3 of both plants showed insignificant flea mortality compared to single plant extracts.

DISCUSSION

Insecticidal activity of ethno-veterinary medicine against several insects has been demonstrated (Isman, 2008) and the deleterious effects of these on insects can be manifested in several ways, which include toxicity, mortality and repellency (Jbilou et al., 2006). In this study, *T. minuta* exhibited both insecticidal and insect-repellent properties. This is in agreement with Cestari et al. (2004) who also documented that *T. minuta* possesses both insecticidal and repellent properties, though the Insecticidal activity of *T. minuta* was insignificant in the current study. The insecticidal activity of *T. minuta* has been reported against stored product pests (Sarin, 2004) and human head lice (Cestari et al., 2004). Its efficacy could be attributed to its composition as the main oil constituents recorded α-terpineol and ocimenone (Liza-lopez, 2008). The α-terpineol and ocimenone in *T. minuta* are responsible for the toxic and repellent effects reported in flies (Njoronge and Bushmann, 2006). Generally, oil extracts are more effective than the aqueous (Sadad, 2011) because water extracts which is different from other solvents do have myriads of compounds that act antagonistically in their overall activities (Boussaada et al., 2008). Nevertheless, in this study, *T. minuta* did not record significant flea mortality. This may be attributed to the fact that aqueous leaf extracts, and not the oil were used. However, it is easy for resource-limited farmers to use aqueous extracts than *T. minuta* oil.

In this current study, *T. minuta* showed significant repellency against fleas. This could be attributed to the pungent smell which repelled fleas away from the test material. The decrease of *T. minuta* repellency efficacy with time could be due to the fact that it contains volatile substances. Zorloni (2007) reported hexane extract of *T. minuta* having a repellency efficacy of 82% against ticks for 2 h at a low concentration of 10%, and this may be attributed to the fact that hexane not aqueous extracts were used. According to Fradin and Day (2002), the minimum repellency efficacy of commercial insecticides is 100% for 2 h and the natural repellent products is 100% for the first 30 to 60 min. *Tagetes minuta* at 50, 75 and 100% and Jeyes fluid at 76.8 and 100% had a repellency efficacy of 100% for more than 1 h, and certainly these conform to the requirement by Fradin and Day (2002). This duration compares well with the required 2 h as there is no significant difference between the two durations and as such these could be used by resource limited farmers as alternative flea repellents. Reapplication of the test material at 2 h interval is needed for the test materials to be effective for several hours. This requires more material, laborious, labour intensive, and consumes more time.

Jeyes fluid is a commercial product used as a household disinfectant. It contains mainly tar acids-13%m/m (carbolic acid) and sodium hydroxide (1%) (Henriette, 2012). It showed some insecticidal activities both through contact and repellency in the study. The effectiveness of Jeyes fluid in killing fleas is attributed to poisonous carbolic acid (Henriette, 2012) and its efficacy compared well with the positive control Karbadust (carbaryl 5%). Jeyes fluid has been used for many years as an alternative method to control ticks (Hlatshwayo and Mbati, 2005). It is accessible to resource-limited farmers and they perceive it cheaper and easy to use compared to the commercial insecticides (Hlatshwayo and Mbati, 2005); this could be the reason why rural farmers use it to control fleas. So far, no reports about its repellency properties against fleas have been documented. The Jeyes fluid is very toxic to aquatic organisms, as it alters the water pH. This may cause long term effects in the aquatic environment if water bodies are contaminated (Henriette, 2012). Its use should be avoided or used with caution to avoid contamination of water bodies.

In this study, used engine oil exhibited significant flea mortality. It contains toxic metals such as aluminium, chromium, lead, and manganese (Delistraty and Stone, 2007) and its efficacy could be due to these toxic substances contained. The toxic and carcinogenic substances contained in used engine oil can cause negative effects to the ecosystem (Delistraty and Stone, 2007). Despite the negative effects previously mentioned, Hlatshwayo and Mbati (2005) reported that used engine oil is being used as an alternative tick control by resource-limited farmers because of its accessibility. However, despite its wide use as an alternative remedy in the control of external parasites it could be used with caution. In addition, paraffin also showed some significant mortality against fleas, however, little or no literature has documented its efficacy against external parasites. Previous studies reported its use by farmers to control fleas (Muchadeyi et al., 2004) and ticks (Hlatshwayo and Mbati, 2005). Paraffin is composed of sulphur, olefins, aromatics and aliphatics (Dlamini and Gqaleni, 2006). Aromatics are associated with high toxicity to skin tissues (Dlamini and Gqaleni, 2006) and they induce necrosis on the skin. These include toluene, ethylbenzene, p-xylene and phenylmethane, which are commonly found as industrial solvents in the manufacturing of paints, chemical, pharmaceuticals and rubber (Broberg et al., 2008). Although it is widely used by resource-limited farmers in most developing countries, it has to be used with caution.

In this study, *C. pulchella* had the highest flea mortality percentage among plants. However, information is lacking on its previous use against fleas and its insecticidal properties. It has previously been reported to be used in drenches for griping pains in calves (Hutchings et al., 1996) and its decoction to treat gall sickness (Masika and Afolayan, 2003). *Calpurnia aurea* also exhibited significant flea mortality at higher concentrations. Some findings by Adedapo et al. (2008) revealed that *C. aurea* (Ait) Benth is used to kill lice. In a study by Zorloni (2007), *C. aurea* exhibited high tick mortality. This could be due to the fact that acetone and not aqueous extracts were used. Also, active compounds from *C. aurea* such as quinolizidine alkaloid calpurnine (12β, 13α dihydroxylupanine) and its 13α-(2-pyrrolycarboxylic acid ester) have been reported by Zorloni (2007). These quinolizidine alkaloids are toxic to insects and animals (Adedapo et al., 2008). Their toxicity could have attributed to the mortality of fleas. Insecticidal properties of *C. aurea* have also been recorded by Blum and Bekele (2002). In another study, Regassa (2000) reported the use of juice from crushed leaves of *C. aurea* in the control of ticks. Furthermore, its insecticidal activity against lice in calves has been reported by Heine and Brenzinger (1988) in Kenya. The wide use of *C. aurea* could justify its use by resource-limited farmers of Amatola basin in controlling flea.

This study revealed that combination extracts of *C. pulchella* and *C. aurea* did not exhibit significant repellency against fleas. The result could be attributed to the observation that both plants did not possess repellent properties or possible that there may have antagonistic effects. In contact bioassay, the proportions of 1.1, 8.2 and 9.2 had significant flea mortality at the highest concentration, but it was less than when single extracts were used. This could be attributed to the fact that the compounds in each extract in plant combinations may neutralize each other (Gilani and Rahman, 2005). Also, in some instances the combination may result in antagonistic effects (Adams et al., 2006); this could explain the effect of two extracts being less than that of a single extract. According to Adams et al. (2006), combining two plants extracts increase potency due to synergistic effect. However, it is hard to predict synergistic and antagonistic effect because in this study, the concentration of active ingredient of plant extract was not estimated. Further investigations are, therefore, recommended in this regard. It was also noted that there were no significant differences in mortality when using different proportions of *C. pulchella* and *C. aurea* and this could necessitate finding the compounds that could contribute to this effect.

Amongst the materials used by resource-limited farmers and reported herein, most of them possess insecticidal and repellent properties against fleas. Despite the fact that some materials showed effectiveness against fleas, used engine oil and Jeyes fluid have potential toxic effects in animals and are also environmental contaminants. Hence, it is recommended that those using it should exercise extreme care and caution

Conclusion

The present investigation established that *T. minuta* and Jeyes fluid showed some repellency efficacy against fleas, while *C. pulchella, C. aurea*, paraffin and used engine oil did not exhibit any efficacy at all. At the same time, *C. pulchella, C. aurea,* paraffin, used engine oil and Jeyes fluid showed some flea mortality efficacy. Plant combinations of *C. pulchella* and *C. aurea* in proportions of 6:4 and 7:3 did not exhibit significant flea mortality at low concentrations. This work demonstrated some of the tested materials to be very effective in the repelling and killing of fleas. However, further investigations are recommended to determine flea insecticidal compounds of *C. pulchella* and *C. aurea*. Despite the wide use of used engine oil, it is an environmental contaminant; therefore farmers need to exercise caution when using it.

ACKNOWLEDGEMENTS

The authors are thankful to the National Research Foundation (NRF), South Africa, for financial support and M. Mwale for statistical analysis assistance.

REFERENCES

Abbot WS (1925). A method for computing the effectiveness of an insecticide. J. Econ. Entomol.18: 265-267.

Adams LS, Seeran NP, Hardy ML, Carpenter C, Heber D (2006). Analysis of the interaction of Botanical extracts combinations against the viability of prostate cancer cell lines. Evidence-based Compl. Alt. Medic.3 (1):117-124.

Adedapo AA, Jimoh FO, Koduru SK, Afolayan AJ, Masika PJ (2008). Antibacterial and antioxidant properties of the methanol extract of the leaves and stems of *Calpurnia aurea*. BMC Compl. Altern. Medic.8:5.

Agboola BA, Remi-Adewunmi BD, Agdede RI (2007). Flea infestation in a Nigerian local chicken. Nigerian Vet. J. 28(2):57-59.

Blum A, Bekele A (2000). Storing grains as a survival strategy of small farmers in Ethiopia. J. Int. Agri. Ext. Ed. 9(1):777-831.

Boughton RK, Atwell JW, Schoech SJ (2006). An introduced parasite, the sticktight flea *(Echidnophaga gallinacean)* and its pathology in the threatened Florida scrub-jay (*Aphelocoma coerulescens*). J. Parasitol. 92(5): 941-948.

Boussaada O, Ammar S, Saidana D, Chriaa J, Chraif I, Daami M, Helal AN, Mighri Z (2008). Chemical composition and antimicrobial activity of volatile components from capitula and aerial parts of Rhaponticum acaule DC growing wild in Tunisia. Microb. Res. 163: 87-95.

Broberg K, Tinnerberg H, Axmon A, Warholm M, Rannug A, Littorin M (2008). Influence of genetic factors of touene diisocyanate-related symptoms: evidence from a cross-sectional study. Environ. Health 7:15.

Cestari IM, Sarti SJ, Waib CM, Castello BA (2004). Evaluation of the potential insecticidal activity of *Tagetes minuta* (Asteraceae) essential oil against the head lice Pediculus human uscapitis (Phthiraptera: Pediculidae). Neo. Entomol. 33 (6): 805-807.

Delistraty D, Stone A (2007). Dioxins, metals, and fish toxicity in ash residues from space heaters burning used motor oil, Chemosphr.68 (5): 907-914.

Dlamini N, Gqaleni N (2006). Chemical composition and toxicity of South African paraffin. Centre for Occupation and Environmental Health, University of KwaZulu Natal. http://www.phasa.za.org/conf2006/present/DlaminN2.pdf. Downloaded 13/06/12.

Dominguez E, Pitchtel J, Coughlin M (2004). Phytoremediation of soil contaminated with used engine oil: Enhanced microbial activities from laboratory and growth chamber studies. Environ. Eng. Sci. 21: 157-168

Fradin MS, Day JF (2002). Comparative efficacy of insect repellents against mosquito bites. N. Engl. J. Med. 347(1):13-18.

Gilani AH, Rahman A (2005). A Trend in ethnopharmacology. J. Ethnopharmacol. 100(1-2): 43-49.

Gyimesi ZS, Hayden ER, Greiner EC (2007). Sticktight flea (*Echidnophaga gallinacea*) Infestation in a Victoria crowned pigeon (Goura Victoria). J. Zoo Wildl. Med. 38(4): 594-596.

Heine B, Brenzinger M (1988). Plant concepts and plant use. In: An ethnobotanical survey of the semi-arid lands of East Africa, Part 1VB. Heine, Editor, Plants of the Borana (Ethiopia and Kenya), Fort. Lauderdale, Breitenbach. 205.

Henriette K (2012). http://www.henriettesherbal.com-1995-2008Henriette Kress. Accessed 13/05/2012.

Hlatshwayo M, Mbati PA (2005). A survey of tick control methods used by resource-poor farmers in the Qwa-Qwa area of the Eastern Free State Province, South Africa. Ondersterpoort J. Vet. Res. 72: 245-249.

Hutchings A, Scott AH, Lewis G, Cunningham AB (1996). Zulu Medicinal Plants: An inventory. University of Natal Press, Pietermaritzburg.

Isman MB (2008). Botanical insecticides: for richer or poorer. Pest Manag. Sci. 64: 8-11.

Jbilou R, Ennabili A, Sayaha F (2006). Insecticidal activity of four medicinal plants extracts against Triboliumcastaneum (Herbst) (*ColeopteraTenebrionidae*). Afr. J. Biotech. 5(10): 936-940.

Liza-lopez M, Norma EB, Julio AZ (2008). Allelopathic potential of *Tagetes minuta* terpenes by a chemical anatomical and phytotoxic approach. Bioch. Systmatics Ecol. 36 (12): 882-890.

Luseba D, Van der Merwe D (2006). Ethno-veterinary medicine practices among Tsonga speaking people of South Africa. Onderstepoort J. Vet. Res.73: 115-122.

Macan J, Varnai VM, Turk R (2006). Health effects of pyrethrins and pyrethroids. ArhHigRadaToksikol 57: 237-243.

Maheswaran R, Sathish S, Ignacimuthu S (2008). Larvicidal activity of *Leucasaspera* (Wild) against the larvae of *Culex quinque fasciatus Say* and *Aedesaegypti L.* Int. J. Integr. Biol. 2 (3): 214 -217.

Masika PJ, Afolayan AJ (2003). An ethnobotanical study of plants used for the treatment of livestock diseases in the Eastern Cape Province,

South Africa. Pharm. Biol. 41(1): 16-21.
Moyo B, Masika PJ (2009). Tick control methods used by resource-limited farmers and the effects of ticks on cattle in rural areas of the Eastern Cape province, South Africa. Trop. Anim. Health Prod. 41: 517-523.
Muchadeyi FC, Sibanda S, Kusina NT, Kusina J, Makuza S (2004). The village chicken production system in Rushinga district of Zimbabwe. Livestock Res. Rural Dev. 16 (4).
Mustapha D, Stephen PF, Daniel S (2006). *Insect repellents: Principles, Methods and Uses*. Bacon Raton: CRC Press. On line at http:books.google.co.za. (Downloaded 13/06/12).
Njoronge NG, Bussmann RW (2006). Herbal usage and information consensus in ethnoveterinary management of cattle diseases among the Kikuyus (Central Kenya). J. Ethnopharmacol.108: 332-339.
Pieters L, Vlietinck AJ (2005). Bioguided isolation of pharmacology active plant components, still a valuable strategy for the finding of new lead compounds? J. Ethnopharmacol.100: 57-60.
Power LE, Sudakin DL (2007). Pyrethrin and pyrethroid exposures in the United States, a longitudinal analysis of incidents reported to poison centers. J. Med. Toxicol. 3:94-99.
Regassa A (2000). The use of herbal preparations for tick control in Western Ethiopia. J. S. Afr. Vet. Assoc.71 (4): 240-243.
Sarin RC (2004). Insecticidal activity of callus culture of *Tagetes erecta*, *Fitoterapia*.75: 62-64.
Snyder DE, Meyer J, Zimmermann AG, Qiao M, Gissendanne SJ, Cruthers LR, Young DR (2007). Preliminary studies on the effectiveness of the novel pulicide, spinosad, for treatment and control of fleas on dogs.Vet. Parasitol.150: 345-351.
Statistical Analysis Systems Institute (2002). SAS user's guide: statistics, version 6. SAS Institute Inc., Cary, NC, USA.
Usavadee T, Apiwat T, Payu B, Prapaid W, Thidarat B, Jaree B, Pranee C, Narumon K, Padet S, Mir SM (2007). Repellent of essential oils against cockroaches (Dictyoptera: Blattidae, Blattellidae and Blaberidae) in Thailand. Southeast Asian. J. Trop. Med. Public Health 38 (4): 663-673.
World Health Organization (1977). Technical Report Series, Instructions for determining the susceptibility or resistance of fleas to insecticides. 443:100-105.
Yin S, Kwok YO (2005). Detection technologies for chemical warfare agents and toxic vapour.Technology and Engineering.CRC Press, USA.
Zorloni A (2007). Evaluation of plants used for the control of animal ectoparasites in southern Ethiopia (Oromiya and Somalia regions). MSc Thesis, University of Pretoria.

Analgesic and anti-inflammatory effects of *Cyphostemma vogelii* (Hook. f.) Desc. root extract in mice

Udegbunam R. I.[1], Udegbunam S. O.[1] and Anosa G. N.[2]

[1]Department of Veterinary Surgery, Faculty of Veterinary Medicine, University of Nigeria, Nsukka, Enugu State, Nigeria.
[2]Department of Veterinary Medicine, Faculty of Veterinary Medicine, University of Nigeria, Nsukka, Enugu State, Nigeria.

Cyphostemma vogelii **(family: Vitaceae) is a herbaceous plant which grows in Obukpa town in South Eastern Nigeria. *C. vogelii* is said to be generally medicinal though no specific medicinal activity was mentioned in the literature for this plant. The analgesic effect of *C. vogelii* was evaluated using acetic acid-induced writhing and formalin-induced nociception tests. The extract was also screened for anti-inflammatory activity using carrageenan-induced and kaolin-carrageenan-induced paw edema tests. *C. vogelii* extract dose dependently inhibited acetic acid-induced pain in mice. The extract at 200 and 400 mg/kg significantly inhibited inflammatory and neurogenic pain induced by formalin. The effects of 200 mg/kg on formalin induced pain were similar to those of aspirin, while 400 mg/kg produced more inhibitory effects. At 2, 3 and 5 h post carrageenan injection, 200 and 400 mg/kg extract significantly inhibited paw edema. The extract dose dependently suppressed kaolin-carrageenan-induced edema from 3 h post treatment for up to 24 h. The effect of 400 mg/kg on kaolin-carrageenan-induced edema was similar to that of aspirin, while 100 mg/kg showed the least activity. The data obtained from this study showed that the methanol extract of *C. vogelii* exhibited mild analgesic activity as well as anti-inflammatory activity.**

Key words: Analgesic, anti-inflammatory, mice, *Cyphostemma vogelii,* nociception.

INTRODUCTION

Pain is defined as an unpleasant sensory or emotional experience associated with actual or potential tissue damage (IASP, 1979). Pain occurs when nociceptors of afferent neurons are exposed to noxious stimuli such as trauma or surgery (Busch et al., 2006). In addition to noxious stimulation, nociceptors can be sensitized by algogens released during the inflammatory process (Jones and Hamm, 1977; Snow, 1981). These chemical mediators of inflammatory pain include bradykinin, prostaglandins, substance P, histamine and serotonin (Hughes and Lang, 1983; Boothe, 1984; Dray, 1995).

In modern medical practice, non steroidal anti-inflammatory drugs (NSAIDs) are considered the drugs of choice in the treatment of inflammatory pain (Hosking and Welchew, 1984). NSAIDs inhibit cyclo-oxygenase, the enzyme responsible for the conversion of arachidonic acid to prostaglandin (Lees et al., 2004; Choi and Kwang, 2004). Prostaglandins are known to cause hyperaemia, modulate inflammation and sensitize pain receptors (Snow, 1981). Thus, by reducing the amount of prostaglandins, NSAIDs reduce inflammation and the amount of pain felt by animals and humans (Hosking and Welchew, 1984).

However, suppression of prostaglandin synthesis may lead to the gastrointestinal bleeding (Sparkes et al., 2010), acute renal failure (Weir, 2002) and delayed wound healing (Haws et al., 1996; Dvivedi et al., 1997). Therefore, plant derived medicines are gaining acceptance as safer alternatives in the management of painful inflammatory conditions (Choi and Kwang, 2004).

Cyphostemma vogelii (family: *Vitaceae*) is a herbaceous climber seen in rain forests and wooded savannah (Verdecourt, 1993; Burkill, 2000). In Obukpa town, an Igbo speaking area of Enugu state, Nigeria, this plant grows widely and is popularly called "Okoho". The inhabitants of Obukpa use powders from its root in the preparation of a native soup. Although, *C. vogelii* is said to be generally medicinal, literature search did not reveal any specific medicinal use of this plant. However, several medicinal activities were recorded for a closely related species (*Cyphostemma adenocaule*) which grows in Ghana, Tanzania, Kenya and Uganda (Katende et al., 1999; Bosch, 2004). In Tanzania, the leaves of *C. adenocaule* are used to prepare medicines used in the treatment of sore throat, cough and pneumonia (Kokwaro, 1993; Bosch, 2004). In Ghana, Gabon and East Africa, a paste made from its root is applied to treat abscess and inflammation (Bosch, 2004).

Despite ethno botanical reports suggesting that plants in the family *Vitaceae* possess anti-inflammatory effects, there is yet no scientific study conducted to ascertain the anti-inflammatory effect of *C. vogelli*, thus the need for this study. In this study, we investigated the analgesic and anti-inflammatory effects of the methanol root extract of *C. vogelii* using experimental models of pain and inflammation.

MATERIALS AND METHODS

Fresh roots of *C. vogelii* were collected from Obukpa in Nsukka Local Government Area of Enugu State, Nigeria in March, 2011. Samples were authenticated by a taxonomist in the International Centre for Ethnomedicine and Drug Development, Nsukka. The voucher specimen was deposited in the herbarium of the centre.

Preparation of extract

The roots were cut into small pieces, air dried and powdered. The plant materials (500 g) were defatted with n-hexane for 48 h at room temperature, followed by filtration. The residue were dried and subsequently macerated in 80% methanol for 48 h followed by filtration. The filtrate was evaporated to dryness under reduced pressure to obtain the extract of *C. vogelii* (yield = 6.32% w/w). At each time of use, the extract was dissolved in 10% Tween 80 and distilled water to obtain the required concentration of the test solution.

Phytochemical screening

Phytochemical screening was performed using *C. vogelii* extract to qualitatively investigate the presence of alkaloids, tannins, flavonoids, saponin and glycosides (Harborne, 1998). *C. vogelii* extract (1000 mg) was dissolved with 10 ml of distilled water to form a 100 mg/ml test solution which was used for the phytochemical assays.

Acute toxicity

Acute toxicity study was carried out as described by Lorke (1983). Mice weighing 18 to 22 g were randomly assigned to five groups (n = 5) and dosed orally with the extract (100, 200, 400, 800 and 1600 mg/kg). These mice were observed for 48 h for symptoms associated with toxicity such as convulsion, ataxia and diarrhea.

Analgesic studies

Acetic acid-induced abdominal writhing test

The acetic acid induced writhing test was performed as described by Koster et al. (1959). Mice in six experimental groups (n = 5) were treated with *C. vogelii* extract (100, 200 and 400 mg/kg, p.o.), aspirin (200 mg/kg, p.o.) and normal saline (1 ml/kg, p.o.) 1 h before 0.6% v/v acetic acid solution (10 ml/kg) was administered intraperitoneally (i.p.). The numbers of abdominal writhings observed in each mouse were counted for 20 min and recorded (Couto et al., 2011).

Formalin induced nociception test

The test was carried using the method described by Dubuisson and Dennis (1977). Mice were treated with *C. vogelii* extract (100, 200, 400 mg/kg, p.o.), aspirin (200 mg/kg, p.o) and normal saline (1 ml/kg, p.o.). Pain was induced 1 h post treatment by intra plantar injection of 0.05 ml 2.5% v/v formaldehyde solution (Couto et al., 2011). The time spent in licking the paws were recorded in the early phase (0 to 5 min) and late phase (15 to 30 min) after formalin injection.

Anti-inflammatory studies

Carrageenan induced paw edema test

Acute inflammation was induced in the paw of mice by sub plantar injection of 0.1 ml carrageenan (1%), 1 h after administration of *C. vogelii* extract (100, 200 and 400 mg/kg, p.o.), aspirin (200 mg/kg) and normal saline (1 ml/kg, p.o.) as described by Winter et al. (1962). The paw thickness of each mouse was measured using a venire caliper before edema induction and at 1, 2, 3 and 5 h post edema induction. Edema and percentage edema inhibition were calculated as described by Zhang et al. (2008).

Kaolin-carrageenan induced paw edema test

Sub acute inflammation was induced by injecting 0.1 ml of a mixture containing kaolin (20%) and carrageenan (1%) into the hind paw of mice (Hajare et al., 2001). By 18 h post edema induction, the extracts (100, 200 and 400 mg/kg, p.o.), aspirin (200 mg/kg, p.o.) and normal saline (1 ml/kg) were administered to mice in the treatment groups. The treatments were repeated at 3 and 7 h after the first dosing. The paw thicknesses were measured before commencing the treatments (0 min) and at 3, 6 and 24 h after the first treatments were administered. Paw edema was calculated.

Table 1. Phytochemical constituents identified in *C. vogelii.*

Phytochemical constituent	Result
Alkaloids	-
Flavonoids	-
Tannins	+
Glycosides	-
Saponins	+

- (Negative); + (positive).

Table 2. Effect of *C. vogelii* on pain induced by acetic acid.

Treatment (ml/kg)	Number of writhing	Percentage inhibition
Normal saline, 1	106.3±3.7[d]	0
Aspirin, 200	37.3±7.5[a]	64.4[a]
Extract, 100	83.5±3.8[c]	21.4[b]
Extract, 200	73.2±2.6[bc]	31.1[b]
Extract, 400	62.5±5.5[b]	41.2[b]

Different superscripts [a,b] in a column show significant difference between group means.

Statistical analysis

Data obtained were presented as mean ± S.E.M. Mean number of writhing, licking time and paw edema in control and extract groups were compared using one way analysis of variance (ANOVA) followed by Duncan multiple range test. The differences were considered significant at $p < 0.05$.

RESULTS

Phytochemical screening

Phytochemical screening showed that the extract contained saponins and tannins, while flavonoids, alkaloids and glycosides were absent (Table 1).

Acute toxicity test

No mortality was recorded in all treatment groups. Also, symptoms such as convulsion, ataxia and diarrhea suggestive of toxicity were not observed.

Acetic acid induced pain

The extract significantly reduced the number of painful writhing induced by acetic acid (Table 2). The inhibitory effect was dose dependent with inhibition of 21.4% recorded for 100 mg/kg, 31.1% for 200 mg/kg and 41.2 % for 400 mg/kg (Table 2).

Formalin induced pain

The extract at 200 and 400 mg/kg significantly inhibited inflammatory and neurogenic pain induced by formalin (Table 3). While the effects of 200 mg/kg on formalin induced pain were similar to those of aspirin, 400 mg/kg produced more inhibitory effects.

Carrageenan induced edema

The paw edema in the treatment groups are shown in Table 4. As shown in this table, treatment of mice with 400 mg/kg extract significantly inhibited paw edema by 1, 2, 3 and 5 h post carrageenan injection. At 2, 3 and 5 h, 200 mg/kg significantly inhibited paw edema while 100 mg/kg inhibited paw edema at 3 and 5 h.

Kaolin-carrageenan-induced edema

As shown in Table 5, the extract dose dependently suppressed kaolin-carrageenan-induced edema from 3 h post treatment for up to 24 h. The effect of 400 mg/kg on kaolin-carrageenan induced edema was similar to that of aspirin, while 100 mg/kg showed the least activity.

DISCUSSION

This study investigated the analgesic effect of methanol root extract of *C. vogelii* using acetic acid-induced writhing test and formalin-induced nociception test. The acetic acid test is often used to assess the peripheral analgesic effect of medicinal plants (Couto et al., 2011). When injected peritoneally, acetic acid stimulates local peritoneal receptors with subsequent release of PGE_2 and $PGF_{2\alpha}$ (Bentley et al., 1983; Deraedt et al, 1980). This test is however a non specific model for detecting peripheral analgesic activity because centrally acting analgesics also positively inhibits pain induced by acetic acid (Trongsakul et al., 2003). Thus, to further investigate the analgesic effect of *C. vogelii,* formalin-induced nociception test, a more satisfactory test for clinical pain was performed (Abbott et al., 1981). The formalin-induced nociception test is used as a model for tonic (Coderre et al., 1990) and inflammatory pain (Tjolsen et al., 1992; Hong and Abbott, 1994). Formalin injection provokes two phases of responses in animals (Lee and Jeong, 2002). Pain in the early phase occurs due to direct stimulation of nociceptors while in the latter phase, pain is due to inflammation (Shibata et al., 1989). Centrally acting analgesics inhibit both phases while peripherally acting analgesics inhibit the second phase (Abram and Olson, 1994; Rosland et al., 1990; Yamamoto et al., 2002). Thus, the ability of 200 and 400 mg/kg *C. vogelii* extract to inhibit acetic acid induced pain as well as both phases of formalin induced nociception

Table 3. Effect of *C. vogelii* on formalin induced pain in mice.

Treatment (ml/kg)	0 - 5 min		15 - 30 min	
	Licking time	Inhibition (%)	Licking time	Inhibition (%)
Normal saline, 1	80.3±7.1^{c}	0	169.3±8.4^{c}	0
Aspirin, 200	54.8±1.9^{b}	31.8^{ab}	89.7±0.6^{ab}	47.0^{ab}
Extract, 100	61.5±7.4^{b}	23.4^{ab}	127.8±9.3^{bc}	24.5^{a}
Extract, 200	45.5±4.8^{ab}	43.3^{ab}	89.8±9.0^{ab}	46.9^{ab}
Extract, 400	36.0±4.9^{a}	55.1^{b}	47.8±9.8^{a}	71.8^{b}

Different superscripts ^{a,b,c} in a column show significant difference between group means.

Table 4. Effect of *C. vogelii* on carrageenan induced paw edema.

Treatment	Paw edema (mm)			
	1 h	2 h	3 h	5 h
Normal saline 1 ml/kg	0.14±0.01^{b}	0.15±0.03^{c}	0.16±0.02^{c}	0.18±0.01^{c}
Aspirin 200 mg/kg	0.11±0.03^{b}	0.07±0.03^{a}	0.06±0.01^{a}	0.05±0.01^{a}
Extract 100 mg/kg	0.13±0.01^{b}	0.12±0.03^{b}	0.12±0.03^{b}	0.10±0.01^{b}
Extract 200 mg/kg	0.13±0.03^{b}	0.11±0.03^{b}	0.11±0.01^{b}	0.09±0.01^{b}
Extract 400 mg/kg	0.06±0.03^{a}	0.08±0.03^{a}	0.06±0.01^{a}	0.06±0.01^{a}

Different superscripts ^{a,b,c} in a column show significant difference between group means.

Table 5. Effect of *C. vogelii* on kaolin-carrageenan-induced paw edema.

Treatment	Dose (mg/kg)	Edema before treatment	Edema post treatment		
			3 h	6 h	24 h
Control	-	0.19±0.01	0.22±0.02^{d}	0.17±0.03^{b}	0.12±0.05^{c}
Aspirin	200	0.17±0.03	0.08±0.02^{a}	0.06±0.02^{a}	0.01±0.01^{a}
Extract	100	0.18±0.03	0.15±0.01^{c}	0.12±0.02^{c}	0.07±0.03^{ab}
Extract	200	0.18±0.02	0.13±0.01^{bc}	0.11±0.02^{c}	0.06±0.01^{ab}
Extract	400	0.18±0.01	0.10±0.01^{ab}	0.08±0.01^{a}	0.04±0.01^{a}

Different superscripts^{a,b,c} in a column show significant difference between group means.

suggests that it exhibited both peripheral and central analgesic effect (Couto et al., 2011).

Carrageenan-induced and kaolin-carrageenan induced paw edema tests were performed to evaluate the effect of *C. vogelii* extract on acute and subacute inflammatory processes (Muruganandan et al., 2001). The injection of carrageenan induces three phases of chemical mediator release which occur in an orderly sequence (Di Rosa, 1972). The initial phase which takes place within 1 to 2 h is mediated by serotonin and histamine, while the intermediate phase is mediated by bradykinin (Zhang et al., 2008). The final phase which occur 2.5 to 6 h post carrageenan injection is presumed to be mediated by PGs (Zhang et al., 2008). Injection of kaolin-carrageenan on the other hand produces inflammation which lasts up to 24 h (Hajare et al., 2001). Edema post kaolin-carrageenan injection is mediated by kinins and activation of kallikrein (Northover and Subramanian, 1961; Bonta and DeVos, 1965). The ability of *C. vogelii* extract to dose dependently inhibit paw edema post carrageenan and kaolin-carrageenan injections shows that it was able to suppress the release of chemical mediators of acute and subacute inflammation (Murunganandam et al., 2001; Hajare et al., 2001; Sini et al., 2010; Udegbunam et al., 2012). Earlier, it has been documented that *Cyphostemma* species contained compounds with anti-inflammatory activity (Bosch, 2004).

Phytochemical screening of the extract of *C. vogelii* revealed the presence of tannins and saponins. Previously, tannins and saponins isolated from medicinal plants exhibited analgesic and anti-inflammatory activities (Thomas et al., 1985; Owoyele et al., 2010; Choi et al., 2005). We thus suggested that the tannins and saponins present in *C. vogelii* extract were responsible for its analgesic and anti-inflammatory activities.

In conclusion, data obtained from this study showed

that the methanol extract of *C. vogelii* exhibited mild analgesic activity as well as anti-inflammatory activity.

REFERENCES

Abbott FV, Franklin KB, Ludwick RJ, Melzack R (1981). Apparent lack of tolerance in the formalin test suggest a different mechanism for morphine analgesia in different types of pain. Pharmacol. Biochem. Behav. 15:637-640.

Abram SE, Olson EE (1994). Systemic opioids do not suppress spinal sensitization after subcutaneous formalin in rats. Anaesthesiology 80:1114-1119.

Bentley GA, Newton SH, Starr J (1983). Studies on the anti-nociceptive action of alpha agonist drugs and their interaction with opioid mechanisms. Br. J. Pharmacol. 79:125-134.

Bonta IL, DeVos CJ (1965). Presence of slow-contraction inducing materials in fluid collected from the rat paw edema induced by serotonin. Experientia 21:34-39.

Boothe DM (1984). Prostaglandins: Physiology and clinical complications. Compend. Cont. Edu. Pract. Vet. 6(11):1010-1021.

Bosch CH (2004). *Cyphostemma adenocaule* (steud ex. A. Rich) Wild and R.B. Drumm.In: Plant resources of Tropical Africa 2: Vegetables. Grubben GJH, Denton OA (eds.) PROTA foundation/Backhuys publishera/CTA Wageningen, Netherlands. pp. 279-300.

Burkill HM (2000). The useful plants of West Tropical Africa, 2nd. Ed. Vol.5, Families S-Z Addenda. Royal Botanic Garden, Kew, United Kingdom. 686 pp.

Busch S, Chambaliss M, Raffel T, Shaffran N (2006). The preoperative patient In: Busch SJ (ed). Small animal surgical nursing: Skills and concepts Elsevier Mosby, St. Louis, Missouri. pp. 3-48.

Choi E, Kwang J (2004). Anti-inflammatory, analgesic and anti-oxidant activities of the fruits of *Foeniculum vulgare*. Fitoterapia 75:557-565.

Choi J, Jung HJ, Lee KT, Park HJ (2005). Antinociceptive and anti-inflammatory effects of saponins and sapogenins obtained from the stem of *Akebia quinata*. J. Med. Food 8(1):78-85.

Coderre TJ, Vaccarino AI, Melzack R (1990). Central nervous system plasticity in the tonic pain response to subcutaneous formalin injection. Brain Res. 535:155-158.

Couto VM, Vilela FC, Dias DF, Dos Santos MH, Soncini R, Nascimento CGO, Giusti-Paiva A (2011). Anti-nociceptive effect of extract of *Emilia sonchifolia* in mice. J. Ethnopharmacol. 134:348-353.

Deraedt R, Jouquey S, Delevallee F, Flahaut M (1980). Release of prostaglandins E and F in an algogenic reaction and its inhibition. Eur. J. Pharmacol. 61:17-24.

Di Rosa M (1972). Biological properties of carrageenan. J. Pharm. Pharmacol. 24:89-102.

Dray A (1995). Inflammatory mediators of pain. Br. J. Anesth. 75:25.

Dubuisson D, Dennis SG (1977). The formalin test, a quantitative study of the analgesic effects of morphine, meperidine and brain stem stimulation in rats and cats. Pain 4:161-174.

Dvivedi S, Tiwari SM, Shama AA (1997). Effect of ibuprofen and diclofenac sodium on experimental wound healing. Indian J. Exper. Biol. 35:1243-1245.

Hajare SW, Chandra S, Sharma J, Tandan SK, Lal J, Telang AG (2001). Anti-inflammatory activity of *Dalbergia sissoo* leaves. Fitoterapia 72:131-139.

Harborne JB (1998). Phytochemical methods: A guide to modern techniques of plant analysis, 3rd edition. Chapman and Hall, London. pp. 1-302.

Haws MJ, Kucan JO, Roth AC, Suchy H, Brown RE (1996). The effects of chronic ketorolac tromethamine (toradol) on wound healing. Ann. Plast. Surg. 37:147-151.

Hong Y, Abott FV (1994). Behavioural effects of intraplantar injection of inflammatory mediators in rats. Neuroscience 63:827-836.

Hosking J, Welchew E (1985). Assessment of pain In: Post-operative pain, Understanding its nature and how to treat it. Faber and Faber, London. pp. 44-55.

Hughes H, Lang M (1983). Control of pain in dogs and cats In: Kitchell R, Erickson H (eds.) Animal pain. Baltimore Waverly press. pp. 207-218.

IASP (1979). Pain terms. A list of definitions and notes on usage recommended by the International Association for the Study of Pain subcommittee on taxonomy. Pain 6:249.

Jones EW, Hamm D (1977). Steroidal and non-steroidal anti-inflammatory drugs for wounds and traumatic inflammation. New Zealand Vet. J. 25(11):317-321.

Katende AB, Ssegawa P, Birnie A (1999). Wild food plants and mushrooms of Uganda. Technical handbook No 19. Regional land management unit, SIDA, Nairobi Kenya. 490 pp.

Kokwaro JO (1993). Medicinal plants of East Africa. 2nd ed. Kenya literature Bureau, Nairobi, Kenya. 401 pp.

Koster RM, Anderson M, De Beer EJ (1959). Acetic acid analgesic screening. Fed. Proc. 18:412-417.

Lee I, Jeong Y (2002). Effects of different concentrations of formalin on paw edema and pain behaviours in Rats. J. Korean Med. Sci. 17:81-85.

Lees P, Giraudel J, Landoni MF, Toutain PL (2004). PK-PD integration and PK-PD modeling of non-steriodal anti-inflammatory drugs. Principles and applications in Veterinary Pharmacology. J. Vet. Pharmacol. Therapeut. 27:491-502.

Lorke D (1983). A new approach to practical acute toxicity testing. Arch. Toxicol. 54:275-287.

Muruganandan S, Srinivasan K, Chandra S, Tandan SK, Lal J, Raviprakash V (2001). Anti-inflammatory activity of *Zyzygium cumini* bark. Fitoterapia 72:369-375.

Northover BJ, Subramanian G (1961). Analgesic and anti-pyretic drugs as inhibitors of kallikrein. Br. J. Pharmacol. 17:107-115.

Owoyele BV, Ngede MN, Olaniran SO, Onasinwo SA, Oguntoye SO, Sanya JO, Onyeleke SA, Ibidapo AP, Soladoye AO (2010). Analgesic and anti-inflammatory effects of aqueous extract of *Zea mays* husk in male wistar rats. J. Med. Food 13:343-347.

Rosland JH, Tjolsen A, Maehle B, Hole DK (1990). The formalin test in mice. Effect of the formalin concentration. Pain 42:235-242.

Shibata M, Ohkubo T, Takahashi H, Inoki R (1989) Modified formalin test: Characteristic biphasic response. Pain 38:347-352.

Sini JM, Yaro AH, Ayanwuyi LO, Aiyelero OM, Mallum SM, Gamaniel L (2010). Anti-nociceptive and anti-inflammatory activities of the aqueous extract of the root bark of *Combretum sericeum* in rodents. Afr. J. Biotechnol. 9(51):8872-8876.

Snow D (1981). Non-steroidal anti-inflammatory agents in horses. In Pract. 3(5):24-31.

Sparke AH, Heiene R, Lascelles DH, Malik R, Sampiero LR, Robertson S, Sherk M, Taylor P (2010). ISFM and AAFP consensus guidelines. Long terrn use of NSAIDs in cats. J. Feline Med. Surg. 12:521-538.

Thomas MLRM, Filho JMB (1985). Anti-inflammatory action of tannins isolated from the bark of *Anacardwn occidentale*. J. Ethnopharmacol. 13:289-300.

Tjolsen A, Berge OG, Hunskaar S, Rosland JH, Hole K (1992). The formalin test: an evaluation of the method. Pain 51:5-17.

Trongsakul SA, Panthong D, Kanjanapothi D, Taesotikul T (2003). The analgesic, antipyretic and anti-inflammatory activity of *Diospyros variegates* kruz. J. Ethnopharmacol. 85:221-225.

Udegbunam RI, Nwamkpa OK, Udegbunam SO, Nwaehujor CO, Offor GE (2012). Evaluation of anti-inflammatory activities of the extracts of *Stephania dinklagei* (Engl.) Diels. roots in mice. Afr. J. Pharm. Pharmacol. 6(11):834-839.

Verdecourt B (1993). Vitaceae In: Polhill RM (ed) Flora of tropical East Africa AA Balkema, Rolterdam Netherlands. 149 pp.

Weir M (2002). Renal effects of non-selective NSAIDs and Coxibs. Cleve. Clin. J. Med. 69 (suppl. I):s153-158.

Winter CA, Risley EA, Nuss GW (1962). Carrageenan induced edema in the hind paw of the rat as an assay for anti inflammatory drugs. Proceedings of Society of Experimental Biology and Medicine, 111:544-547.

Yamamoto T, Nozaki-Taguchi N, Chiba T (2002). Analgesic effect of intrathecally administered orexin-A in the rat formalin test and in the rat hot plate test. Br. J. Pharmacol. 137:170-176.

Zhang G, Huang X, Wang H, Leung AK, Chan C, Fong DWF, Yu Z (2008). Anti-inflammatory and analgesic effects of the ethanol extract of *Rosa multiflora* Thunb. hips. J. Ethnopharmacol. 118:290-294.

Physicochemical and sensory qualities of spiced soy-corn milk

Ajala, Lola[1*], Ologunde, M. O.[2] and Adetuyi, F. O.[3]

[1]Food Science and Technology Department, Rufus Giwa Polytechnic P.M.B 1019, Owo, Ondo State, Nigeria.
[2]Ladoke Akintola University of Technology, Ogbomosho, Oyo State, Nigeria.
[3]Ondo State University of Technology, Okitipupa, Ondo State, Nigeria.

Soy-corn milk type was produced from a blend of soybean milk and corn milk extract at a ratio of 3:1. The soy-corn milk type was spiced with ginger and garlic extract respectively to improve the taste. Total dissolved solid (TDS), total titrable acidity (TTA) specific gravity (SG), apparent colloidal stability, pH and sensory evaluation of the spiced soy-corn milk type were determined. The results show that the specific gravity of all milk types was not significantly (P>0.05) different from each other. The spiced milk types were relatively stable under refrigerated condition for 72 h while corn milk, soy-corn milk was not stable even under refrigeration. There was no significant (P>0.05) difference in the pH of the milk type at preparation and the value ranged between 6.8 and 6.5. The spiced milk types became more acidic after 24 h. Spicing improved the colloidal stability of the soy-corn milk type and its acceptability to the consumer, but has no significant difference (P>0.05) on the other physicochemical properties investigated.

Key words: Corn milk, soybean milk, soy-corn milk, colloidal stability.

INTRODUCTION

Maize, also known as corn (*Zea mays* L) a common staple tropical crop is mainly starch with very low protein and poor amino acid profile. For corn to be efficient in ameliorating malnutrition, it must be supplemented with high grade protein. Soybean (*Glycine max*) can provide a useful complement to cereals and tubers to give a balanced diet that could prevent protein energy malnutrition (PEM). The protein content of soybeans is particularly high though low in the sulphur containing amino acids. Soymilk is a potential substitute for cow's milk, and could be used for solving malnutrition problems in developing countries (Wei et al., 1985). Iwe and Agu (1993) reported the use of natural flavourants to improve soymilk acceptability. Owing to soymilk's immense health benefits, research targeted at improving its acceptability should be undertaken. Omueti and Ashaye (1998) reported that soy-corn milk was more acceptable than soymilk.

Spices have been used all over the world through history and for many centuries, have been an important trading commodity. They are used in food preparation whole, crushed or powdered. They generally have a strong aromatic flavor. Garlic (*Allium sativum* L.) is the second most widely cultivated Allium after onion. Raw garlic no doubt has the ideal flavor and provides desirable textural and water retaining properties when incorporated into food product, bulk of raw garlic is water and lacks flavor and aroma. Garlic is used practically all over the world for flavouring various dishes (Pandey, 2001). The spice ginger is obtained from the underground stems or rhizomes of *Zingiber officinale* (Rosc.), an herbaceous tropical perennial belonging to the family Zingiberaceae. The whole plant is refreshingly aromatic, but it is the underground rhizome, raw or processed, that is valued as spice. Fresh ginger is outstanding for flavouring as it contains the full note of the spice compared to other products from it (Vasala, 2001).

The aim of this work was to produce spiced soy-corn milk by the addition of garlic and ginger and to determine

*Corresponding author. E-mail: lola_ajala2006@yahoo.co.uk.

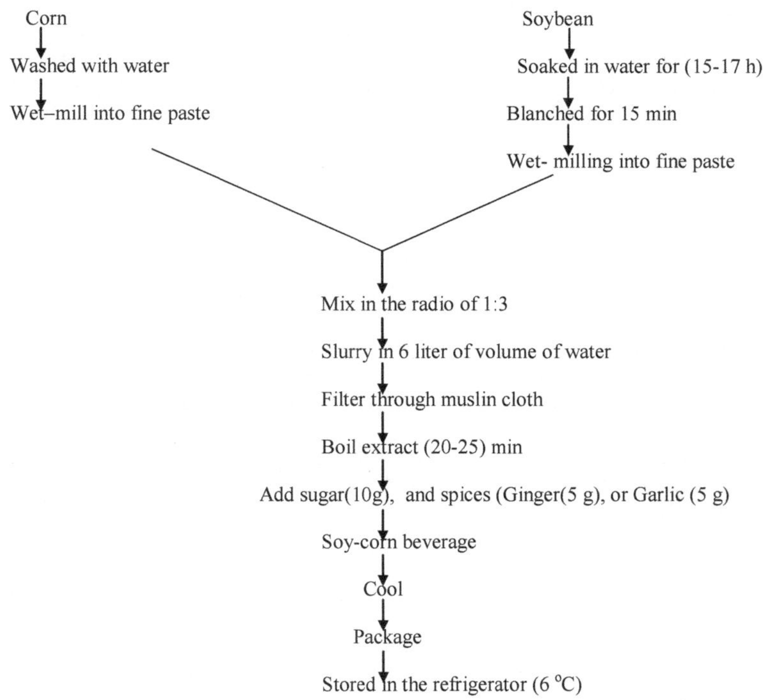

Figure 1. Flow diagram for the production of spiced soy-corn milk.

the physicochemical and sensory quality of the spiced soy-corn milk, aiming to improve the acceptability and hence diversity of the food resources of low-income families, especially in developing countries.

MATERIALS AND METHODS

The soybeans, maize (or corn) and spices were bought from Ikoko market in Owo, Ondo State, Nigeria.

Preparation of spiced milk types extracts.

Freshly harvested maize grains (1 kg) were separated from the corm and cleaned to remove silk hairs and other extraneous materials. The grains were wet milled into fine paste and filtered through muslin cloth to obtain the corn milk.

Healthy and unbroken soybean seed were soaked overnight (16 to 17 h) drained and added directly into boiling water. The soybeans were allowed to boil for 15 min after which they were transferred into cold tap water, dehulled and drained. A mass (1.5 kg) of the seeds were milled. The soymilk slurry was mixed with water and sieved using muslin cloth to get the filtrate which is soymilk. The soymilk was boiled for 20 to 25 min and cooled at room temperature to (30 ± 1°C). The filtrate from corn grains and soybeans cotyledon was measured separately in a ratio of 1:3 (corn to soybean). The mixture was carried out according to the method of Omueti and Ajomale (2005) slightly modified by the addition of spices (Figure 1).

Chemical analysis

Total dissolved solid (TDS), total titrable acidity (TTA) and pH were determined according to the method described by Egan et al. (1981). Apparent colloidal stability was measured by the method of Omueti and Ashaye (1998). Specific gravity determination was done according to the AOAC (1990) method.

Sensory evaluation

The sensory qualities of the soymilk, soy–corn milk, corn milk, soy–corn ginger milk and soy–corn garlic milk namely appearance, taste, flavour, thickness and overall acceptability, were analyzed by 10 panellists familiar with the taste of soymilk. The nine point hedonic scale was used to determine preference of the panellists from 1 to 5 (1 – dislike to 5-extreme likeness).

Statistical analysis

An analysis of variance (ANOVA) was performed on the results for each quality variable to determine the significance of each treatment. Means were compared by the least significance difference (LSD) test; significance was indicated at P>0.05.

RESULTS AND DISCUSSION

The result of the chemical attributes of the various milk types is presented in Table 1. The ginger and garlic spiced soy-corn milk variant had a total solid content of 8.58 and 8.40%, respectively; this value was not significantly different (P>0.05) from the total solid content reported for soymilk and soy-corn milk (9.44 and 9.48%, respectively). These values were comparable to the value reported by Omueti and Ajomole (2005) where a total solid content of soymilk and soy-corn milk were reported as 8.84 and 9.48%, respectively. The total solid content measured for soy-corn milk in this work was higher than that reported by Omueti and Ajomole (2005) and this could be due to higher level of suspended particles in the soy-corn milk types (Nelson et al., 1976).

Table 1. Attributes of various soy-corn milk types.

Milk type	Total solid (%)	Total acidity	Specific gravity
Soymilk	9.44[a]	0.016[b]	1.02[a]
Soy-corn ginger	8.58[a]	1.011[a]	1.02[a]
Soy-corn garlic	8.40[a]	1.015[a]	1.02[a]
Corn milk	1.00[b]	1.014[a]	1.03[a]
Soy-corn milk	9.48[a]	1.015[a]	1.01[a]

Values are average of three determinations. Values with the same letter in the same column are not significantly different (p = 0. 05).

Table 2. Duration of refrigerated storage on apparent colloidal stability (mL) of various soy-corn milk types.

Time (h)	Soymilk	Soy-corn ginger	Soy-corn garlic	Soy-corn milk	Corn milk
0	15.00[a]	15.00[a]	15.00[a]	15.00[a]	15.00[a]
24	14.70[a]	14.90[a]	14.80[a]	12.30[b]	13.00[a]
48	14.50[a]	14.80[a]	14.70[a]	8.59[b]	13.00[a]
72	14.50[a]	14.60[a]	14.60[a]	8.59[b]	12.70[a]

Values are average of three determinations. Values with the same letter in the same column are not significantly different (p = 0. 05).

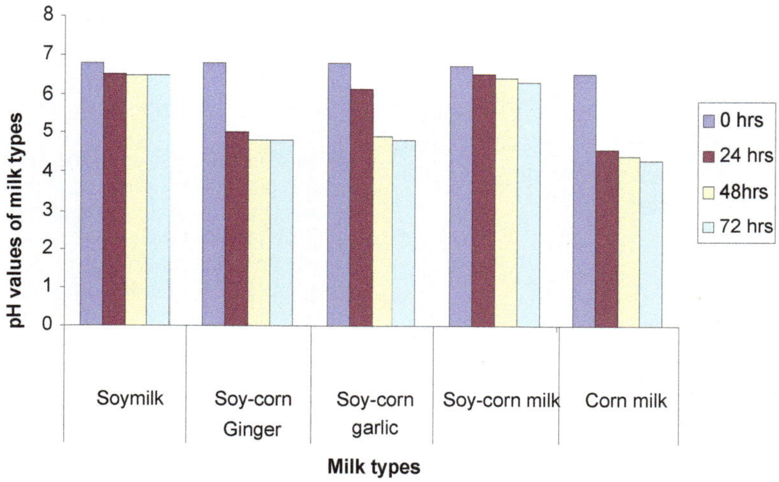

Figure 2. pH of various soy-corn milk type under refrigerated storage.

The total acidity of the ginger and garlic spiced milk types were found to be 1.01 and 1.02, respectively. These values were higher than the values reported for soy-corn milk (0.063 and 0.057 in white and yellow maize, respectively) by Omueti and Ajomole (2005). Shojaei and Yoidoilahi (2008) reported a total acidity of 0.16 and 0.15 for pasteurized and ultra-heated temperature milk respectively. The acidic nature of ginger and garlic could be responsible for the higher total acidity of the spiced soy-corn milk types

The specific gravity of all the milk types were not significantly different (P>0.05). This result could be attributed to uniform dispersion of solutes. The values were comparable to the value reported by Omueti et al. (2000) in nutritional evaluation of home-level prepared soy-corn milk.

The result of the colloidal stability of the milk types under refrigeration is presented in Table 2. The apparent colloidal stability of all milk types decreased when stored under refrigerated condition for 72 h, however, with the exception of soy-corn milk the decrease was not statistically significant (P>0.05) from the value measured at 0 h. this significant decrease in colloidal stability of soy-corn milk within 72 h under refrigerated condition, was also observed by Omueti and Ajomole (2005) where the colloidal separation of the product was reported.

The result of the pH of the samples under refrigeration for 72 h is presented in Figure 2. There was no significant difference (P>0.05) in the pH of the milk type at preparation, while the value ranged between 6.8 and 6.5 which is within the standard (Shojaei and Yodollahi 2008). The spiced

Table 3. Sensory score of various soy-corn milk types.

Milk type	Appearance	Taste	Flavour	Thickness	Overall acceptability
Soymilk	3.4[a]	3.7[a]	3.5[a]	3.1[a]	3.9[a]
Soy-corn ginger	2.9[a]	3.3[a]	3.1[b]	3.2[a]	3.3[b]
Soy-corn garlic	3.0[a]	2.7[a]	2.7[b]	2.9[a]	2.4[c]
Corn-milk	2.2[b]	2.3[b]	2.4[b]	3.2[a]	2.3[c]
Soy-corn milk	3.6[a]	3.3[a]	3.5[a]	2.9[a]	2.5[b]

Values are average of three determinations. Values with the same letter in the same column are not significantly different (p = 0. 05).

The spiced milk types became more acidic even after 24 h. This may be due to the presence of high level of acidic amino acids in corn and the acidic nature of the spices.

Taste panelists gave significantly lower (P>0.05) sensory score to corn milk in appearance, taste, flavour and overall acceptability while soymilk was rated best in all the sensory parameters; soy-corn milk type tasted better than the corn milk types meanwhile addition of ginger improved the sensory quality of the soy-corn milk while taste panelist rated the garlic spiced corn milk very low (Table 3).

Conclusion

Spicing improves the colloidal stability of the soy-corn milk type and its acceptability to the consumer.

REFERENCES

AOAC (1990). Official methods of analysis. 15th Edn Association of Official Analytical Chemists. Arlington, VA.

Egan H, Kirk RS, Sawyer R (1981). Pearson's chemical analysis of foods, Longman Scientific and Technical. 8th Ed.

Iwe MO, Agu IF (1993). Use of natural flavourants to improve soy milk acceptability. Nig. Food J. 11:16- 24

Nelson AI, Steinberg MP, Wei LS (1976). Illinois process for preparation of soymilk. J. Food Sci. 41:57-61.

Omueti O, Ashaye AO (1998). Home –Level preparation, nutrient content and acceptability of soy-corn milk, in Chainuvati C, Sarobol NP (Eds). Proceedings, World Soybean Research Conference V (Supplement to 1997). pp. 579-583.

Omueti O, Oguntona EB, Jaiyeola O, Ashaye OA (2000). Nutritional evaluation of home-level prepared soy-corn milk: a protein beverage. Nutr. Food Sci. 30:128-132

Omueti O, Ajomole K (2005). Chemical and sensory attributes of soy-corn milk types. Afr. J. Biotechnol. 4(6):847-851

Pandey UB (2001). Garlic in: Handbook of herbs and spices edited by Peter K.V. Woodhead Publishing Ltd. pp. 196-210

Shojaei ZA, Yadollahi A (2008). Physicochemical and microbiological quality of raw- pasteurized and UHT milks in shops. Asian J. Sci. Res. 1:532-538.

Vasala PA (2001). Ginger In: Handbook of herbs and spices edited by Peter KV. Woodhead Publishing Ltd. pp. 211- 222

Wei LS, Urbanski GE, Steinberg MP, Nelson AJ (1985). Feasibility of shelf-stable soybean beverage, In Proceeding of World Soybean Research Conference III. Shibles R (Ed.) Westview Press Boulder, Colorado, pp. 191-198.

Characterization of a newly isolated *Rubrivivax benzoatilyticus* PS-5 with self-flocculation property and optimization pathway for 5-aminolevulinic acid production

Supalak Sattayasamitsathit[1] and Poonsuk Prasertsan[1,2*]

[1]Department of Industrial Biotechnology, Faculty of Agro-Industry, Prince of Songkla University, Hatyai 90112 Thailand.
[2]Palm Oil Product and Technology Research Center, Faculty of Agro-Industry, Prince of Songkla University, Hatyai 90112 Thailand.

Six strains of phototrophic purple non-sulfur bacteria were isolated from paddy soil, lake water and pond sediment samples. They showed the ability to produce 5-aminolevulinic acid under anaerobic-light (3,000 lux) condition. A selected strain, designated as PS-5, produced the highest 5-aminolevulinic acid (ALA) concentration (45.10 μM) and productivity (0.94 μM h^{-1}) in glutamate-malate (GM) medium as well as possessed self-flocculating ability. Based on physiological and 16S rRNA sequence, the isolate PS-5 was identified to be *Rubrivivax benzoatilyticus*. The optimum concentration of C_4 pathway precursors was 7.5 mM glycine and 10 mM succinate with supplementation of 10 mM levulinic acid. This resulted in 4.2 fold increase of 5-aminolevulinic acid production to 190.74 μM and 2.7-fold increase of its productivity to 2.57 μM h^{-1} compared to the control. In addition, the optimum concentration of C_5 pathway precursor was 50 mM glutamate with the maximum 5-aminolevulinic acid production of 84.12 μM and productivity of 1.56 μM h^{-1}. Therefore, *R. benzoatilyticus* PS-5 could produce 5-aminolevulinic acid using both C_4 and C_5 pathway precursors and preferred C_4 pathway over C_5 pathway as it produced more than 2.3 folds higher 5-aminolevulinic acid concentration.

Key words: 5-Aminolevulinic acid, photosynthetic bacteria, self-flocculation, optimization, *Rubrivivax benzoatilyticus*.

INTRODUCTION

Photosynthetic bacteria (PSB) are gram-negative prokaryotes that convert light energy into chemical energy by the process of anoxygenic photosynthesis. They contain photosynthetic pigments, bacteriochlorophylls and carotenoids, and can grow autotrophically with CO_2 as the sole carbon source (Pfennig, 1969). PSB has been used for production of 5-aminolevulinic acid commercially with various applications, such as function as a natural photodynamic compound effective as a biodegradable herbicide, growth-promoting factor and a stress tolerance enhancer for plants (Tanaka et al., 1992). Applications of 5-aminolevulinic acid in medical and pharmaceutical field include photodynamic medicine for treating superficial skin cancers, actinic keratoses (Dijkstra et al., 2001), psoriasis (Zeitouni et al., 2003), cutaneous T-cells (Leman et al., 2002), basal cell carcinoma (De Rosa et al., 2000), squamous cell carcinoma (Lee et al., 2010) induce fluorescence diagnosis of pleural malignant tumor (Ali et al., 2011) and the suppression of the inflammatory

*Corresponding author. E-mail: poonsuk918@yahoo.com.

Abbreviations: ALA, 5-Aminolevulinic acid; **GM,** glutamate-malate.

response to coronary and iliac injuries (Sasaki et al., 2005). As a second-generation photodynamic medicine, its outstanding advantages include its minimal side effect, strong curative effect, high penetrability, and low price (Nishikawa et al., 1999). 5-Aminolevulinic acid is nontoxic to humans and animals and it is readily biodegradable, thus causing no adverse effects on the environment (Rebeiz et al., 1988).

Two major pathways for the biosynthesis of 5-aminolevulinic acid have been described (Sasikala et al., 1994). In the Shemin pathway (C_4 pathway), 5-aminolevulinic acid is formed by the condensation of glycine and succinyl-CoA that is catalyzed by aminolevulinic acid synthase (ALAS). This pathway is found in animal cells, yeasts, fungi, and certain bacteria. In the C_5 pathway, 5-aminolevulinic acid is formed in three steps from glutamate (Beale and Weinstein, 1990) and found in plant chloroplasts, cyanobacteria, anaerobic archaebacteria, *Escherichia coli*, and *Salmonella typhimurium* (Beale, 1999).

In general, the homogeneous characteristic of photosynthetic bacteria in liquid medium make harvesting of biomass relatively high cost. To reduce the production cost, many alternative methods of bacterial cells recovery are proposed such as addition of chemical or biological flocculating agents (Kurane and Matsuyama, 1994) and using flocculating microorganism (Chernyanovskaya, 1990).

Using flocculating strain of photosynthetic bacteria was expected to give three advantages. Firstly, cell mass can be easily harvested by decantation within a few minutes. Secondly, the cells flocs can be used directly in the repeated culture.

Finally, the living flocculating cells can be used for cleaning up the polluted water. Therefore, this work aimed to isolate the flocculating photosynthetic bacteria and optimize the production of 5-aminolevulinic acid from the selected strain.

MATERIALS AND METHODS

Medium

The *Rhodospirillum* medium (RM) used for enrichment and isolation contained (gl^{-1}): 4, D,L malate; 0.1, yeast extract; 0.12, $MgSO_4.7H_2O$; 0.076, $CaCl_2.2H_2O$; 0.02, Na_2EDTA; 0.59, KH_2PO_4; 0.39, K_2HPO_4; 0.0065, $FeSO_4.7H_2O$; 0.0003, $MnCl_2.4H_2O$; 0.01, $ZnSO_4.7H_2O$; 0.02, $CoSO_4.7H_2O$; 0.03, H_3BO_3; 0.003, Na $MoO_4.2H_2O$; 0.001, $CuSO_4.5H_2O$ (Madukasi et al., 2011).

Modified glutamate-malate medium (GM) without NaCl was used to maintain the bacterial cells and to prepare a starter culture. The medium contained (g l^{-1}) 2.7, malate; 3.7 g l^{-1}, monosodium glutamate; 0.8, $(NH_4)_2SO_4 \cdot 7H_2O$; 0.5, yeast extract; 0.5, KH_2PO_4; 0.5, K_2HPO_4; 0.2, $MgSO_4 \cdot 7H_2O$; 1.0×10^{-3}, thiamine-HCl; 1.0×10^{-3}, of nicotinic acid and 1.0×10^{-5}, biotin. The initial pH of this medium was adjusted to 7.0 with 5 M NaOH (Tangprasittipap et al., 2007).

Basal medium was used for optimization studies on 5-aminolevulinic acid production. It had the same composition as the modified GM medium but without carbon source.

Isolation, selection and identification of photosynthetic bacteria

Samples were collected from paddy soil, lake water and pond sediment mud from Songkhla Province, Thailand. The samples (2 g soil or 2 ml) were added into 40 ml *Rhodospirillum* medium (RM) and mixed well. Sterile liquid paraffin (5 ml) was added on top of the medium to create anaerobic condition. Cultivation was conducted at room temperature (32±2°C) under illumination (3,000 lux) with a 100 W incandescent lamp, until the medium turned red. The samples were streaked on modified GM agar plates and incubated under anaerobic-light (3,000 lux) condition. The reddish colonies were purified by subculture several times until pure cultures were obtained. They were kept on the modified GM agar at 4°C until used.

Starter cultures were prepared by transferring a loopfull of cells from the modified GM slant to 100 ml modified GM medium (pH 7.0) and incubated under anaerobic-light (3,000 lux) condition at room temperature (32±2°C) for 24 h. They were harvested by centrifugation (8,000 x *g* for 15 min), and resuspended in the basal medium to a concentration of 0.2 g wet cell weight per 10 ml medium. The cell suspension was transferred to 120 ml serum bottle, each containing 90 ml basal medium and cultivated for 48 h under the same condition as above. Samples were taken for determination of pH and 5-aminolevulinic acid (Mauzeall and Granick, 1956).

Selection was conducted by cultivating all isolates in the modified GM medium plus 10 mM levulinic acid at 24 h (Sasaki et al., 1991) under anaerobic-light (3,000 lux) conditions at room temperature. Preliminary results indicate that without addition of levulinic acid, the isolates could not produce 5-aminolevulinic acid. The highest 5-aminolevulinic acid producing strain was selected and identified according to Bergey's Manual of Systematic Bacteriology (Staley et al., 1989) and the classification was confirmed using 16S rRNA sequencing (Achenbach et al., 2001).

Morphological and physiological characterization of the selected strain was conducted on modified GM medium. Colony pigmentation, margin, and elevation of the colony were recorded while size and shape were determined using Scanning Electron Microscopy (SEM) (JEOL JSM-5800LV) (Harley and Prescott, 2007). Besides, SEM could also illustrate biofilm surface structures at high-resolution by the acquisition of three dimensional images of the surface, revealing details about 1 to 5 nm in size. Cell pigment scans and absorption spectra of the living cells suspension were performed using cell pellets resuspended in 60% sucrose (Truper and Pfennig, 1981). Carbon utilization test was carried out as described by Kantachote et al. (2005).

Identification of the selected strain using molecular technique was also conducted. The primer set 27F-1492R was used in PCR amplification of the 16S rRNA gene fragment of the selected strain. Sequence similarity searches were conducted using the National Center for Biotechnology Information BLAST network service (nucleotide blast). Similar 16S rRNA gene sequences were downloaded from GenBank and manually checked for ambiguous sites using bioedit 7.0.1 software. Alignments were then performed against the 16S rRNA gene sequence of the selected strain, where the pair-wise deletion option for gaps was employed. Phylogenetic tree was constructed from multiple sequence alignment, with related rRNA by the Neighbor Joining analysis with 1000 bootstrap replicates (MEGA version 5.1; Arizona State University, USA), UPGMA, Minimum Evolution or Maximum Parsimony methods (Ramana et al., 2006).

Effect of precursors on 5-aminolevulinic acid (ALA) production from the selected PSB

Fermentation was carried out in a basal medium (90 ml) containing (g l^{-1}) 0.5, yeast extract; 0.8, $(NH_4)_2SO_4{\cdot}7H_2O$; 0.5, KH_2PO_4; 0.5, K_2HPO_4; 0.2, $MgSO_4{\cdot}7H_2O$; $1.0{\times}10^{-3}$, thiamine-HCl, $1.0{\times}10^{-3}$, nicotinic acid and $1.0{\times}10^{-5}$, biotin, at pH 7.0, and incubated under anaerobic-light (3,000 lux) condition at room temperature (32±2°C). The effect of C_4 pathway precursors (glycine, succinate) and C_5 pathway precursor (glutamate or malic acid) at various concentrations plus 10 mM levulinic acid at 24 h (Sasaki et al., 1987) were investigated by supplementing each precursor into the medium.

Effect of inhibitor on ALA production

Various levulinic acid concentrations (2.5-12.5 mM) were added into the medium at 24 h to study their effect on 5-aminolevulinic acid production. The fermentation was conducted in a basal medium as described above with the supplementation of either C_4 or C_5 pathway precursors.

Analytical methods

Samples taken during the fermentation runs were centrifuged (at 8,000×*g* for 15 min) and the supernatant was used for determination of 5-aminolevulinic acid by a colorimetric method modified from the method of Mauzerall and Granick (1956). Specifically, 0.5 ml of either sample or standard was mixed with 1 ml of 1M sodium acetate buffer (pH 4.7), 25 µl of acetylacetone (2,4-pentanedione) was added, and then the mixtures were boiled in a water bath for 15 min. After cooling, 1.75 ml of freshly prepared modified Ehrlich's reagent was added, and 15 min later, the absorbance at A_{553} nm was measured. The acetic acid, propionic acid, butyric acid, succinic acid and levulinic acid concentrations were determined using High Performance Liquid Chromatography (HPLC) (Agilent 1200) equipped with a Bio-Rad HPX-87H (300 mm × 7.8 mm) column (Hercules, CA, USA) and a refractive index detector. The samples were diluted with deionized water, filtered through 0.22 µm, 13 mm Nylon membrane filter (Sartorius, German) and then injected in the chromatograph under the conditions: column temperature at 65°C, 5 mM sulfuric acid as mobile phase at a flow rate of 0.8 ml min^{-1}, and an injection volume of 20 µL (Kang et al., 2011). The concentration of these compounds was calculated using calibration curves obtained from the standard solutions.

RESULTS AND DISCUSSION

Isolation, selection and identification of photosynthetic bacteria

Thirty (30) samples collected from paddy soil, lake, and pond sediment mud were enriched in the RM medium and incubated under anaerobic-light (3,000 lux) condition for growth of photosynthetic bacteria. The RM medium contained high malate that could enhance the growth of photosynthetic bacteria as malate is a photosynthetic reductant that plays a critical role in the noncyclic photophosphorylation of *Rhodospirillum rubrum* (Stenn, 1968). Selective enrichment in liquid media is still an appropriate method for isolation and incubation time of five or more days are required for intense color of purple non-sulfur bacteria. Six colonies, different in forms and colors on the agar plates, were purified by sub-cultured. The six selected isolates producing red, pink, reddish brown, orange brown and pale pink colonies were designated as PS-1, PS-2, PS-3, PS-4, PS-5 and PS-6, respectively. The occurrence of different colors was dependent on the species of photosynthetic bacteria that contain specific photopigments in their cells particularly carotenoid types (Imhoff, 1988). All six isolates were tested for 5-aminolevulinic acid production in modified GM medium with the addition of 10 mM levulinic acid at 24 h. Levulinic acid is an analogue of 5-aminolevulinic acid and can be used to inhibit ALA dehydratase activity in the biosynthesis of tetrapyrrols by competitive inhibition. Since levulinic acid could retard the growth of photosynthetic bacteria, it should be added in the middle log phase to increase the extracellular 5-aminolevulinic acid concentration (Sasaki et al., 1987). It had been utilized to enhance 5-aminolevulinic acid production effectively in various organisms such as *Rhodobacter sphaeroides* (Sasaki et al., 1991), *Rhodovulum* sp. (Noparatnaraporn et al., 2000), *Chlorella regularis* (Ano et al., 2000), *Methanobacterium thermoautotrophicum* (Lin et al., 1989), *Rhodopseudomonas palustris* (Saikeur et al., 2009). Among the six isolates, the strain PS-5 showed the highest 5-aminolevulinic acid production (45.10 µM) at 48 h cultivation with the productivity of 0.94 µM h^{-1} (Figure 1). This maximum value was higher than that from *Rhodopseudomonas palustris* KG31 (38.24 µM) after 48 h cultivation in GM medium supplemented with 5 mM levulinic acid at 24 h cultivation under anaerobic-light condition (Saikeur et al., 2009).

The special property of the isolate PS-5 was self-flocculating ability (Figure 2A) that would benefit the harvesting process using decantation method instead of commonly used energy-intensive centrifugation method. The characteristic of flocculant growth has two forms, compact floc (tightly attachment of cells in flocs) and friable flocs (loosely cells attachment). The SEM of the strain PS-5 illustrated that flocculated cells polarly attached to each other and embedded in gum-like substance, which found covering on the surface of the flocs (Figure 2B). Some flocculating photosynthetic bacteria was reported to produce extracellular polymeric substances (EPS) on the surface of cells (Noparatnaraporn et al., 2000). The arrangement of numerous cells in flocs or pellets of photosynthetic bacteria is relatively uniform in each pattern and seem to be the clusters or layers. In contrast, the cells arranged in multispecies biofilm formed highly complex distribution of microorganisms and structure (Beer et al., 1993). Several flocculating mechanisms were proposed such as bridge formation between the EPS and cells (Urbain et al., 1993), production of extracellular nucleic acids and form structured communities of cells called flocs (as in the genus Rhodovulum) (Ando et al., 2006; Watanabe et al., 1998) and production of RNAs (contain mainly fully mature-sized tRNAs and fragments of 16S and 23S rRNAs) (as in the genus Rhodovulum) (Ando et al., 2006) and extracellular soluble DNA (*Rhodovulum*

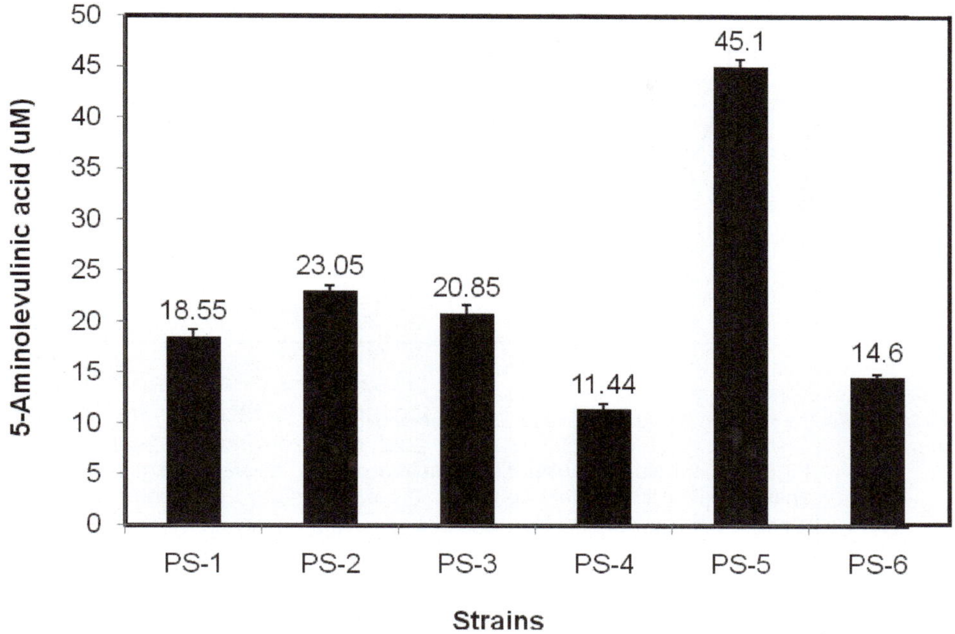

Figure 1. 5-Aminolevulinic acid production from cultivation of the isolated strains in modified glutamate-malate (GM) medium with addition of 10 mM levulinic acid at 48 h under anaerobic-light (3,000 lux) condition.

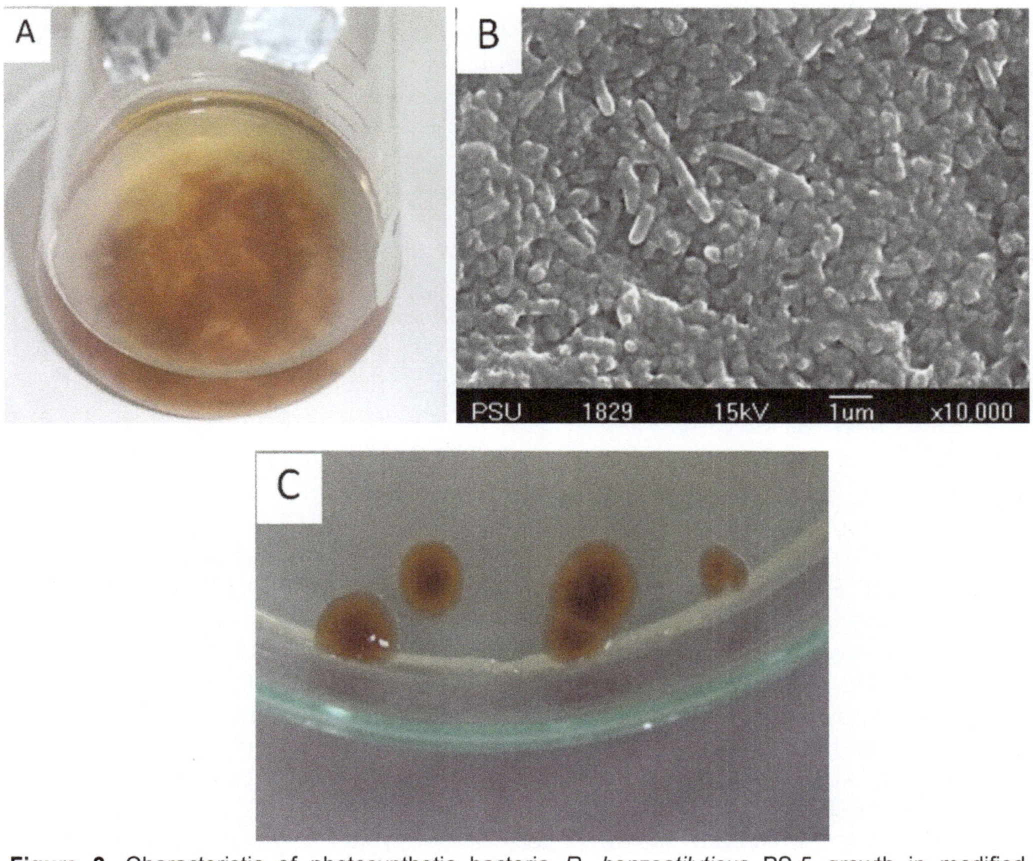

Figure 2. Characteristic of photosynthetic bacteria *R. benzoatilyticus* PS-5 growth in modified glutamate-malate (GM) medium (A), Scanning electron micrographs of biofilms formed (B) magnification x 10,000 and colony on GM agar (C).

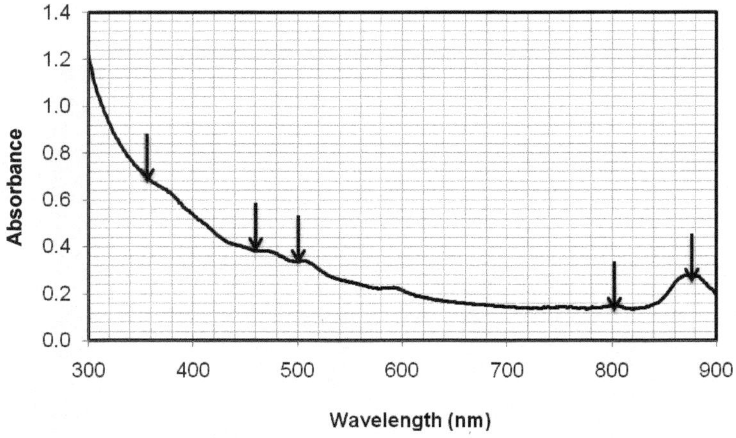

Figure 3. Whole-cell absorption spectrum of the isolate PS-5 under anaerobic-light (3,000 lux) condition. Wavelengths (nm) of absorption maxima are shown at the tops of peaks.

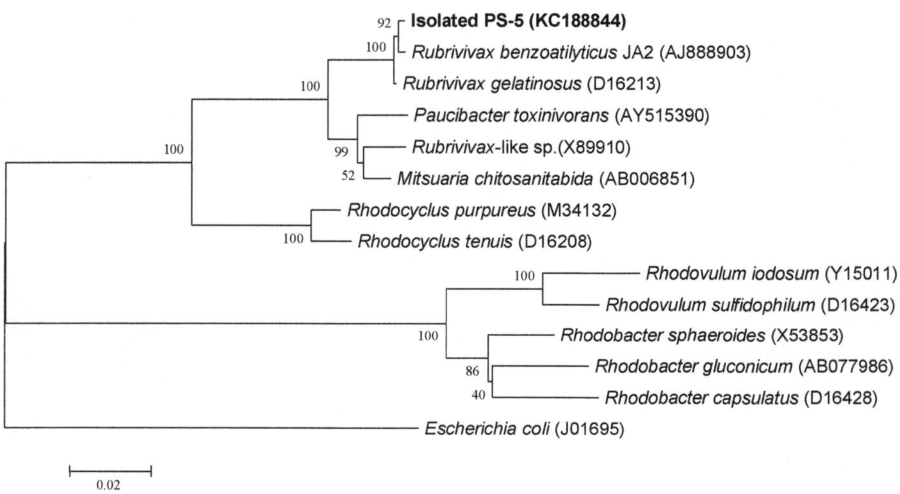

Figure 4. Phylogenetic relationship between the isolate PS-5 and other phototrophic purple non-sulfur bacteria, constructed by the Bootstrap percentages that are written on internal branches, neighbor-joining method using the MEGA 5.1 program. Accession numbers for the sequences are shown in parentheses following each strain designation.

sulfidophilum) (Suzuki et al., 2009). The isolate PS-5 was chosen for 5-aminolevulinic acid production because it gave the highest 5-aminolevulinic acid and self-flocculation.

Morphology of the isolate PS-5 was rod-shaped, 0.7-1.0 μm wide and 1.5 μm long, and cells multiply by binary fission. Colony appearance on agar plates was round, convex and smooth, with dark orange-brown color (Figure 2C). After growing under anaerobic photoheterotrophic conditions, the living cell suspensions became orange and the absorption spectra showed maxima at 370, 475, 520, 800 and 875 (Figure 3). The main peaks at 800 and 875 nm, which are closely related to bacteriochlorophyll a and b as well as carotenoid, are characteristics of photosynthetic bacteria (Madigan et al., 2000). The isolate PS-5 were tested to determine the range of suitable electron donors and carbon sources for photosynthetic growth (Table 1). The strain grew photoorganotropically on a wide variety of organic compounds, including short-chain fatty acids (lactic, formic, acetic, propionic, butyric acid), sugars (glucose, fructose, maltose) and amino acids (glutamic, aspartic) and glycerol. Moreover, this strain showed excellent growth on starch, indicating its production of enzyme amylase. The ability of the isolate PS-5 to assimilate lactic acid but not tartaric acid was opposite to that of *Rubrivivax benzoatilyticus* JA2 previously reported (Ramana et al., 2007).

A typical phylogenetic tree for the relationship of 16S rRNA of the isolate PS-5 to the 16S rRNA of the other phototrophic bacterial strains is shown in Figure 4. It showed that the isolate PS-5 was most closely related to

Table 1. Utilization of electron donors and carbon sources used by the isolate PS-5 and *Rubrivivax benzoatilyticus* JA2[a] .

Donor/source	PS-5	*R. benzoatilyticus* JA2
Organic acids		
Formic acid	+[b]	+
Acetic acid	+	ND
Propionic acid	+	+
Valeric acid	-	-
Butyric acid	+	ND
Pyruvic acid	++	+
Citric acid	-	-
Lactic acid	+	-
Malic acid	+	+
Succinic acid	+	ND
Tartaric acid	-	+
Fumaric acid	+	+
Benzoic acid	+	+
Sugars and alcohols		
Glucose	++	+
Sorbital	-	ND
Manital	-	-
Sucrose	-	ND
Fructose	++	+
Maltose	++	+
Glycerol	+	+
Ethanol	-	ND
Methanol	-	-
Amino acids		
Glutamic acid	+	+
Aspartic acid	++	ND
Glutamine	-	ND
Glycine	-	ND
Carbohydrate		
Starch	+++	ND
Chemolithoautotrophic substrates		
Sulfide	-	-
Sulfur	-	-
Thiosulfate	-	-

[a]Data for *R. benzoatilyticus* JA2 were obtained from Ramana et al. (2007). All cultures were incubated at room temperature in the presence of light (3,000 lux) for 48 h. [b]+++, very good growth; ++, good growth; + poor growth; -, no growth; ND, not determined; +++, Excellent growth (A_{660nm} > 1.0); ++, good growth (A_{660nm} = 0.5-1.0); +, moderate growth (A_{660nm} = 0.2-0.5); (+), utilized; ±, some strains only; -, no growth; nt, not determined.

Rubrivivax benzoatilyticus (GenBank accession No. AJ888903.1) that was isolated from the rhizosphere of a flooded paddy field near Eluru, Andhra Pradesh, India (Ramana et al., 2007). The accession numbers of the gene sequences submitted to GenBank was KC188844.

Effect of C_4 pathway precursor

5-Aminolevulinic acid is formed by the C_4 pathway, called Shemin pathway, in some bacteria including photosynthetic bacteria such as *R. sphaeroides,*

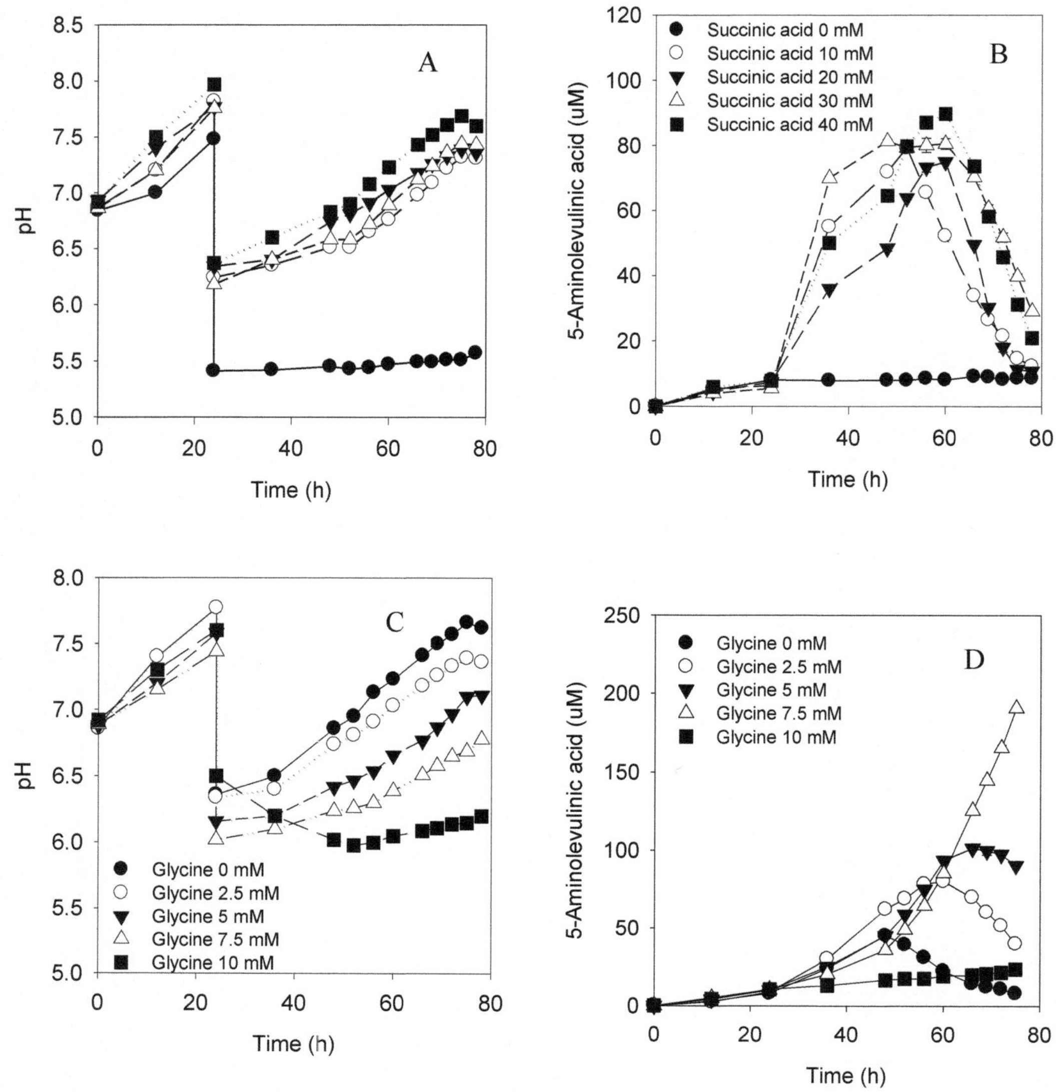

Figure 5. Time course of pH change (A, C) and ALA production (B, D) at various succinic acid concentration with 2.5 mM glycine and various glycine concentrarion with 10 mM succinic acid and adding 10 mM LA at 24h during cultivation of *R. benzoatilyticus* PS-5 under anaerobic-light (3,000 lux) condition at room temperature (32±°C).

Rhodopseudomonas sp.. Biosynthesis of 5-aminolevulinic acid proceeds from C_4 pathway precursors, glycine and succinyl-CoA through a combined action of the host enzymes, succinyl-CoA synthase and ALA synthase (Sasaki et al., 2002). In addition, substrate inhibition occurred at high initial concentrations of glycine and succinate and could not enhance the production of 5-aminolevulinic acid (Fu et al., 2007). Therefore, the optimum concentration of each precursor in the basal medium is anticipated to be necessary for effective 5-aminolevulinic acid production. In this study, succinic acid was used instead of the expensive succinyl-CoA to reduce the production cost and being another carbon source for cell growth as an indirect precursor. The effect of succinic acid concentration on growth and 5-aminolevulinic acid production from *Rubrivivax benzoatilyticus,* PS-5 was tested in the range of 0 to 10 mM (0 to 1.2 gl^{-1}) and fix concentration of glycine at 2.5 mM (0.19 gl^{-1}). Due to self-flocculating property, cell growth was detected indirectly via pH changes during cultivation. The pH increased from 6.8 to 7.3 to 7.8 during growth in the first 24 h cultivation. Addition of levulinic acid resulted in a rapidly drop of pH to 5.4 to 6.3 (Figure 5A), then increased linearly to 7.3 to 7.6 along with a continuous decrease of levulinic acid concentration during 78 h cultivation. It was interesting to find that this strain could not use glycine without succinate, therefore, there was no growth (no pH change) and no consumption of levulinic acid concentration. On the other hand, succinate of 2.5 to 7.5 mM was

Table 2. Effect of precursors and levulinic acid concentration.

Parameter	Final pH	Maximum ALA (μM)	Succinic acid consumption (%)	Levulinic acid consumption (%)
C4 pathway				
Succinic acid (mM) + 2.5 mM glycine				
0	5.57	9.26 ± 0.07	0	5.34
2.5	7.31	74.95 ± 1.04	100	100
5	7.43	81.32 ± 0.49	100	73.39
7.5	7.6	87.06 ± 1.11	100	76.86
10	7.3	90.59 ± 0.69	78	66.95
Glycine (mM) + 10 mM succinic acid				
0	7.62	39.26 ± 1.04	81.24	80
2.5	7.36	80.49 ± 1.66	75.38	68.8
5	7.11	101.76 ± 1.11	62.7	47.97
7.5	6.78	190.74 ± 0.90	48.8	7.32
10	6.2	25.49 ± 0.97	26.56	2.17
C5 pathway				
Glutamate (mM)				
10	7.17	58.73 ± 1.04	-	96.92
20	7.32	71.08 ± 1.66	-	95.65
30	7.44	80.88 ± 1.46	-	76.98
40	7.4	84.12 ± 1.39	-	69.37
50	7.39	85.29 ± 0.21	-	51.22
Malic acid (mM)				
10	7.36	35.29 ± 1.53	-	87.5
20	7.77	22.06 ± 0.69	-	89.51
30	7.89	19.12 ± 0.55	-	93.55
40	7.98	17.35 ± 0.35	-	90.76
50	7.85	13.53 ± 0.21	-	91.8
Inhibitor ALA dehydratase				
Levulinic acid (mM)				
2.5	7.63	57.74 ± 1.73	87.7	90.2
5	7.55	96.18 ± 1.25	78.87	73.91
7.5	7.33	120.00 ± 1.39	73.98	61.27
10	7	186.50 ± 1.11	34.15	25.52
12.5	5.35	73.82 ± 0.83	27.64	2.27

completely consumed at 70 h cultivation whereas 10 mM succinate was consumed about 79.2% (Table 2). 5-aminolevulinic acid production increased ($P<0.05$) as succinic acid concentration increased and it was secreted after addition of levulinic acid at 24 h (Figure 5B). 5-aminolevulinic acid production was highest ($P<0.05$) when succinic acid was supplemented at 10 mM. The maximum 5-aminolevulinic acid concentration and productivity were 89.8 μM and 1.5 μM h^{-1}, respectively, which was 9.7 times higher than that without succinate. This indicates that appropriate succinate concentration could enhance both cell growth and 5-aminolevulinic acid production. This agreed with the report that succinate could serve as another precursor for 5-aminolevulinic acid production, as well as a carbon source for cell growth (Qin et al., 2006).

Glycine is another important precursor for 5-aminolevulinic acid production (Choi et al., 1999; Sasaki et al., 2002), therefore, its concentration in the range of 2.5 to 10 mM plus optimum concentration of succinic acid (10 mM) on growth and 5-aminolevulinic acid production was investigated. The pH increased from 6.8 to 7.5 to 7.8 at 24 h and dropped to 6.0 to 6.5 when levulinic acid was added, then increased to 7.3 to 7.6 during 78 h cultivation at low glycine concentration (2.5 to 5.0 mM) (Figure 5C). However, in the presence of high glycine concentration (10 mM) the pH was dropped to 5.98 to 6.2 after levulinic acid addition, and slightly changes thereafter. In addition, only 3.84% of levulinic acid and 26.56% succinate were consumed (Table 2). Therefore, the substrate inhibition was occurred.

Glycine concentration had an effect (P>0.05) on 5-aminolevulinic acid production. The optimum glycine concentration was found to be 7.5 mM, giving the maximum 5-aminolevulinic acid production of 190.74 μM and productivity of 2.54 μM h^{-1} (Figure 5D) which was 2.4 and 1.9 folds higher than those at 2.5 and 5.0 mM, respectively. The negative effect of high concentrations of glycine (10 mM) on 5-aminolevulinic acid production (25.49 μM) was caused by its inhibitory effect on cell growth from ammonia generated during glycine metabolism (Sasaki et al., 2002). The results agreed with that occurred to *Rhodopseudomonas palustris* KG31 (Saikeur et al., 2009). Moreover, high amount of glycine in the medium can inhibit porphyrin formation (Neuberger, 1961). Optimum concentrations of glycine and succinic acid added to the culture medium were different from those in other experiment, that is, 30 mM glycine and 90 mM succinate in a LB complex medium for *Escherichia coli* transformed with ALA synthhase gene of *Bradyrhizobium japonicum* (Choi et al., 1999). Besides, the optimum glycine concentration of *Rubrivivax benzoatilyticus* PS-5 (7.5 mM) was lower than that of *Rhodobacter sphaeroides* CR-606 (60 mM) (Nishikawa et al., 1999). Therefore, it depends on bacterial strain, culture condition, and medium used, which may have an effect on the ALA synthase (ALAS) activity levels (Chung et al., 2005).

Effect of C_5 pathway precursor

Glutamate is the precursor of 5-aminolevulinic acid biosynthesis in higher plants and most green algae via the light-dependent C_5 pathway in the plastid, called Beale pathway (Reinbothe and Reinbothe, 1996). Phytoflagellate *Euglena gracillis* (Weinstein and Beale, 1983) and the green alga *Scenedesmus obliquus* (Dreschler-Theilmann et al., 1993) have both ALA-forming pathways. Effect of glutamate on growth (pH change) and 5-aminolevulinic acid production from *Rubrivivax benzoatilyticus* PS-5 was investigated. The change of pH was similar to those of using succinate as a precursor. The pH increased from neutral to alkaline (6.92 to 7.93) at 24 h cultivation and decreased rapidly to acidic pH (pH 5.6) after adding 10 mM levulinic acid, then increased to 7.3 at the end of cultivation (102 h) (Figure 6A). This indicated that glutamate could enhance cell growth.

The optimum glutamate concentration for 5-aminolevulinic acid production was found to be 50 mM, giving the highest 5-aminolevulinic acid production of 84.12 μM at 54 h with the productivity of 1.56 μM h^{-1} (Figure 6B). This may be due to the degradation of 5-aminolevulinic acid into pyrazine compounds and tetrapyrroles (Bunke et al., 2000) or generation of ammonia from assimilation of glutamate that caused the reduction of ALA synthase activity or induced ALA dehydratase activity. Moreover, ALA in aqueous solution was found to be unstable since it is dependent on pH and other factors (Gadmar et al., 2002) such as temperature, and ALA concentration.

It was previously reported that photosynthetic bacteria could not produce 5-aminolevulinic acid via C_5 pathway (glutamate) (Sasikala et al., 1994). However, this work clearly indicated that *R. benzoatilyticus* PS-5 could produce 5-aminolevulinic acid from both C_4 and C_5 pathways precursors in which the product from C_4 pathway was 2-fold higher than that from C_5 pathway. Therefore, it is a special property of *R. benzoatilyticus* PS-5 that it may have the mechanism for changing glutamate to 5-aminolevulinic acid *via* transfer RNA-dependent five-carbon route and formed by ALA synthase-catalyzed condensation of succinyl coenzyme A and glycine. The biosynthetic source of succinyl coenzyme A of *R. benzoatilyticus* PS-5 may contain α-ketoglutarate dehydrogenase and use succinyl-coenzyme A as a tricarboxylic acid cycle intermediate (Mayer and Beale, 1991).

Malic acid is a precursor of 5-aminolevulinic acid biosynthesis via C_5 pathway (malate → fumarate → succinate + CO_2 → oxoglutarate → glutarate) (Sasikala et al., 1994). Effect of malic acid concentration (10 to 50 mM) with the addition of 10 mM levulinic acid at 24 h on 5-aminolevulinic acid production was studied. The culture pH increased from neutral to alkaline pH (from 7.0 to 8.42) within 24 h cultivation and decreased rapidly to acidic pH (pH 6.5) after adding 10 mM levulinic acid then increased from 6.92 to about 8.3 at the end of cultivation (102 h) (Figure 6C). Malic acid concentration had no effect on 5-aminolevulinic acid production (Figure 6D) as the values were low (P<0.05) at all malic acid concentrations tested. The 5-aminolevulinic acid production reached the maximum 5-aminolevulinic acid of 35.29 μM obtained at the minimum concentration of malic acid (10 mM) and decreased thereafter (P<0.05). This might be due to the increase of pH resulted from the assimilation of malic acid (Gadmar et al., 2002). Levulinic

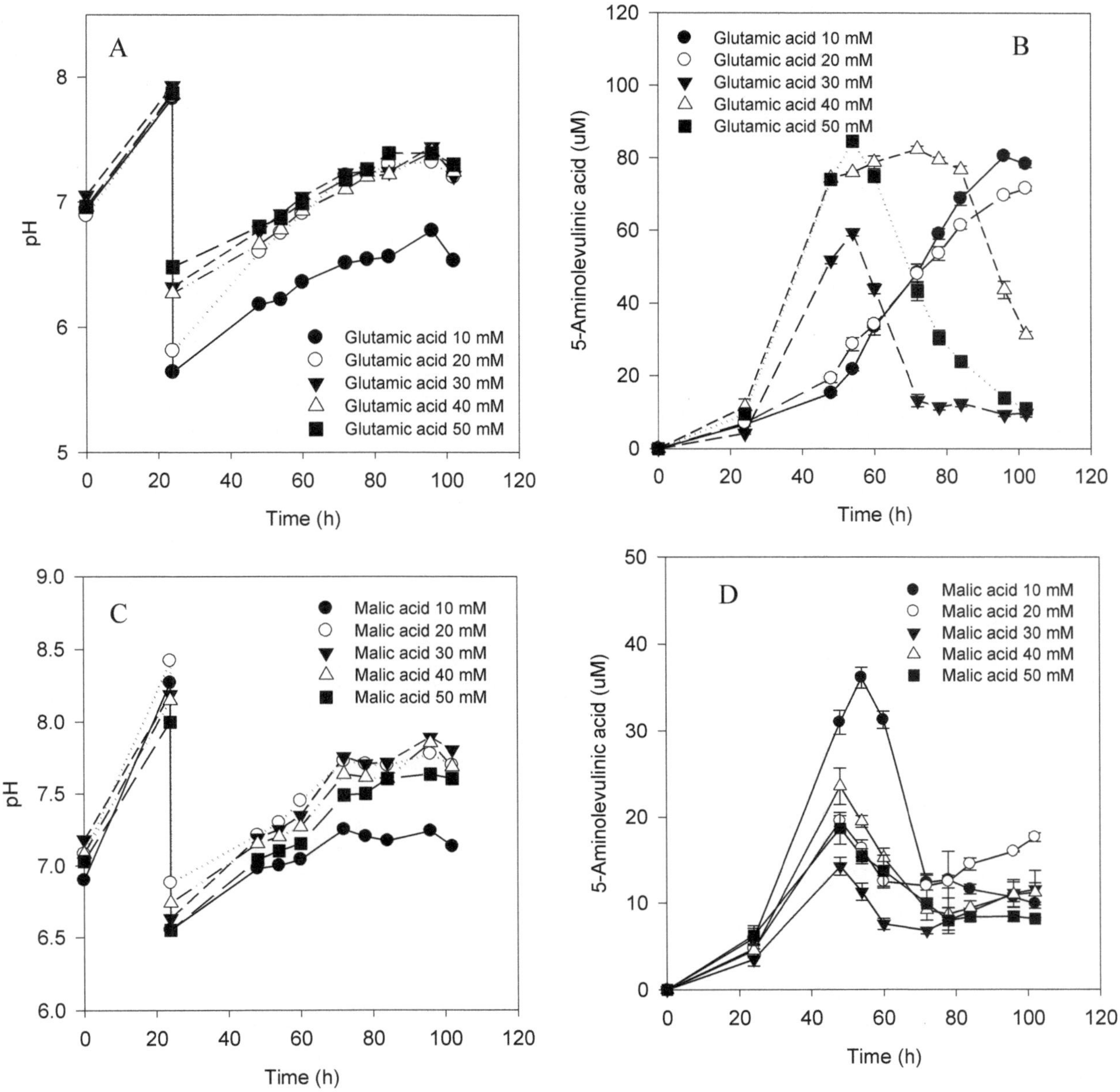

Figure 6. Time course of pH change (A, C) and ALA production (B,D) at various glutamic acid concentration and malic acid concentration and adding 10 mM LA at 24h during cultivation of *Rubrivivax benzoatilyticus* PS-5 under anaerobic-light (3,000 lux) condition at room temperature (32±°C).

acid consumption was in the range of 87.5-91.8% at all malic acid concentrations. Possibly, levulinic acid uptake was stimulated in the presence of malic acid. In addition, malic acid at 30, 40, and 50 mM could be converted to succinic acid at 7.62, 10.75 and13.04 mM, respectively (data not shown).

The presence of C_4 precursors could enhance 5-aminolevulinic acid production more effectively than C_5 pathway precursors. This is the first report that *R. benzoatilyticus* could produce 5-aminolevulinic acid via the consumption of both C_4 (glycine and succinic acid) and C_5 (glutamate or malic acid) pathway precursors.

Effects of LA concentration on ALA production

Levulinic acid is structurally similar to 5-aminolevulinic acid except for the absence of the amino group. The undissociated form of levulinic acid can enter the cell and function as a competitive inhibitor of ALA dehydratase (Sasaki et al., 1997). Since levulinic acid could retard the growth of photosynthetic bacteria, it was recommended to add the middle log phase to increase the extracellular 5-aminolevulinic acid concentration (Sasaki et al., 1987). In this study, levulinic acid was added at 24 h. Effect of levulinic acid concentration (2.5 to 12.5 mM) with

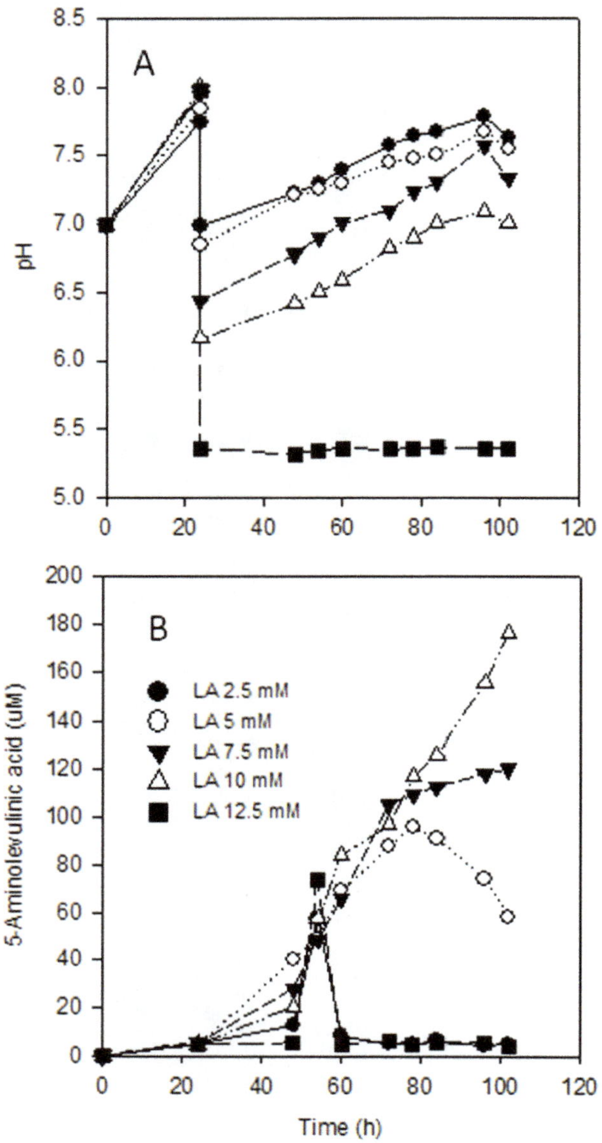

Figure 7. Time course of pH change and ALA production at adding various levulinic acid (LA) concentration at 24h during cultivation of *Rubrivivax benzoatilyticus* PS-5 under anaerobic-light (3,000 lux) condition at room temperature (32±°C).

optimum precursors (7.5 mM glycine plus 10 mM succinate) on 5-aminolevulinic acid production from *R. benzoatilyticus* PS-5 were investigated (Figures 7A and B). Increase levulinic acid concentration from 2.5 to 10 mM caused the decrease of pH from 7.63 to 7.0 at 96 h. At 2.5 mM levulinic acid, the consumption of levulinic acid and succinic acid was 90.20 and 87.70%, respectively (Table 2). This indicates that the lowest concentration of levulinic acid could be uptake more than high concentration (12.5 mM). Moreover, levulinic acid could be metabolized to acetate and propionate and gave the maximum values of 0.23 and 0.34 g l^{-1}, respectively under low levulinic acid concentration (2.5 mM) (data not shown). These results agreed with the report that levulinic acid incorporated by a photosynthetic bacterium (*Rhodopseudomonas* sp.) was metabolized to acetate and propionate under anaerobic-light condition (Okayama et al., 1990). Levulinic acid would be incorporated into the main metabolism, which must first pass through glycolysis to form pyruvate and acetyl-CoA (Jaremko and Yu, 2011).

Levulinic acid supplementation did affect (P>0.05) 5-aminolevulinic acid production as it increased from 57.74 to 176.50 μM with increasing levulinic acid concentration from 2.5 to 10 mM. The decrease at 12.5 mM levulinic acid was due to the drop of pH to 5.3 that strongly suppressed the growth of *R. benzoatilyticus* PS-5 as observed by cells decolorization (data not shown). In addition, levulinic acid also affects the pigment synthesis of the blue-green alga *Agmenellum quadruplicatum* PR-6 chlorophyll, c-phycocyanin, and ALA syntheses (Kipe-nolt and Stevens, 1979). Moreover, the drop of pH that resulted in the decrease of ALA production might be due to the decrease of enzyme activities under low pH conditions (Sasaki et al., 1991). For example, at pH 6.0, ALAD activity was relatively inhibited by levulinic acid addition, however, ALAS activity was also low (Noparatnaraporn et al., 2000).

Conclusion

R. benzoatilyticus PS-5 was isolated from pond sediment and found to be a novel 5-aminolevulinic acid producer with self-flocculating property. The optimum precursors of C_4 pathway for 5-aminolevulinic acid production were 7.5 mM glycine plus 10 mM succinic acid and addition of 10 mM levulinic acid at 24 h cultivation. Moreover, the optimum precursors of C_5 pathway was 85.29 μM with the addition of 10 mM levulinic acid at 24 h. The strain preferred C_4 to C_5 pathway. Maximum 5-aminolevulinic acid production was 190.74 μM and the productivity was 2.54 μM h^{-1}. Further optimization of the fermentation process should improve the 5-aminolevulinic acid production to even higher level.

ACKNOWLEDGEMENT

This research was supported by the Postdoctoral Fellowship from Prince of Songkla University.

REFERENCES

Achenbach LA, Carey J, Madigan M (2001). Photosynthetic and phylogenetic primers for detection of anoxygenic phototrophs in natural environments. Appl. Environ. Microbiol. 67:2922-2926.

Ali AHK, Takizawa H, Kondo K, Matsuoka H, Toba H, Nakagawa Y, Kenzaki K, Sakiyama S, Kakiuchi S, Sekido Y, Sone S, Tangoku A (2011). 5-Aminolevulinic acid-induced fluorescence diagnosis of pleural malignant tumor. Lung Cancer 74:48-54.

Ando T, Suzuki H, Nishimura S, Tanaka T, Hiraishi A, Kikuchi Y (2006). Characterization of extracellular RNAs produced by the marine photosynthetic bacterium *Rhodovulum sulfidophilum*. J. Biochem. 139:805-811.

Ano A, Funahashi H, Nakao K, Nishizawa Y (2000). Effects of levulinic acid on 5-aminolevulinic acid production in heterotrophic cultures of *Chlorella regularis* YA-603. J. Biosci. Bioeng. 89:176-180.

Beale SI (1999). Enzymes of chlorophyll biosynthesis. Photosynth. Res. 60:43-73.

Beer DD, Stoodley P, Roe F, Lewandowski Z (1993). Effects of biofilm structure on oxygen distribution and Mass transport. Biotechnol. Bioeng. 43:1131-1138.

Bunke A, Zerbe O, Schmid H, Burmeister G, Merkle HP, Gander B (2000). Degradation mechanism and stability of 5-aminolevulinic acid. J. Pharm. Sci. 89:1335-1341.

Chernyanovskaya EA (1990). Flocculation of *Methylococcus capsulatus* cells by biopolymers. Appl. Biochem. Microbiol. 25:690-695.

Choi C, Hong BS, Sung HY, Lee HS, Kim JH (1999). Optimization of extracellular 5-aminolevulinic acid production from *Escherichia coli* transformed with ALA synthhase gene of *Bradyrhizobium japonicum*. Biotechnol. Lett. 21:551-554.

Chung SY, Seo KH, Rhee JI (2005). Influence of culture conditions on the production of extra-cellular 5-aminolevulinic acid (ALA) by recombinant *E. coli*. Process Biochem. 40:385-394.

De Rosa FS, Marchetti JM, Thomazini JA (2000). A vehicle for photodynamic therapy of skin cancer: influence of dimethylsulphoxide on 5-aminolevulinic acid *in vitro* cutaneous permeation and *in vivo* protoporphyrin IX accumulation determined by confocal microscopy. J. Control Release 65:359-366.

Dijkstra AT, Majoie IM, van Dongen JW (2001). Photodynamic therapy with violet light and topical 6-aminolaevulinic acid in the treatment of actinic keratosis, Bowen's disease and basal cell carcinoma. J. Eur. Acad. Dermatol. Venereol. 5:550-504.

Dreschler-Theilmann B, Dornemann D, Senger H (1993). Synthesis of protoheme via both the C5 and the Shemin pathway, in the pigment mutant C-2A' of *Scenedesmus obliquus*. Z. Naturforsch. 48:584-589.

Fu WQ, Lin JP, Cen PL (2007). 5-Aminolevulinate production with recombinant *Escherichia coli* using a rare codon optimizer host strain. Appl. Microbiol. Biotechnol. 75:777-782.

Gadmar OB, Moan J, Scheie E, Ma LW, Peng Q (2002). The stability of 5-aminolevulinic acid in solution. J. Photochem. Photobiol. 67:187-193.

Harley JP, Prescott LM (2007). Laboratory Exercises in Microbiology. 7th ed. USA : McGraw- Hill.

Imhoff JF (1988). Anoxygenic phototrophic bacteria. In Austin B (ed) Methods in Aquatic Bacteriology, New York, John Wiley and Sons, pp. 207-240.

Jaremko M, Yu J (2011). The initial metabolic conversion of levulinic acid in *Cupriavidus necator*. J. Biotechnol. 155:293-298.

Kang Z, Wang Y, Gu P, Wang Q, Qi Q (2011). Engineering *Escherichia coli* for efficient production of 5-aminolevulinic acid from glucose. Metab. Eng. 13: 492-498.

Kipe-Nolt JA, Stevens SE (1979). Effect of levulinic acid on pigment biosynthesis in *Agmenellum quadruplicatum*. J. Bacteriol. 137:146-152.

Kurane R, Mutsuyama M (1994). Production of a bioflocculant by mixed culture. Biosci. Biotechnol. Biochem. 58:1589-1594.

Lee WR, Shen SC, Pai MH (2010) Fractional laser as a tool to enhance the skin permeation of 5-aminolevulinic acid with minimal skin disruption: a comparison with conventional erbium: YAG laser. J. Control Release 145:124-133.

Leman JA, Dick DC, Morton CA (2002) Topical 5-ALA photodynamic therapy for the treatment of cutaneous T-cell lymphoma. Clin. Exp. Dermatol. 27:516-518.

Lin D, Nishio N, Nagai S (1989). Production of 5-aminolevulinic acid by methanogens. J. Ferment. Bioeng. 68:88-91.

Madigan MT, Jung DO, Woese CR, Achenbach LA (2000). *Rhodoferax antarcticus* sp. nov., a moderately psychrophilic purple non-sulfur bacterium isolated from an Antarctic microbial mat. Arch. Microbiol. 173:269-277.

Madukasi EI, Chunhua H, Zhang G (2011). Isolation and application of a wild strain photosynthetic bacterium to environmental waste management. Int. J. Environ. Sci. Technol. 8:513-522.

Mauzerall D, Granick S (1956). The occurrence and determination of 5-aminolevulinic acid and porphobilinogen in urine. J. Biol. Chem. 219:435-446.

Mayer SM, Beale SI (1991). δ-Aminolevulinic acid biosynthesis from glutamate in *Euglena gracilis* photocontrol of enzyme levels in a chlorophyll-free mutant. Plant Physiol. 97:1094-1102.

Neuberger A (1961). Aspects of the metabolism of glycine and porphyrins. Biochem. J. 78:1-10.

Nishikawa S, Watanabe K, Tanaka T, Miyachi N, Hotta Y, Murooka Y (1999). *Rhodobacter sphaeroides* mutants which accumulate 5-aminolevulinic acid under aerobic and dark conditions. J. Biosci. Bioeng. 87:798-804.

Noparatnaraporn N, Watanabe M, Sasaki K (2000). Extracellular formation of 5-aminolevulinic acid by intact cells of the marine photosynthetic bacterium *Rhodovulum* sp. under various pH conditions. World J. Microbiol. Biotechnol. 16 :313-315.

Okayama A, Fuji S, Miura R (1990). Optimized fluorometric determination of urinary 5-aminolevulinic acid by using pre-colume derivatization and identification of the derivative. Clin. Chem. 36:1494-1497.

Pfennig N (1967). Photosynthetic bacteria. Ann. Rev. Microbiol. 21:285-324.

Qin G, Lin J, Liu X, Cen P (2006). Effects of medium composition on production of 5-aminolevulinic acid by recombinant *Escherichia coli*. J. Biosci. Bioeng. 102:316-322.

Ramana ChV, Sasikala Ch, Arunasri K, Anil Kumar P, Srinivas TNR, Shivaji S, Gupta P, Su ling J, Imhoff JF (2007). *Rubrivivax benzoatilyticus* sp. nov., an aromatic, hydrocarbon-degrading purple betaproteobacterium. Int. J. Syst. Evol. Microbiol. 56:2157-2164.

Rebeiz CA, Juvik JA Rebeiz CC (1988) Photodynamic insecticides. I. Concept and phenomenology Pesticide Biochem. Physiol. 30:11-27.

Reinbothe S, Reinbothe C (1996). Regulation of chlorophyll biosynthesis in angiosperms. Plant Physiol. 111:1-7.

Saikeur A, Choorit W, Prasertsan P, Kantachote D, Sasaki K (2009). Influence of precursors and inhibitor on the production of extracellular 5-aminolevulinic acid and biomass by *Rhodopseudomonas palustris* KG31. Biosci. Biotechnol. Biochem. 73:987-992.

Sasaki K, Ikeda S, Nishizawa Y, Hayashi M (1987). Production of 5-aminolevulinic acid by photosynthetic bacteria. J. Ferment. Technol. 65:511-515.

Sasaki K, Tanaka T, Nishizawa Y, Hayashi M (1991). Enhanced production of 5-aminolevulinic acid by repeated addition of levulinic acid and supplement of precursors in photoheterotrophic culture of *Rhodobacter sphaeroides*. J. Ferment. Bioeng. 71:403-406.

Sasaki K, Watanabe M, Nishio N (1997). Inhibition of 5-aminolevulinic acid (ALA) dehydratase by undissociated levulinic acid during ALA extracellular formation by *Rhodobacter sphaeroides*. Biotechnol. Lett. 19:421-424.

Sasaki K, Watanabe M, Tanaka T, Tanaka T (2002). Biosynthesis, biotechnological production and applications of 5-aminolevulinic acid. Appl. Microbiol. Biotechnol. 58:23-29.

Sasikala C, Ramana CV, Rao PR (1994). 5-aminolevulinic acid:A potential herbicide/insecticide from microorganisms. Biotechnol. Prog. 10:451-459.

Staley JT, Bryant MP, Pfennig N, Holt JG (1989). Bergy's Manual of Systematic Bacteriology. 3rd ed. The Willams and Wilkins Co, Baltimore.

Stenn KS (1968). Cation transport in a photosynthetic bacterium. J. Bacteriol. 96:862-864.

Suzuki H, Umekage S, Tanaka T, Kikuchi Y (2009). Extracellular tRNAs of the marine photosynthetic bacterium *Rhodovulum sulfidophilum* are not aminoacylated. Biosci. Biotechnol. Biochem. 73:425-427.

Tangprasittipap A, Prasertsan P, Choorit W, Sasaki K (2007). Biosynthesis of intracellular 5-aminolevulinic acid by a newly identified halotolerant *Rhodobacter sphaeroides*. Biotechnol. Lett. 29:773-778.

Truper HG, Pfennig N (1981). Isolation of members of the families Chromatiaceae and Chlorobiaceae. In Starr et al. (eds) The Prokaryotes: a Handbook on Habitats, Isolation, and Identification of Bacteria, Baltimore, Springer, pp. 279-289.

Urbain V, Block JC, Manem J (1993). Bioflocculation in activated

sludge: an analytic approach. Wat. Res. 27:829-838.
Watanabe M, Sasaki K, Nakashimada Y, Kakizono T, Noparatnaraporn N, Nishio N (1998). Growth and flocculation of a marine photosynthetic bacterium *Rhodovulum* sp. Appl. Microbiol. Biotechnol. 50:682-691.
Weinstein JD, Beale SI (1983). Separate physiological roles and subcellular compartments for two tetrapyrrole biosynthetic pathways in *Euglena gracilis*. J. Biol. Chem. 258:6799-6807.
Zeitouni NC, Oseroff AR, Shieh S (2003). Photodynamic therapy for nonmelanoma skin cancers. Current review and update. Mol. Immunol. 39:1133-1136.

15

Optimization of cultivation conditions for extracellular lipase production by *Yarrowia lipolytica* using response surface method

Gonçalves, F. A. G.[1] , Colen, G.[1], Takahashi, J. A.[2]

[1]Faculdade de Farmácia, Universidade Federal de Minas Gerais, Belo Horizonte, MG, 31270-901, Brazil.
[2]Departamento de Química, Instituto de Ciências Exatas, Universidade Federal de Minas Gerais, Belo Horizonte, MG, 31270-901, Brazil.

A wild strain of *Yarrowia lipolytica*, identified as LMI 91, was isolated from an oleaginous Brazilian fruit called *pequi* (*Caryocar brasiliense* Camb), which is a native species of the Brazilian savannah and which acts as a potent lipase producer. The present study examined the effects of certain factors, such as the combination of casein and meat peptones, as well as the initial pH of a culture medium in lipase produced by *Y. lipolytica* LMI 91. Experiments were conducted within a low concentration of olive oil, at a temperature of 30°C and a fermentation time of 60 h, under orbital shaking. The response surface methodology was used to determine the optimum concentration of the constituents' composition. A 2^2 experimental design with four axial points and three central points was employed in the optimization experiments, and interactive regression analysis was performed to obtain the optimum concentration. The experimental values were found to be in accordance with the predicted values, presenting a correlation coefficient of 0.866. The maximum lipase activity (13.0 U/mL) was obtained by peptone mixture of 70 g/L (1:1) and an initial pH of 5.0.

Key words: *Yarrowia lipolytica*, wild strain, extracellular lipase, fermentation, optimization.

INTRODUCTION

The growing demand for lipases to be incorporated into several biotechnological processes has been stimulating research on novel microbial lipases. As such, the present research focuses on both the screening of microorganisms from different sources and the development of better conditions to obtain higher enzyme production with lower costs (Sharma et al., 2001; Colen et al., 2006; Salihu et al., 2012). Among these the most microorganisms are yeasts, mainly *Candida* sp, which is commonly reported lipase producer in the literature (Salihu et al., 2012), but whose use is limited because it is a human pathogen. On the other hand, *Yarrowia lipolytica,* a non-pathogenic yeast with a Generally Recognized as Safe (GRAS) status, secretes several proteins with high molecular weight, depending on its growth conditions (Glover et al., 1997; Madzak et al.,

2004). Intracellular, membrane-bound lipases (carboxylesterases) and extracellular lipases (triacylglycerol hydrolases) can all be detected in *Y. lipolytica* (Fickers et al., 2006). These differ from lipases produced by other yeasts through their amino acid compositions, protein chain lengths and enzymatic properties (Pignède et al., 2000; Zhao et al., 2011).

Lipase production depends largely on the microorganisms' source, culture type (liquid or solid fermentations), medium composition and bioreactor design (Salihu et al., 2012).

The knowledge on how wild strains become domesticated represents a common research priority for microbiologists. The diversity of microbial communities and their ecological and metabolic functions have the potential for remarkable scientific, social and economic impacts (Suzzi, 2011). In fact, the evaluation of strain diversity provides a great chance to increase the knowledge of metabolic functions and regulation, as well as to speed up the improvement process of a targeted strain (Wittmann and Heinzle, 2002).

In the development of an industrial fermentation, fermentation medium designing is of critical importance, as the medium's composition can significantly affect the concentration, yield, and volumetric productivity of the product. In addition, the cost of the medium substantially affects the value of overall process (Kennedy and Krouse, 1999). Optimization through factorial design and RSM analysis is a common practice in biotechnology. Various researchers have applied this approach, especially to optimize process parameters, including the pH, temperature, aeration, among others (Gupta et al., 2004). The best conditions for the production of a microbiological enzyme can be gathered by experimental design using response surface methodology (Bornscheuer et al., 2002; Burkert et al., 2004; Rajendran and Viruthagiri, 2007). This methodology consists of a group of techniques used to investigate the relationship among the experimental variables and their responses, in turn allowing for a reduction in the number of experiments without neglecting the interaction among variable parameters (Myers, 1971).

Y. lipolytica is known for its ability to grow in hydrophobic substrates like oils (Pignède et al., 2000). Considering the consumption of fatty acids, it seems that *Y. lipolytica* preferably consumes unsaturated fatty acids (Kar et al., 2012). Unsaturated oleic acid has been pointed out as a potent inducer of *Y. lipolytica* lipase production (Fickers et al., 2004). The strain used in the present work was isolated from a unique biological resource, a typical Brazilian fruit called *pequi* (*Caryocar brasiliensis* Camb.). This fruit is rich in palmitic and oleic acids, containing 442 and 517 g kg^{-1} of each fat acid, respectively (Segall et al., 2006); consequently, a singular lipase production was expected from this wild strain.

Through fractioned factorial planning, the present study investigated the variables that affect the production of extracellular lipase by the wild strain of *Y. lipolytica* LM 91. The influence of different growing conditions (peptones, yeast extract, olive oil concentrations and the initial pH of the culture medium) could be observed at 48, 60 and 72 h of fermentation. The aim of the present work was to determine the best combination of casein and meat peptones, as well as the initial pH of the culture medium's lipase production in the presence of low concentrations of olive oil, during 60 h of fermentation. A response surface methodology was used to compare and optimize the medium's components for lipase production. A 2^2 central composite design matrix was used in the optimization experiments, and interactive regression analysis was performed to obtain the optimum concentration.

MATERIALS AND METHODS

Microorganism, culture media and chemicals

A *Y. lipolytica* LMI 91 strain was isolated from the *pequi* pulp (*C. brasiliense* Camb), an oleaginous fruit that is a native species of the Brazilian savannah (Ferreira, 2007). The yeast was placed in GYMP agar, consisting of 20 g/L dextrose, 5 g/L yeast extract, 10 g/L malt extract and 2 g/L NaH_2PO_4, 20 g/L agar, with a pH of 4.6 at 4°C, and was cultivated in Sabouraud Agar (SA) for production storage. Slants were incubated at 30°C for 72 h.

This experiment used casein peptone, meat peptone, yeast extract, malt extract and agar, all of the BD brand (Becton, Dickinson, and Company, USA). The commercial olive oil (1% acidity) used in this study was from GALLO (Lisboa, Portugal). All reagents were of analytical grade.

Production of lipase by submerged fermentation

A cell suspension containing 10^7 cells/mL, prepared by washing 3 day-old SA slants with sterile distilled water, was used (1 mL) to inoculate each 125 mL Erlenmeyer flask containing 25 mL of the liquid growth medium. Cultivations were carried out in an orbital shaker (Marconi, Rio Piracicaba, SP, Brazil) at 150 rpm and 30°C. The composition of the medium varied according to the design matrix. The basal liquid medium contained $NaNO_3$ (1.0 g/L), KH_2PO_4 (1.0 g/L), $MgSO_4.7H_2O$ (0.5 g/L) and chloramphenicol (0.2 g/L). The initial pH and the concentration of other components (g/L) used in the medium's composition are presented in Table 1. The cultures of each flask were centrifuged (2000 rpm/15 min), and the supernatants were analyzed. Product formation was measured at 48, 60 and 72 h. In the optimization step, the composition of the medium varied according to the matrix design and was made up of: a mixture (1:1) of meat and casein peptones (varying from 55.9 to 84.1 g/L), olive oil (5.0 mL/L) (Table 2) and the basal medium described above. The initial pH was adjusted to 6.0. The lipase activities of cultures after 60 h of incubation were determined in the supernatants after separation of the cells by centrifugation.

Table 1. Experimental values and coded levels for the fractional factorial design and lipase activity during *Y. lipolytica* fermentation.

Experiment	Coded variables (concentration)				Y_1	Y_2	Y_3
	X_1	X_2	X_3	X_4	(48 h)	(60 h)	(72 h)
1	-1 (30)	-1 (0)	-1 (3)	-1 (5)	0.00	3.68	0.80
2	+1 (70)	-1 (0)	-1 (3)	+1 (7)	17.90	4.00	0.00
3	-1 (30)	+1 (2)	-1 (3)	+1 (7)	0.00	0.00	0.48
4	+1 (70)	+1 (2)	-1 (3)	-1 (5)	15.03	4.96	1.28
5	-1 (30)	-1 (0)	+1 (7)	+1 (7)	0.00	2.40	1.28
6	+1 (70)	-1 (0)	+1 (7)	-1 (5)	2.40	9.91	11.67
7	-1 (30)	+1 (2)	+1 (7)	-1 (5)	5.75	1.60	0.80
8	+1 (70)	+1 (2)	+1 (7)	+1 (7)	1.76	0.00	0.00
9	0 (50)	0 (1)	0 (5)	0 (6)	8.47	2.44	4.88
10	0 (50)	0 (1)	0 (5)	0 (6)	4.64	1.48	5.62
11	0 (50)	0 (1)	0 (5)	0 (6)	6.71	1.33	4.14

X_1 Peptones (g/L): Mixture of meat and casein peptones (1:1); X_2 yeast extract (g/L); X_3 olive oil (mL/L); X_4 pH; $Y_{1,2,3}$ Lipase activity (U/mL) at 48, 60 and 72 h of fermentation, respectively.

Table 2. Variables and their coded levels and concentrations used in the first factorial at 60 h of fermentation and lipase activity (U/mL).

Run	Peptones (g/L)	pH	Coded levels		Y^c (U/mL)
			X_1^a	X_2^b	
1	50	4	-1	-1	0.00
2	90	4	+1	-1	10.69
3	50	6	-1	+1	2.99
4	90	6	+1	+1	1.26
5	41.8	5	-1.41	0	0.00
6	98.2	5	+1.41	0	2.51
7	70	3.59	0	-1.41	5.03
8	70	7.41	0	+1.41	5.66
9	70	5	0	0	2.67
10	70	5	0	0	3.77
11	70	5	0	0	2.83

[a] Concentration of mixture of meat and casein peptones (g/L); [b] pH; [c] lipase activity (U/mL).

Lipase assay

The lipase activity was determined using a titration method (Watanabe et al., 1977). The reaction mixture contained 5.0 mL of 25% (v/v) olive oil emulsion in a 2% (v/v) polyvinyl alcohol solution, prepared at 13.500 rpm for 5 min, using an ultra-turrax dispersor T25 (Ika, Wilmington, NC, USA); 5.0 mL of 0.1 M Tris-HCl buffer, pH 8.0; and 1.0 mL of the enzyme supernatant. The reaction mixture was incubated at 30°C in a reciprocal shaker (Solab, Piracicaba, SP, Brazil) at 45 cycles/min for 10 min. The emulsion was immediately disrupted after incubation by adding 10 mL of an acetone/ethanol (1:1, v/v) solution. The released free fatty acids were determined by titration with 0.05 N NaOH solution, using thymolphthalein as an indicator. A supernatant culture, boiled for 10 min, was used as the control. One unit of lipase was defined as the amount of enzyme required to release one µmol of fatty acid per minute under the specific conditions.

Experimental design

A 2^{4-1} fractional factorial design with three central points was used to calculate the main effects on a desired response. Four independent variables (concentration of peptones, yeast extract, olive oil, and pH) were used, the levels and values of which are presented in Table 1. Eleven experimental runs were carried out, which included eight factorial points and three central points. The lipase activity after 48, 60 and 72 h of fermentation was studied as a response.

Next, the effects of the peptone mixture and the initial pH of the culture medium on the lipase production were determined by modulating the variables according to a two-factor, five-level central composite rotatable design (CCRD). The equivalence of the design and levels are shown in the Tables 2 (first factorial design) and 3 (second factorial design).

A 2^2 experimental design with star (4 axial points) and 3 central

Table 3. Variables and their coded levels and concentrations used in the second factorial at 60 h of fermentation and lipase activity (U/mL).

Run	Peptones (g/L)	pH	Coded levels X_1[a]	Coded levels X_2[b]	Y^c (U/mL)
1	60	4	-1	-1	2.22
2	80	4	+1	-1	3.40
3	60	6	-1	+1	0.00
4	80	6	+1	+1	0.89
5	55.9	5	-1.41	0	5.32
6	84.1	5	+1.41	0	2.44
7	70	3.59	0	-1.41	4.29
8	70	7.41	0	+1.41	2.66
9	70	5	0	0	13.53
10	70	5	0	0	12.57
11	70	5	0	0	12.61

[a] Concentration of mixture of meat and casein peptones (g/L); [b] pH; [c] lipase activity (U/mL).

Table 4. Analysis of effects on lipase activity as determined by fractional factorial planning at 48, 60 and 72 h of fermentation.

Factor	Effect			Standard Error			t			P		
	48 h	60 h	72 h	48 h	60 h	72 h	48 h	60 h	72 h	48 h	60 h	72 h
Peptones (g/L)	7.83	2.80	2.40	3.86	1.18	2.29	2.03	2.38	1.04	0.09	0.05	0.34
Yeast extract (g/L)	0.56	-3.36	-2.80	3.86	1.18	2.29	0.15	-2.85	-1.22	0.89	0.03	0.27
Olive oil (mL/L)	-5.75	0.32	2.80	3.86	1.18	2.29	-1.49	0.27	1.22	0.19	0.80	0.27
pH	-0.88	-3.44	-3.20	3.86	1.18	2.29	-0.23	-2.92	-1.39	0.83	0.03	0.21

points, totaling 11 experiments, was carried out to obtain a second order model. The distance of the axial points was 1.414, calculated from Equation 1, where α is the distance of the axial points; k is the number of independent variables.

$$\alpha = (2^k)^{1/4} \qquad \text{(Equation 1)}$$

The results of CCRD were used to fit a second-order polynomial equation represented as

$$Y = \beta_0 + \beta_1 X_1 + \beta_2 X_2 + \beta_{12} X_1 X_2 + \beta_{11} X_1^2 + \beta_{22} X_2^2 \qquad \text{(Equation 2)}$$

In Equation 2, Y is the predicted response; β_0, the intercept; β_1, β_2, the linear coefficients; β_{12}, β_{11}, β_{22}, the squared coefficients.

Statistical analysis

The Statistica 8.0 software was used to determine the regression and graph analyses of the obtained data. The statistical significance of the aforementioned model equation was determined by Fisher's test value and the proportion of variance explained by the polynomial model was given by the multiple coefficient of determination, R squared (R^2) value. Analysis of variance (ANOVA) was also performed, and the level of significance was set at 0.1.

RESULTS AND DISCUSSION

The present study evaluated several parameters involved in lipase produced by a wild strain of *Y. lipolytica*, including the effect of the culture medium composition, given that sources and concentrations of nutrients directly influence both cell growth and enzyme production. Parameters selected as variables and lipase activity measured after 48, 60 and 72 h of fermentation are shown in Table 1.

Fractioned factorial planning was used as a screening tool to select the variables that affected lipase production. For optimization, other factorial designs with 2 or 3 variables were necessary, depending on the fractioned planning results. Table 4 shows the effect of the variables on lipase activity in *Y. lipolytica* LMI 91.

The parameters were considered significant when respective *p* values were lower than 10% ($p < 0.1$). In this manner, results obtained after 48 h of fermentation indicated that only the dependent variable peptone ($p = 0.09$) was significant in lipase production (Table 4). Therefore, determining the optimal concentration of peptone was required to reach a maximum enzymatic activity. By contrast, the results indicated that any value within the range of the tested pH, as well as all yeast extracts and olive oil concentrations, can be used without interfering in enzyme production.

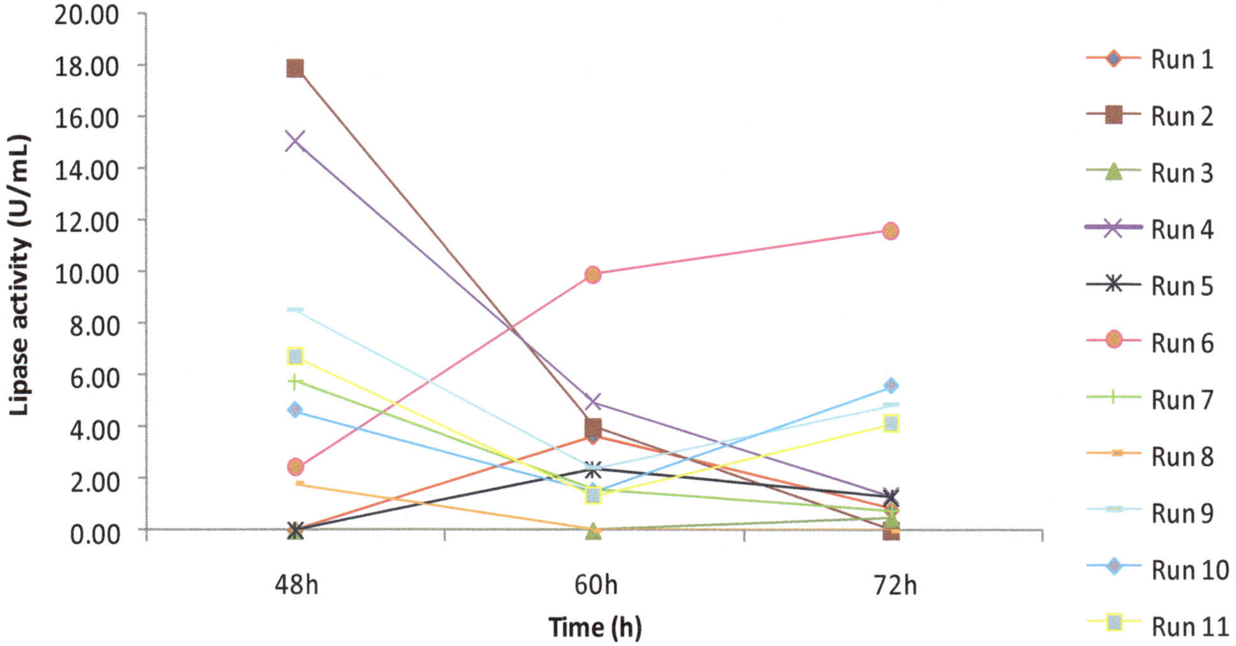

Figure 1. Lipase activity in function of fermentation time in the fractionated factorial planning.

A change in the influence of factors affecting lipase production could be observed after 60 h of fermentation (Table 4). At this time, in addition to the increase in peptone concentration, two other parameters presented statistical significance ($p < 0.1$): a decrease in the concentration of the yeast extract and a lower pH. By contrast, variations in the amount of olive oil present in the medium did not affect lipase production at 60 h of incubation. Since the yeast extract did not show statistically significant effects, this ingredient can be removed from the medium's composition without decreasing lipase production, given that level -1 represents an absence of yeast extract in the culture.

According to the statistical analyses, lipase activity was not affected by any of the studied variables at 72 h of fermentation (Table 4). According to Figure 1, lipase activity was higher at 48 h of fermentation (assays 2 and 4), decreasing abruptly at 60 h, and reaching nearly zero at 72 h. Assay 6 shows a high lipase activity from 48 to 60 h, at which time it reached 9.9 U/mL, remaining almost constant from then on.

Rathi and collaborators (2002) showed that the decrease in lipase activity may well be due to increased protease activity in the post-exponential phase of microorganism growth due to the decrease of oil availability or to the accumulation of fatty acids. Corzo and Revah (1999) also agree that lipase activity is strongly inhibited by the presence of oleic acid. Dalmau et al. (2000) reported that lipase activity decreases quickly after reaching the maximum production due to lipase proteolysis catalyzed by proteases produced during cell growth. *Y. lipolytica* is a dimorphic yeast that secrets several large proteins, such as alkaline proteases and at least one acid protease.

The first step in the process of seeking optimum conditions is to identify the input variables that have the greatest influence on the experimental response. According to Kennedy and Krouse (1999), there are many challenges associated with medium design. Designing the medium is a laborious, expensive, open-ended, often time-consuming process involving many experiments. Many constraints occur during the design process, and an industrial scale must be kept in mind when designing the medium. The present study verified that the most important variables for lipase production by the wild strain after 60 h of fermentation in orbital shake flasks were peptone concentrations and the initial pH of the culture medium and that variations in the amount of olive oil did not affect lipase production, despite prior reports (Fickers et al., 2004) describing that oleic acid enhanced lipase production in *Y. lipolytica*.

According to Gupta et al. (2004), the emphasis has shifted towards medium optimization using response surface methodology (RSM). The factorial design of a limited set of variables is advantageous when compared with the conventional method of manipulating a single parameter per trial, as the latter approach frequently fails to locate the optimal conditions for the process due to a

Table 5. ANOVA for the first factorial design.

Source of variation	Sum of square	Degrees of freedom	Mean square	F ratio	P
Regression	58.09	2	29.05	7.18	< 0.02
Residual	32.39	8	4.05		
Total	90.48	10	9.05		

R^2 = 64.2; F tab. 2; 8; 0.1 = 3.11.

Table 6. Regression coefficient for lipase activity (Y) for the second factorial design.

Factors	Regression coefficient	Standard error	t(5)	P	Estimate interval (90%)	
					Lower limit	Upper limit
X_1 (L)	-0.3	0.75	-0.3	0.7500	-1.75	1.25
X_1 (Q)	-4.5	0.89	-5.1	0.0038	-6.31	-2.74
X_2 (L)	-0.9	0.75	-1.2	0.2913	-2.38	0.62
X_2 (Q)	-4.7	0.89	-5.3	0.0031	-6.51	-2.94
$X_1 X_2$	-0.1	1.05	-0.1	0.9468	-2.20	2.05

L, Q: Linear and quadratic terms.

failure to consider the effect of possible interactions between factors. Moreover, the factorial design makes it possible to take advantage of practical knowledge about the process during the final RSM analysis.

Lipid carbon sources appear to be essential to obtaining a high lipase yield. However, according to Tamilarasan and Dharmendira (2011), higher oil concentration can affect the oxygen transfer in the medium. Although the lipid carbon source is essential, Treichel et al. (2010) claim that nitrogen sources and essential micronutrients should also be carefully considered for growth and production optimization. These nutritional requirements for microbial growth are fulfilled by several alternative media, including those based on defined compounds (synthetic medium), like sugars, oils and complex components, such as peptone, yeast extract, malt extract media, as well as agroindustrial residues containing all the components necessary for microorganism development.

According to Table 5, which shows the values of the variance analysis for the first factorial planning (Table 2), it could be observed that, although the F_{calc} for regression (7.18) is significantly greater than the F_{tab} (3.11), the percentage of variance (R^2) explained by the template proved to be unsatisfactory, approximately 64.2%. Moreover, there was a lack of adjustment for the model, since the F_{calc} in relation to the pure error (14.92) was greater than the F_{tab} (9.33). In this manner, since a response surface was impossible to obtain, a second factorial planning was required to determine new levels for the variable peptones (Table 3). The results obtained by means of CCRD (Table 3) were used to calculate the regression coefficients shown in Table 6, which gave rise to Equation 3, in which the variables are coded values, represented as: lipase activity (Y) as a function of the peptone mixture (variable X_1) and the initial pH (variable X_2); a second order regression equation (Equation 3), which shows the dependency of the medium's components for lipase activity. Only the quadratic terms proved to be significant model terms for the response ($p < 0.1$).

$$Y = 11.90 - 4.52 X_1^2 - 4.73 X_2^2 \quad \text{(Equation 3)}$$

The analysis of variance (Table 7) indicated that the second-order polynomial model (Equation 3) was highly significant and adequate enough to represent the relation between the response (lipase production) and the variables studied, with a very small p-value and a satisfactory coefficient of determination ($R^2 = 0.866$). This provides proof that the model accurately describes the studied region.

The response surface described by the model equation (Y) is represented in Figure 2a and by the contour diagrams in Figure 2b. The maximum lipase activity (13.0 U/mL) was obtained using 70 g/L of peptones and an initial pH of 5.0.

The results obtained in prior literature are diverse. Under optimum conditions, Muralidhar et al. (2001) reported the use of response surface methodology to optimize lipase production by *Candida cylindraceae*. Although, this study examined a different microorganism, the carbon source used was the same, olive oil, but in a much larger amount (33.7 g/L), resulting in a higher

Table 7. ANOVA for the second factorial design.

Source of variation	Sum of square	Degrees of freedom	Mean square	F ratio	P
Regression	186.88	2	93.44	25.839	<0.0001
Residual	28.93	8	3.62		
Total	215.81	10	21.58		

R^2 = 0.866 ; F tab. 2; 8; 0.1 = 3.11.

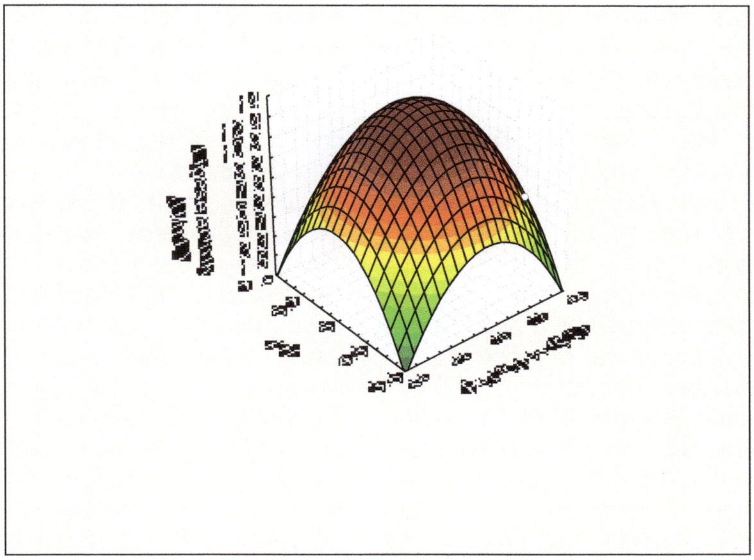

(a)

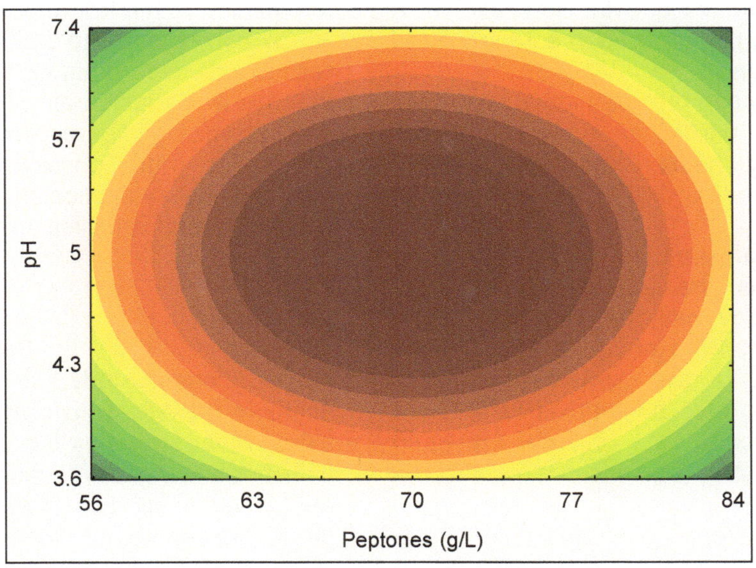

(b)

Figure 2. (a) Response surface plot showing the effect of concentration of peptones mixture and initial pH on lipase production from *Y. lipolytica* wild strain under conditions predicted by the model; (b) Contour diagrams for the lipase activity as a function of concentration of peptones mixture and initial pH of the culture medium.

activity (47.25 U/mL). Rajendran and Viruthagiri (2007) also optimized the conditions of culture mediums for lipase production by another yeast, *Candida rugosa* NCIM 3462, using response surface methodology, and reached maximum activity of 5.95 U/mL in 50 h of fermentation. Burkert et al. (2004), optimizing the culture medium conditions using olive oil as a carbon source, reached the maximum activity of 17 U/mL of lipase produced by the fungus *Geotrichum* sp. In another study, Papagora et al. (2013) optimized lipase production, using a *Debaryomyces hansenii* strain isolated from dry-salted olives, and reached a productivity of 7.44 U/mL.

To optimize the culture medium, peptones were used as a nitrogen source, as can be seen in the works of Muralidhar et al. (2001) and Rajendran and Viruthagiri (2007). Rajendran and Viruthagiri (2007) used only peptones and observed increased activity upon increasing the peptone concentrations, while Muralidhar et al. (2001) used peptone in a much smaller concentration, together with yeast and malt extracts. Another study on the production of lipase by a species of *Bacillus* corroborated with the finding that peptone is the preferred source of organic nitrogen (Gupta et al., 2004). Almeida et al. (2012) evaluated the effects of inorganic (ammonium chloride, sulphate, and nitrate) and organic (peptone and tryptophan yeast extract) nitrogen sources in the growth and production of lipase by the yeast *Candida viswanathii*. These authors reported that ammonium nitrate was the source that most influenced both enzyme productivity and cell growth. However, yeast extract was considered the better nitrogen source for cell growth and lipase production.

Organic nitrogen sources can provide many cell growth factors and amino acids, which are required for cell metabolism and enzyme synthesis (Darvishi et al., 2009; Tamilarasan and Dharmendira, 2011; Almeida et al., 2012).

Lipase production is substantially increased when working with two situations: the use of mutant strains obtained by molecular biology methods (Tan et al., 2003; Fickers et al., 2006; Bussamara et al., 2010) and the increase in the oxygen transfer rate, usually as a result of shaking the cultures, which also contributes to a better access of microorganisms to culture medium components, thus accelerating and increasing enzyme production.

Improvements in microbial strains for the overproduction of industrial products have been the hallmark of all commercial fermentation processes. Such improved strains can reduce the cost of the processes through increased productivity and may also possess some specialized and desirable characteristics (Karanam and Medicherla, 2008).

Despite the smaller productivity of wild strains, the screening of microorganisms from new sources, such as that used in the present work, is of utmost importance due to the industrial demand for new sources of lipases with different catalytic characteristics that need to be isolated from new strains (Sharma et al., 2001; Colen et al., 2006). In the present study, the target strain was isolated from the *pequi* fruit, an oil-rich fruit typically found in the Brazilian savannah (Segall et al., 2006).

Analyzing the results, it was possible to observe the behavior of *Y. lipolytica* LMI 91 towards different compositions of culture media and the influence of fermentation time on the parameters of cultivation conditions for lipase production. The composition of the medium and cultivation time interfere directly with cell growth, as well as in lipase production, since this production increases and reaches maximum levels in the growth phase and begins to decrease in the mid-stationary phase (Fickers et al., 2006; Almeida et al., 2012). It is also interesting to note the influence that peptone concentrations had on enzyme production, regardless of the time of fermentation.

By contrast, yeast extract, another nitrogen source used in fractional factorial planning, did not influence lipase activity. Therefore, it was possible to remove this component from the growth medium, making it less costly and easier to be controlled, since a variable had been removed. In addition to the yeast extract, olive oil also showed little interference in the production of the enzyme, even if it is the main and only source of cell energy; therefore, this study opted for a medium formulation with a low concentration of olive oil (5.0 mL/L, which corresponds to approximately 4.5 g of oil). The results obtained in this work made it possible to consider that this wild yeast strain of *Y. lipolytica* LMI 91 is a good and promising lipase-producing strain, indicating its importance for possible future work aimed at biotechnological applications, considering that it was able to reach a relatively high lipase activity.

Conclusion

In the present study, using orbital shake flasks, the optimum conditions to obtain a high-level production of extracellular lipase by the wild strain of *Y. lipolytica* LMI 91 included 70 g/L of a casein and meat peptone mixture and a culture medium with an initial pH of 5.0, with the basal experimental conditions presenting a low concentration of olive oil (5.0 mL/L). Under these conditions, an enzyme activity of 13 U/mL was achieved. According to the results found in this study, *Y. lipolytica* LMI 91 is a promising lipase producer. Since this is a wild strain, yield improvement can be achieved by increasing the scope of further developments and applications using *Y. lipolytica* LMI 91 for lipase production to supply the growing biotechnological needs for the supply of lipase in the world market.

ACKNOWLEDGEMENTS

We are thankful for the finantial support provided by FAPEMIG (Fundação de Amparo à Pesquisa do Estado de Minas Gerais, Brazil) and CNPq (Conselho Nacional de Desenvolvimento Científico e Tecnológico, Brazil).

REFERENCES

Almeida AF, Taulk-Tomisielo SM, Carmona EC (2012). Influence of carbon and nitrogen sources on lipase production by a newly isolated *Candida viswanathii* strain. Ann. Microbiol. doi:10.1007/s13213-012-0580-y.

Bornscheuer UT, Cornelius B, Ramisetti S, Hari Krishna S (2002). Optimizing lipases and related enzymes for application. Trends Biotechnol. 20:433-437.

Burkert JFM, Maugeri F, Rodrigues MI (2004). Optimization of extracelullar lipase production by *Geotrichum* sp. using factorial design. Bioresour. Technol. 91:77-84.

Bussamara R, Fuentefria AM, Oliveira ES, Broetto L, Simcikova M, Valente P, Schrank A, Vainstein MH (2010). Isolation of a lipase-secreting yeast for enzyme production in a pilot-plant scale batch fermentation. Bioresour. Technol. 101:268-275.

Colen G, Junqueira RG, Moraes-Santos T (2006). Isolation and screening of alkaline lipase-producing fungi from Brazilian savanna soil. World J. Microbiol. Biotechnol. 22:881-885.

Corzo G, Revah S (1999). Production and characteristics of the lipase from of the lipase from *Yarrowia lipolytica* 681. Bioresour. Technol. 70:173-180.

Dalmau E, Montesinos JL, Lotti M, Casas C (2000). Effect of different carbon source on lipase production by *Candida rugosa*. Enzyme Microb. Technol. 26:657-663.

Darvish F, Nahvi I, Zarkesh-Esfahani H, Momenbeik F (2009). Effect of plant oils upon lipase and citric acid production in *Yarrowia lipolytica* yeast. J. Biomed. Biotechnol. 2009: doi:10.1155/2009/562943.

Ferreira LC (2007). Aspectos microbiológicos da conservação de polpas de pequi (*Caryocar brasiliense* Camb.): qualidade, higiene, adaptação de bactérias ao estresse ácido e isolamento de microorganismos com potencial para bioconservação (PhD Thesis, Universidade Federal de Minas Gerais).

Fickers P, Nicaud JM, Gaillardin C, Destain J, Thornart P (2004). Carbon and nitrogen sources modulated lipase production in the yeast *Yarrowia lipolytica*. J. Appl. Microbiol. 96:742-749.

Fickers P, Ongena M, Destain J, Weekers F, Thonart P (2006). Production and down-stream processing of an extracellular lipase from the yeast *Yarrowia lipolytica*. Enzyme Microbiol. Technol. 38:756-759.

Glover DJ, McEwen RK, Thomas CR, Young TW (1997). pH-regulated expression of the acid and alkaline extracellular proteases of *Yarrowia lipolytica*. Microbiol UK. 143:3045-3054.

Gupta R, Gupta N, Rathi P (2004). Bacterial lipases: an overview of production, purification and biochemical properties. Appl. Microbiol. Biotechnol. 64:763-781.

Kar T, Destain J, Thonart P, Delvigne F (2012). Scale-down assessment of the sensitivity of *Yarrowia lipolytica* to oxygen transfer and foam management in bioreactors: investigation of underlying physiological mechanisms. J. Ind. Microbiol. Biotechnol. 39:337-346.

Karanam SK, Medicherla NR (2008). Enhanced lipase production by mutation induced *Aspergillus japonicus*. Afr. J. Biotechnol. 7:2064-2067.

Kennedy M, Krouse D (1999). Strategies for improving fermentation medium performance: a review. J. Ind. Microbiol. Biotechnol. 23:456-475.

Madzak C, Gaillardin C, Beckerich JM (2004). Heterologous protein expression and secretion in the non-conventional yeast *Yarrowia lipolytica*: a review. J. Biotechnol. 109:63-81.

Muralidhar RV, Chirumamila RR, Marchant R, Nigam P (2001). A response surface approach for a comparison of lipase production by *Candida cylindracea* using two different carbon sources. Biochem. Eng. J. 9:17-23.

Myers RH. Response surface methology. Boston (1971) apud Essamri M, Deyris V, Comeau L (1998). Optimization of lipase production by *Rhizopus oryzae* and study on the stability of lipase activity in organic solvents. J. Biotechnol. 60:97-103.

Papagora C, Roukas T, Kotzekidou P (2013). Optimization of extracellular lipase production by *Debaryomyces hansenii* isolates from dry-salted olives using response surface methodology. Food Bioprod. Process. doi:10.1016/j.fbp.2013.02.008.

Pignède G, Wang GHJ, Fudalej F, Gaillardin C, Seman M, Nicaud J (2000). Characterization of an extracellular lipase encoded by LIP2 in *Yarrowia lipolytica*. J. Bacteriol.182:2802-2810.

Rajendran A, Viruthagiri T (2007). Optimization of medium composition for lipase production by *Candida rugosa* NCIM 3462 using response surface methodology. Can. J. Microbiol. 53:643-655.

Rathi P, Goswami VK, Sahai V, Gupta R (2002). Statistical medium optimization and production of hyperthermostable lipase from *Burkholderia cepacia* in a bioreactor. J. Appl. Microbiol. 93:930-936.

Salihu A, Alam Md Z, Abdukarim MI, Salleh HM (2012). Lipase production: an insight in the utilization of renewable agricultural residues. Resour. Conserv. Recycl. 58:36-44.

Segall SD, Artz WE, Raslan DS, Ferraz VP, Takahashi JA (2006). Triacylglycerol analysis of pequi (*Caryocar brasiliensis* Camb.) oil by electrospray and tandem mass spectrometry. J. Sci. Food Agric. 86:445-452.

Sharma R, Chisti Y, Banerjee UC (2001). Production, purification, characterization, and applications of lipases. Biotechnol. Adv.19:627-662.

Suzzi G (2011). From wild strain to domesticated strain: the philosophy of microbial diversity in foods. Front. Microbiol. 2: doi:10.3389/fmicb.2011.00169.

Tamilarasan K, Dharmendira KM (2011). Optimization of médium components and operating conditions for production of solvent tolerant lipase by *Bacillus sphaericus* MTCC 7542. Afr. J. Biotechnol. 10:15051-15057.

Tan T, Zhang M, Wang B, Ying C, Deng L (2003). Screening of high lipase producing *Candida* sp. and production of lipase by fermentation. Process Biochem. 39:459-465.

Treichel H, Oliveira D, Mazutti MA, DI Luccio M, Oliveira JV (2010). A review on microbial lipases production. Food Bioprocess Technol. 3:182-196.

Watanabe N, Ota Y, Minoda Y, Yamada K (1977). Isolation and identification of alkaline lipase producing microorganisms, cultural conditions and some properties. Agric. Biol. Chem. 41:1353-1358.

Wittmann C, Heinzle E (2002). Genealogy profiling through strain improvement by using metabolic network analysis: metabolic flux genealogy of several generations of lysine-producing corynebacteria. Appl. Environ. Microbiol. 68:5843-5859.

Zhao H, Zheng GL, Wang X, Liu Y, Xu L, Yan Y (2011). Cloning, expression and characterization of a new lipase from *Yarrowia lipolytica*. Biotechnol. Lett. 33:2445-2452.

Physiological responses of donkeys (*Equus asinus*, Perissodactyla) to work stress and potential ameliorative role of ascorbic acid

AYO, J. O., OLAIFA, F. H.* and AKE, A. S.

Department of Veterinary Physiology, Ahmadu Bello University, Zaria, Nigeria.

The review examines donkey as a draught animal, used mainly for carrying load from one place to another in Africa. Although, motorized transport has been thriving rapidly in the twentieth century, the use of donkeys for transport has not experienced a proportionate decline. Donkeys are often overused, neglected and exposed to various stressors, which eventually lead to their poor performance. This review also examines the effect of packing on the physiological parameters of donkeys and the ameliorative role of ascorbic acid in donkeys subjected to work stress. In conclusion, a better understanding of the pathophysiologic mechanism of stress in donkeys may enhance their work output under the unfavourable environmental conditions of the tropics.

Key words: Donkeys, ascorbic acid, packing, stress, physiological parameters.

INTRODUCTION

The modern domestic donkey (*Equus asinus, Perissodactyla*) was bred from the wild ass (*Equus africanus*) in north-eastern Africa during the predynastic period of Egypt, about 6 000 years ago. Two wild ass sub-species are thought to have had a role in the development of the modern donkey. They are the Nubian (*Equus africanus africanus*) and Somali (*Equus africanus somaliensis*) donkeys. Recent DNA analysis suggested that only the Nubian ass contributed genetically to the domestic donkey (Kimura et al., 2010). The early history of the donkey in Africa is difficult to reconstruct because of the sparse archaeological data. Draught animals, the principal being donkeys, play an important role in agricultural production and transport in sub-Saharan Africa. Although draught animal power has been superseded by tractors on many large commercial farms in Africa, it is a relevant farm technology in small-scale agriculture for economic and agro-ecological reasons (Pearson and Vall, 1998). Nigeria, with a donkey population of one million is ranked third in Africa after Ethiopia and Egypt (Blench, 2003).

*Corresponding author. E-mail: folashadeakanmu@yahoo.com.

Donkeys are preferred as draught and transport animals because they are friendly, hardy, quiet and more economical, compared to horses and oxen. They can be maintained on local farm produce, and are easy to train, intelligent and patient while working (Inns, 1979). Donkeys as pack animals can carry a load that is up to 100 kg or that is 50% of their body weight (Pal et al., 2002). It has also been reported that donkeys can carry a heavy load twice the weight a person can carry for a longer distance (Martin and Smith, 2005). Donkeys are major assets to human lives in arid areas, but their physical needs are always ignored due to their hardiness. It is important to protect donkeys from harsh elements of the weather in order to improve their health (Kataria and Kataria, 2010). As donkeys become a more popular choice of work animals for small-holder farmers, specific management practices need to be devised in order to fully maximize their natural survival advantages (Smith and Pearson, 2005). The improvement in health and welfare will enhance their productivity and provide their owners with a more reliable source of income (Martin et al., 2005). A better understanding of the pathophysiologic mechanism of stress in donkeys may enhance their work output under the unfavo-

unfavorable environmental conditions of the tropics. From a welfare point of view, the use of ascorbic acid may provide a less expensive and readily available alternative to other anti-stress agents. This is particularly important, if they can be shown to remove or at least mitigate the harmful consequences of stress responses in the donkey.

ADAPTATION OF THE DONKEY

Origin of the wild ass in Africa is controversial, but it may have originated outside the sub-saharan regions of Africa. However, the true home of the donkey is the arid and semi-arid areas of North-East and Western Asia (Fielding and Krause, 1998). According to Fielding and Krause (1998), donkeys adapt well to hot-dry desert environment through body temperature control, water metabolism, and special nutritional and anatomic features. Donkeys in the tropics are able to maintain homeothermy by compensatory mechanisms in order to keep their physiological values within the established normal range (Minka and Ayo, 2007). This despite the prevailing high ambient temperature (AT) (Minka and Ayo, 2007a), is established to induce heat stress in other livestock (Ayo et al., 1996, 2008).

The most limiting factor for survival in semi-arid and arid areas and during drought is water. Donkeys survive better than cattle in these areas and during drought (Homer 1993; Smith and Pearson, 2005). They have the ability to tolerate thirst and this allows them to have access to more remote source of forage, inaccessible to cattle in rangeland (Smith and Pearson, 2005). Nengomasha et al. (2000) reported that donkeys with limited access to water (2 to 3 days) loose less water through faeces than their counterparts with *ad libitum* access to water. This is because faecal water loss can account for 50% of all water lost from the body. Donkeys are also adapted to low-quality, high-fibre feed, which contributes to their ability to eat and survive on very little quantity (Yousef, 1991; Nengomasha et al., 2000). Donkeys do very well in selecting their feed and it is better to feed them *ad libitum* when given poor-quality cereal straw diets and stovers to improve the feeding value (Pearson and Meritt, 1991). Pearson et al. (1998), however, reported that donkeys have the tendency to overeat when they have access to good quality roughages. Fielding and Krause (1998) reported that the positioning of the stomach avoids food from being washed out of it by water when donkeys drink. They have tough, compact hooves that have successfully adapted to walking on sandy and rocky terrain. Donkeys have wooly hair which insulates them from desert heat and cold and a lean body mass that is fuel-efficient and easily cooled. Socio-economically, owing to their single-purpose role and their low social status, maintenance of a low sustainable population of donkeys is possible (Smith and Pearson, 2005).

SPECIAL PHYSIOLOGICAL FEATURES OF THE DONKEY COMPARED TO THE HORSE

The physiology of the donkey differs in some aspects from horses and mules. Normal base-line values for most hematology and biochemical indices have been published for donkeys and are slightly different from those of horses (Grosenburgh et al., 2011). For instance, there is absence of reticulocytes in peripheral circulation, even after severe hemorrhage in donkeys. Donkeys usually have fewer, but larger erythrocytes.

Hypothermia in donkeys is defined as having rectal temperature below 35°C during cold period and may not be secondary to other diseases related to diet or management. Donkeys are thermolabile, and so their temperature may increase more in a hot climate, or following exercise than for a normal horse. Normal values of resting heart rates are higher in donkeys than for a normal horse, and heart rate is a very good indicator of stress or pain, even when it is not noticed physically in donkeys. The donkey's fluid balance and water compartment partitioning differ from those of the horse, thus they tolerate dehydration better than the horse and recover quickly (Grosenburgh et al., 2011). The hematocrit increase secondary to dehydration does not occur in donkeys until they are 30% dehydrated.

USES OF DONKEYS FOR TRACTION

Animal traction can be defined as the employment of domestic animals for tillage or transport (Starkey, 1992). After cattle, the main work animals world-wide are equids. The real importance of the donkeys relates to their use as pack animals because they are adapted to the desert and can carry loads through arid lands. Donkeys have been employed for packing over 5 000 years ago (Clutton-Brock, 1992). Donkeys as pack animals, are used to carry people, water, manure, firewood, farm inputs and produce, and to relieve women of much of their time and energy-consuming work (Pearson, 2000; Minka and Ayo, 2007a; Ayo et al., 2008).

In Africa, while motorized transport has been thriving rapidly since the twentieth century, the use of animal power for local transport has not experienced a proportionate decline. Donkeys have been important in the history of salt trade in Ethiopia (Wilson, 1991). In West-Africa, donkeys were not widespread until about 1 500 years ago, and ever since then, they have been used mostly as pack animals (Blench, 1993). The use of donkeys for transportation continues, especially in rural areas with rugged terrain, absence of well-developed modern transport network and prevailing poverty (Mengistu et al., 2005) and the contribution of donkeys in facilitating agricultural production is very high (Youseph et al., 2005). Animal traction is socially accepted with no taboos and there is a supplementary and complementary connection between the crop sector and draught animals (Djang-Ford jour et al., 2003). It has been predicted that animal traction will still spread in most sub-Saharan Africa even in the foreseeable future.

STRESS IN DONKEYS

Donkeys in Nigeria are used for transport purposes in different parts of the country, especially in places with difficult terrain. Under this condition, donkeys are subjected to different stressors, which may compromise their health and consequently, result in poor performance (Gupta et al., 1999). Stress generally affects the host-defense system and increases its susceptibility to infections (von Borell, 2001). The different conditions by which donkeys are subjected to stress are enumerated below.

Work stress

Work is produced when animals by physical exertion use energy to pull an implement through the soil or to carry a load over a distance (Pearson and Vall, 1998). The amount of work an animal can do depends on the speed at which it works, and the draught force generated. For a particular draught force, it is the speed that determines the power output of the animal. Therefore, as draught force increases, speed correspondingly decreases (Francis, 1988). Adverse working conditions, including high ambient temperature and relative humidity resulting in heat stress, badly fitted harness and difficult ground conditions, are stressors that reduce the rate of work (Pearson and Vall, 1998). Thus, work output is not just a function of the speed of the animal itself, but is influenced by the environment and soil conditions (Pearson and Vall, 1998). The oxidation of chemical energy to provide work produces heat, which has to be lost by the animal, if it is to continue to work and maintain homeostasis. However, donkeys can still work while withstanding severe dehydration (Youseph, 1991). They do this by reducing water intake, sweating rate and water excretion, and by maintaining feed-intake. They also maintain plasma volume by drawing substantial fluid reservoir in the hind-gut, especially the caecum and ventral colon. Mild chronic dehydration experienced by the donkey produces morphological and physiological adaptations in the hind-gut, which enhance fermentation and absorptive capacity of the hind-gut (Sneddon et al., 2006). With their good adaptability, in hot climate, heat loss in donkeys has been observed. Olaifa et al. (2012a) observed an increase in rectal temperature in donkeys after a 4 h load carrying. Rose (1986) reported that horses exercising over a distance of 100 km may lose 30 to 40 L of water in sweating during the course of the exercise. Pal et al. (2002) observed signs of fatigue, which include unwillingness to continue, uncoordinated legs and excitement after work, and concluded that continuous working of donkeys causes increased stress.

Thermal environmental stress

Meteorological stressors that directly and adversely affect livestock performance and health care, especially in tropics and sub-tropical Africa are AT, relative humidity (RH), thermal radiation and wind speed (Ayo et al., 2011). Smith et al. *(*1995) shows that the meteorological stressors exert adverse effects on animal health, growth, production and overall performance. AT is positively correlated with the rectal temperature of donkeys during the harmattan season (Olaifa et al., 2012a) and this is because exposure to high ambient temperature induces the animal to attempt to balance the excessive heat load by using thermoregulatory mechanisms to dissipate their latent heat. The temperature begins to rise if the mechanism fails. Of all the meteorological stress factors adversely affecting livestock, heat stress factors have been shown to be the most severe in the tropics (Ayo et al, 1996). Heat stress is always a potential problem, which causes obvious discomfort, tremors and rapid shallow breathing in donkeys (Fielding and Krause, 1998). Heat stress causes cascade of changes in the physiological mechanisms of animals. In donkeys, Kataria and Kataria (2010) reported an increase in prolactin and cortisol levels, suggestive of stress adaptation to unfavorable environment.

Although there is paucity of information on the effect of cold stress on the donkey, especially in the tropics, stress due to cold environment or low ambient temperature may be a potential problem too, especially during the harmattan season, in the early hours of the day (Ayo et al., 1998). It may be reflected in an elevated metabolic rate and increased production of reactive oxygen species (Heise et al., 2003). Cold stress has been shown to disrupt the balance in the oxidant/antioxidant system, and it causes oxidative damage to several tissues by altering the enzymatic and non-enzymatic antioxidant status, protein oxidation and lipid peroxidation (Sahin and Gusmulu, 2004).

Poor management

Improper handling is considered as a major stressor, adversely affecting farm animals (Knowles and Warris, 2000; Minka and Ayo, 2009). It exerts deleterious effects on health, well-being, behavior, performance and production quality (Geverink et al., 1998). Improper housing and handling expose donkeys to wounds, which are commonly not attended to and leading to poor performance. Donkeys are exposed to wound infections mostly during the hot-dry and rainy seasons (Asha et al., 2006). They are commonly fed with poor-quality feed, where proper feeding of donkeys enables them to resist better to disease, live longer and have a higher rate of reproducetion to provide replacement animals. Besides, donkeys need supplements, especially throughout the dry season, when they are increasingly used for traction (Olaifa et al., 2012a).

BIOMARKERS OF STRESS IN DONKEYS

Rectal temperature, respiratory rate and heart rate

It has been established that rectal temperature, respiratory rate and heart rate are relevant body parameters for immediate evaluation of health status and adaptability of animals (Ayo et al., 1996, 2008; Minka and Ayo 2007a; Dey et al., 2010). They can be measured easily, especially during transportation in rural areas (Minka and Ayo, 2007c). Continuous packing for 4 h has been reported to induce a transient increase in rectal temperature (Olaifa et al., 2012a) and respiratory rate in donkeys (Pal et al., 2002). High body temperature initiates compensatory and adaptive mechanisms to establish the body's homeostasis and homeothermy (Stott, 1981). Donkeys adapt well to different environmental conditions by compensatory increase in values of rectal temperature, respiratory rate and heart rate (Minka and Ayo, 2007a). Increase in these physiological parameters above the normal range indicates that the animals are stressed. This has been confirmed in sheep (Srikandakumar et al., 2003), donkeys (Pal et al., 2002; Minka and Ayo, 2007a) and goats (Ayo et al., 1996). In donkeys, rectal temperature, respiratory rate and heart rate are stress quantifiable parameters, used to define stress levels under continuous work (Singh, 2000). The standard ranges used to define levels of stress according to Singh (2000) are as follows:

Rectal temperature (°C) = T_0 + (1.0 – 3.0)
Heart rate (beats per minute) = H_0 + (15 – 45)
Respiratory rate (breaths per minute) = R_0 + (15 - 50);

Where T_0, H_0 and R_0 parameters are recorded at rest. In an experimental study carried out by Olaifa et al. (2012a), a significant increase in rectal temperature was observed in donkeys following packing. The result support the hypothesis that donkeys subjected to packing are often stressed.

BEHAVIORAL RESPONSES

Behavioral responses are the first line of defense to environmental challenges (Ayo et al., 2002; Cable et al., 2007). When the brain perceives a stressful experience, it initiates physiologic and behavioral responses that lead to allostasis and adaptation (Armario et al., 2008). The hypothalamic-pituitary-adrenal axis and the sympathetic adrenomedullary system have a crucial function in the metabolic and cardiovascular preparation of the body to exhibit certain behavioral responses (Sapolsky et al., 2000). Stress responses in donkeys are always conducted on the basis of irregular behavioral phenomena that may be difficult to interprete as observed in previous works (Minka and Ayo, 2007a; Ayo et al., 2008). The perception of stress is influenced by experience, genetics and behavior. A major problem of applied stress studies in animals relates to the methodology of stress assessment or to the non-specificity of the selected variable (von Borell, 2001). Laedwig (2000) reveals that animals, like donkeys, under intensive housing conditions, exhibit chronic intermittent stress response, occurring in situations such as standing up or lying down in inadequately designed cubicle or tether stalls. The response of the donkey to a stressor also depends on temperament and early experience. It has been debated that farm animals develop distinct coping strategies as demonstrated in laboratory animals. Grandin (1997) demonstrated that excitability of animals depend on their temperament. The more the animals are handled, the aggressive they become and conversely, frequent handling of calm animals result in increase in their aggression (von Borell, 2001). In a study by Minka and Ayo (2007a), unshaded donkeys subjected to work during the hot-dry season exhibited behavioral values; lower frequency of lying down and lower speed compared to shaded donkeys.

HEMATOLOGY AS A STRESS BIOMARKER

Carrying load over a long distance is a strenuous exercise for donkeys (Pearson and Vall, 1998) thus; this can influence their hematologic parameters due to work stress (Lorena et al., 2006). The sympathetic nervous system generally plays an important role in the mediation of response to exercise by modifying cardiovascular function (Cuniberti et al., 2010), and thereby affecting the responses that are observed in blood parameters. Changes occurring in hematologic parameters induced by physical work have been well documented (Hindchiff et al., 2002; Pal et al., 2002; Lorena et al., 2006).

Erythrocytes

Erythrocytes are naturally prone to oxidative stress because of the high content of polyunsaturated fatty acid in their membranes and high amount of hemoglobin-bound iron (Agrawal and Sharma, 2010). Erythrocytes are an ideal model for studying the effects of physical stress since they lack nucleus and, have a short lifespan (Vani et al., 2010). Many *in vitro* and *in vivo* studies have demonstrated that several parameters of erythrocytes are negatively affected by increased oxidative stress (Brzezinska-Slebodzinska, 2003; Bhatti and Shaikh, 2007; Pandey and Risvi, 2009), produced following physical exercise, which is a form of stress. Bhatti and Shaikh (2007) reported an increase in erythrocyte count following physical exercise and slight increase in erythrocyte sedimentation rate. They may increase apparently, due to inflammatory components and release of acute phase protein in the liver cells. Increase in total number of circulating erythrocytes, hematocrit and hemoglobin concentration has been observed (Boning et al., 2010; Cuniberti et al., 2010) and may be due to splenic contractions during exercise. The spleen acts as a reservoir for erythrocytes, storing a large volume of blood, about 40 to 50% of the entire circulatory population, and the release of erythrocytes is under the influence of sympathetic nervous system and circulating catecholamines (Cuniberti et al., 2010).

Erythrocyte osmotic fragility

Erythrocyte osmotic fragility is also a good indicator of work stress (Hesta et al., 2008). Increase in erythrocyte osmotic fragility has been reported in donkeys after subjecting them to packing (Olaifa et al., 2012b). Changes in the erythrocyte osmotic fragility during exercise in equine are an important indicator of intravascular hemolysis (Hanzawa and Watanabe, 2000). Erythrocyte osmotic fragility is used to determine the integrity of erythrocytes (Asala et al., 2011). It is reported that heavy exercise increases erythrocyte osmotic fragility while light exercise decreases it because light exercise has antioxidant properties (Cuniberti et al., 2010). The decrease in blood pH and increase in lactate and peroxide caused by anaerobic exercise promote erythrocyte osmotic fragility. Oyewale et al. (2011) reported that donkey erythrocyte is less resistant to osmotic hemolysis than that of other equids.

Leukocytes

Responses of leukocytes and their sub-populations to physical work are well established (McCarthy and Dale, 1988; Gabriel et al., 1992; Pederson and Hoffman-Goetz, 2000; Bhatti and Shaikh, 2010). Pederson and Hoffman-Goetz (2000) demonstrated that neutrophils rose during and after physical exercise, but a fall below basal values results after a prolonged exercise. Lorena et al. (2006) also reported an increase in leukocyte counts after marathon race in humans. On the contrary, Cuniberti et al. (2010) did not observe any change in total leukocyte counts after prolonged exercise, although they recorded neutrophilia, lymphopenia and monocytosis similar to the findings of Robson et al. (2003) in horses. The change in differential leukocyte count has been attributed to a rise in the concentration of hormones (that is adrenaline, cortisol and growth hormone) during exercise and increased cardiac output, which increases mobilization of leukocytes from marginated pool (Tonnesen et al., 1987). Exercise-induced leukocytosis has been often compared to inflammation-like reaction. After strenuous work, the leukocyte count is almost double, the baseline similar to the response against physiological insults to the immune system (Bhatti and Shaikh, 2007).

Neutrophil : lymphocyte ratio as an index of stress

Neutrophil : lymphocyte ratio is a good indicator of stress (Altan et al., 2003) as the ratio increases in stressed individuals (Plyaschenko and Sidorov, 1987). Minka et al. (2004) demonstrated an increase in neutrophil/lymphocyte ratio in pullets following road transportation stress due to inflammatory responses. Similarly, in donkeys, an increase in neutrophil/lymphocyte ratio was observed after work (Olaifa et al., 2012c).

BIOCHEMICAL RESPONSES

The parameters are usually evaluated to determine the extent of stress include electrolytes (Na^+, K^+, Cl^-, H^+) and serum enzymes such as aspartate aminotransferase, alkaline phosphatase, creatinine kinase, superoxide dismutase, catalase, glutathione peroxidase (GSH) and molecules of urea, glucose and total protein (Ambali et al., 2007; Lemma and Morges, 2009). Acids and bases are usually added continuously to the body fluids as a result of either ingestion or production during cellular metabolism (Srikandakumar et al., 2003). However, environmental and animal factors as well as thermoregulatory mechanisms exert significant effect on the energy exchange between the animal and the environment leading to changes, occurring in the normal acid-base balance. In order to combat these changes, the body utilizes three basic mechanisms, which are chemical buffering and respiratory adjustment of blood carbonic acid as well as excretion of H^+ or HCO_3^- by the kidneys (Srikandakumar et al., 2003). Heat stress has been known to cause respiratory alkalosis as a result of hyperventilation and compensation, which often result in urinary HCO_3^- loss in an attempt to balance the ratio of carbonic acid to bicarbonate in the blood (Benjamin, 1978).

Na^+ and K^+ have significant effects on the hydration status of the body system. Increased extracellular fluid osmolality (increased sodium concentration) or decreased K^+ plasma concentration indicates dehydration, which may eventually stimulate drinking mechanism (Hindchiff et al., 2002). During heat stress or dehydration, there is reduced supply of blood to the kidneys, because heat stress causes peripheral vasodilation in order to lose heat (Srikandakumar et al., 2003). All serum biochemicals, except total protein, have been reported to be significantly affected by workload in donkeys (Lemma and Moges, 2009). During exercise, water, Na^+, K^+ and Cl^- are lost in sweat and this may lead to electrolyte imbalance (Hinchdiff et al., 2002). Na^+, Cl^-, K^+ and bicarbonate are responsible for maintaining electrical neutrality and extracellular fluid by exerting enough osmotic pressure during low water intake. Donkeys subjected to stress due to feed deprivation did not show any significant change in these ions (Gupta et al., 2011). In the study conducted by Gupta et al. (2011), the response of donkeys was low ingestion of water and feed and was less severe in donkeys than the mules because of the lighter body weight and lesser nutrient requirement in the donkeys compared to the mules. Donkeys subjected to different workload and different categories of body condition have been known to show variations in concentrations of most serum biochemical compounds. The variations have been attributed to differences in metabolic and electrolyte regulatory mechanisms that control their concentrations in plasma and extracellular fluid (Lemma and Morges, 2009).

Malondialdehyde

Malondialdehyde concentration is a biomarker for lipid peroxidation. Lipid peroxidation is the oxidative deterioration of polyunsaturated lipids, which leads to the production of a degraded product called malondialdehyde (Belge et al., 2003). Malondialdehyde is a by-product of lipid peroxidation. Lipid peroxidation has been shown to cause profound alterations in the structural organization and function of cell membranes, including decreased membrane fluidity, increased membrane permeability, inactivation of membrane-bound enzymes and loss of essential fatty acids.

Lipid peroxidation is a complex process which leads to membrane damage and cell death (Belge et al., 2003). Malondialdehyde is formed from metabolism of arachidonic acid to endoperoxides by fatty acid cyclo-oxygenase. These endoperoxides because of their unstable property are rapidly converted into other products, including malondialdehyde. It has been established that increased malondialdehyde values can be due to stress (Belge et al., 2003; Ambali et al., 2010a, b, c; Asala et al., 2011). Draper and Hadley (1990) assert that the generation of malondialdehyde *in vivo* increases with exposure to environmental oxidants. The lipoperoxidative alteration in the structural and functional components of the erythrocyte membranes compromises their integrity, making it prone to increased erythrocyte fragility (Ambali et al., 2010c). Olaifa et al. (2012b) observed an increase in malondialdehyde after subjecting the donkeys to packing. Thus tissue malondialdehyde concentration may be used as biomarker of oxidative stress in donkeys subjected to stress due to packing. This requires further investigation.

Antioxidants as anti-stress agents

Antioxidants are substantial defense network that combat oxidative stress imposed by several stressors, including physical exercise (Asha et al., 2005). Antioxidants provide cells with protection collectively or individually (Ghiselli et al., 2000) against reactive oxygen and nitrogen species. Cells are equipped with an effective and complex antioxidant system, including protective enzymes and biological antioxidants such as superoxide dismutase, catalase, glutathione peroxidase, glutathione, ascorbic acid and vitamin E (Vani et al., 2010). Glutathione peroxidase and catalase remove H_2O_2, while superoxide dismutase catalyzes the dismutation of O_2^- to form H_2O_2 and O_2 (Aruoma, 1998). Natural antioxidants are very important in maintaining high growth levels, reproduction and immunocompetence (Surai, 2006). Several compounds possessing antioxidant properties, which are capable of neutralizing free radicals include fat-soluble (vitamin E and carotenoids) (Ajakaiye et al., 2011) and water-soluble (Ambali et al., 2010a) compounds. Thus antioxidant defense has been considered to be a key element in preventing lipid peroxidation (Henkel, 2011).

ASCORBIC ACID AS AN ANTIOXIDANT AND ANTI-STRESS AGENT

Ascorbic acid is a potent antioxidant compound used to mitigate adverse effects of stress in livestock (Ayo and Sinkalu, 2007), including the donkey (Olaifa et al., 2012b). It has been used as a dietary supplement to pets, even if they are capable of ascorbic acid synthesis (Wang et al., 2001) because ascorbic acid is depleted from the body in stress situations (Hesta et al., 2008). Ascorbic acid, as an antioxidant, detoxifies the harmful reactive oxygen species by giving free molecules of hydrogen, especially when the body's natural antioxidants are exhausted and overwhelmed (Minka and Ayo, 2007b). Ascorbic acid recycling by its regeneration from its oxidized forms such as semidehydroascorbyl radical and dehydroascorbic acid has key functions in maintaining redox homeostasis. It is perhaps the most important water-soluble antioxidant in plasma and plays a fundamental role in the antioxidant defense, both as a scavenger and possibly in the regeneration of other antioxidants (Frei et al., 1989; Urban-Chmeil et al., 2009). It has been observed that ascorbic acid reduces the adverse effects of stress on goats when administered before the commencement of transportation (Minka and Ayo, 2010). Similarly, ascorbic acid has been shown to ameliorate chlorpyrifos-induced erythrocyte fragility (Ambali et al., 2010b), and stress induced by handling, loading, transportation and meteorological factors (Minka et al., 2009; Minka and Ayo, 2010).

Ghanem et al. (2007) also show that ascorbic acid may be used easily by farmers to alleviate water deprivation stress in the semi-arid or arid regions, where scarcity of water is a common challenge imposed by harsh weather and long road transportation. Ascorbic acid as a dietary substance is effective in the amelioration of transportation stress in livestock, including pullets (Minka and Ayo, 2008) and goats (Minka and Ayo, 2011).

Ascorbic acid has the ability to sequester the singlet oxygen radical, stabilize the hydroxyl radical, and regenerate reduced vitamin E back to its active state, which functions to end peroxidation of cellular lipid membranes (Kaminski and Boal, 1992; Harrison and May, 2009).

Ascorbic acid has been shown to induce a lower-frequency fatigue (indicates less muscle damage) (Jakeman and Maxwell, 1993). It has also been demonstrated that ascorbic acid can control the rise of reactive oxidant species, formed during exercise (Shephard et al., 1974; Tauler et al., 2003). If not controlled, reactive oxygen species have the ability to react with cell membranes and damage them, by initiating lipid peroxidation. Kaminski and Boal (1992) examined the relationship between ascorbic acid given to 19 men for three days before exercise (and seven days after) and the muscle damage induced by two bouts of eccentric exercise.

The authors concluded that ascorbic acid reduced muscle damage; however, they did not measure the indices of oxidative stress other than muscle soreness. Supplementation with two 500 g dosages of ascorbic acid for one day is associated with a decreased shift from pre-exercise to post-exercise pro-oxidant activity, when compared to a placebo. The same dosage given over a two-week period did not elicit such great change in peroxidation activity as a one-day supplementation provided (Alessia and Blasi, 1997). This difference may be related to the fact that a one-day dose of ascorbic acid, helped to regenerate other antioxidants in the body such as vitamin E. The authors hypothesize that a two-week period of ascorbic acid supplementation may replenish other antioxidants and then lead to its pro-oxidant properties within the body (Alessia and Blasi, 1997), likely via the Fenton reaction. Thus, there is an obvious lack of research addressing antioxidants and exercise, particularly ascorbic acid. Although the preliminary information on the obtained data seems promising, there is the need for further research in this area.

THE USE OF ASCORBIC ACID IN DONKEYS

Dey et al. (2010) reported a case of heat stress in donkeys and the ameliorative effect of ascorbic acid. Although there is paucity of information in the literature to support the prophylactic use of ascorbic acid in donkeys, its use has been demonstrated in horses, their counterparts (Snow et al., 1987). The Equidae family is known to synthesize ascorbic acid; thus ascorbic acid is a non-essential vitamin. It has been shown that when all animals are subjected to different forms of stressors, there is usually a significant depletion of ascorbic acid. They would, therefore, need to be supplemented with this vitamin in order to ensure balance between oxidants and antioxidants, thereby preventing oxidative stress.

Dey et al. (2010) show that ascorbic acid could serve as a curative agent in donkeys during heat stress. Olaifa et al. (2010a and b) reports that ascorbic acid ameliorated the effect of work stress on rectal temperature and erythrocyte osmotic fragility in donkeys. The uses of antioxidants have not been fully elucidated in donkeys, and there is a need to evaluate its prophylactic use in work stress, especially in those that are subjected to work and adverse environmental conditions.

CONCLUSION

Available information in the literature has shown that using donkeys as pack animals is a cheap and affordable means by which farm produce can be transported in many tropical regions of the world, including Africa. Furthermore, because of the nature of the donkeys' major role in traction and the general view of the people of their 'supposed adaptation', they are often overused and exposed to various stressors, such as work stressors, thermal environment stress and poor management, which eventually can lead to poor performance. Stress experienced in the donkeys may lead to increased production of free radicals. Ascorbic acid, a potent antioxidant, has the ability to ameliorate the effect of free radicals. The possible oxidative effect of packing and the ameliorative role of ascorbic acid in donkeys are yet to be explored. Therefore, it is proposed that more work should be done on the effects of ascorbic acid, and, indeed, other potent antioxidants on work performance of donkeys, reared in the tropical regions of the world, especially in Africa.

REFERENCES

Agrawal A, Sharma B (2010). Pesticides induced oxidative stress in mammalian systems: a review. Int. J. Biol. Med. Res. 1: 90-104.

Ajakaiye JJ, Cuesta-Mazorra M, Garcia-Diaz JR (2011). Vitamin C and E can alleviate adverse effects of heat stress on liveweight and some egg quality profiles of layer hens. Pakistan Vet. J. 31: 1-5.

Alessia HM, Blasi ER (1997). Physical activity as a natural antioxidant booster and its effect on a healthy lifestyle. Res. Quart. Ex. Sport. 68: 292-302.

Altan O, Pabuccuoglu A, Koyanliogu S, Bayraktav H (2003). Effect of heat stress on oxidative stress, lipid peroxidative and some stress parameters in broilers. Br. Poult. Sci. 44: 545-550.

Ambali S, Akanbi D, Igbokwe N, Shittu M, Kawu M, Ayo J (2007). Evaluation of sub-chronic poisoning on haematological and serum biochemical changes in mice and protective effect of vitamin C. J. Toxicol. Sci. 32: 111-120.

Ambali SF, Akanbi D, Shittu M, Giwa A, Oladipo OO, Ayo JO (2010a). Chlorpyrifos-induced clinical, haematological and biochemical changes in Swiss Albino mice: mitigating effect by co-administration of vitamins C and E. Life Sci. J. 7: 37-44.

Ambali SF, Ayo JO, Ojo SA., Esievo KAN (2010b). Ameliorative effect of vitamin C on chronic chlorpyrifos-induced erythrocyte osmotic fragility in Wistar rats. Hum. Exp. Toxicol. 30: 19-24.

Ambali SF, Ayo JO, Ojo SA, Esievo KAN (2010c). Vitamin E protects Wistar rats fron chlorpyrifos-induced increase in erythrocyte fragility. Food Chem. Toxicol. 48: 3477-3480.

Armario AA, Escorihuela RM, Nadal R (2008). Long-term neuroendocrine and behavioural effects of a single exposure to stress in adult animals. Neurosci. Biobehav. Rev. 32: 1121-1135.

Aruoma OI (1998). Free radicals, oxidative stress and antioxidants in human health diseases. J. Am. Oil Chem. Soc. 75: 199-213.

Asala OO, Ayo JO, Rekwot PI, Minka NS, Omoniwa DO, Adenkola AY (2011). Effect of ascorbic acid administration on erythrocyte osmotic fragility of pigs transported by road during the hot-dry season. Vet. Res. Commun. 35: 245-254.

Asha DS, Subramanyam MVV, Vani R, Jeevaratnam K (2005). Adaptation of the antioxidant system in erythrocytes of trained adult rats: impact of intermittent hypobaric-hypoxia at two altitudes. Comp. Biochem. Physiol. 140: 59-67.

Asha ME., Salim MO, Amel OB, Ibrahim AA (2006). The aerobic bacteria of equine wound infection associated with type of animal use and sampling. Sudan J. Vet. Sc. Anim. Husb. 45: 1-7.

Ayo JO, Dzenda T, Zakari FO (2008). Individual and diurnal variations in rectal temperature, respiration and heart rate of pack donkeys during the early rainy season. J. Equine Vet. Sci. 28: 281-289.

Ayo JO, Obidi JA, Rekwot PI (2011). Effect of heat stress on the well-being, fertility and hatchability of chickens in the Northern Guinea Savannah zone of Nigeria: A review, ISRN Vet. Sc. 2011, Art. #838606, 10pages.

Ayo JO, Oladele SB, Fayomi A (1996). Effects of heat stress on livestock Production: A review. Nig. Vet. J. (Special Edition). 1: 58-68.
Ayo JO, Oladele SB, Fayomi A (2002). Diurnal fluctuations in rectal temperature of the Red Sokoto goat during the harmattan season. Res. Vet. Sc. 66: 7-9.
Ayo JO, Sinkalu VO (2007). Effect of ascorbic acid on diurnal variations in rectal temperature of Shaver Brown pullets during the hot-dry season. Int. J. Poult. Sci. 6: 642-646.
Belge F, Cinar A, Selcuk M (2003). Effects of stress produced by adrenocorticotropin on lipid peroxidation and some antioxidants in vitamin C treated and nontreated chickens, S. Afr. J. Anim. Sci. 33:201-205.
Benjamin MM (1978). Fluids and electrolytes. Outline of Veterinary Clinical Pathology. Iowa state University Press, Ames, IO, Pp. 213-214.
Bhatti R, Shaikh DM (2007). The effect of exercise on blood parameters. Pakistan J. Physiol. 3:1-3.
Blench RM (1993). Ethnographic and linguistic evidence for the pre-history of African ruminant livestock, horses and ponies. In: Shaw T, Sindar P, Andah B, Okpoko A (eds), The Archaeology of Africa. Food, metals and Towns. Routledge, London, UK, pp. 71-103.
Boning D, Massen N, Pries A (2010). The haematocrit paradox - how does blood doping really work? Int. J. Sports. 5: 16-22.
Brzezinska-Slebodzinska E (2003). Species differences in the susceptibllity of erythrocytes exposed to free radicals in vivo. Vet. Res. Comm. 27: 211–217.
Cable NT, Drust B, Gregson WA (2007). Impact of altered climatic conditions and altitude on circadian physiology. Physiol. Beh. 90: 267-273.
Clutton-Brock J (1992). Horse power: A history of the horse and the donkey in human societies. Havard University Press, Cambridge, Massachusetts, USA. 192P.
Cuniberti B, Badino P, Odore R, Girardi C, Re G (2010). Effects induced by exercise on lymphocyte β-adrenergic receptors and plasma catecholamine levels in performance horses. Res. Vet. Sc. 92: 116-120.
Dey S, Dwivedi SK, Malik P, Panisup AS, Tandon SN, Singh BK (2010). Mortality associated with heat stress in donkeys in India. Vet. Rec. 166: 143-145.
Draper HH, Hadley M (1990). A review of recent studies on the metabolism of exogenous and endogenous malondialdehyde. Int. Society Study Xenobiot. 9: 901-907.
Dyang-Fordjour KT, Asare-Mantey G, Bediako JA, Otchere EO (2003). The uses and management practices for draught animals in the northern zone of Ghana. Draft Anim. News. 39: 2-5.
Fielding D, Krause P (1998). The Tropical Agriculturist-Donkeys. Edited by Coste R, Smith AJ, Macmillian Education Ltd., London and Babingstoke, 119 pp.
Frei B, England L, Ames BN (1989). Ascorbate is an outstanding antioxidant in human blood plasma. Proc Nat Acad Sc USA. 86: 6377-6381.
Francis PA (1988). Ox draught power and agricultural formation in northern Zambia. Agric. Syst. 27:15-28.
Gabriel H, Schwarz L, Steffens G, Kinderman W (1992). Immunoregulatory hormones circulatory leucocyte and lymphocyte subpopulations before and after exercise. Eur. J. Appl. Physiol. 65: 164-170.
Geverink NA, Kappers A, van de Burgwal JA, Lambooij E, Blokhuis HJ, Wiegant VM (1998). Effects of regular moving and handling on the behaivoural and physiological responses of pigs to pre-slaughter treatment and consequences for subsequent meat quality. J Anim. Sci. 76: 1806-1811.
Ghanem AM, Jabel LS, Abi SM, Barbour EK, Hamadeh SK (2007). Physiological and chemical responses in water deprived Awasi ewes treated with vitamin C. J. Arid Environ. 72: 141-149.
Ghiselli A, Serafini M, Natella F, Scaccini C (2000). Antioxidant capacity as a tool to assess redox status: critical view and experimental data. Free Rad. Biol. Med. 29: 1106-1114.
Grandin T (1997). Assessment of stress during handling and transportation. J. Anim. Sci. 75: 249-257.
Grosenburgh DA, Reinmeyer CR, Figueiredo MD (2011). Pharmacology and therapeutics in donkeys. Eq. Vet. Ed. 23: 523-530.
Gupta K, Mamta YP, Yadav MP (1999). Effect of feed deprivation on biochemical indices of equids. J. Eq. Sci. 10: 33-38.
Hanzawa K, Wantanabe S (2000). Changes in osmotic fragility of erythrocytes during exercise in athletic horses. J. Eq. Sci. 11: 51-61.
Harrison FE, May JM (2009). Vitamin C function in the brain: vital role of the ascorbate transporter SVCT2. Free Rad. Biol. Med. 46: 719-730.
Henkel RR (2011). Leucocytes and oxidative stress: dilemma for sperm function and male fertility. Asian J. Androl. 13: 43-52.
Heise K, Puntarulo S, Portner HO, Abele D (2003). Production of reactive oxygen species by isolated mitochondria of the antartic bivalve, *Laternula elliptica* (King and Broderip) under heat stress. Biochem. Physiol. 134: 79-90.
Hesta M, Ottermans C, Krammer-Lucas S, Zentek J, Helweg P, Buyse J, Janssens GPS (2008). The effect of vitamin C supplementation in healthy dogs on antioxidative capacity and immune parameters. J. Anim. Physiol. Anim. Nut. 93: 26-34.
Hinchdiff KW, Geor RJ, Pagan E (2002). Equine exercise physiology. Eq. Vet. J. 34: 23-26.
Homer BS (1993). Grain and Protein Supplement for Beef Cattle on Pasture. Published by University of Missouri, Pp. 122-125.
Inns FM (1979). Animal Power in Agricultural Production Systems. Agricultural Mechanization Group of Agriculture Services Division, FAO, Iringa, 143 pp.
Jakeman P, Maxwell S (1993). Effect of antioxidant vitamin supplementation on muscle function after eccentric exercise. Eur. J. Appl. Physiol. 67: 426-430.
Kaminski M, Boal R (1992). An effect of ascorbic acid on delayed-onset muscle soreness. Pain, 50: 317-321.
Kataria N, Kataria AK (2010). Assessement of stress due to hot ambience in donkeys from arid tracts in India. J. Stress Physiol. Biochem. 6: 12-17.
Kimura B, Marshall FB, Chen S, Rosenbom S, Moelham PD, Tuross N, Sabin RC, Peters J, Barich B, Yohaness H (2010). Ancient DNA from Nubian and Somali wild ass provide insights into donkey ancestry and domestication. Proc. Royal Soc. B: Biol Sci.
Knowles TG, Warris, PD (2000). Stress physiology during transport. In: Grandin, T. (Ed.), Livestock Handling and Transport. CAB International, Wallingford, Oxon, U.K., Pp. 385-407.
Laedwig J. (2000). Chronic intermittent stress: a model for the study of long term stressors. In: Moberg GP, Mench JA (Eds.), The Biology of Animal Stress. CAB International, Wallingford, Oxon, UK, Pp. 159-169.
Lemma A, Moges M (2009). Clinical, haematological and serum biochemical reference values of working donkeys (Equus asinus) owned by transport operators in Addis Ababa. Livest. Res. Rural. Dev. 21: 89-92.
Lorena D, Da Ponte A, Cozzi M, Carbone A, Pomati M, Nava I, Capellini MD, Fiorelli G (2006). Changes in erythropoiesis, iron metabolism and oxidative stress after half-marathon. Int. J. Emerg. Med. 1: 30-34.
Martin CM, Smith DG (2005). The impact of donkey ownership on livelihoods of female peri-urban dwellers in Ethiopia. Trop. Anim. Health Prod. 37: 67-86.
Martin CM, Feseha G, Smith G (2005). The impact of access to animal health services on donkey health and livelihoods in Ethiopia. Trop. Anim. Health Prod. 37: 47-65.
McCarthy DA, Dale MM (1988). The leucocytosis of exercise. A review and model. Sports Med. 6: 333-336.
Mengistu A, Smith DG, Yoseph S, Nega T, Zewdie W, Kassahun WG, Taye B, Firew T (2005). The effect of providing feed supplementation and antihelminthic to donkeys during late pregnancy and lactation on liveweight and survival of dams and their foals in Central Ethiopia. Trop. Anim. Health Prod. 37: 21-33.
Minka NS, Ayo JO (2007a). Effect of shade provision on some physiological parameters, behaviour and performance of pack donkeys (Equus asinus) during the hot-dry season. J. Eq. Sci. 18: 39-46.
Minka NS, Ayo JO (2007b). Physiologlcal responses of goats treated with ascorbic acid during the hot-dry season. Anim. Sci. J. 78: 164-172.
Minka NS, Ayo JO (2007c). Road transportation effect on rectal temperature, respiration and heart rates of ostritch (Struthio camelus) chicks. Vet. Arhiv. 77: 39-46.

Minka NS, Ayo JO (2008). Haematology and behaviour of pullets transported by road and administered with ascorbic acid during the hot-dry season. Res. Vet. Sci. 85: 389-393.
Minka NS, Ayo JO (2009). Physiological responses of food animals to road transportation stress. Afr. J. Biotech. 8: 7415-7427.
Minka NS, Ayo JO (2010). Physiological responses of erythrocytes of goats to transportation and modulatory role of ascorbic acid. J. Vet. Med. Sci. 72: 875-881.
Minka NS, Ayo JO (2011). Modulating role of Vitamin C and E against transport-induced stress in pullets during the hot-dry conditions. ISRN Res. Vet. Sc. 2011, Art. #497138, 10 pages.
Minka NS, Ayo JO, Sackey AKB, Adelaiye AB (2009). Assessment and scoring of stresses imposed on goats during handling, loading, road transportation and unloading, and the effect of pretreatment with ascorbic acid. Livest. Sci, 125: 275-282.
Minka NS, Fayomi A, Ayo JO (2004). Effect of road transportation and ascorbic acid on the haematological parameters of pullets during the hot-dry season. Proceedings of the 38th Ann Conf Agric Society Nigeria, Lafia, Pp. 653-659.
Nengomasha EM, Pearson RA, World GA (2000). Empowering people through donkey power into the next millennium. In: Kaumbutho PG, Pearson RA, Simalenga TE (Eds.). Empowering Farmers with Animal Traction, Proceedings of a Workshop of the Animal Traction Network for Eastern and Southern Africa, held 20-24 September, 1999, Mpumalanga, South Africa, 344 pp.
Olaifa F, Ayo JO, Ambali SF, Rekwot PI, Minka NS (2012a). Rectal temperature responses of donkeys subjected to load-carrying (packing) during the harmattan season in Nigeria. Trop. Anim. Health. Prod. Doi:10.1007/S11250-012-0242-x.
Olaifa F, Ayo JO, Ambali SF, Rekwot PI (2012b). Effect of packing on erythrocyte osmotic fragility in donkeys administered with ascorbic acid. Onderstepoort J. Vet. Res. (In Press).
Olaifa F, Ayo JO, Ambali SF, Rekwot PI (2012c). Haemato-biochemical responses of pack donkeys administered ascorbic acid during the harmattan season. J. Vet. Med. Sc. (In Press).
Oyewale JO, Dzenda T, Yaqub L, Akanbi D, Ayo J, Owoyele O, Minka N, Dare T (2011). Alterations in the osmotic fragility of camel and donkey erythrocytes caused by temperature, pH and storage. Vet. Arhiv. 81: 459-470.
Pal Y, Kumar S, Gupta AK (2002). Blood gases, acid-base and physiological indices in donkeys as pack animals. Draught Anim. News, 37: 27-33.
Pandey KB, Risvi SI (2009). Protective effect of resveratrol on markers of oxidative stress in human erythrocytes subjected to in vitro oxidative insult. Phytotherapy Res. doi: 101002/ptr.2853.
Pearson RA (2000). Introduction to the project use and management of donkeys by poor societies in Peri-urban areas in Ethiopia. In: Smith DG, Agajie T, More L (Eds.), Alleviating Poverty in Peri-urban Ethiopia by Improving the Health, Welfare and Management of Donkeys, CTVM, Edinburgh, Pp. 2-5.
Pearson RA, Dijkman JT, Krecek RC, Wright P (1998). The effect of density and weight of load on the energy cost of carrying loads by donkeys and ponies. Trop. Anim. Health. Prod. 30: 67-78.
Pearson RA, Meritt J (1991). Intake, digestion and gastro-intestinal transit time in resting donkeys and ponies and exercised donkeys given ad libitum hay and straw diets. Eq. Vet. J. 23: 339-343.
Pearson RA, Vall E (1998). Performance and management of draught animals in agriculture in sub-Saharan Africa: A review. Trop. Anim. Health. Prod. 30: 309-324.
Pederson BK, Hoffman-Goetz L (2000). Exercise and the immune system: Regulation, integration and adaptation. Physiol. Reviews. 80: 1055-1081.
Plyaschenko SI, Sidorov VT (1987). Stress in Farm Animals. Agropromizdat, Moscow, 192 pp (in Russian).
Robson PJ, Altson TD, Myburgh KH (2003). Prolonged suppression of the innate system in the horse following an 80 km endurance race. Equine Vet. J. 35: 133-137.
Rose RC (1986). Ascorbic acid transport in mammalian kidney. American J. Physiol. 250: 627-632.
Sahin E, Gusmulu S (2004). Cold-stress-induced modulation of antioxidant defence: role of stressed conditions in tissue injury followed by protein oxidation and lipid peroxidation. Int. J. Biometeorol. 48: 165-171.
Sapolsky A, Wortsman J, Thomas L, Paus R, Solomon S (2000). Corticotropin releasing hormone and proopiomelanocortin involvement in the cutaneous response to stress. Physiol. Reviews, 80: 979-1020.
Shephard RJ, Campbell R, Rimm P (1974). Vitamin E, exercise and the recovery from physical activity. Eur. J. Appl. Physiol. 33: 119-126.
Singh G (2000). Empowering farmers through animal traction in India. Proceedings of the workshop of the Animal Traction Network for Eastern and Southern Africa held 20-24 September, 1999, ISBN 0907146104.
Smith BL, Jones JH, Hornof WJ, Miles JA, Longworth KE, Willits NH (1995). Effects of road transport on indices of stress in horses. Eq. J. Vet. Sci. 30: 9-12.
Smith DJ, Pearson RA (2005). A review of the factors affecting the survival of donkeys in semi-arid regions of sub-Saharan Africa. Trop. Anim. Health Prod. 37: 1-19.
Sneddon JC, Boomker E, Howard CV (2006). Mucosal surface area and fermentation activity in the hind gut of hydrated and chronically dehydrated working donkeys. J. Anim. Sci. 84: 119-124.
Snow DH, Gash SP, Cornelius J (1987). Oral administration of ascorbic acid to horses. Equine Vet. J. 19: 520-523.
Srikandakumar A, Johnson EH, Maghoub O (2003). Effect of heat stress on respiratory rate, rectal temperature and blood chemistry in Omani and Australian Merino sheep. Small Rumin. Res. 49: 193-198.
Starkey P (1992). Changes in animal traction in Africa and Asia: implications for development. In: den Hartog G, van Huis JA (eds), The role of Draught animal Performance in rural Development. Proceedings of an international seminar held 2-12 April 1990, Edinburgh, Scotland, The Netherlands, pp 11-24.
Stott GH (1981). What is animal stress and how is it measured? J. Anim. Sci. 52: 150-153.
Surai PF (2006). Natural antioxidants in poultry nutrition: New developments. 16th European Symposium on Poultry Nutrition, Pp. 669-676.
Tauler P, Aguilo A, Gimeno I, Fuentespina E, Tur JA, Pons A (2003). Influence of vitamin C diet supplementation on endogenous antioxidant defence during exhaustive exercise. Eur. J. Physiol. 446: 658-664.
Tonnesen E, Christensen NJ, Brinklov MM (1987). Natural killer cell activity during cortisol and adrenaline infusion in healthy volunteers. Eur. J. Clin. Invest. 17: 497-503.
Urban-Chmeil R, Kankofer M, Wernicki A, Albera E, Puchalski A (2009). The influence of different doses of α-tocopherol and ascorbic acid on selected oxidative stress parameters in *in vitro* culture of leucocytes isolated from transported calves. Livest. Sci. 124: 89-92.
Vani R, Shiva Shankar RCS, Asha Devi S (2010). Oxidative stress in erythrocytes: a study of an effect of antioxidant mixtures during intermittent exposures to high altitude. Int. J. Biometeorol. 54: 553-562.
von Borell EH (2001). The biology of stress and its application to livestock handling and transportation assessement. J. Anim. Sci. 79: 260-267.
Wang S, Berge, G, Sund R (2001). Plasma ascorbic concentrations in healthy dogs. Res. Vet. Sci. 71: 33-35.
Wilson TR (1991). Equines in Ethiopia. In: Fielding, D. and Pearson, R. A. (eds), Donkey, mules and horses in tropical agricultural development. Proceedings of colloquium held 3-6 September, Edinburgh, UK, pp 366.
Yousef MK (1991). Physiological responses of donkeys to heat stress. In Fielding D, Pearson RA (Eds.). Donkeys, Mules and Horses Tropical Agricultural Development CTVM: Edinburgh, P. 96 (Abstract).
Youseph S, Smith DG, Mengistu A, Teklu F, Firwe T, Betere Y (2005). Seasonal variation in the parasite burden and body condition of working donkeys in East Shewa and West Shewa Region of Ethiopia. Trop. Anim. Health Prod. 37:35-45.

17

Biosorption of cadmium (ii) ion from aqueous solution by *Afzelia africana*

Onwu, F. K.[1] , Ogah, S. P. I.[2] and Ngele, S. O.[2]

[1]Department of Chemistry, Michael Okpara University of Agriculture Umudike, P.M.B 7267, Umuahia, Abia State, Nigeria.
[2]Department of Industrial Chemistry, Ebonyi State University, P.M.B 053, Abakaliki, Ebonyi State, Nigeria.

The batch adsorption of cadmium (II) ion from aqueous solution using low-cost adsorbent of biological origin, *Afzelia africana* shell under different experimental conditions was investigated in this study. The influences of initial Cd (II) ion concentration, initial pH, contact times and temperature were reported. Adsorption of Cd (II) was found to be pH dependent and the results indicate that the optimum pH for its removal from aqueous solution was 5. No marginal effect on the biosorption of cadmium was detected for temperatures between 298 and 313K as observed from their Langmuir isotherm constants (q_{max}) at different temperatures. q_{max} values obtained at the different temperatures of 298, 303 and 313K were 13.59, 14.04 and 14.27 mg g^{-1}, respectively. The adsorption equilibrium shows that the process followed both Freundlich and Langmuir models with Freundlich giving a better fit for the adsorption data in comparison to the Langmuir model. The fit of the adsorption data into Freundlich model shows that the adsorption process was predominantly a physisorption. The results reveal that cadmium (II) was considerably adsorbed on the *A. africana* shell and could serve as an economical method for the removal of cadmium from aqueous solutions.

Key words: Adsorption isotherms, adsorption kinetics, *Afzelia africana,* biosorbent, biosorption.

INTRODUCTION

The presence of considerable amounts of heavy metals in industrial wastewaters poses serious environmental threats. The physico-chemical methods such as precipitation, filteration, evaporation, ion-exchange, membrane technologies and electrochemical methods, which have been employed in the sequestration of these metal ions, have been found to be capital intensive (Matheickal and Yu, 1999). These conventional methods in addition have been found to be less efficient and ineffective, especially when low concentrations of these heavy metal ions (in the range of 1- 100 mg L^{-1}) are involved (Volesky, 1990). Some of these techniques sometimes generate sludge and other toxic waste products that require secondary treatment and when released into the environment culminates into increased environmental pollution load

Researches into the use of agricultural by-products as biosorbents for the removal of heavy metals from aqueous solutions have been on the increase. The increased interest in the use of biomass is their abundant occurrence in nature, which makes them less costly as compared with conventional adsorbents like activated carbon. Some of the major advantages of biosorbents include: Competitive performance, heavy metal selectivity, cheapness, ease of regeneration, known process equipment, possible metal recovery and no sludge generation. These merits in the use of biosorbents have led numerous researches into the use of various plants/agricultural by-products in the removal of heavy metals from industrial effluents. The binding mechanisms of heavy metals by biosorption have been explained to result from different physical and chemical interactions between cell wall ligands (active sites) and the adsorbates (metal ions) by ion-exchange, complexation, coordination and microprecipitation (Gardea-Terresdey et al., 1996; Gang and Weixing, 1998). The diffusion of metals from the bulk solution to active sites of biosorbents occurs predominantly by passive transport mechanism (Veglio and Beolchini, 1997) and various functional groups such as carboxyl, hydroxyl, amino and phosphate present on the cell wall of biosorbents can bind the heavy metals (Avery and Tobin, 1993). The use of various plant biomass for adsorption of heavy metals in solution has been reported in literature and some of these include: African white star apple shell (Anusiem et al., 2010), maize leaf (Babarinde et al., 2006); unmodified and modified maize cob (Igwe and Abia, 2007), rice husk (Ong et al., 2007), shear butter seed husks (Eromosele and Otitolaye, 1994), sago waste (Quek et al., 1998), pomelo peel (Saikaew et al., 2009), husk of bengal gram (Ahalya et al., 2005), groundnut husks (Okieimen et al., 1991), brown seaweed (Antunes et al., 2003), cassava waste biomass (Horsfall et al., 2004), tobacco stems (Wei et al., 2008), chemically modified *Rhizopus nigrigans* (Bai and Abraham, 2002), cone biomass of *Thuja orientalis* (Nuhoglu and Oguz, 2003), use of submerged aquatic plant *Ceratophyllum demersum*, (Keskinkam et al., 2004), modified pine tree (Argun et al., 2005) among others.

The current work focused on the use of an economically cheaper adsorbent of natural origin, *Afzelia africana* in the adsorption of cadmium (II) ion from aqueous solutions. Cadmium which is a widely used material in the industries is extremely toxic even in relatively low dosages and is one of the major heavy metals responsible for causing kidney damage, renal disorder, high blood pressure, bone fraction and destruction of red blood cells (Drash, 1993). In this work, the influence of different parameters on adsorption of cadmium such as initial pH, initial cadmium concentration, contact time and temperature were investigated. The kinetic and equilibrium of the biosorption process were evaluated and maximum adsorption capacity of the adsorbent was determined using the Langmuir isotherm model.

MATERIALS AND METHODS

Adsorbent preparation

The *A. africana* seeds used in this work were basically sourced from local market in Abakaliki in Ebonyi State of Nigeria. The shells were extensively washed with distilled water to remove particulate material from their surface. They were cut into small pieces and air-dried and finally oven-dried at 80°C for 12 h in preparation for the adsorption analysis. The oven-dried shells were crushed with manual blender to smaller particles and sieve analysis was performed using the sieve screen to obtain final sample size of 180 μm. Thereafter, 200 g of the screened (180 μm size) adsorbent was further soaked in 500 mL of 0.3 M HNO_3 solution, stirred for 30 min and allowed to stay for 24 h. It was then filtered through a filter paper and washed with excess deionised water. The rinsed biosorbent was later air-dried for 6 h. The treatment of the adsorbent with 0.3 M HNO_3 solution was to remove any soluble biomolecules that might interact with the metal ions during the adsorption process and to open-up the micropores of the adsorbent in readiness for the adsorption process.

Biosorption studies

All reagents used were of analytical reagent grades and double distilled deionized water was used in the sample preparation. 100 mg L^{-1} stock solution of cadmium (II) from $CdSO_4.8H_2O$ were prepared at a pH 5.0 and temperature of 30°C. From the stock solutions, working solutions of 10 - 60 mg L^{-1} of each of the adsorbates was obtained by serial dilution.

The effect of concen-tration on the adsorption of the metal ion was studied by transferring 100 mL of the different concentrations of the adsorbate (cadmium ion) into different 250 mL Erlenmeyer flasks while maintaining the pH of the solutions at 5.0. Thereafter, 0.4 g of the adsorbent was added into the flasks, corked and labelled. The flasks were placed in a rotary shaker and agitated for 120 min to attain equilibrium.

At the end of the adsorption process, the content of each flask was filtered, centrifuged and the residual metal ion concentrations (C_e) analyzed using buck scientific atomic absorption spectrophotometer (AAS) model 210 VGP. The concen-trations of the metal ions adsorbed at the different concentrations were calculated by difference.

The effect of pH on the adsorption of the metal ion was carried out by adding 0.4 g of the adsorbent into different 250 mL Erlenmeyer flasks containing 100 mL of the 60 mg L^{-1} solutions of the metal ion and then adjusting the pH of the solutions to different pH ranging from 1.0 to 6.0 using 0.1M HCl or 0.1 M NaOH. The flasks were placed in a rotary shaker and agitated for 120 min to attain equilibrium. The concentrations of the metal ions adsorbed at the different pH were calculated by difference.

Kinetic studies were also carried out by adding 0.4 g of the adsorbent into different 250 mL Erlenmeyer flasks containing 100 mL of the 60 mg L^{-1} solutions of the metal ion and agitated in a thermostated versal shaker at different contact times (10, 20, 30, 40, 50, 60, 70, 90, and 120 min). At the end of agitation time, the content of each flask was filtered with a filter paper, centrifuged and the residual concentration of the metal ions at the different contact times analyzed using AAS. The amount of metal ion adsorbed was then calculated by difference. Each of these analysis was carried out in triplicates and mean residual concentration analyzed. Blank solutions were also prepared and analyzed.

Metal uptake

The initial metal ion concentration and metal concentration at any given time and at equilibrium, respectively, were measured and the metal uptake (amount of Cd (II) ion adsorbed in mg g^{-1}) was evaluated from the mass balance expression as shown:

$$q_e = \left(C_o - C_e\right)\frac{V}{m} \qquad (1)$$

Where, V is the volume of the solution in dm^3, m the dry mass of the adsorbent in g while C_o and C_e are the initial and equilibrium concentrations of the cadmium ion in mg L^{-1}.

RESULTS AND DISCUSSION

The results of the experimental investigations for the adsorption of Cd (II) ion on the adsorbent *A. africana* are thus presented. These results are presented as amount

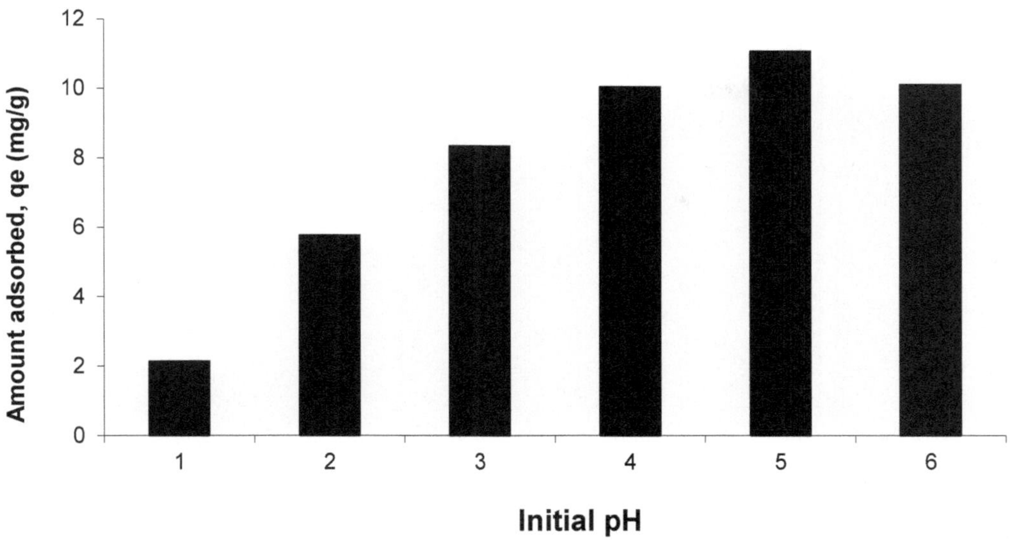

Figure 1. Effect of the initial pH on equilibrium cadmium ions adsorption by the adsorbent.

cadmium (II) ion adsorbed against the various experimental conditions which include initial concentration, initial pH, temperature and time employed.

Influence of Initial pH

pH is an important parameter in the biosorption of metal ions from aqueous solutions since the pH of a medium affects the solubility of metal ions and the concentration of the counter ions on the functional groups of the biomass cell walls. The effect of initial pH on the adsorption of Cd (II) by the adsorbent is shown in Figure 1. The experiments were not conducted beyond pH 6.0 to avoid possible hydroxide precipitation. The amount adsorbed increased with increasing initial pH showing a plateau in the range of 4.0 to 6.0 and having a peak at pH 5.0. This shows that the optimum pH in the removal of Cd (II) from the aqueous solution using the adsorbent is 5.0. The lower uptake at pH below 4.0 may possibly result from competition between hydroxonium ions and Cd (II) ions on the active sites of the adsorbent. The increased presence of these hydroxonium ions (H_3O^+) at lower pH restrict the access of the adsorbates (Cd^{2+} ions) due to repulsive forces to the adsorbent active sites and thereby leading to decreased uptake. Therefore, at lower pH values, the functional groups on the cell walls of the adsorbents are closely associated with hydroxonium ions, but when the pH is increased, the hydroxonium ions are gradually dissociated and the positively charged cadmium ions will be associated with the free binding sites. This may have led to the decreased adsorption at low pH values. Similar findings have been reported by other researchers (Antunes et al. 2003, Benguella and Benaissa, 2002).

Effect of initial metal ion concentration

The effect of concentration on the adsorption of the cadmium (II) ion is displayed as a variation of amount adsorbed with initial cadmium ion concentration (Figure 2). Results show that the equilibrium sorption capacity of the adsorbent increased with increase in initial concentration of the metal ions. The increased amount adsorbed resulting from increase in initial concentration is supported by the distribution coefficients shown in Table 1. The distribution coefficient is the concentration of cadmium in the adsorbent phase divided by the concentration in the continuous (bulk) phase. It expresses the relative effectiveness of the adsorbent in binding the metal ion. It is observed (Table 1) that for all the initial metal ion concentrations used, the distribution coefficients were all greater than unity, thus indicating an uptake of more than 50%. Results generally show that the amount adsorbed increased with initial cadmium ion concentration.

Adsorption isotherms

To study the relationship between the amounts of metal ions adsorbed (q_e, mg g^{-1}) and the equilibrium concentrations left after adsorption, adsorption isotherm models are generally applied for fitting the data. Equilibrium adsorption isotherm is fundamental for the industrial application of biosorption. This is because it gives information for comparison among different biomaterials under different operational conditions and for designing and optimizing operating procedures (Benguella and Benaissa, 2002). In this study, the Langmuir and Freundlich isotherm models were applied

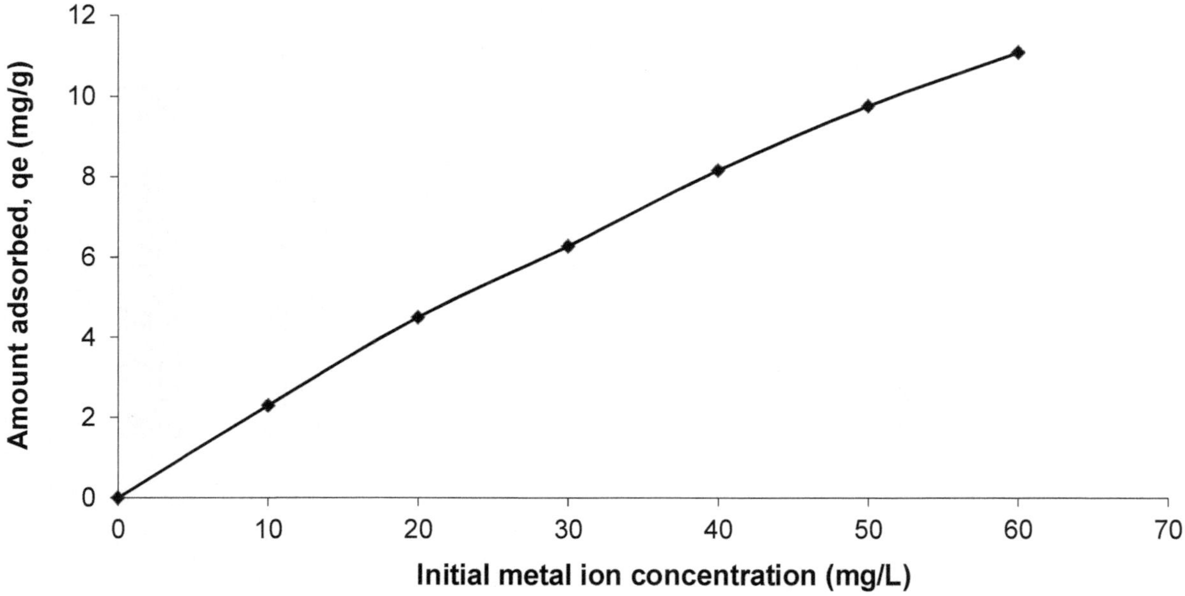

Figure 2. Effect of concentration on the biosorption of cadmium (II) ion by the adsorbent.

Table 1. Distribution coefficients for adsorption of the cadmium (II) ions by the adsorbent.

Initial concentration, C_o (mg L^{-1})	Distribution coefficient
10	11.5000
20	9.0000
30	5.0852
40	4.3908
50	3.5455
60	2.8168

to correlate our experimental data. The linear form of Langmuir adsorption isotherm is represented as:

$$\frac{C_e}{q_e} = \frac{1}{q_{\max} K_L} + \frac{C_e}{q_{\max}} \quad (2)$$

Where, C_e is the equilibrium concentration of the metal ion (mg L^{-1}), q_{max} (mg g^{-1} adsorbent) is the maximum adsorption capacity of the adsorbent corresponding to monolayer coverage or sites saturation, K_L is the equilibrium Langmuir isotherm constant, which is related to the affinity of the binding sites. K_L represents the ratio of the adsorption to desorption constants and gives information about the intensity of the adsorption process while q_{max} allows the comparison of adsorption performance, especially in the cases where the adsorbent did not attain its full saturation in experimental processes. The values of q_{max} and K_L were calculated from the slope and intercept of the Langmuir plot of C_e versus C_e/q_e (Figure 3).

On the other hand, the Freundlich equation is given as:

$$\ln q_e = \ln K_F + \frac{1}{n} \ln C_e \quad (3)$$

Where, K_F and n are the Freundlich constants characteristic of the systems, which expresses the adsorption capacity and adsorption intensity, respectively. Values of these constants were calculated from the intercept and slope of the plot (Figure 4) and are presented in Table 2. Parameters, K_F and n affect the adsorption process. Larger values of K_F and n indicate higher adsorption capacity.

The Langmuir and Freundlich constants and their corresponding coefficients of correlation (R^2) values are presented in Table 2. The high coefficients of correlation obtained from both isotherms show that both isotherms could correlate the experimental data with Freundlich giving a better fit since it presents a higher correlation coefficient. According to Karthikeyan et al. (2004), the fit of experimental data to Freundlich isotherm model indicates that the forces of adsorption by the adsorbents are governed by physical adsorption (physisorption). Therefore, the fit of the experimental data into the Freundlich model suggests that the mode of bonding of the cadmium (II) ions onto the adsorbent was by physisorption. Furthermore, the Langmuir isotherm constants were also evaluated at different temperatures ranging from 298 to 313 K and results obtained show that increase in temperature in the range studied did not significantly affect the biosorption of cadmium (II) on the adsorbent employed. Figure 5 shows the Langmuir plots at different temperatures.

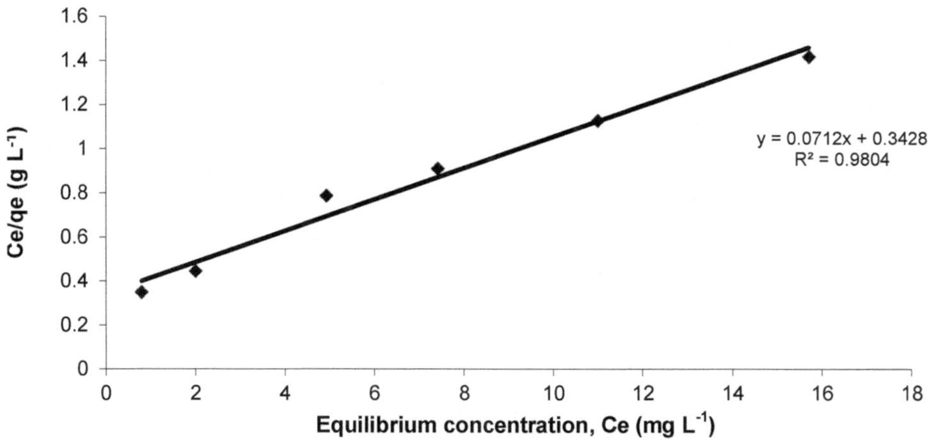

Figure 3. Langmuir isotherm for the adsorption of Cadmium (II) ion from the aqueous solution by the adsorbent at 30°C.

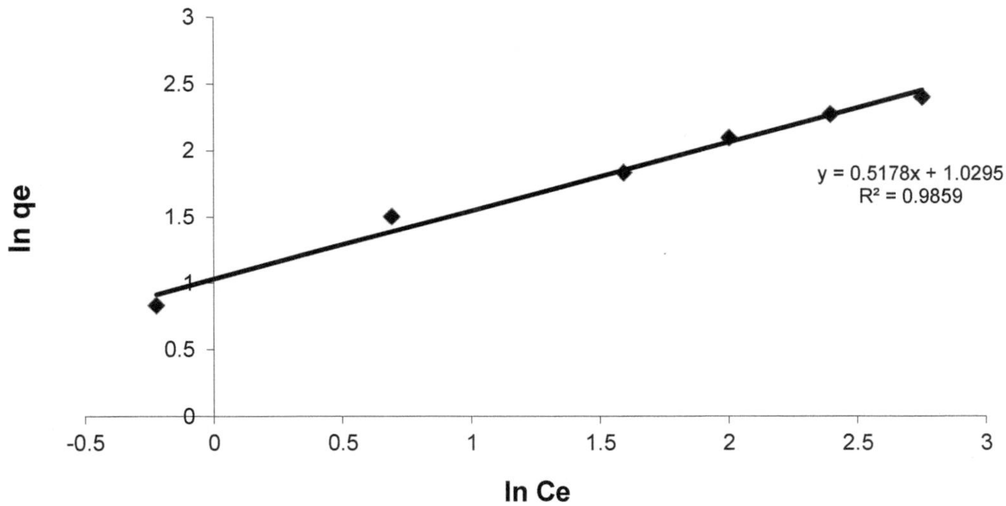

Figure 4. Freundlich isotherm for the adsorption of Cadmium (II) ion from the aqueous solution by the adsorbent at 30°C.

Table 2. Langmuir and Feundlich model's regression constants for the adsorbent at 30°C.

Type of Isotherm	Parameter	Value
	q_{max} (mg g^{-1})	14.04
Langmuir isotherm	K_L (L mg^{-1})	2.078 x 10^{-1}
	R^2	0.980
	K_F	2.800
Freundlich isotherm	N	1.932
	R^2	0.986

Heat of adsorption

The heat of adsorption was evaluated using the Langmuir parameter, K_L, and this parameter can be expressed as a function of temperature using the Arrhenius type relationship:

$$K_L = K_0 \exp\left[-\frac{\Delta H}{RT}\right] \quad (4)$$

Where, K_0 is a constant, ΔH heat of adsorption (kcal mol^{-1}), R is the gas constant and T is the absolute temperature (K). If K_L values are known for various temperatures, the heat of adsorption can be calculated from the plots of log K_L versus 1/T (Ozer and Ozer, 2003). The variation of ln K_L with reciprocal of temperature, (1/T), is shown in Figure 6 and the value of biosorption heat of adsorption obtained was 4.163 kcal mol^{-1} (17.419 kJ mol^{-1}), while the regression coefficient

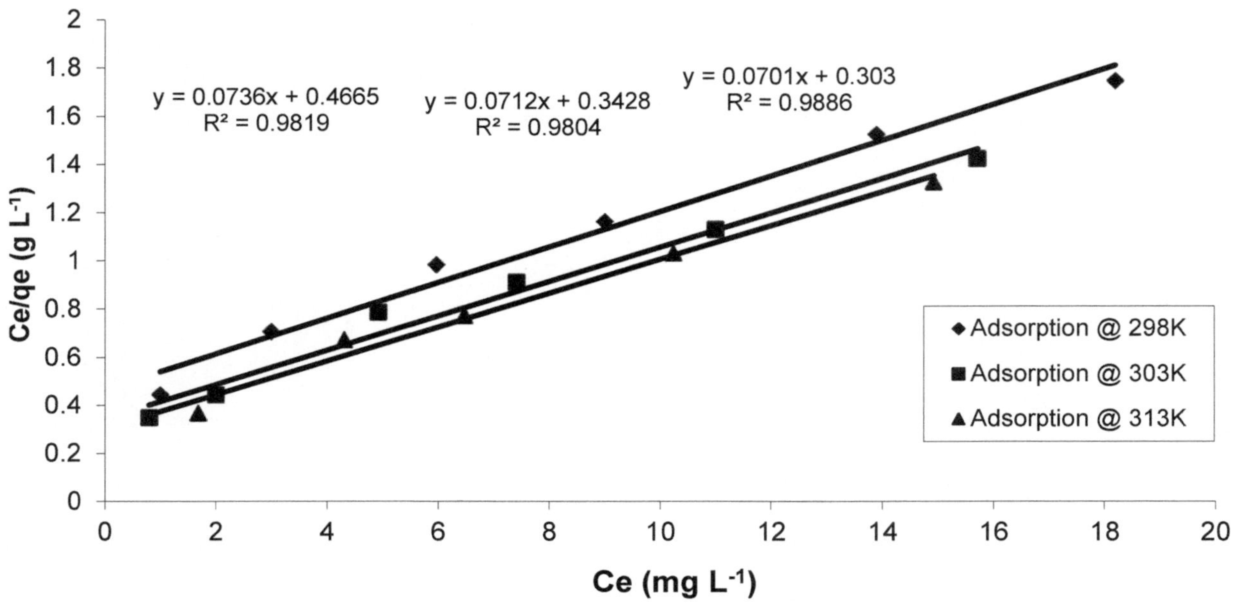

Figure 5. Langmuir isotherm plots at different temperatures.

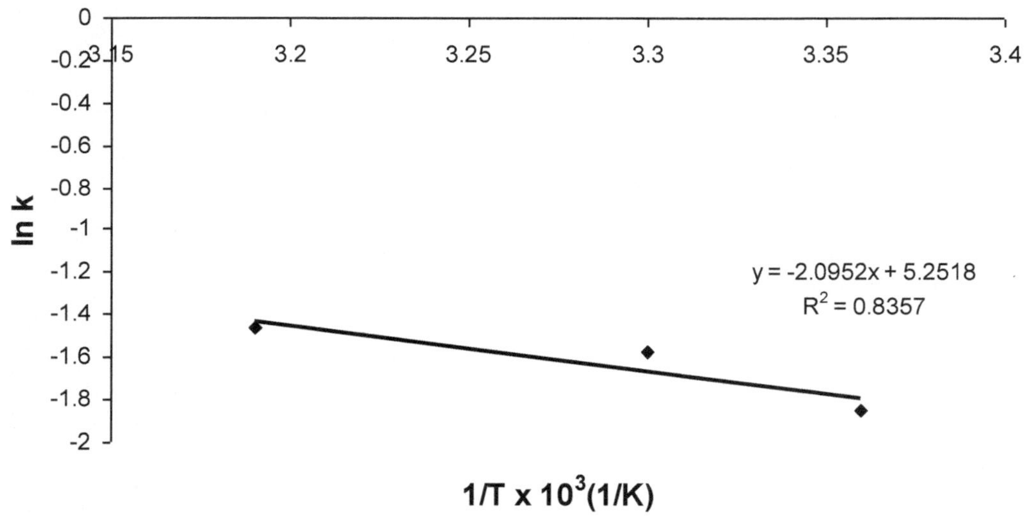

Figure 6. Variation of Langmuir constant, K_L with temperature (Arrhenius type plot) for the adsorption of the cadmium (II) ion by the adsorbent.

obtained is 0.8357. The value of biosorption heat for the cadmium (II) ion onto the biosorbent shows that the reaction is endothermic. Similar results were reported for biosorption of cadmium and lead by spent grain (Low et al., 2000).

Maron and Lando (1974) had stated in general, that the heat of physical adsorption is of the order of 10 kcal mol^{-1} or less and that of chemical adsorption is in the range of 20-100 kcal mol^{-1}. Since the heat of adsorption in our study is 4.163 $kcalmol^{-1}$, we could conclude that the mode of adsorption of cadmium (II) ion by the adsorbent is physisorption. Kim et al., 2005, had reported that the biosorption of lead, copper and cadmium by waste brewery yeast was an endothermic reaction and heat of biosorption values obtained showed that the mode of adsorption was between physisorption and chemisorption.

Effect of contact time

Figure 7 shows the effect of contact time on the degree of adsorption of cadmium (II) ions by the adsorbent. The

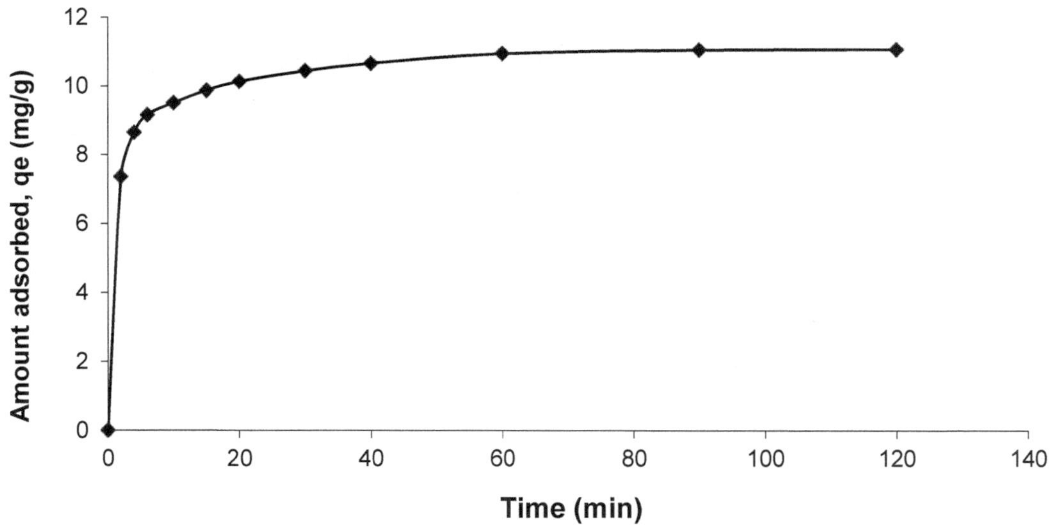

Figure 7. Effect of contact time on cadmium (II) ion adsorption from solution by the adsorbent.

rate of adsorption of the cadmium (II) ions was extremely fast in the first 20 min, but decreased afterward and attained equilibrium in about 60 min. Maximum Cd (II) removal was achieved within 60 min after which Cd (II) concentration in the test solution became constant. This may be explained by the fact that the initial rapid biosorption was due to the availability of a large number of vacant sites, which slowed down later due to saturation of remaining surface sites and repulsive force between solute molecule and bulk phase. The rapid adsorption kinetics observed is also typical for biosorption of metals involving no energy-mediated reactions, where metal adsorption from solution is due to purely physico-chemical interactions between biomass and metal solution (Aksu, 2001).

Sorption kinetic

To investigate the internal diffusion mechanism in the adsorption of cadmium ions onto the biosorbent, the intra-particle diffusion equation has been used. Gerente et al. (2007) stated that adsorption is usually controlled by an external film resistance and/or mass transfer which is controlled by internal or intra-particle diffusion. The intra-particle diffusivity model was employed in our kinetic study. The intra-particle diffusion is characterized by the relationship between specific sorption (q_t) and the square root of time ($t^{1/2}$) and is represented as:

$$q_t = k_{id}\, t^{1/2} \qquad (5)$$

Where, k_{id} is the intra-particle rate constant (mg g^{-1} $min^{-1/2}$), t is the time (min), q_t is the specific sorption that is amount of cadmium adsorbed per unit mass of the adsorbent (mg g^{-1}) at a given time.

According to Weber and Morris (1963), if the intra-particle diffusion is the rate-limiting step in the adsorption process, then the graph of q_t vs. $t^{1/2}$ should yield a straight line and particle diffusion would be the only rate controlling mechanism if this line passed through the origin. McKay and Allen (1980) suggested that three linear sections on the plot q_t vs. $t^{1/2}$ can be identified. This means that two or three (multilinear) steps can occur. The first portion represents external surface adsorption or an instantaneous adsorption stage. The second portion is a gradual adsorption stage, where the intra-particle diffusion is the controlling factor. The third portion is a final equilibrium stage where the intra-particle diffusion starts to decelerate due to extremely low solute concentrations in the solution (Wu et al., 2000). Figure 7 shows the intra-particle diffusivity plot for the adsorption of cadmium (II) ion by the adsorbent. A deviation of straight lines from the origin indicates that intra-particle transport may not be the only rate-limiting step. From Figure 8, it can be seen that the intra-particle diffusion of the cadmium (II) ion on the adsorbent occurred in 2 stages. The linear portion in phase 1 is attributed to the macropore diffusion and the second linear portion due to micro-pore diffusion (phase II) (Allen et al., 1989). The intra-particle diffusion constants for these 2 stages are presented as kid_1 = 0.949 and kid_2 = 0.171. The results indicate that the cadmium (II) ion diffused quickly among the particles at the beginning of the adsorption process, and then intra-particle diffusion slowed down and stabilized. In the two stages observed, kid_1 was found to be greater than kid_2.

Conclusion

The results obtained in this study reveal that initial cadmium (II) ion concentrations, initial pH, contact time

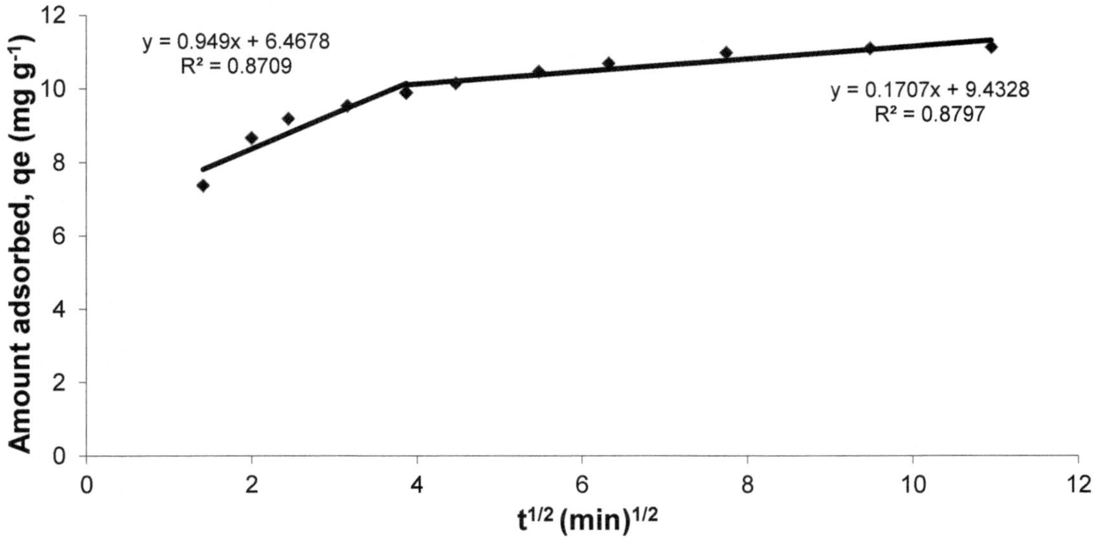

Figure 8. Weber-Morris intra-particle diffusion plot for the adsorption of the cadmium (II) ion by the adsorbent.

and temperature influenced the adsorption capacity of the biosorbent, *A. africana*. Maximum cadmium (II) removal from the solution occurred at a pH 5.0. The equilibrium adsorption showed that the system followed both Freundlich and Langmuir models, with Freundlich giving a better fit for the adsorption data in comparison to the Langmuir model. The temperature did not affect the equilibrium adsorption significantly, since q_{max} values obtained from Langmuir plots vary only slightly at the different temperatures studied. The results revealed that the biosorbent could be a potential biomaterial for the removal of cadmium (II) ions from aqueous solutions having q_{max} values of 14.04 mg g^{-1}.

REFERENCES

Ahalya N, Kanamadi RD, Ramachandra TV (2005). Biosorption of Chromium (VI) from aqueous solutions by the husk of Bengal gram (*Cicer arientinum*). Electron. J. Biotechnol. 8(3):258-264.

Aksu Z (2001). Equilibrium and Kinetic modelling of cadmium (II) biosorption by *C.vulgaris* in a batch system. Sep. Purif. Technol. 21:285-294.

Allen SJ, McKay G, Khader KYH (1989). Intra-particle diffusion of a basic dye during adsorption onto Sphagnum peat. Environ. Pollut. 56:39-50.

Antunes WM, Luna SA, Henriques CA, Costa AC (2003). An evaluation of Copper biosorption by a brown seed weed under optimized conditions. Electron. J. Biotechnol. 6(3):174-184.

Anusiem ACI, Onwu FK, Ogah SPI (2010). Adsorption Isotherm Studies of Ni (II), Cd (II) and Pb (II) ions from aqueous solutions by African white star apple shell. Int. J. Chem. 20(4):265-274.

Argun ME, Dursun S, Gur K, Ozdemir C, Karatas M, Dorgan S (2005). Nickel adsorption on the modified Pine Tree materials. Environ. Technol. 26:479-488.

Avery SV, Tobin JM (1993). Mechanism of Adsorption of Hard and Soft Metal ions to *Saccharomyces cerevisiae* and influence of Hard and Soft ion. Appl. Environ. Michrobiol. 59:2852.

Babarinde NAA, Babalola JO, Sanni RA (2006). Biosorption of Lead (II) ions from aqueous solution by maize leaf. Int. J. Phys. Sci. 1(1): 023-026.

Bai RS, Abraham TE (2002). Studies on the enhancement of Cr (VI) biosorption by Chemically modified biomass of *Rhizopus nigricans*. Water Reserves 36:1224-1236.

Benguella B, Benaissa H (2002). Cadmium removal from Aqueous solutions by Chitin: Kinetic and equilibrium studies. Water Res. 36:2464.

Drash GA (1993). Increase of cadmium body burden for this century. Sci. Total Environ. 67:75-89.

Eromosele IC, Otitolaye OO (1994). Binding of iron, Zinc and Lead ions from aqueous solution by Shea Butter (*Butyrospermum parkii*) Seed Husks. Bull. Environ. Contam. Toxicol. 52:530-537.

Gang S, Weixing S (1998). Sunflower stalk as adsorbents for the removal of metal ions from waste water. Ind. Chem. Res. 37(4):1324-1328.

Gardea-Torresday JL, Tang L, Salvador JM (1996). Copper adsorption by esterified and unesterified fractions of sphagnum peat moss and in different humic substances. J. Hazard. Mater. 48:191-206.

Gerente C, Lee VKC, Cloirec PL, McKay G (2007). Application of Chitosan for the Removal of Metals from Wastewaters by Adsorption - Mechanisms and Models Review. Crit. Rev. Environ. Sci. Technol. 37:41-127.

Horsfall MJ, Spiff AI, Abia AA (2004). Studies on the influence of mercaptoacetic acid (MAA) modification of cassava (Manihot sculentacranz) waste Biomass on the adsorption of Cu^{2+} and Cd^{2+} from aqueous solution. Bull. Korean Chem. Soc. 25(7):969-976.

Igwe JC, Abia AA (2007). Adsorption isotherm studies of Cd (II), Pb (II) and Zn (II) ions bioremediation from aqueous solutions using unmodified and EDTA- modified maize cob. Eclet. Quim.32(1):33-42.

Karthikeyan G, Anbalagan K, Muthulakshmi A (2004). Adsorption dynamics and equilibrium studies of Zn (II) on chitosan. J. Chem. Sci. 116 (2):119-127.

Keskinkam O, Goksu MZL, Basibuyuk M, Forster CF (2004). Heavy metal adsorption properties of a submerged Aquatic Plant (*Ceratophyllum demersum*). Bioresour. Technol. 92:197-200.

Kim T, Park S, Cho S, Kim H, Kim S (2005). Adsorption of Heavy Metals by Brewery Biomass. Korean J. Chem. Eng. 22 (1): 91-98.

Low KS, Lee CK, Liew SC (2000). Sorption of cadmium and lead from aqueous solutions by Spent Grain. Process Biochem. 36:59.

Maron SH, Lando JB (1974). Fundamentals of Physical Chemistry. Revised edition. Collier Macmillan International Publishers, Canada. 754-755.

Matheickal JT, Yu Q (1999). Biosorption of lead (II) and copper (II) from aqueous solution by pre-treated biomass of Australian marine algae. Bioresour. Technol. 69(3):223-229.

McKay G, Allen SJ (1980). Surface Mass Transfer Processes using Peat as an adsorbent for dyestuffs. Can. J. Chem. Eng. 58:521.

Nuhoglu Y, Oguz E (2003). Removal of copper (II) from aqueous solution by biosorption of cone biomass of *Thuja orientalis.* Process Biochem. 38:1627-1631.

Okieimen FE, Okundia EU, Ogbeifun DE (1991). Sorption of cadmium and lead ions on modified groundnut (*Arachis hypogea*) husks. J. Chem. Technol. Biotectnol. 51:97-103.

Ong S, Seng C, Lim P (2007). Kinetics of Adsorption of Cu (II) and Cd (II) from aqueous solution on Rice Husk and modified Rice Husk. Electron. J. Environ. Agric. Food Chem. 6(2):1764-1774.

Ozer A, Ozer D (2003). Comparative study of the Biosorption of Pb(II), Ni (II) and Cr (VI) ions onto *S. cerevisiae*: Determination of Biosorption Heats, J. Harzad. Mater. B100:219.

Quek SY, Waste DAJ, Torster CF (1998). The use of Sago waste for the sorption of lead and copper. Water SA. 24(3):251-256.

Saikaew W, Kaewsarn P, Saikaew W (2009). Pomelo Peel: Agricultural Waste for Biosorption of Cadmium Ions from Aqueous solution. World Acad. Sci. Eng. Technol. 56:287-290.

Veglio F, Beolchini F (1997). Removal of metals by biosorption: a review. Hydrometallurgy 44: 301-316.

Volesky B (1990). Removal and recovery of heavy metals by biosorption. In: Biosorption of heavy metals. Boston, USA, CRC Press. 7-43.

Weber WJ, Morris JC (1963). Kinetics of adsorption on carbon from solution. J. Sanit. Div. Am. Soc. Civ. Eng. 89: 31-60.

Wei L, Libo Z, Jinhui P, Ning L, Shimin Z, Shenghui G (2008). Tobacco Stems as low cost adsorbent for the removal of Pb (II) from wastewater: Equilibrium and Kinetic Studies. Ind. Crops Prod. 28:294-302.

Wu FC, Tseng RL, Juang RX (2000). Comparative Adsorption of Metal and Dye on Flake- and Bead-Types of Chitosans prepared from Fishery Wastes. J. Hazard. Mater. B73:63-75.

18

Evaluation of physicochemical, functional and textural properties of rainbow trout (*Oncorhynchus mykiss*) stored at low temperatures

Arturo Silva-Ríos[1], Octavio Dublán-García[3], Baciliza Quintero-Salazar[2], Aurelio Dominguez-Lopez[1], Leobardo Manuel Gómez-Oliván[3], Leticia Xochitl López-Martínez[3] and Abdel-Fattah Z. M. Salem[4,5]

[1]Facultad de Ciencias Agronómicas, Universidad Autónoma del Estado de México, El Cerrillo Piedras Blancas, Toluca, Edo. de México. CP 50090, México.
[2]Facultad de Turismo y Gastronomía, Campus "el Rosedal", Universidad Autónoma del Estado de México, Carretera Toluca-Atlacomulco Km. 14.5, Toluca, C.P. 50200 México.
[3]Facultad de Química, Universidad Autónoma del Estado de México, El Cerrillo Piedras Blancas, Toluca, Edo. de México. CP 50090, México.
[4]Facultad de Medicina Veterinaria y Zootecnia, Universidad Autónoma del Estado de México, El Cerrillo Piedras Blancas, Toluca, Edo. Mexico. CP 50090, Mexico.
[5]Faculty of Agriculture (El-Shatby), Alexandria University, Egypt.

The physicochemical and functional behaviour of rainbow trout stored at 4 and -20°C for nine days and six months, respectively, were investigated. The combined effect of proteolytic activity and physicochemical changes in the muscle (pH, solubility and proteolytic activity) altered the molecular structure of the proteins. SDS-PAGE showed the degradation of myosin and the formation of peptides of low molecular weight, causing an increase in solubility, which is correlated with changes in the functional properties of meat (water retention, capacity of emulsification and ability of gelling). In a gel obtained from the stored samples, the solubility had an effect on the distribution of the water gel, and significant differences (*P<0.05*) in colour and texture parameters were observed during storage. Regarding correlation analysis, during cooling, all variables studied were dependent on each other, however, in freezing, the correlation diminished in variables such as hardness, solubility and CRA. Surplus meat of trout according to the time of cold storage, can be incorporated into technological processes up to two days at a storage temperature of 4°C and three months at - 20°C, thus could contribute to the functional and textural properties of commercial meat products, generating a value liked trout producers. After this time of storage, the trout flesh can be consumed in the conventional way.

Key words: Rainbow trout, refrigeration, freezing, texture.

INTRODUCTION

Rainbow trout (*Oncorhynchus mykiss*) is a species whose production through aquaculture is growing in countries such as France, Italy, Denmark, Germany and Spain and recently in Chile, which ranks first in production for export (FAO, 2005). In Latin American countries and the Caribbean, such as Ecuador, Mexico, Cuba, Brazil and Colombia, is similar and has attained great importance due to the high nutritional and economic value it provides to some regions of these countries. However, this form of marketing is limited. So it is important to seek alternatives to the storage of the surplus using traditional methods such as low temperature to ensure not only the preservation of the functional (gelling capacity, water holding capacity, emulsifying capacity) and textural properties of muscle of trout but also of its derivatives, such as gels produced from its myofibrillar protein, for possible use in food technological processes. Generally, the conservation of aquaculture species at low temperatures can cause changes to occur in the lipids and myofibrillar proteins of trout, and they can denature during frozen storage (Tejada et al., 2003; Michalczyk and Surówka, 2007; Chen et al., 2007), which leads to loss of integrity of the muscle, modification of the protein solubility (Mohan et al., 2006; Baron et al., 2007) and loss of texture, indicators and factors of functionality such as gellification, emulsification and water holding capacity. Additional studies on trout have demonstrated that the change in texture also resulted from the incorporation of additives (Moreno et al., 2010) and the role of endogenous proteolytic enzymes, specifically the *Lysosomal cathepsins* B, L and D, as well as cytosolic calcium-activated calpains (Godiksen et al., 2009). Both of these enzymes are important proteolytic systems in the softening of the post-mortem meat, and their action involves the breakdown of the myofibrillar structure and is directly correlated with the texture (Bahuaud et al., 2010; Chéret et al., 2007; Delbarre et al., 2006; Yamashita and Konagaya, 1991). With the loss of the protein functional characteristics, the quality of the products obtained from species stored at low temperatures could be affected. Physicochemical characteristics, for example, the colour, are related to the fat content, diet and age, have also been studied to attempt to explain the changes occurring during storage with controlled coatings (Sun et al., 2012; Choubert and Baccaunaud, 2006; Helge et al., 2006). Many studies on rainbow trout involve physicochemical characteristics and proteolytic or oxidation of lipids and proteins; this work aims to assess the effect of storage under cooling and freezing on the physicochemical and functional properties of trout fillets as well as on the textural properties of gel-type surimi obtained from the protein myofibrillar of this species.

MATERIALS AND METHODS

Preparation of the sample

Whole rainbow trout were from a farm located in the municipality of Malinalco, State of Mexico (Mexico). The general appearance of the species was good with a firm flesh texture. Twenty specimens (24 h post-mortem; taking into account the freshness of the fish such as: skin: bright iridescent pigmentation; eye: transparent cornea; gills: bright colour; flesh (cut from abdomen): bluish translucent, smooth, ahining) (Huss, 1995) of approximately 2.5 kg were eviscerated, cleaned, deboned and cut into fillets, and the mean weight were300 ± 10 g (mean ± S.D.). The fish muscle samples were placed in polyethylene food bags, each containing 3 frozen samples. The bags were vacuum packed (Vacuum Packing Machine, Stainless Steel, TVPM-400, Penang, Malaysia). Two batches of samples were divided into 2 lots: 1 was stored at 4°C, and other was stored at -20°C; the latter were stored after a rapid freezing pre-treatment dip in solid-state CO_2 for 20 s. Analysis of the initial sample was performed the day following their capture. The samples were stored at 4°C were analysed every three days for 9 days, and the samples stored at -20°C were analysed each month for 6 months.

Physicochemical and functional properties

pH

The pH measurement was performed according to Owen et al. (1982) as follows: 10 g of meat was homogenised using 90 mL distilled water for 1 min in a domestic blender. After removing the connective tissue, a digital potentiometer (Hanna instrument pH 210, Ukraine) was used to measure the pH.

Fillet water-holding capacity (WHC)

The WHC was determined on 5 g minced muscle as described by Hamm (1975). The storage samples were collected in a centrifuge tube to which 8 mL of 0.6 M NaCl was added. The samples were spun at 1,643 x *g* for 15 min in a centrifuge (5810 R Eppendorf, Hamburg, Germany). The supernatant was decanted, and the unretained volume was measured. The WHC was expressed as the volume of retained 0.6 M NaCl per 100 g of muscle.

Capacity emulsifier (EC)

The EC was measured using the method proposed by Zorba et al. (2006) after homogenisation of 25 g of minced sample with 100 mL of 0.4 M NaCl at pH 6.6 (2 to 4°C). Subsequently, 6 g of the homogenised sample was mixed with 18 mL 0.4 M NaCl at pH 6.6 for 5 s at low speed, and 30 mL of sunflower oil was added to the mixture. The oil was added at a rate of 0.9 - 1.0 mL/s with the help of a homogeniser (Proctor Silex 59735, China), until breakup of the emulsion. The EC was expressed as the average number of mL necessary to affect the change of phase.

Protease extraction

Protease extraction was performed according to the method reported by Okomato et al. (1993) with modifications; the sample was cut finely and homogenised (1:1 w/v) with phosphate buffer 25 mM, pH 7 (with 0.25 M NaCl). Extract centrifugation was performed at 1,643 x *g* for 20 min at 4°C, the supernatant, or enzyme extract (EE), was collected and stored at 4°C until use.

Extraction of acid and alkaline protease and analysis of proteolytic activity

The muscle was homogenised with a phosphate buffer solution 25 mM, pH 7 (with NaCl 0.025 M) at a 1:1 ratio (w/v) and then centrifuged at 1,643 x *g* for 20 min at 4°C (5810 R Eppendorf, Hamburg, Germany). The supernatant was collected and subjected to fractionation with 20 to 60% ammonium sulfate saturation; the precipitate was resuspended in phosphate buffer 25 mM, pH 7 (with NaCl 0.025 M) and dialyzed overnight against the same buffer using a 12,000 kDa cutoff dialysis tube (Sigma-Aldrich, Steinheim, Germany) after this time, was maintained at 4ºC until the subsequent determination of proteolytic activity.

Proteolytic activity

The proteolytic activity was evaluated using the method of Kunitz (1947), with 1% casein in phosphate (20 mM, pH 7) with 0.9% NaCl as a substrate, and by the method of Anson as described in Ouali and Valin (1980), using 1% haemoglobin in a 25 mM universal buffer (phosphoric acid: boric acid: acetic acid; 1:1:1). The substrates were incubated at 35°C for 10 min in a temperature-controlled bath before addition of the enzyme extract. This mixture was then incubated for 10 min to perform hydrolysis. The reaction was halted by adding 0.25 mL of 50% trichloroacetic acid. The reaction mixture was cooled for 10 min at 4°C and centrifuged at 10,822 x *g* for 15 min at 4°C to remove the protein that was insoluble in acid. The soluble protein was determined by measuring the supernatant absorbance at 280 nm using a spectrophotometer (Espectrofotometer Thermo Spectronic 4001/4, USA). We used pH of 2.7 or 7.3 at 35°C as optimal parameters of enzymatic activity present in different aquatic and seafood species according to previous studies (Dublan-Garcia et al., 2006; Gómez-Guillén et al., 1996; Kolodziejska et al., 1994; Konno and Fukazawa, 1993; Wasson et al., 1992; Leblanc and Gill, 1982).

Residual activity was measured after incubation of enzyme extracts with substrate solution pH (2.7 or 7.3) at 35°C; 100% of enzyme activity is the activity of enzyme without storage, and the residual activity is the 100% of enzyme activity less enzyme activity measured during the different sampling intervals. One unit of enzyme activity is defined as the amount of enzyme that produces a change in optical density of 0.001 at 280 nm in 1 min under the conditions established by Yamaguchi et al. (1982) using the following equation (Kunitz, 1947; Ouali y Valin, 1980):

$$PA = (abs*dilution) / ((0.001)\ (time)\ (mL\ of\ extract)) \qquad (1)$$

Protein extraction

Myofibrillar protein extraction was performed according to the method described by Ngapo et al. (1992), with modifications. Trout muscle (25 g) was homogenised with ice and cold water in a 1:2:1 ratio (w:w:v) using an Oster commercial blender (Sumbeam Mexican S.A. de C.V. 405-42, Mexico) until the tissue and water were completely integrated. This extract was mechanically shaken for 10 min in an ice bath and then passed through a sieve to remove the connective tissue.

Two volumes (200 mL) of cold water were added, and the sample was shaken once more for 15 min. The homogenate was centrifuged (Eppendorf 5810 R centrifuge, Germany) at 1,643 x *g* for 25 min at 4°C, and the precipitate was re-suspended in a phosphate buffer (25 mM, pH 7, 0.6 M NaCl). After different sampling intervals, the myofibrillar extracts were analysed using SDS PAGE.

Protein content

The protein content was obtained by applying the biuret technique reported in Gornall et al. (1949), which consisted of taking 1 mL of extracted protein and then adding 3 mL of reagent biuret (per liter final volume: 1.5 g copper sulphate x $5H_2O$ ($CuSO_4{\cdot}5H_2O$), 6.0 g sodium potassium tartrate x $4H_2O$ ($KNaC_4H_4O_6{\cdot}4H_2O$) all dissolved in order in 300mL 10% NaOH before to final volume). The mixture was allowed to rest for 30 min in darkness, and the absorbance was subsequently measured at 540 nm (Espectrofotometer, Thermo Spectronic, 4001/4, USA). The reading obtained correlated to a curve pattern of seroalbumin bovine with a concentration of 0-10 mg/mL.

Solubility

The solubility was determined as proposed by Pilosof (2000). The myofibrillar protein extract at a concentration of 5 mg/mL and pH 7 was centrifuged (Eppendorf 5810 R centrifuge, Germany) at 1,207 x *g* for 20 min at 4°C. The protein content was determined in the supernatant, and the solubility was expressed as the ratio between the protein content in the supernatant and the protein content in the uncentrifuged sample, multiplied by 100. The protein content was determined by the biuret method as described by Alanís-García (1997). 3 mL of the biuret reagent was added to 1 mL of the myofibrillar protein extract. This mixture was stirred gently for a few seconds to homogenise the mixture and was left at room temperature for 30 min in the dark. The absorbance was read at 540 nm using a Beckman spectrophotometer. The results obtained were interpolated on a bovine albumin type curve with a maximum concentration of 10 mg/mL.

SDS-PAGE profile of protein molecular weights

The samples were analysed with SDS-PAGE using the procedure described by García et al. (1993). In total, 15 µL of the sample and 10 µL of the molecular weight markers (10-200 kDa range) (Promega V8491, Madison, USA) were analysed at a constant current of 200 V (Biolec, TC50, Mexico) using Mini-Protean II Cell electrophoresis equipment (Cole Palmer 2857-00, USA) at 4°C. After analysis, the gels were disassembled and stained with 0.1% Coomassie Brilliant Blue R-250 (glacial acetic acid: methanol: water) for 30 to 40 min. They were then placed in a destaining solution (glacial acetic acid: methanol: water; 10:40:50), making the necessary changes (approximately 5 times) in this solution for the protein bands to be clearly observed.

Colour

The colour was measured in the gels based on proteins of rainbow trout. The reflected colours of the gel samples were measured using a Hunter Lab colorimeter (Chroma Meter CR-200, 80025393, Tokyo, Japan) model D25-PC2 using the method described by Little (1975) in the Lab System. The instrument was calibrated using a standard white tile (L = 97.38, a = 0. 17, b = 1.94), to obtain luminosity values (L*), a* and b*; Luminosity (L) is a variable from 0 to 100; where 0 is black and 100 white. Furthermore, for a better assessment of the colour, the two coordinates (a, b) were combined to obtain the chroma (McLaren, 1980). The values of a* and b* serve to determine the hue angle (tonality) (McGuire, 1992) values:

$$\text{Hue angle } (H°ab) = Tan^{-1}\ [b^*/a^*] \qquad (2)$$

$$\text{Chromaticity (Chroma)} = (a^2 + b^2)^{1/2} \qquad (3)$$

Texture properties

Texture profile analysis (TPA)

The protein extract was adjusted to a concentration of 40 mg/mL and was placed in 3-cm diameter, and 4-cm high glass bottles, according to "semi-infinite geometry" (Klettner, 1995). The tubes were immersed in a water bath and were gradually heated at 1°C/min from room temperature to 70°C. Once a temperature of 70°C was attained, the tubes were placed in an ice bath for 30 min and stored refrigerated for 12 h.

The textural profile analysis (TPA) was performed at room temperature with a texturometer (Texture Technologies Corp., Scarsdale, NY, USA) using the programme Texture Expert Exceed v2.3 (Stable Micro Systems Ltd, Surrey, England). The conditions of the analysis included: cell load: 5 Kg; type of physical evidence: % compression; speed of tube: 1 mm/s, and instrumental of test: 2.5 cm long x 1 cm diameter acrylic cylinder. Hardness (H) corresponds to the maximum force (N) required to compress the sample. Cohesiveness (A2/A1) is defined as the extent to which the sample could be deformed before rupturing, where A1 is the total energy required for the first compression, and A2 is the total energy required for the second compression. Elasticity (T2/T1) is the ability of the sample to recover its original shape after a force of deformation, where T1 is the first compression peak duration and T2 is the second compression peak duration. Finally, masticability is a parameter obtained from the product of hardness x cohesiveness x elasticity.

Statistical analysis

One-way analysis of variance (ANOVA) was performed. Pearson's correlation coefficients between the means of the parameters studied were also calculated, the correlations were presented as R^2. ANOVAs showing significant differences were followed by a Duncan multiple range tests using the package Statgraphics Plus for Windows 4.1 (Statistical Graphics Corp. 1994-1999, USA). All analysis was performed in triplicate. In all statistical analyses, significance was accepted when $P < 0.05$.

RESULTS AND DISCUSSION

Physicochemical and functional properties

The measurement of these parameters is crucial in ensuring the degree of freshness of the species. In this study, the rainbow trout exhibited a pH of 6.68, coinciding with the value reported by García et al. (2004), who reported a pH of 6.65 for fillets of this species. During storage at 4°C, a significant increase in pH was observed ($P< 0.05$) from the sixth day (Figure 2); this effect could result from the degradation of trimethylamine oxide in the muscle post mortem, as this compound is reduced to trimethylamine by bacterial or enzymatic action (Quitral et al., 2003) and is also responsible for the decarboxylation of amino acids that generates low molecular weight amines (Paczkowski and Schütz, 2011). On the other hand, significant difference was observed in the storage at -20°C, ($p< 0.05$) with a decrease in pH from the second month (Figure 2); this could be explained by changes in glycogen, lactate accumulation and production of inosinic acid phosphorylated as demonstrated in a study conducted in cod stored at 0°C (Fennema, 2000). One of the parameters which presented a good correlation with respect to pH was the WHC, which as shown in Figure 2, increased during storage at 4°C ($p< 0.05$); 67 to 107 mL NaCl 0.6 M per each 100 g of muscle. The amount of water immobilized within muscle tissue depends on how the myofibrillar actin and myosin proteins are organized. The change of the organization of the proteins in storage at 4°C is a reflection of the variation of pH and the action of proteases during the period of conservation, both favor hydrolysis by inducing increased distance between the filaments (Huff and Lonergan, 2005) and exposure of the domain of myosin by induction of depolymerization proteins and promoting the interaction of fractions of the myosin, which results in an increase of the WHC (Moreno et al., 2010). The behaviour of the WHC in storage at -20°C has the same trend during refrigeration storage for 67 to 73 mL (0.6 M NaCl per 100 g of muscle) ($P< 0.05$); however, there was no correlation with respect to pH or the capacity of water retention according to a statistical analysis of correlation at -20°C, which could result from freezing. Another type of chemical interactions, including those between protein carbonyl groups and macromolecules such as carbohydrates and lipids (Dalle et al., 2001; Bader and Grune, 2006; Baron et al., 2007), contributes to the reduction of the holding capacity of water in proportion to the status of storage.

Proteolytic activity and SDS-PAGE profile of protein molecular weights

The endogenous proteolytic activity in the enzyme extract of trout, evaluated at pH of 2.7 and 7.3 at 35°C (Table 1) indicates that there was enzyme activity in both acid proteases, similar to that occurring in neutral proteases. The effect of storage at 4 and -20°C on the proteolytic activity resulted in a significant decrease ($P<0.05$) from 100 to 78 and 81%, respectively, at pH 2.7 and 100 to 57 and 37%, respectively, at pH 7.3 with a linear behaviour with respect to time (Table 1), coinciding with reports by Michalczyk and Surówka (2007), who observed a decrease in cathepsin activity in trout stored at low temperatures. In addition, in some studies on mackerel surimi (*Scomber scombrus*) stored under refrigeration (1 to 2°C) and freezing (-2°C), the proteolytic activity decreased from the first month (Fik et al., 1988; Ming et al., 2000). The decrease of this activity could be due to several factors such as temperature or partial oxidation of thiol groups in the centre of the active enzyme (Fik et al., 1988) or their interactions with salt or sugar in the muscle itself (Reddi et al., 1972). According to the analysis of the electrophoretic profile of the proteins, we observed that although the enzyme activity decreased during storage, the enzyme activity continued during this time, as there was a decrease of the intensity of the band molecular weight 225 kDa, corresponding to myosin. In addition, a thickening and increase in intensity of the 43-kDa actin, 39.5-kDa β-tropomyosin and 37.9-kDa Troponin T bands occurred with increasing time at a storage temperature of 4°C (Figure 3), indicating that hydrolysis of the protein myosin may have occurred. According to Godikse et al.

Table 1. Proteolytic activity in stored trout fillets at 4 and -20°C and pH 2.7 and 7.3.

Sample	Period of storage	pH 2.7 (U/mL)	Residual activity (%)	pH 7.3 (U/mL)	Residual activity (%)
Day					
	0	376.45±1.61^{a}	100	343.03±0.59^{a}	100
	2	335.35±2.64^{b}	89	312.84±3.67^{b}	91
Trout at 4°C	4	305.87±2.64^{c}	81	303.74±1.93^{c}	89
	6	301.51±0.84^{cd}	80	287.99±1.43^{d}	84
	9	295.96±4.99^{d}	79	279.37±5.07^{e}	81
Month					
	0	376.45±1.61^{a}	100	343.03±0.59^{a}	100
	1	328.86±4.28^{b}	87	286.65±1.83^{b}	84
	2	287.01±4.63^{c}	76	255.96±4.70^{c}	75
Trout at -20°C	3	268.60±4.27^{d}	71	226.32±2.19^{d}	66
	4	242.08±3.19^{e}	64	185.20±4.23^{e}	54
	5	217.72±2.54^{f}	58	152.17±1.94^{f}	44
	6	215.36±0.63^{f}	57	127.41±0.56^{g}	37

Values with the different letters in the same column are statistically different at ($p < 0.05$). Different letters (a, b, c, d, e, f, g) indicate significant difference ($p < 0.05$) in each treatment with storage time.

(2009), who observed that degradation of the electrophoretic profile of the proteins appeared to be similar in trout muscle extract mixed with commercial cathepsins B, L and D compared with an extract without additives. This result might suggest that the enzymes act *in situ*, followed by the hydrolysis of muscle proteins.

Solubility

The solubility in samples stored at 4°C positively correlated with the WHC and the EC; in contrast, a negative correlation was observed with luminosity, hardness and chewiness. A significant linear increase (*P<0.05*) in solubility with respect to time was observed (Table 2). For samples stored at -20°C, no correlation was observed. The solubility increased from the second month and decreased from the third (Figure 2). The change in solubility of the proteins was closely associated with the change in their secondary and tertiary structures (Michalczyk and Surówka, 2007), which were themselves affected by various factors such as pH and proteolytic action. According to Mohan et al. (2007), by moving away from the isoelectric point pH amending the electric charge of the protein, the resulting electrostatic repulsion between the molecules contributed to the solubility of the proteins, coupled with the hydrolysis of the myofibrillar proteins derived from the endogenous proteolytic activity; as a result, low molecular weight peptides, which are more apt to dissolve, are produced. Furthermore, in terms of the effect of storage at -20°C on the solubility of the myofibrillar proteins, freeze denaturation exposes the carbonyl groups of the proteins and generates aggregation reactions, such as those occurring with lipids (Dalle et al., 2001), which are compounds observed in high proportions in trout. As mentioned above, these reactions diminish the WHC and the solubility of proteins, such that these interactions generate insoluble products in saline solutions and generate non-linear solubility behaviour with respect to time (Table 2).

Capacity emulsifier (EC)

The emulsification capacity of extracted proteins (Figure 2) increased with storage time at 4°C (*P<0.05*), most likely due to dissociation of muscle protein complexes (Farouk et al., 2003), which would have allowed increasing molecular flexibility and hydrophobicity, favouring solubility. The latter plays an important role in emulsifying properties because a greater quantity of soluble protein increases the ability of emulsification stabilisation coupled to it by the effect of pH (Fennema, 2000), as it is located away from the isoelectric point of protein emulsifiers. The effect of the emulsifying capacity upon storage at -20°C decreased significantly (*P<0.05*) over time. This result could be due to the decreased pH and protein solubility (Figure 2), or, according to Ramirez et al. (2000), to the formation of intramolecular disulphide bridges resulting the conformational changes that occur during the storage period, such as the formation of disulphide, hydrogen and hydrophobic bonds, which associate with the oil added in what is known as the aggregation phenomena (Benjakul and Bauer, 2000).

Colour

At a storage temperature of 4°C, the values of light and shade (Table 3) indicate a significant difference from the

Table 2. Proteolytic activity, colour, texture and physicochemical variables compared with time of storage at 4 and -20°C.

Sample	Value	pH 2.7 (U/mL)	pH 7.3 (U/mL)	L*	H°ab	Chroma	Hardness (N)	Elasticity	Cohesiveness	Chewiness (N)	pH	Solubility	WHC	EC
Trout at 4°C	Ordinate	358.89	333.83	81.41	79.14	8.93	6.04	0.70	0.48	1.90	6.65	2.92	63.81	103.26
	Slope	-8.54	-6.77	-0.55	0.39	-0.07	-0.39	0.00	0.00	-0.12	0.02	0.73	4.33	3.78
	R^2	-0.89	-0.95	-0.95	0.56	-0.55	-0.91	-0.96	0.11	-0.98	0.92	0.98	0.97	0.87
Trout at -20°C	Ordinate	356.99	330.95	82.41	78.66	7.92	6.49	0.70	0.46	2.08	6.70	3.45	64.43	92.33
	Slope	-26.80	-35.24	-1.83	1.28	0.73	-0.16	0.00	0.00	-0.07	-0.03	0.21	1.38	-1.43
	R^2	-0.97	-1.00	-0.94	0.77	0.77	-0.77	-0.76	-0.14	-0.84	-0.96	0.57	0.87	-0.91

Table 3. Colour and texture parameters in trout gel stored at 4 and -20°C.

Sample	Period of storage	L*	H°ab	Chroma	Hardness (N)	Elasticity	Cohesiveness	Chewiness (N)
Day								
	0	81.99±0.09^{a}	80.29±0.52^{abc}	8.58±0.17^{a}	6.87±0.09^{a}	0.69±0.04^{a}	0.42±0.03^{a}	1.98±0.13^{a}
	2	80.04±0.54^{ab}	78.28±0.56^{a}	8.89±0.57^{a}	4.43±0.11^{b}	0.69±0.02^{a}	0.53±0.11^{a}	1.61±0.32^{ab}
Trout at 4°C	4	79.01±0.78^{bc}	82.53±0.70^{bc}	9.24±0.21^{a}	4.15±0.20^{b}	0.68±0.01^{a}	0.53±0.02^{a}	1.31±0.06^{b}
	6	77.27±0.09^{c}	78.81±2.63^{ab}	8.36±0.01^{a}	3.71±0.05^{c}	0.66±0.01^{a}	0.50±0.01^{a}	1.22±0.02^{bc}
	9	77.13±1.90^{c}	83.97±1.77^{c}	8.13±1.00^{a}	2.88±0.12^{d}	0.65±0.01^{a}	0.46±1.04^{a}	0.85±1.08^{c}
Month								
	0	81.99±0.09^{a}	80.29±0.52^{ab}	8.58±0.17^{ab}	6.87±0.09^{a}	0.69±0.04^{a}	0.42±0.03^{a}	1.98±0.13^{a}
	1	80.76±1.08^{a}	80.41±1.74^{ab}	9.09±0.53^{bc}	6.09±0.047^{ab}	0.70±0.02^{a}	0.50±0.03^{b}	2.16±0.36^{a}
	2	79.95±0.25^{ab}	79.80±0.21^{a}	9.49±0.03^{c}	5.88±0.33^{b}	0.68±0.01^{a}	0.48±0.01^{ab}	1.92±0.07^{a}
Trout at -20°C	3	77.80±0.84^{b}	82.37±0.89^{b}	9.18±0.26^{bc}	5.82±0.33^{b}	0.69±0.01^{a}	0.48±0.05^{ab}	1.94±0.30^{a}
	4	72.78±0.07^{c}	79.47±0.13^{a}	8.30±0.04^{a}	5.93±0.35^{ab}	0.66±0.02^{a}	0.43±0.03^{ab}	1.68±0.19^{a}
	5	71.92±0.29^{c}	87.57±0.10^{c}	12.81±0.04^{d}	6.08±0.42^{ab}	0.67±0.02^{a}	0.40±0.01^{a}	1.64±0.18^{a}
	6	73.18±1.50^{c}	87.58±0.53^{c}	13.28±0.32^{d}	5.34±0.25^{b}	0.67±0.01^{a}	0.48±0.02^{ab}	1.72±0.02^{a}

Values with the different letters in the same column are statistically different at ($p < 0.05$). Different letters (a, b, c, d) indicate significant difference ($p < 0.05$) in each treatment with storage time.

second day (*P<0.05*). Because the amount of light determines the white component in the sample, when the values obtained were decreased from 82 to 77, the chromaticity did not differ.

For storage at -20°C, the values of brightness, hue and chroma indicate a difference (*P<0.05*) with respect to the values of the initial sample from the second month. The brightness decreased to 72. These variations may be attributed to the distribution of water in the sample, as water from the surface will cause the sample to appear whiter. Qiao et al. (2001) reported that according to a correlation analysis, this particular variable is an excellent value for predicting the responses of physicochemical and functional variables. In this study, this variable correlated with the luminosity of most variables upon storage at a temperature of 4°C, while at

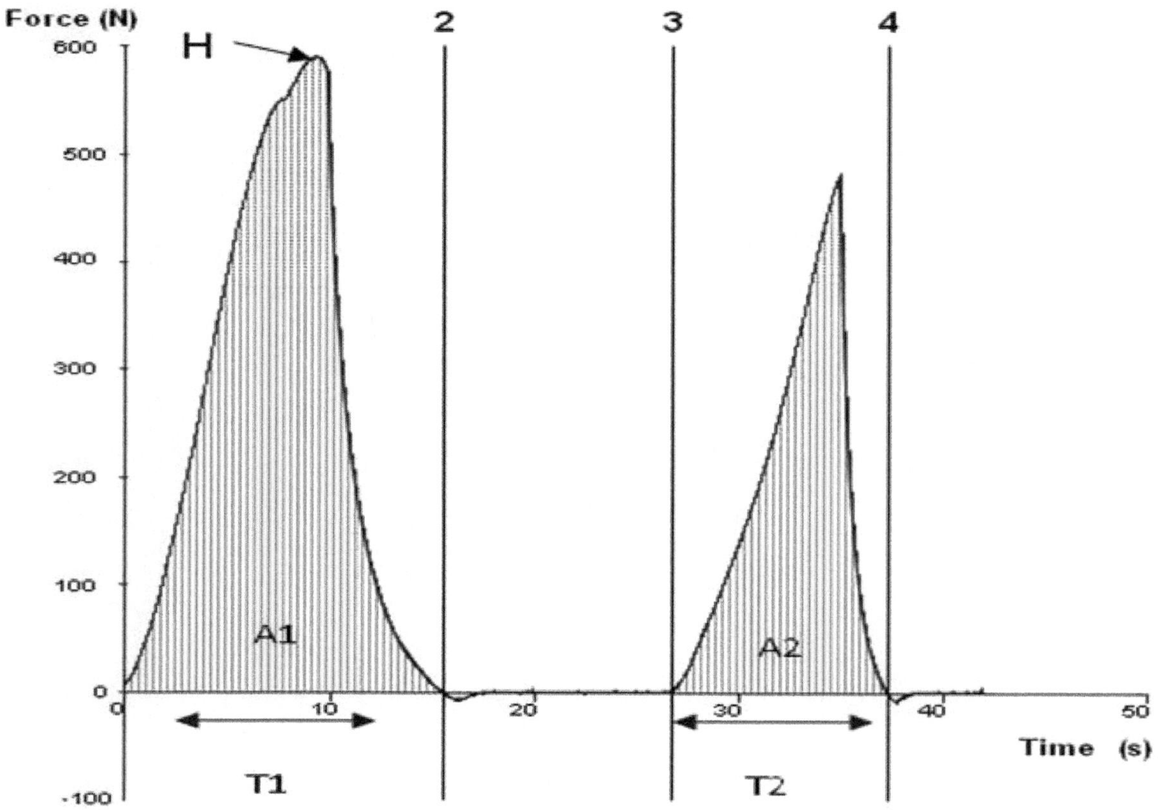

Figure 1. Texture profile analysis curve. Hardness=H; cohesiveness=A2/A1; elasticity=T2=T1; chewiness= hardness x cohesiveness x elasticity.

-20°C, the variables only correlated with pH, WHC, EC, elasticity and chewiness. The variation of the colour with respect to the key variable is expressed in degrees, where 0° = red, 90° = yellow, 180° = green and 270° = blue (McGuire, 1992). In the present study, for both storage temperatures, 78 values were obtained with respect to 88°. Notably, the values greater than 88° correspond to samples from storage at -20°C because at this temperature, other types of reactions occur, such as oxidation of lipids and interaction with the pigments that colour. According to Hunt (1980), in meat products, it is desirable to have values less than 90°; higher values indicate a poor meat. The increase is attributed to the key changes that exist in the pigments myoglobin and oxymyoglobin to metmyoglobin (Choubert and Baccaunaud, 2006). The chromaticity (chroma) is a dimensionless value that expresses the intensity of a hue (H°ab). In the present work, the values of this intensity (Table 3) are in a range that is considered tenuous, with significant differences (*P<0.05*) occurring at a storage temperature of -20°C, with higher values in months 5 and 6.

Textural properties

The TPA analysis parameters are presented in Table 3. A linear decrease was observed in the hardness (Table 2) until the end of storage time for both temperatures (*P<0.05*), indicating that the muscle becomes softer; this result could be due to endogenous enzyme activity (Figure 1), which may act on structural proteins such as myosin (Bahuaud et al., 2010) that provide firmness in meat products (Godiksen et al., 2009). A downward trend was also observed for chewiness, demonstrating a significant difference (*P<0.05*) in storage at 4°C but not for storage at -20°C; this result occurs because chewiness is a derived parameter of hardness, which decreased 22.3% upon storage at -20°C.

For samples stored at 4°C, a decrease in hardness of 58.1% occurred, indicating a direct relationship with chewiness. No significant difference was observed in cohesiveness (*P<0.05*). Upon storage at 4°C, the obtained value is near 0.5, which indicates that the gel recovered 50% of its original structure after the first compression. Upon storage at -20°C, a significant difference (*P<0.05*) was observed, taking values near 0.5.

The elasticity parameter was not changed significantly during the storage time for both temperatures, with average values of 0.7, indicating that the product has a good degree of recovery after compression. It follows that the changes in texture and functionality could be mainly

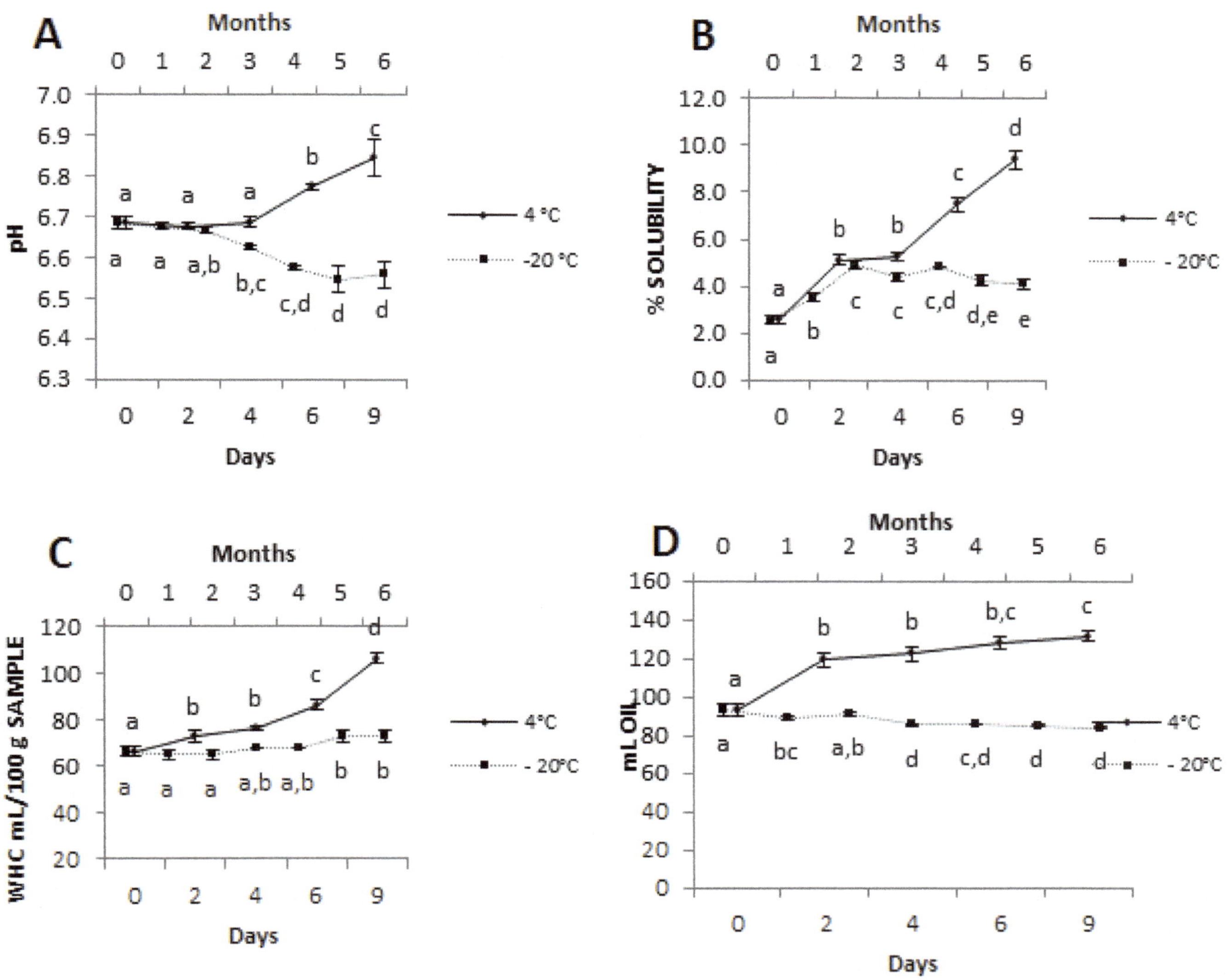

Figure 2. Physicochemical and functional properties as a function of time. A) pH, B) % solubility, C) and WHC D) emulsification capacity. Different letters (a, b, c, d, e) indicate a significant difference ($p < 0.05$) in each treatment with storage time.

attributed to changes in myofibrillar proteins due to the action of proteases; as a result, physicochemical changes changes are produced by the effect of cold storage, triggering the behaviour of the variables described in this investigation.

Conclusions

The study demonstrates that both storage temperatures significantly affected the physicochemical, functional and textural properties of beef and trout. Once the product is captured, the recommended storage is for up to six days; however, the use of this technology for processing meat stored at 4°C may be more feasible to two days and a temperature of -20°C is possible for storing in a period of six months and the technology can be used to at least three months.

After this time, it may be possible to use conventional consumption. Because during the study, some properties were observed to be maintained and others, such as the WHC and EC, were improved, one could say that the storage can be beneficial depending on the desired application that is, longer storage, which increases the EC, although altering the myosin, could be used in emulsified foods such as sausages. The functional and textural properties, such as the emulsifying capacity and gel strength, of the trout suggest its incorporation into commercial meat products.

ACKNOWLEDGEMENTS

Arturo Silva-Ríos thanks the National Council of Science and Technology (Consejo Nacional de Ciencia y Tecnología, Mexico) for a graduate scholarship.

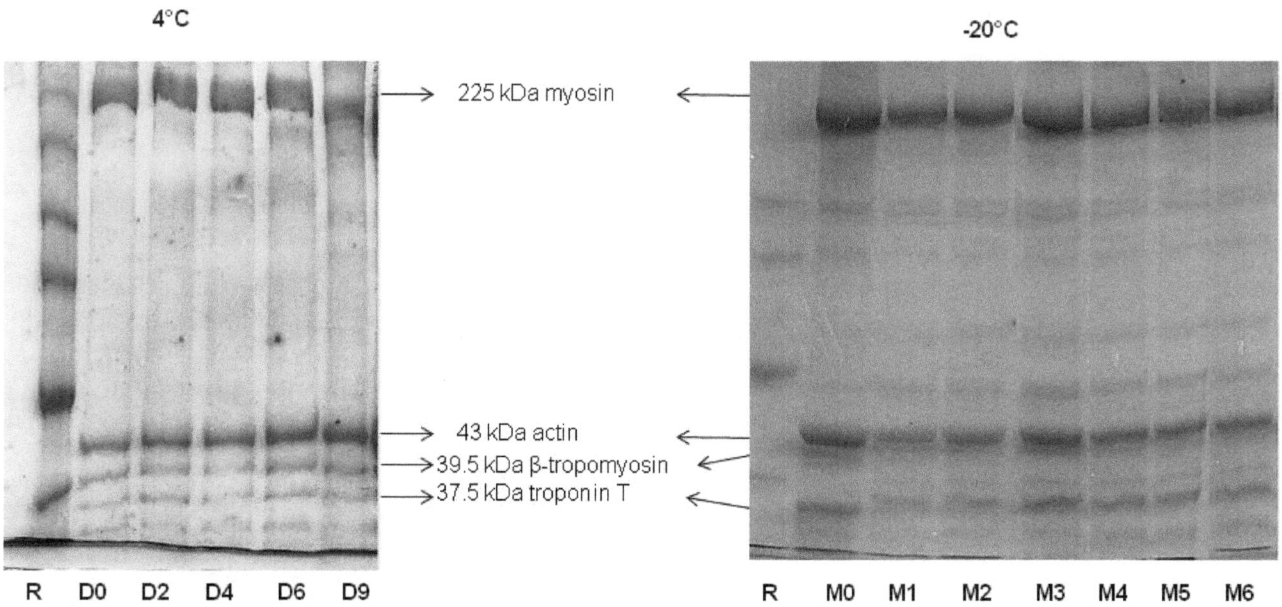

Figure 3. Electrophoretic profile extract of trout stored at 4°C and -20°C. R=reference, D= day and M= month.

REFERENCES

Bader N, Grune T (2006). Protein oxidation and proteolysis. Biol. Chem. 387:1351-1355.

Bahuaud D, Gaarder M, Veiseth KE, Thomassen M (2010). Fillet texture and protease activities in different families of farmed Atlantic salmon (*Salmo salar* L.). Aquaculture 310:213-220.

Baron PC, KjÆrsgard VHI, Flemming J, Charlotte J (2007). Protein and Lipid Oxidation during Frozen Storage of Rainbow Trout (*Oncorhynchus mykiss*). J. Agric. Food Chem. 55:8118-8125.

Benjakul S, Bauer F (2000). Physicochemical and enzymatic changes of cod muscle proteins subjected to different freeze-thaw cycles. J. Sci. Food Agric. 80(8):1143-1150.

Chen YC, Nguyen J, Semmens K, Beamer S, Jaczynski J (2007). Physicochemical changes in ω-3-enhanced farmed rainbow trout (*Oncorhynchus mykiss*) muscle during refrigerated storage. Food Chem. 104:1143-1152.

Chéret R, Delbarre LC, de Lamballerie AM, Verrez BV (2007). Calpain and cathepsin activities in post mortem fish and meat muscles. Food Chem. 101:1474-1479.

Choubert G, Baccaunaud M (2006). Colour changes of fillets of rainbow trout (*Oncorhynchus mykiss* W.) fed astaxanthin or canthaxanthin during storage under controlled or modified atmosphere. LWT- Food Sci. Technol. 39:1203-1213.

Dalle DI, Rossi R, Giustarini D, Gagliano N, Lusini L, Milzani A, Di Simplicio P, Colombo R (2001). Actin carbonylhdation: from a simple marker of protein oxidation to relevant signs of severe functional impairment. Free Radical Biol. Med. 31:1075-1083.

Delbarre lC, Chéret R, Taylor R, Verrez BV (2006). Trends in postmortem aging in fish: Understanding of proteolysis and disorganization of the myofibrillar structure. Critical Reviews in Food Sci. Nutr. 46:409-421.

Dublan-Garcia O, Cruz-Camarillo R, Guerrero-Legarreta I, Ponce-Alquicira E (2006). Effect of refrigerated storage on proteolytic activity, physicochemical and microstructural properties of giant squid (*Dosidicus Gigas*) mantle muscle. J. Muscle Foods 17(3):142-159.

FAO (2005-2011). Programa de información de especies acuáticas. Oncorhynchus mykiss. Programa de información de especies acuáticas. Texto de Cowx, I. G. In: Departamento de Pesca y Acuicultura de la FAO [en línea]. Roma. http://www.fao.org/fishery/culturedspecies/Oncorhynchus_mykiss/es#tcNA00ED.

Farouk MM, Wieliczko KJ, Merts I (2003). Ultra-fast freezing and low storage temperatures are not necessary to maintain the functional properties of manufacturing beef. Meat Sci. 66:171-179.

Fenema O (2000). Química de los Alimentos, 2a ed., Ed. Acribia, España, pp. 451:1077-1078.

Fik M, Surówka K, Leszczynska-Fik A (1988). Application of superchilling to prolong the keeping time of rainbow trout. Nahrung, 32(3):291-300.

García CFL, Dimes LE, Haard NF (1993). Substrate-gel electrophoresis for composition and molecular weight of proteinases or proteinaceous proteinase inhibitors. Analytic. Biochem. 214:65-69.

García MJA, Nuñez GFA, Rentería MAL, Jiménez CGA, Espinoza HMR (2004). Calidad de cabal y carne de tres variedades de trucha arcoiris oncorhynchus mykiss (*Walbaum*). Hidrobiologia 16(001):11-22.

Godiksen H, Morzel M, Hyldig G, Jessen F (2009). Contribution of cathepsins B, L and D to muscle protein profiles correlated with texture in rainbow trout (Oncorhynchusmykiss). Food Chem. 113:889-896.

Gómez GC, Solas T, Borderías AJ, Montero P (1996). Ultraestructural and rheological changes during gelation of giant squid (*Dosidicus gigas*) muscle. Z. Lebensm-Unters-Forsh.. 202:215-220.

Gornall AJ, Bardawill CJ, David MM (1949). Determination of serum proteins by means of the biuret reaction. Biol. Chem. 177:751-766.

Hamm R (1975). Water holding capacity of meat. En Meat. Editores D.J.A., Cole, R.A. Lawrie. Butterworths. Londres, Reino Unido, pp. 321-337.

Helge SL, Manne F, Ruohonene K, Kause A, Rungruangsak-Torrissen K, Kiessling A (2006). Automated image analysis as a tool to quantify thecolour and composition of rainbow trout (*Oncorhynchus mykiss* W.) cutlets. Aquaculture 26:695-705.

Huff LE, Lonergan S (2005). Mechanisms of water-holding capacity of meat: The role of postmortem biochemical and structural changes. Meat Sci. 71:194-204.

Hunt MC (1980). Meat Color Measurements 33d. Annual Reciprocal Meat Conference Perdue University, Weat Lafayette 33:41-46

Huss, HH (1995). Quality and quality changes in fresh fish. FAO Fisheries Technical Paper. Rome. 348.

Klettner PG (1995). Meat and meat products. Measuring texture and consistency with testing machines. Fleischwirtsch international, 36-36.

Klettner PG (1995). Meat and meat products. Measuring texture and consistency with testing machines. Fleischwirtsch international 36-36.

Kolodziejka I, Niecikowska C, Sikorski ZE (1994). Dimethylamine and formaldehyde in cooked squid *Illex argentinus* muscle extract and mantle. Food Chem. 50:281-283.

Kono K, Fukazawa C (1993). Autolysis of squid mantle muscle protein as affected by storage e-conditions and inhibitors. J. Food Sci. 58 (6):1198-1202.

Kunitz M (1947). Crystaline soybean trypsin inhibitor, general properties. J. General Physiol. 30:291-310.

Leblanc ELY, Gill TA (1982). Comparative study of proteólisis in short-finned (*Illex illecebrousus*) and long-finned (*Loligo pealei leseur*) squid. Comp. Biochem. Physiol. 73B:201-210.

Little AC (1975). Off on tangent. J. Food Sci. 40:410-411.

McGuire RG (1992). Reporting of objective color measurements. Hortscience 27(12):1254-1255.

Michalczyk M, Surówka K (2007). Changes in protein fractions of rainbow trout (*Oncorhynchus mykiss*) gravads during production and storage, Food Chemistry, 104:1006-1013.

Ming LH, Gen HC, Shann TJ (2000). Effect of mackerel cathepsins L and L-like, and calpain on the degradation of macrelsurimi. Fisheries Sci. 66(3):558-568.

Mohan M, Ramachandran D, Sankar TV, Anandan R (2007). Influence of pH on the solubility and conformational characteristics of muscle proteins from mullet (*Mugil cephalus*). Process Biochem. 42:1056-1062.

Moreno MH, Borderías AJ, Baron PC (2010). Evaluation of some physicochemical properties of restructured trout and hake mince during cold gelation and chilled storage. Food Chem. 120:410-417.

Ngapo T, Wilkinson B, Chong R, Haisman D (1992). Gelation of bovine myofibrillar protein induced by 1,5 Gluconolacone. En 38th International Congress of Meat Science and Technology, Clermont-Ferrand, Francia. pp.1095-1098.

Okomato Y, Fuchino O, Horiuchi S, Tamaya T, Matsumoto JJ, Tsuchiya T (1993). Purification and caracterization of two matalloproteinases from squid mantle muscle, myosinasa I and miosinase II. Biochem. Biophys. Acta. 1161:97-104.

Ouali A, Valin C (1980). Effect of muscle lysosomal enzymes and calcium activated neutral proteinase myofibrillar ATPase activity relationship with ageing changes. Meat Sci. 5(3):233-245.

Owen JE, Nuñez FA, Arias MT, De CANO, Ríos O (1982). Manual De Prácticas De Cursos De Tecnología De La Carne. Facultad de Zootecnia, Universidad Autónoma de Chihuahua, Chihuahua, México.

Pilosof AM (2000). Solubilidad. En Caracterización Funcional y estructural de proteínas. Pilosof, A.M.R y Bartholomai, G.B. editors. Edit. Ceudeba. CYTED (Programa Iberoamericano de Ciencia y Tecnología para el Desarrollo). pp. 60-75.

Qiao M, Fletcher DL, Smith DP, Northcutt JK (2001). The Effect of Broiler Breast Meat Color on pH, Moisture, Water-Holding Capacity, and Emulsification Capacity. Poult. Sci. 80:676-680.

Quitral RV, Abugoch JL, Vinagre LJ, Guarda MA, Larraín BMA, Santana RG (2003). Efecto de tratamientos térmicos sobre las características químicas de carne de jaiba mora (*Homalaspis plana*). Archivos Latinoamericanos de Nutr. 53(1):90-95.

Ramirez JA, Marín-Polo MO, Bandam E (2000). Fish myosin aggregation asaffect by freezing and initial physical state. J. Food Sci. 65:556-561.

Reddi PK, Constantinides SM, Dymsza HA (1972). Catheptic activity of fish muscle. J. Food Sci. 37(5):643-648.

Tejada M, Huidobro A, Mohamed GF (2003). Comparison of gilthead seabream (*Sparus aurata*) and hake (*Merluccius merluccius*) muscle proteins during iced and frozen storage. J. Sci. Food Agric. 83:113-122.

Wasson DH, Babbit JK, French JS (1992). Characterization of a heat stable protease from arrowtooth flounder; atheresthes stomias. J. Aquat. Food Prod. Technol. 1:167-182.

Yamashita M, Konagaya S (1991). Hydrolytic action of salmon cathepsin-B and Cathepsin-L to muscle structural proteins in respect of muscle softening. Nippon Suisan Gakkaishi 57:1917-1922.

Zorba Ö, Kurt S (2006). Optimization of emulsion characteristics of beef, chicken and turkey meat mixtures in model system using mixture design, Meat Sci. 73:611-618.

19

Physicochemical and antioxidant characteristics of *Baccaurea angulata* fruit juice extract

Darina Ibrahim[1], Norazlanshah Hazali[1], NurhazniKhir Jauhari[1], Muhammad Nor Omar[2], Mohd Noor Adros Yahya[3], Idris Adewale Ahmed[1], Maryam Abimbola Mikail[1] and Muhammad Ibrahim[1]

[1]Department of Nutrition Science, Kulliyyah of Allied Health Sciences, International Islamic University Malaysia, Jalan Istana, Bandar Indera Mahkota, 25200 Kuantan, Pahang, Malaysia.
[2]Department of Biotechnology, Kuliyyah of Science, International Islamic University Malaysia, Jalan Istana, Bandar Indera Mahkota, 25200 Kuantan, Pahang, Malaysia.
[3]Malaysian Agriculture Research and Development Institute (MARDI), Sarawak, Malaysia.

The purpose of this study was to explore the physicochemical and antioxidant characteristics of *Baccaurea angulata* fruit juice extract. Freeze-dried whole fruit (FDWF), freeze-dried berry (FDB), and freeze-dried skin (FDS) of *B. angulata* were analyzed for total phenolic, total flavonoid, total anthocyanin, and antioxidant activities. FDS recorded the highest moisture and ash content, protein, total fat, and water activity, compared to FDWF and FDB. FDS also contained the highest total phenolic, total flavonoid, and total anthocyanin, while FDWF recorded the highest in scavenging xanthine oxidase (35.9%) and ferric reducing activity (44.9 µM TE/g). FDS, however, showed the highest DPPH (102.66 mg AA/100 g) and TEAC (847.46 mg TE/100 g) values. DPPH, TEAC and FRAP were strongly correlated with total phenol (r = 0.979; 0.948; 0.997) and total flavonoid (r = 0.987; 0.960; 0.992). Total anthocyanin had no correlation with DPPH and TEAC, but moderately with FRAP (r = 0.734). Physicochemical and antioxidant characteristics of *B. angulata* may indicate that this fruit may impart health benefits when consumed and can be suggested as a good source for nutraceutical beverages.

Key words: Antioxidant properties, *Baccaurea angulata*, freeze-dried fruit, nutritional composition, physicochemical.

INTRODUCTION

Reactive oxygen species (ROS) is as important as anti-oxidant to human health. However, due to imbalance of free radicals and antioxidant in our body, development of chronic diseases such as hyperlipidemia, cancer, diabetes, and many others may occur (Liu et al., 2010). There are many conventional therapies and pharmacological drugs available to combat these diseases. Despite this, the treatment is quite expensive; taking synthetic drugs in a long run may also impose a threat to the human`s health. Therefore, medications, or any products derived from natural sources such as plants and fruits have a great demand from the public (Leong and Shui, 2002).

Epidemiological studies showed that there is an inverse association between fruit and vegetable consumption and the risk of developing degenerative diseases (Oliveira et al., 2009). This is because fruits and vegetables contain antioxidant compounds, such as flavonoids, carotenoid, and polyphenols, which have protective effects to human body (Ashraf et al., 2010). These compounds help our body to balance out the ratio with free radical by scavenging free radicals, decomposing peroxides, and making complex of redox-catalytic metal ions (Zibadi et al., 2007).

Malaysia has plenty of nutritious fruits and medicinal plants due to constant exposure to 6-8 h of daily sunlight throughout the year. Therefore, tropical fruits in Malaysia are rich in antioxidants. Interestingly, there are some fruits which are commonly consumed by the locals but yet to be discovered by researchers in terms of nutritional as well as health benefits. *Baccaurea angulata*, locally known as 'belimbing dayak' or 'belimbing hutan', is one of the underutilized fruits from Borneo island of Malaysia. *B. angulata* belongs to the *Euphorbiaceae* family. The tree is about 8-10 m tall. It has thick and broad leaves, and the leaves are ellipse in shape. This fruit is seasonal, and the fruit grows on stems and branches of the tree (Rukayah, 2002). The review of literature showed that very limited studies have been conducted using this fruit in terms of physical and chemical characteristics. In view of this, the determination of antioxidant content and antioxidant activities of this fruit is carried out in this work.

MATERIALS AND METHODS

Sample preparation

Fresh *B. angulata* fruits were collected from Bau, Sarawak, Malaysia. These fruits were separated into three portions, whole fruit, fruit skin, and berries (edible portion). The fruits were rinsed with distilled water, sliced, and blended with ratio of 1:1 (v/v) of edible portion to water. For berries, they were blended with ratio of 1:3 (v/v) of edible portion to water. These juices were filtered and transferred into separated containers. The samples were freeze-dried at -52°C, 0.63 mbar (Freezone 18 Liter Console Freeze Dry System, Labconco, USA). Freeze-dried powder retrieved was stored in an air-tight container at 4°C for further analysis.

Sample extraction of freeze-dried *Baccaurea angulata*

Extraction of powdered sample was carried according to a method by Velioglu et al. (1998) with slight modification. About 200 mg of sample was extracted with 2 mL of methanol at room temperature for 2 h with continuous stirring. Then, the mixture was centrifuged at 1000 *g* for 15 min. The supernatant was collected and kept at -20°C before the determination of total phenolic content, total flavonoid, total anthocyanin, and antioxidant capacity study.

Physical characteristics of *B. angulata* fruits

The fruit was randomly selected for physical measurement which includes average fruit, skin, and berry weight (g), fruit and berry length (cm), fruit and berry width (cm). Fruit weight was measured using calibrated weighing scale. The length and width of the fruit were measured using vernier caliper. All the measurements were done in triplicate and the average of the reading was reported.

Nutritional composition of *B. angulata* freeze-dried extract

Analysis of the moisture, ash, protein, total fat, crude fiber, dietary fiber, and carbohydrate contents were carried out according to the Association of Official Analytical Chemist (AOAC) (2003) method. Gross energy was calculated according to FAO/WHO/UNU (1981) method. Water activity was measured using AquaLab CX-2 water activity equipment.

Determination total phenolic content

Total phenolic content was determined according to the method of Lim et al. (2006) with slight modifications. About 1 mL of extract was mixed with 5 mL of Folin-Ciocalteu reagent and allowed to stand at room temperature for 5 min. 4 mL of 7.5% sodium carbonate solution was then added into the mixture. The solution was incubated for 40 min. The absorbance was recorded at 765 nm. Calibration curve was constructed using different concentration of gallic acid range between 0.001 to 0.02 mg/mL. Results were expressed as mg gallic acid equivalent in 1 g of dried sample (mg GAE/g).

Determination of total flavonoid content

Total flavonoid content was determined by colorimetric method as described by Abu Bakar et al. (2009). Briefly, 0.5 mL of extract was mixed with 2.25 mL of distilled water and 0.15 mL of 5% $NaNO_2$ solution. The mixture was allowed to stand for 6 min before 0.3 mL of 10% $AlCl_3.H_2O$ solution was added and incubated for another 5 min. 1 mL of NaOH was then added into the mixture, vortexed, and the absorbance was immediately measured at 510 nm. A linear standard calibration curve was constructed using quercetin range from 0.01- 0.1 mg/mL. Results were expressed as mg quercetin equivalents per gram of dried sample (mg QE/g).

Determination of total anthocyanin content

Determination of total anthocyanin content was based on pH differential method according to the modified method by Lee et al. (2005) and Abu Bakar et al. (2009). About 3.5 mL of 0.025 M potassium chloride buffer pH 1.0 was added to 0.5 mL of the sample extract and mixed thoroughly using vortex. The mixture was allowed to stand for 15 min before the absorbance was read at 515 and 700 nm. The extract was then treated similarly with 0.4 M sodium acetate buffer pH 4.5 and the absorbance was measured at the same wavelengths. The total anthocyanin concentration was calculated using the following equation and expressed as mg of cyanidine-3-glucoside equivalent in 100 g of dried sample (mg c-3-gE/ 100g dried sample).

$$\text{Total anthocyanin concentration} = \frac{A \times MW \times DF \times 103}{\varepsilon \times 1} \quad (1)$$

Table 1. Physical characteristics of whole fruit, berries, and skin of *Baccaurea angulate*.

Physical characteristic	Parameter
Whole fruit	
Weight (g)	22.69±9.05
Length (cm)	4.61±0.59
Circumference (cm)	11.12±1.61
Width (cm)	3.02±0.52
pH	3.34±0.03
Whole berry	
Weight (g)	7.73±4.55
pH	4.05±0.53
Skin	
Weight (g)	15.95±7.03
pH	3.13±0.07

Values were the means standard deviations of three replicates analysis.

Where, A= (A_{515} nm – A_{700} nm) pH 1.0 – (A_{515} nm – A_{700} nm) pH 4.5; molecular weight (MW) = 449.2 g/mol for cyanidin-3-glucoside (cyd-3-glu); DF = dilution factor of samples; 1 = path length in cm; ε= 26,900 molar extinction coefficient, in L x mol-1 x cm-1 for cyd-3-glu; 103 = factor for conversion from g to mg.

Determination of antioxidant capacity

Xanthine oxidase dismutase (XOD)/superoxide scavenging assay (SOD)

Xanthine oxidase dismutase (XOD) assay was conducted according to the method of Vimala et al. (2003) with slight modification. About 4.1 mM/l 4-nitroblue tetrazolium chloride (NBT) solution was prepared by adding 3.15 g TrisHCl, 0.1 g $MgCl_2$, 15.0 mg 5-bromo-4-chloro-3-indolyl phosphate and 34.0 mg NBT to 100 mL of distilled water. Reaction mixture was prepared by dissolving 0.53 g of Na_2CO_3 (pH10.2), 4.0 mg ethylenediaminetetra acetate (EDTA), and 5.0 mg xanthine before the addition of 10 mL of NBT solution prior to use and kept at 4°C. 5 µL of the sample extract, 995 µL of the reaction mixture and 0.1 µL of XO were added into a cuvette, mixed thoroughly and the absorbance was taken at 560 nm for 120 s. Absorbance of 1000 µL of the reaction mixture and 0.1 µL XO of the same spectrophotometer condition was considered as control.

1,1-Diphenyl, 2-picryl hydrazyl (DPPH) free radical scavenging assay

Radical scavenging activity of Freeze-dried whole fruit (FDWF), freeze-dried berry (FDB), and freeze-dried skin (FDS) was determined according to the method described by Molyneux (2004) with slight modification. About 1 mL of sample was mixed with 2 mL of DPPH solution. The mixture was shaken vigorously and left to stand in a dark room for 30 min. After incubation, the absorbance of the mixture was measured using UV/Vis spectrophotometer at 517 nm. A linear standard calibration curve was constructed using ascorbic acid range from 0.001 - 0.007 mg/mL (r^2 = 0.996). Results were expressed as mg ascorbic acid equivalents per 100 g sample (mg AA/100 g).

Trolox-equivalent antioxidant capacity (TEAC)/ 2,2-azino-bis (3-ethylbenzthiazoline-6-sulfonic acid *(ABTS) assay*

The improved method adopted for ABTS radical scavenging activity was described by Re et al. (1999). The stock solution was prepared which include 7 mM of ABTS and 2.45 mM of potassium persulfate. To prepare the working solution, both solutions were mixed in equal quantities and left to stand in a dark room for 12-16 h. Then, about 1 mL from the working solution was diluted with 60 mL methanol to obtain an absorbance of ±0.70 at 734 nm using spectrophotometer. For every assay, fresh $ABTS^{\cdot +}$ solution was prepared. About 150 µL of the methanol extract sample was mixed with 2850 µL of $ABTS^{\cdot +}$ and left to react for 6 min. A linear standard calibration curve was constructed using ascorbic acid range from 0.5 – 18 µM/mL (r^2 = 0.998). Results were expressed as mg trolox equivalents per 100 g sample (mg TE/100g).

Ferric reducing /antioxidant power (FRAP) assay

FRAP assay was conducted based on Benzie and Strain (1996) with slight modification. The fresh working solution of FRAP reagent was prepared by mixing 300 mM acetate buffer pH 3.6, 10 mM 2,4,6-tripyridyl-s-triazine (TPTZ), and 10 mM $FeCl_3.6H_2O$ in a 10:1:1 ratio and warmed to 37°C in water bath prior to use. Briefly, 3 mL of FRAP reagent was taken into a cuvette and a blank reading was measured at 593 nm and considered as blank reading. Then, 100 µL of the sample extract and 300 µL of distilled water were added into the cuvette and the second reading was performed after 4 min incubation in dark condition at the same wavelength. The change in absorbance was compared with a standard curve. The standard curve was linear between 0.002 to 0.2 µM trolox (r^2 = 0.993). Results were expressed in micromolar trolox equivalent per gram of dried sample (µM TE/g dried sample).

RESULTS AND DISCUSSION

Physical characteristics of *B. angulata*

Physical characteristics of *B. angulata* which include weight of the whole fruit, berry, and skin, the length and perimeter of whole fruit are shown in Table 1. The shape of this fruit is like star fruit (*Averhhoa carambola*), from which the name "belimbing" was derived. The ripe fruit is bright red in colour. It is round at the base of the fruit and pointed at the end. Each fruit contains 1 – 4 cloves of berry which are white in colour and sticky.

The average length and width of *B. angulata* was 4.61±0.59 cm and 3.02±0.52 cm compared to diameter of *B. ramiflora* (Burmese grape) which was 2.5 – 3 cm (Singh, 2007). About 70.3% of whole fruit weight of *B. angulata* was contributed by fruit skin. The edible portion was about 34.1% fraction of fruit weight which is lesser than *B. ramiflora* by 12% (Haque et al., 2009). Furthermore, the average weight of berry of *B. angulata* was 7.73±4.55 g which is heavier when compared to *B. ramiflora* berry (2 g) (Singh, 2007). pH of edible portion of *B. ramiflora* was more acidic (pH 2.4) than *B. angulata* (pH 4.05) (Haque et al, 2009).

Table 2. Nutritional and chemical composition of FDWF, FDB, FDS.

Nutrient composition	FDWF	FDB	FDS
Moisture (%)	19.63±0.04	20.71±0.54	21.04±0.20
Ash (%)	4.74±0.25	5.04±0.05	10.04±0.97
Protein (%)	1.33±0.05	1.28±0.16	2.12±0.21
Total fat (%)	0.11±0.03	1.73±1.76	2.09±1.89
Carbohydrate (%)	74.12±0.00	70.85±0.00	64.58±0.00
Crude fiber (%)	0.07±0.06	0.36±0.21	0.13±0.11
Gross energy (kcal/100g)	302.79±0.00	304.09±0.00	285.61±0.00
Total dietary fiber (%, w/w)	6.3	3.6	6.3
Water activity (W_A)	0.210	0.384	0.467

Values were the means standard deviations of three replicates analysis.

Table 3. Antioxidant content and antioxidant capacities of FDWF, FDB, and FDS.

Sample	TPC (mgGAE/g)	TFC (mg QE/g)	TAC (mgc-3-g/100 g)	XOD (%)	DPPH (mgAA/100g)	TEAC (mg TE/100g)	FRAP (µM TE/g)
FDWF	7.91±0.05	12.74±0.24	0.42±0.28	35.9±1.95	60.96±4.24	412.39± 33.33	44.91±0.73
FDB	4.78 ±0.04	7.86 ±0.10	0.25±0.02	22.9± 2.90	56.31± .32	406.83± 75.93	15.54±0.49
FDS	16.58±0.05	31.05±0.52	0.72±0.13	20.6± 5.35	102.66± 2.12	847.46± 10.57	112.88±3.86

Values are presented in mean ± standard deviations of three replicates analyses of freeze-dried samples.

Nutritional and chemical composition of *B. angulata*

Nutritional and chemical compositions of freeze-dried samples are shown in Table 2. From the results, almost 20% of freeze-dried powder from all samples contained moisture. Furthermore, water activity (a_w) recorded for these samples ranges from 0.2 to 0.5. Moisture and water activity are essential properties as they can determine the shelf life of food (Ruzainah et al, 2009). Presence of high moisture content can facilitate microbial growth in food products. However, in food with water activity that ranges from 0.2 to 0.7 a_w, microbial proliferation will not be observed (Fontana and Campbell, 2004).

Percentage of ash represents the mineral content that is present in the sample. Therefore, FDS recorded the highest ash content (10.04 %), protein content (2.12 %) and total fat (2.09%) compared to FDB (5.04%; 1.28%; 1.73%) and FDWF (4.74%; 1.33%; 0.11%), respectively. Furthermore, FDB has higher crude fiber content (0.36%) compared to FDS and FDWF with 0.13 and 0.07%, respectively. Carbohydrate contributed the major composition for FDWF, FDB and FDS which was about 65 - 75% and gross energy for all three samples were about 300 kcal/100 g. FDS and FDWF contained about 6.3% of total dietary fiber which is higher than orange juice (NDB No: 09206) and apple juice (NDB No: 09016) by 6.1% (USDA National Nutrient Database for Standard Reference, 2011). Dietary fiber further adds nutritional benefits to FDS because it is known to have many positive effects such as lowering the risk of cardiovascular disease, obesity, and gastrointestinal diseases (Ibrahim et al., 2010; Rosamond, 2002).

Antioxidant constituent and capacity of samples

From the results (Table 3), FDS showed the highest total phenolic content (TPC) of 16.58 mg GAE/g, compared to FDB (4.78 mg GAE/g) and FDWF (7.91 mg GAE/g). The result was expressed in gallic acid equivalent because, in plant, gallic acid is considered as one of the major phenolic compounds (Norshazila et al., 2010).

Natural phenolic compounds are largely found in teas, fruits, and vegetables. The most common and widely distributed plant phenolic compounds are flavonoids that are characterized by benzo-y-pyrone structure (Abu Bakar et al., 2009). From the results, FDS contained flavonoid compound of 31.05 mg QE/g and anthocyanin compound, which is higher compared to other samples. Skin of freeze-dried *B. angulata* was richer in total flavonoid compared to total phenol and anthocyanin. Therefore, flavonoids probably are the most attributable scavenger in it and it has high pharmacological activity as radical scavengers (Abu Bakar et al., 2009). Polyphenols are able to scavenge free radicals in human body such as superoxide anion radicals that cause oxidative stress (Ajila et al., 2007). Thus, phenolic substances in freeze-dried skin of *B. angulata* may exert an antioxidant effect, thus, may prevent the development of atherosclerosis (Gorinstein et al., 1999).

XOD, DPPH, TEAC and FRAP assays were selected to measure the capacity of polyphenols in *B. angulata*. Xanthine oxidase dismutase assay measure the ability of sample to scavenge superoxide free radical anions. From the result, superoxide scavenging (%) by XOD was as follows: FDWF>FDB>FDS. DPPH assay measures the ability of *B. angulata* extract to donate hydrogen to the radicals (Lim et al., 2006). The result for DPPH was in the order of FDS>FDWF>FDB. Reduction capability of DPPH is determined by the degree of discoloration at 517 nm. This indicates the reaction of antioxidant with DPPH by converting it to 2,2-diphenyl-1-picrylhydrazine (Ajila et al., 2007). Ascorbic acid equivalent capacities (AEAC) for FDWF and FDB are less than 70 mg AA_{eq}/100g (Leong and Shui, 2002). These results indicate that FDWF and FDB fall into low antioxidant capacity category. While, FDS, having more than 70 mg AA_{eq}/100g, but less than 200 mg AA_{eq}/100g, falls into moderate antioxidant capacity. Nevertheless, the reducing ability of FDWF, FDB and FDS is strongly correlated with total phenol ($r = 0.979$, $p<0.01$) and total flavonoid ($r = 0.987$, $p<0.01$), while anthocyanin constituent has no correlation with DPPH values.

ABTS scavenging assay is widely adopted to study food substances because this intensely-coloured protonated radical has maximum absorbance at 734 nm and mostly coloured food do not absorb light at 734 nm. $ABTS^+$ is a cation radical that will decrease as the protons are scavenged. From the result, trolox equivalent antioxidant capacities are as follows: FDS > FDWF > FDB. Pearson correlation shows strong correlation between trolox value in studied extract and total phenol ($r = 0.948$, $p<0.01$) and total flavonoid ($r = 0.960$, $p<0.01$), while there is no correlation between anthocyanin and trolox value.

The freeze-dried samples were further evaluated for antioxidant capacities through FRAP assay. This assay was adopted in order to measure the ability of *B. angulata* to reduce Fe^{3+} complex of tripyridyl triazine $Fe(TPTZ)^{3+}$ to the intensely blue coloured Fe^{2+} complex $Fe(TPTZ)^{2+}$ in acidic medium (Lim et al, 2006). FRAP value was highest in FDS, followed by FDWF and FDB. A strong correlation was observed between all samples with total phenol ($r = 0.997$, $p<0.01$) and total flavonoid ($r = 0.992$, $p<0.01$). However, correlation was found moderate with total anthocyanin ($r = 0.734$, $p<0.05$). The presence of polyphenol compounds may contribute to the high antioxidant activities via hydrogen donation or electron donation (Ajila et al, 2007).

Conclusion

The result of this investigation shows that FDWF, FDB, and FDS contain the essential nutrients including dietary fiber that are good in promoting human health. Besides, all freeze-dried powder showed presence of essential amount of antioxidant constituents as confirmed by antioxidant activities. Comparing all the samples, it was noted that FDS has the best potential as nutrient supplement due to its higher antioxidant activity contributed by polyphenol compounds. Therefore, FDS derived from *B. angulata* has the potential to be promoted as functional food and drinks. Further study is suggested to isolate and identify phenolic compounds present in the *B. angulata*, as well as identifying other biomarker compounds such as ascorbic acid and beta-carotene.

ACKNOWLEDGMENT

This work was supported by the Ministry of Higher Education under Fundamental Research Grant Scheme, Project No. 0510-138; Ministry of Science, Technology and Innovation (EScience), Project No. 06-01-08-SF0111; Ministry of Science, Technology and Innovation, EScience Project No. 06-01-08-SF0112 and Research Endowment Grant Scheme of the International Islamic University Malaysia, Project No. EDW B11-065-0543. We would also like to appreciate the technical supports provided by the Kulliyyah of Allied Health Sciences, IIUM.

REFERENCES

Abu Bakar MF, Maryati M, Asmah R, Fry (2009). Phytochemicals and antioxidant activity of different parts of bambangan (*Mangifera pajang*) and tarap (*Artocarpus odoratissimus*). Food Chem. 113:479-483.

Ajila CM, Naidu MKA, Bhat SG, Prasada Rao UJS (2007). Bioactive compounds and antioxidant potential of mango peel extract. Food Chem. 105:982–988.

AOAC (2003). Association of Official Analytical Chemists. Official Methods of Analysi, (17th ed.). USA: AOAC Inc.

Ashraf MA, Maah MJ, Yusoff I (2010). Estimation of antioxidant phytochemicals in four different varieties of durian (*Durio zibethinus murray*) fruit. Middle-East J. Sci. Res. 6(5):465-471.

Benzie IFF, Strain JJ (1996). The Ferric Reducing Ability of Plasma (FRAP) as aMeasure of "Antioxidant Power": The FRAP Assay. Anal. Biochem. 239:70–76.

FAO/WHO/UNU (1981). Energy and Protein Requirements. Technical Reports Series 724, Geneva., 133-138.

Fontana AJ, Campbell CC (2004). Water activity. In L. M. L. Nollet (Ed.). Handbook of food analysis: physical characterization and nutrient analysis. Belgium: Marcel Dekker, Inc.

Gorinstein S, Zemser M, Haruenkit R, Chuthakorn R, Grauer F, Martin-Belloso O, Trakhtenberg S (1999). Comparative content of total polyphenols and dietary fiber in tropical fruits and persimmon. J. Nutr. Biochem.10:367–371.

Haque MN, Saha BK, Karim MR and Bhuiyan MNH (2009). Evaluation of nutritional and physico-chemical properties of several selected fruits in Bangladesh. Bangladesh J. Sci. Ind. Res. 44(3), 353-358.

Ibrahim M, Prasad KN, Ismail A, Azrina A, Azizah AH (2010). Physiochemical composition and antioxidant activities of underutilized Mangifera pajang fruit. Afr. J. Biotechnol. 9(28):4392-4397.

Lee J, Durst RW, Wrolstad RE (2005). Determination of total monomeric anthocyanin pigment content of fruit juices, beverages, natural colorants, and wines by the pH differential method: Collaborative study. J. AOAC Int. 88:1269–1278

Leong LP, Shui G (2002). An investigation of antioxidant capacity of fruits in Singapore markets. Food Chem. 76:69–75.

Lim YY, Lim TT, Tee JJ (2006). Antioxidant properties of guava fruit: comparison with some local fruits. Sunway Acad. J. 9-20.
Liu YT, Lu BN, Xu LN, Yin LH, Wang XN, Peng JY, Liu KK (2010). The antioxidant activity and hypolipidemic activity of the total flavonoids from the fruit of *Rosa laevigata Michx*. Nat. Sci. 2(3):175-183.
Molyneux P (2004). The use of the stable free radical diphenyl-picrylhydrazyl (DPPH) for estimating antioxidant activity. *Songklanakarin*. J. Sci. Technol. 26(2):211-219.
Norshazila S, Syed Zahir I, Mustapha Suleiman K, Aisyah MR, Kamarul Rahim K (2010) Antioxidant levels and activities of selected seeds of Malaysian tropical fruits. Mal. J. Nutr. 16(1):149-159.
Oliveira AC, Valentim IB, Silva CA, Bechara EJH, Barros MP, Mano CM, Goulart MOF (2009) Total phenolic content and free radical scavenging activities of methanolic extract powders of tropical fruit residues. Food Chem. 115:469–475.
Re R, Pellegrini N, Proteggente A, Pannala A, Yang M, Rice-Evans C (1999). Antioxidant activity applying an improved ABTS racicalcation decolorization assay. Free Radic. Biol. Med. 26(9/10): 1231–1237.
Rosamond WD (2002) Dietary fiber and prevention of cardiovascular disease. J. Am. College Card, 39: 57-59.
Rukayah A (2002). *Buah-buahan Malaysia*. Kuala Lumpur: Dewan Bahasa dan Pustaka.
Ruzainah AJ, Nor ZCM, Vasudevan R (2009). Proximate analysis of dragon fruit (*Hyelecereuspolyrhizus*). Am. J. Appl. Sci. 6(7):1341-1346.
Singh DR (2007). Underutilized fruit crops of Andaman and Nicobar Islands. In K. V. Peter (Ed.). Underutilized and Underexploited Horticultural Crops, Volume 2. New Delhi: New India Publishing. 11-54.
United State Department of Agricultural (USDA) (2011). USDA National Nutrient Database for Standard Reference, Release 24. Retrieved May 17, 2011. Websitehttp://www.nal.usda.gov/fnic/foodcomp/search/
Velioglu YS, Mazza G, Gao L, Oomah BD (1998) Antioxidant activity and totalphenolics in selected fruits, vegetables and grain products. J. Agric. Food Chem. 46:4113-4117.
Vimala S, Adenan MI, Ahmad AR, Shahdan R (2003). Nature`s choice to wellness: antioxidant vegetables/ulam. Kuala Lumpur: Forest Research Institute Malaysia.
Zibadi S, Farid R, Moriguchi S, Lu Y, Foo L, Tehrani P (2007). Oral administration of purple passion fruit peel extract attenuates blood pressure in female spontaneously hypertensive rats and humans. Nutr. Res. 27:408–416.

Effect of ethanol seed extract of *Buccholzia coriacea* (wonderful kola) on the lipid profile of albino rats

Enechi, O. C. , Manyawo, Lewis, N. and Ugwu Okechukwu P. C.

Department of Biochemistry, University of Nigeria, Nsukka, Enugu State, Nigeria.

This study investigates the effect of ethanol extract of *Buccholzia coriacea* seeds on the serum lipid profile of albino rats within a period of two weeks of treatments. Twelve (12) Wistar albino rats weighing between 85 to 130 g purchased and randomly divided into three groups were used for this study. The control group received normal saline while the two other groups received graded doses of 100 and 200 mg/kg body weight of the extract, respectively. Measurement of the serum lipid profile in rats administered with extract indicated a significant decrease ($p<0.05$) in the mean values of low density lipoprotein (LDL) cholesterol but there was a significant increase ($p<0.05$) in high density lipoprotein (HDL) cholesterol levels of the rats administered with the same extract. There was no significant difference ($p>0.05$) between the total cholesterol concentration and triacylglycerol concentrations of rats administered with the same extract compared to the control. This study suggests that ethanol extract of *B. coriacea* seeds have active ingredients that are capable of improving the lipid profile and thus, might be useful in the management of cardiovascular diseases caused by hyperlipidemia.

Key words: *Buccholzia coriacea*, triacylglycerol, high density lipoprotein, low density lipoprotein, total cholesterol.

INTRODUCTION

Immense benefits have been derived by man from using medicinal herbs in disease management because they are relatively safer, more affordable and sometimes offer better therapeutic value than synthetic drugs. The increasing discovery of more medicinal plants have necessitated increased scientific scrutiny of their bioactivity in order to provide data that will help physician and patients make wise decision before using them (Oyewole and Akigbala, 2011).

Atherosclerosis or coronary artery disease is a condition characterized by deposits of lipids, mainly cholesterol on the inner walls of the arteries (Ghasi et al., 2002; Ugwu et al., 2013; Olantunji et al., 2005; Ezekwesili et al., 2008). These deposits narrow the arterial channels and partly block the normal flow of blood through them (Ahman et al., 1992; Ugwu et al., 2013; Ghasi et al., 2002; Olantunji et al., 2005; Ezekwesili et al., 2008). The decrease in blood flow and oxygen can result in stroke, partial paralysis, loss of speech and sometimes death (Olson, 1998).

Atherosclerosis is the main cause of mortality and morbidity in western countries and it is progressively increasing in developing countries (Barter et al., 2007). Low fat diets are often prescribed for the management of atherosclerosis.

Buccholzia coriacea (Wonderful cola) is a tropical plant used in traditional medicine. It is grouped under the genus, *Buccholzia.* Over the years, it has been proven to possess pharmacological and cosmetic effects. The plant is a shrub, evergreen, with a dense crown, large glossy and leathery leaves.

In this study, the effects of *B. coriacea* leaf extract on the lipid profile of female Wistar albino rats were evaluated.

MATERIALS AND METHODS

Plant

Fresh seeds of wonderful kola (*Buccholzia coriacea*) were got from Ibagwa Nkwo market, Igbo-Eze South Local Government Area of Enugu State, Nigeria. The seeds were identified by Mr. A. Ozioko of Bioresource Development and Conservation Programme (BDCP) Research Centre, Nsukka.

Preparation of the extract

Fresh seeds of *B. coriacea* were plucked and washed with distilled water. The sliced seeds were spread on a clean mat in a well-ventilated room with regular turning to enhance even drying and avoid decaying. The sliced seeds were shade-dried for 10 weeks. The shade dried sliced seeds were pulverized with an electric blender and a known weight (800 g) of the pulverized *B. coriacea* seeds were macerated in 70% ethanol (700 ml) and allowed to stand for 24 h. The mixture was separated with Whatman No. 1 filter paper. The filtrate was concentrated using rotary evaporator and water bath. The extract was then weighed and stored in a refrigerator.

Experimental design

Twelve (12) Wistar albino rats were randomly divided into three (3) different groups as follows: Group 1: Normal control rats that were administered 0.2 ml of normal saline; group 2: rats were administered 100 mg/kg body weight of *B. coriacea* extract orally; group 3: rats were administered 200 mg/kg body weight of *B. coriacea* extract orally.

The extracts were administered for two weeks and analyses were done before administration to determine the baselines and also on day 14 of treatment.

Determination of extract yield

The percentage yield of the aqueous extract of *B. coriacea* seeds were determined by weighing the seeds before extraction and the concentrated extract was obtained after extraction and then calculated using the formula:

$$\text{Percentage yields} = \frac{weight\ (g)\ of\ extract}{weight\ (g)\ of\ pulverized\ leaves} X 100$$

Animals

Adult female Wistar rats of 4 to 6 weeks old with average weight of 85 to 130 g were obtained from the Animal House of the Faculty of Veterinary Medicine, University of Nigeria, Nsukka. The animals were acclimatized for seven days under standard environmental conditions, with a 12-h light and dark cycle maintained on a regular feed and water *ad libitum*.

Chemicals/reagents/samples

All chemicals used in this study were of the analytical grade and products of May and Baker, England; BDH, England and Merck, Darmstadt, Germany. Reagents used for all the assays were commercial kits and products of Randox, USA; QCA, Spain; Teco (TC), USA; Biosystem Reagents and Instruments, Spain; Glucose Analyzer, Accu Chek Active of Germany; alloxanmonohydrate, Sigma St. Louis, MO, USA.

Instruments/Equipment

Water Bath (Gallenkamp, England), Chemical Balance (Gallenkamp, England), Conical Flasks (Pyrex, England), Hotbox (Gallenkamp, England), Centrifuge (3,500 rpm, PIC, England), Digital Photo Calorimeter (EI 312 Model, Japan), Adjustable Micropipette (Perfect, U.S.A.), Refrigerator (Kelvinator, Germany), pH Meter (Pye, Unicam 293, England), multi-well microtiter plate reader (Tecan, Austria), and Glucose Analyzer, (Accu Chek Active of Germany) were used for the study.

Determination of high density lipoprotein (HDL)

High density lipoprotein (HDL) cholesterol concentration was determined using QCA commercial kit according to the study of Albers et al. (1978).

Determination of low density lipoprotein (LDL)

Low density lipoprotein (LDL) cholesterol concentration was determined using the method of Assmann et al. (1984) methods.

Determination of total cholesterol

Total cholesterol concentration was determined using QCA commercial kit according to the study of Allain et al. (1976).

Determination of triacylglycerol (TAG)

Triacylglycerol (TAG) was determined using the method of Allain et al. (1976)

Statistical analysis

Data were reported as means ± SEM, where appropriate. Both one- and two-way analyses of variance (ANOVA) were used to analyse the experimental data and Duncan multiple test range was used to compare the group means obtained after each treatment with control measurements. Differences were considered significant when $p \leq 0.05$.

RESULTS

Percentage yield of extract

Percentage yield of the extract was calculated thus:

Weight of pulverized seed = 800 g

Weight of extract = 87.5 g

$$\text{Percentage yield} = \frac{87.5}{800} X \frac{100}{1} = 10.94\%$$

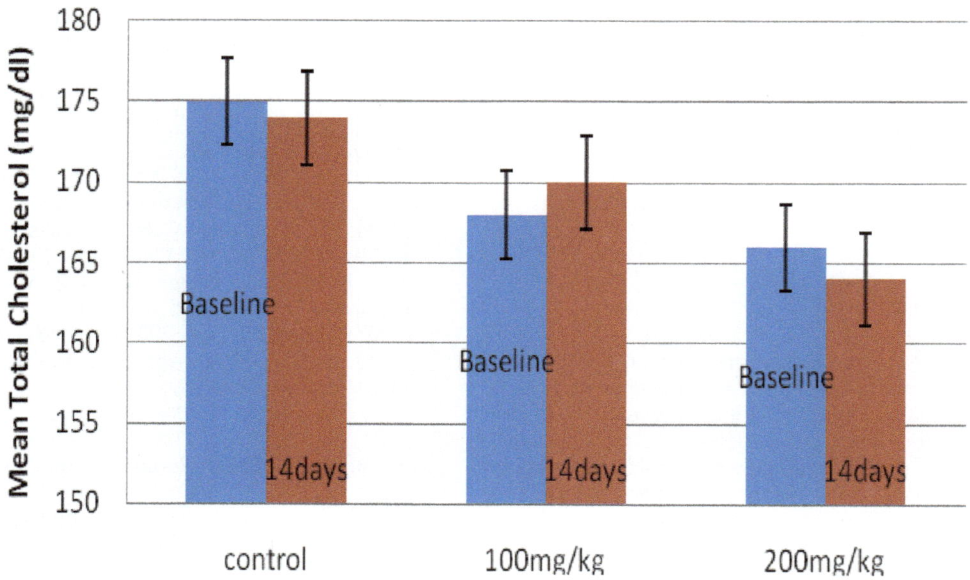

Figure 1. Effect of *Buccholzia coriacea* extract on the total serum cholesterol concentration of rats.

Effect of *B. coriacea* ethanol seed extract on the total serum cholesterol concentration of experimental rats

After daily oral administration of graded doses of 100 and 200 mg/kg b.w. of *B. coriacea* ethanol seed extract for two weeks, there was no significant ($p>0.05$) difference in the total serum cholesterol concentration of the test groups when compared with the control groups as shown in Figure 1.

Effect of *B. coriacea* seed extract on the serum triacylglycerol concentration of experimental rats

After daily oral administration of *B. coriacea* seed extract of 100 and 200 mg/kg b.w. for two weeks, there was no significant ($p>0.05$) difference in the serum triacylglycerol concentration of the test groups when compared with the control groups as shown in Figure 2.

Effect of *B. coriacea* seeds extract on the serum HDL cholesterol concentration of experimental rats

After daily oral administration of *B. coriacea* extract seeds (100 and 200 mg/kg b.w) for two weeks, there was a significant increase ($p<0.05$) in the serum HDL cholesterol concentration of the test groups when compared with the control groups as shown in Figure 3.

Effect of *B. coriacea* seed extract on the serum LDL cholesterol concentration of experimental rats

After daily oral administration of graded doses of *B. coriacea* seed extract (100 and 200 mg/kg b.w.) for two weeks, there was a significant decrease ($p<0.05$) in the LDL cholesterol concentration of the test groups when compared with the control groups as shown in Figure 4.

DISCUSSION

The global world today is challenged with cardiovascular diseases. Some of the key manifestations include coronary heart diseases, stroke and hypertension. Elevated concentrations of plasma lipids are risk factors in cardiovascular problems (Brown and Goldstein, 1992), and important lipids whose elevations are implicated in these conditions are cholesterol and triaglycerides. Lipids are transported in the blood by combination of lipids and proteins complexes called lipoproteins (Nwanjo, 2005).

The main identified determinants of hyperlipidemia are increased LDL-cholesterol and reduced HDL-cholesterol (Ghasi et al., 2002; Ugwu et al., 2013). Thus, any attempt to lower serum concentrations of LDL and increase HDL concentration is considered as one of the strategies that can hinder or delay the on-set of chronic disorders that are associated with hyperlipidemia in humans (Olantunji et al., 2005; Ezekwesili et al., 2008).

In this study, the results obtained on the effect of ethanol extract of *B. coriacea* seed on the lipid profile of normal rats revealed that this plant could improve the lipid profile by increasing the serum HDL-cholesterol level on the 14th day of treatment. This was deduced from the fact that the serum concentration of these fractions in the treated animals were significantly higher ($p>0.05$) on day 14 compared to the baseline values and control animals.

Also, LDL-cholesterol concentration significantly reduced ($p<0.05$). The observed decrease in serum concentration of LDL could be due to the presence of certain

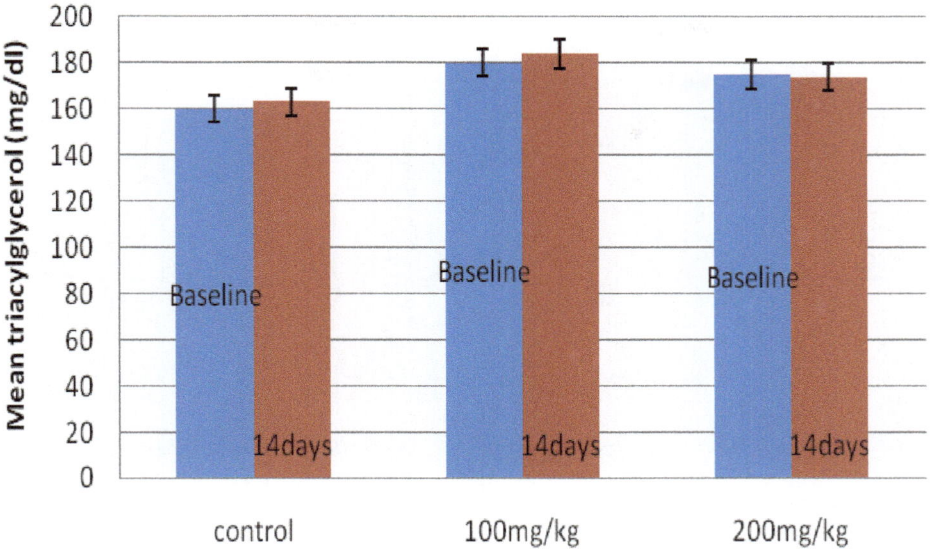

Figure 2. Effect of *Buccholzia coriacea* extract on the triacylglycerol concentration of rats.

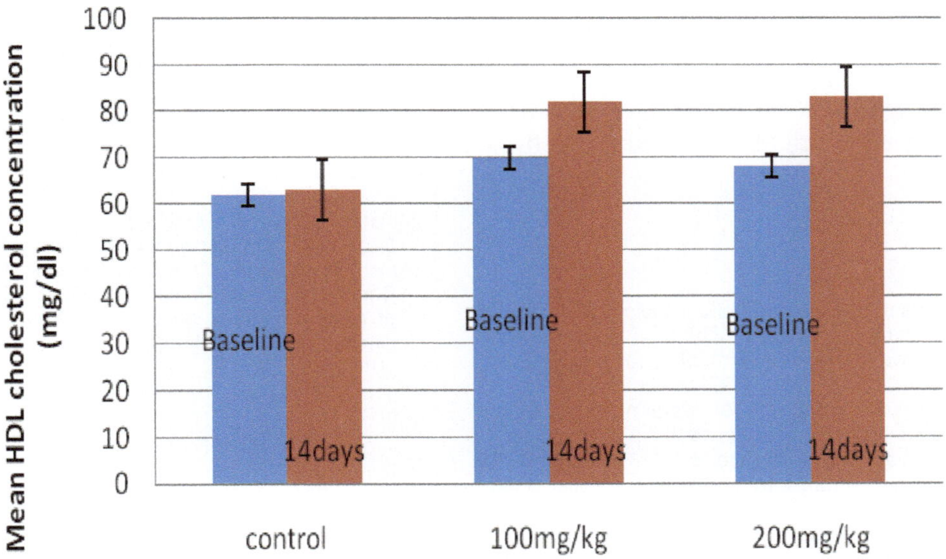

Figure 3. Effect of *Buccholzia coriacea* extract on the HDL cholesterol concentration of rats.

physiologically active constituents in the extract. There was no significant changes ($p>0.05$) in serum concentration of total cholesterol and triacylglycerol. HDL cholesterol has a protective effect against cardiovascular disease as it removes excess cholesterol from circulation and carries it back to the liver where it is degraded or converted into bile acids (Ahmad et al., 1992). The lipid raising effect of some traditional medicinal plants have been reported (Sundram et al., 1999).

The hypocholesterolemic activity of the extract after administration may be due to a number of mechanisms including the inhibtion of rate limiting of enzyme of cholesterol biosynthesis, HMG-CoA reductase, conversion of cholesterol into bile acids and inhibtion of cholesterol absorption from the intestine due to formation of complexes with compounds such as glycosides and saponins (Gaamoussi et al., 2010).

Conclusion

In conclusion, the results obtained from this study showed that oral administration of ethanol extract of *B. coriacea* seeds resulted in a significant improvement on

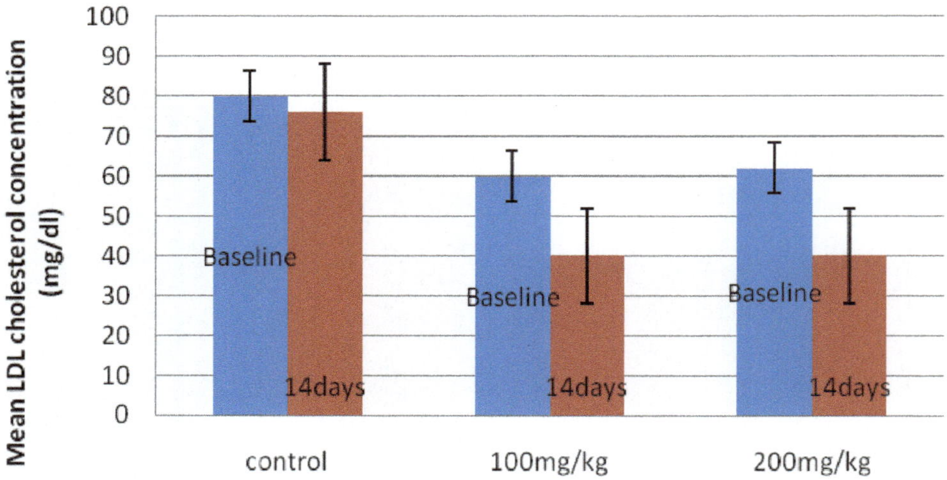

Figure 4. Effect of *Buccholzia coriacea* extract on the LDL cholesterol concentration of rats.

the levels of serum lipid profile by a significant increase in HDL concentration, reducing LDL-cholesterol concentration without any practical effect on triacylglycerol and total cholesterol. This may suggest that the plant extract possess hypolipidemic activity, and may be useful in the management of cardiovascular diseases and thus suggest that further research on the extract should be carried out to elucidate the actual mechanism underlying this effect.

REFERENCES

Ahman M, Saeed MM, Alam H, Ashgar Z (1992). Biological studies of indigenous medicinal plants 11: Effect of *Aplotaxis lappa Dene* on various parameters of liver metabolism in rabbits. J. Islamic Acad. Sci. 5:61-66

Albers JJ, Warmick GR, Cheng MC (1978). Determination of high density lipoprotein (HDL)–cholesterol. Lipids 13:926-932.

Allain CC, Poon LS, Chan CSG, Richmond W, Fu PC (1976). Enzymatic determination of serum total cholesterol. Clinical Chem. 20:470-475.

Assmann G, Jabs HU, Kohnert U, Nolte W, Schriewer H (1984). Determination of low density lipoprotein (LDL)–cholesterol. Clinica Chimica Act. 140:77-83.

Barter P, Gotto L, Rosa AM, Szarek J, Grundy SM, Kastelein JP, Bittner V (2007). HDL Cholesterol, VLDL cholesterol and cardiovascular events. The New England J. Med. 357(13):1301-1309.

Brown SB, Goldstein JL (1992). Drugs used in the treatment of hyperlipoproteinaemias. In Galman, a.g. Rall, T.W., Nies, A.S., Taylor, P. (eds), Goodman and Gilman's pharmacological Basis of therapeutics, volume1. Eight ed , McGraw-Hill, Inc. New York. pp. 874-894.

Ezekwesili CN, Obidoa O, Nwodo OFC (2008). Effects of ethanol Extract of *Acalypha torta* leaves on the Lipid profile and serum Electrolytes of Rabbit, Niger. J. Biochem. Mol. Biol. 23(1):15-19.

Gaamoussi F, Isreali ZH, Lyoussi B (2010). Hypoglycemic and hypolipidemic effects of an aqueous Extract of *Chamaerops humilis* leaves in obese, hyperglycemic and hyperlipidemic *meriones shawi* rats. Pak. J. Pharm. Sci. 23(2):212-219.

Ghasi S, Nwobodo E, Ofili JO (2002) .Hypocholesterolemic effects of crude extract of leaf of *Moringa oleifera* Lam. in high-fat diet Wistar Rats. J. Ethnopharmacol. 69:21-25.

Nwanjo HU (2005). Efficacy of aqueous leaf extract of *Vernonia amygdalina* plasma lipoprotein and oxidative status in diabetic rat models. Niger. J. Physiol. Sci. 20:39-42.

Olantunji LA, Adebayo JO, Oguntoye OB, Olatunde NO, Olantunji VA, Soladoye AO (2005). Effects of aqueous extracts of petals of red and green *Hibiscus sabdariffa* Plasma lipid and hematological variables in rats. Pharm. Biol. 43(5):471-474.

Olson RE (1998). Discovery of the lipoproteins, their role in fat transport and their significance as risk factors. J. Nutr. 128: (2): 439-443.

Oyewole OI, Akigbala PF (2011). Phytochemical Analysis and Hypolipidemic Properties of *Jatropha tanjorensis* Leaf Extract. Eur. J. Med. Plants. 1(4):180-185.

Sundram K, Khor HT, Ong AS (1999). Effect of dietary palm oil and its fractions on rat plasma and high density lipoprotein. Lipids 25:07-13.

Ugwu Okechukwu PC, Nwodo Okwesili FC, Joshua Parker E, Odo Christian E, Ossai Emmanuel C (2013). Effect of Ethanol Leaf Extract of *Moringa oleifera* on lipid profile of mice. Res. J. Pharm. Biol. Chem. Sci. 4(1):1324-1332.

21

Impact of salicylic acid on antioxidants, biomass and osmotic adjustments in *Vigna unguiculata* L. walp. during water deficit stress

Umebese, C. E. and Bankole, A. E.

Department of Botany, P.M.B 1029 UNILAG Post Office, University of Lagos, Akoka-Yaba, Lagos, Nigeria.

Ameliorative impact of salicylic acid (SA) on *Vigna unguiculata* L. (cowpea) cultivar IT93k-452-1 during water deficit stress was investigated. Plants were subjected to water deficit stress (WD) for 7 days at either the vegetative stage (DVS) or the reproductive stage (DRS) and the leaf water potentials (ψ_w) of -1.9 MPa and -2.01 MPa were obtained, respectively. Stress caused reductions in almost all parameters studied. 3 and 5 mM SA foliar treatment caused increases of 27% in leaf ψ_w, 94% in chlorophyll content, 75% in plant biomass, 7% in nitrate reductase activity and 38% in proline content in DVS plants while the impact was much lower in DRS plants. Stress stimulated almost 3-fold increase in antioxidant vitamin B_{12} in DVS plants and 40% increase in vitamin C in DRS plants and SA (3 mM) further increased the latter by 14%. Overall, SA had a protective influence on the water potential and growth of stressed plants accompanied by an increase in production of osmolyte proline and antioxidant vitamin C.

Key words: Antioxidants, growth, salicylic acid, water stress.

INTRODUCTION

Water deficit stress elicits many different physiological responses in plants. It causes reduction in leaf water potential, stomatal conductance, nitrate reduction and inhibits leaf enlargement while osmolytes such as total soluble sugars and proline are increased (Jafar et al., 2004; Adejare and Umebese, 2007). Prolonged water stress affects virtually all metabolic processes and often results in severe reductions in plant productivity and death of plants. The effect of water deficit varies with the variety and the growth stage of the plants.

All abiotic stresses such as water deficit and salt stress cause increased production of reactive oxygen species leading to oxidative damage. Almost all organisms possess antioxidant defense and repair systems that have evolved to protect them against oxidative damage. Majority of the antioxidant activity is from flavonoids, isoflavone, flavones, anthocyanin, catechin, isocatechine, vitamins B, C, E and β-carotene (Bronzetti et al. 2001; Amic et al., 2003). Others are enzymes such as, catalase, peroxidase and superoxide dismutase (Jaleel et al., 2009). Salicylic acid acts as an endogenous signal molecule responsible for the induction of antioxidant responses that protect plants from damage (Senaratna et al., 2000; Hayat et al., 2007, 2010).

Salicylic acid can stimulate flowering, increase flower life, inhibit seed germination, promote ethylene synthesis (Singh and Usha, 2003), enhance photosynthetic rate and growth rate (Khan et al., 2003). Salicylic acid has been shown to produce a protective effect in plants under the action of stress factors of different abiotic nature, such as, heat, chilling, osmotic and salt stress (Borsani et al., 2001; Janda et al., 1999; Singh and Usha, 2003; Wang and Li, 2006). Salicylic acid (SA) is an endogenous growth regulator of phenolic nature, which participates in the regulation of physiological processes in plants such as growth, photosynthesis, nitrate metabolism, ethylene production, heat production and flowering (Hayat et al., 2010) and also provides protection against biotic and abiotic stresses such as salinity (Kaya et al., 2002). An efficient N assimilation is said to be favoured by a high rate of CO_2 assimilation (Larsson et al., 1989). SA induced conservation of water in stressed plants also results in the protection of nitrate reductase activity (NRA) in SA treated stressed plants.

The present study is aimed at evaluating the protective effect of salicylic acid on the development of *Vigna unguiculata* L. (cowpea) seedlings during water deficit stress, thereby considering the leaf water potentials, plant biomass, leaf chlorophyll content, nitrate reductase activity, proline concentrations, vitamine B12 and ascorbate contents at both vegetative and reproductive stages.

MATERIALS AND METHODS

Plant material and planting procedure

Seeds of *V. unguiculata* L. cultivar IT93k-452-1 were collected from the International Institute of Tropical Agriculture (IITA) Ibadan, Oyo State. The experiment was set up at the botanic garden of the University of Lagos, Akoka Lagos. Seeds of cowpea were planted 2-3 cm deep in 48 pots of 30.4 cm diameter (4 treatments) each filled with 2 kg garden topsoil mixed with 2 g NPK fertilizer. After three weeks, seedlings were thinned to 4 plants per pot. Plants were subjected to water deficit for 7 days at vegetative (21st-27th days after sowing) and reproductive (45th - 52nd days after sowing) stages (DVS and DRS plants). At the beginning of the 7 day stress period a batch of 12 plants was treated with 3 mM SA as foliar spray, the other batch of 12 plants was treated with 5 mM SA, the third batch which consist of yet another 12 plants was subjected to water deficit only while the fourth batch of 12 plants served as the control (no stress treatment). All plants were watered daily apart from the period of stress treatment. The plants were arranged in a randomized complete block design with 3 replicates. The pots were kept at the green house with 12 h photoperiod and a relative humidity of 60% during the day and 68% at night. Plants were harvested on the last day of stress treatment at each growth stage, dried in an oven at 80°C for 3 days and the biomass taken.

Determination of leaf water potential

After subjecting plants to 7 days water deficit, the degree of stress in the plants was measured using the tissue weight- change method outlined by Hopkins and Hüner (2004). 0.5 g leaf from the different treatments was placed in each of a graded series of 0.2 – 0.8 molal concentration of sucrose in different test tubes. The leaves were removed after 1 h equilibration and dried between filter papers and reweighed. Osmotic potential (ψ_s) of each sucrose solution was calculated using the van't Hoff equation:

$$\psi_s = - C\gamma RT$$

Where, C is the molal concentration, γ is the activity coefficient (a value of 1 for neutral solutes such as sucrose in dilute solution), R is the gas constant (.00831 kg MPa mol^{-1} $^oK^{-1}$), T is the absolute temperature ($^oK = ^oC + 273$).

The change in weight was calculated as a percentage of the original weight and plotted against ψ_s of sucrose solutions. The leaf water potential (ψ_w) is estimated as equivalent to the ψ_s of the solution in which there is no change in weight (the intercept on the abscissa).

Determination of nitrate reductase activity

Nitrate reductase activity in the leaves was assayed as described by McCashin (2000). 5 ml of the incubation medium (comprising 100 ml 0.1 M phosphate buffer, pH 7.5, 1.5825 g KNO_3 and 1 ml 4% propan-1-ol), was used to incubate 0.5 g sample of leaves in boiling tubes in 3 replicates for a period of 1 h at a room temperature of 30 – 35°C. Then the reaction was stopped by adding 1 ml 1% sulphanilic acid, followed by 1 ml of 0.02 % naphthyl-ethylenediamine dichloride. It was then carefully shaken and left for 20 min for colour development. The blank was prepared by adding 1 ml substrate assay solution, 1 ml 1% sulphanilic acid and 1 ml napthylethylenediamine dichloride together in a boiling tube. After full colour development, the absorbance was measured at 540 nm optical density of the test solution, was determined using a Corning spectrophotometer model 258. A standard curve was prepared using 0-10 μM nitrite ml^{-1}. NRA is proportional to the concentration of nitrite in the reaction medium.

Determination of proline content

Proline content of leaves was determined as described by Bates et al. (1973). 0.5 g of dried ground leaves was homogenized in 10 ml 3% aqueous sulfosalicylic acid and the homogenate filtered through filter paper. 2 ml acid ninhydrin and 2 ml glacial acetic acid were added to 2 ml of the filtrate in a digestion tube and placed in a boiling water bath for 90 min. The reaction was terminated in an ice bath. The reaction mixture was extracted with 4 ml toluene and agitated vigorously for 30 s. The chromophore containing toluene was aspirated from the aqueous phase and the absorbance read at 520 nm using toluene as the blank. The concentration was determined from a standard curve prepared from 1-10 mg l^{-1} of proline and calculated on a dry weight basis.

Estimation of chlorophyll content

The amount of chlorophyll a and b (mg g^{-1} fresh weight) were calculated using the formula of Maclachlam and Zalik (1963).

Determination of the antioxidant vitamins content

Antioxidant vitamins cyanocobalamin (vitamin B_{12}) and ascorbic acid (vitamin C) contents were determined as described by Van (2006). Ten grams of the fresh leaves from different plants were macerated using a ceramic mortar and pestle. The extract was mixed thoroughly using an electric mixer and centrifuged at 1000 g at 5°C. 1 ml aliquot was made up to 10 ml with distilled water for cyanocobalamin (vitamin B_{12}) content while 1 ml of the extract was made up to 100 ml with distilled water for ascorbic acid (vitamin C). After filtration each was injected into the high performance liquid chromatography (HPLC) at a wavelength of 204 nm. The chromatographic conditions consist of the following: Column-zorbax SB

Table 1. Leaf water potential of cowpea subjected to water deficit (WD) and salicylic acid (SA) treatment at the vegetative and reproductive stages of growth.

Treatment	Water potential (MPa) vegetative stage	Reproductive stage
Control	-0.92[a]	-1.35[a]
WD	-1.97[b]	-2.01[b]
WD + 3 mM SA	-1.88[b]	-1.90[b]
WD + 5 mM SA	-1.44[d]	-1.70[c]

Superscripts with same letters at each growth stage are not significantly different at p<0.05 using the Duncan's multiple range tests.

C18 (250 x 4.6 mm, 5 nm), thermostatically regulated at 20°C; the mobile phase was a gradient composition of activities: acetonitrite (A) and 25mM KH_2PO_4 buffer (B) pumped at a flow rate of 0.5 ml min^{-1}. When the gradient elution of A is 8 and 16% and B is 92 and 84% at 0 min and 5 min, respectively, 20 µL of reference standards were injected to determine their respective retention time, and graded concentration of mixed reference standards (vitamin C and B_{12}) were injected. The peak areas (MAU) were acquired and integrated by enhanced integrator for plotting of calibration curve and concentrations given as µg $ml^{-1}$and calculated as mg g^{-1} fresh weight.

All data were analyzed using Statistical Analysis System (SAS) software. When ANOVA showed significant treatment effects, Duncan's multiple range test was applied to compare the means at p<0.05.

RESULTS AND DISCUSSION

Leaf water potential of *V. unguiculata* L. cultivar IT93k-452-1 was significantly reduced when plants were subjected to 7 days water deficit stress at the vegetative and reproductive stages (DVS and DRS respectively) of growth, with the reproductive stages (DRS) showing a greater decrease in water potential (Table 1). This stage dependent response to water deficit was shown in an earlier study on soybean leading to a greater increase in stomatal resistance and reduced stomatal capacity at the reproductive stage (Adejare and Umebese, 2007). This resulted in much reduction in growth and yield of plants (Figure 3). Foliar application of salicylic acid (SA) was effective in raising the water potential of stressed plants by 27% though the values were still lower than the control.

Water deficit did not affect the total chlorophyll content of cowpea at both stages of growth (Figure 1). This was also observed by Vasileva and Llieva (2011), when lucerne was subjected to water deficiency stress. However, the application of 5 mM SA to DVS plants significantly (p<0.05) enhanced the chlorophyll content over the control. This corroborates the report by Rajasekaran and Blake (1999) that SA has a positive effect on photosynthesis. Nitrate reductase activity (NRA) was significantly reduced by water deficit (Figure 2). 5 mM SA raised the NRA level of vegetative stages (DVS) of plants to that of the control while 3 mM SA increased the NRA of DRS plants though it was still lower than that of the control. There is a strong positive correlation between NRA and relative water content (Umebese and Okeola, 1998). SA increased NRA of stressed plants and this corresponded with the increase in the leaf water potential. This agrees with an earlier report that SA increases the conservation of water in tomato and amaranth during water stress and enhances NRA (Umebese et al., 2009).

Plant biomass was significantly reduced by water deficit, corresponding to the low water potential induced by the 7 days water deficit (Figure 3). Energy requiring processes such as NRA and growth are reduced by water stress since CO_2 fixation is limited, reducing the $NADP^+$ regeneration by the Calvin cycle leading to an over-reduced photosynthetic electron transport chain (Krause, 1994). Though treatment with SA significantly increased (p<0.05) plant biomass, values were still lower than the control. In an earlier study, plant biomass was enhanced by SA when tomato and amaranth were subjected to water stress (Umebese et al., 2009). Concentration of SA as low as 0.05 mM increased the level of cell division within the apical meristem of seedling roots which caused an increase in plant growth (Bezrukova et al., 2001).

Proline was significantly decreased in the DVS plants while it was significantly increased in DRS plants. Both concentrations of SA enhanced proline production during water stress at both stages of growth (Figure 4). The accumulation of the osmolyte proline, in plants is a mechanism by which plants resist water stress and develop anti-stress reaction. Water deficit induces accumulation of proline in seedlings and also results in a decline in metabolic activity of plants cells, and this is inevitably reflected in inhibition of their growth (Bezrukova et al., 2001; Umebese et al., 2009). SA induced abscisic acid (ABA)-mediated protective reactions in plants involves increased production of proline by the expression of the gene encoding D'-pyroline-5-carboxylate synthetase (P5CS) in proline biosynthesis. When plants constantly produce high levels of proline, they are capable of surviving drought (Delauney and Verma, 1993). Thus, the increased production of proline by SA in plants subjected to water deficit at the two growth stages may have

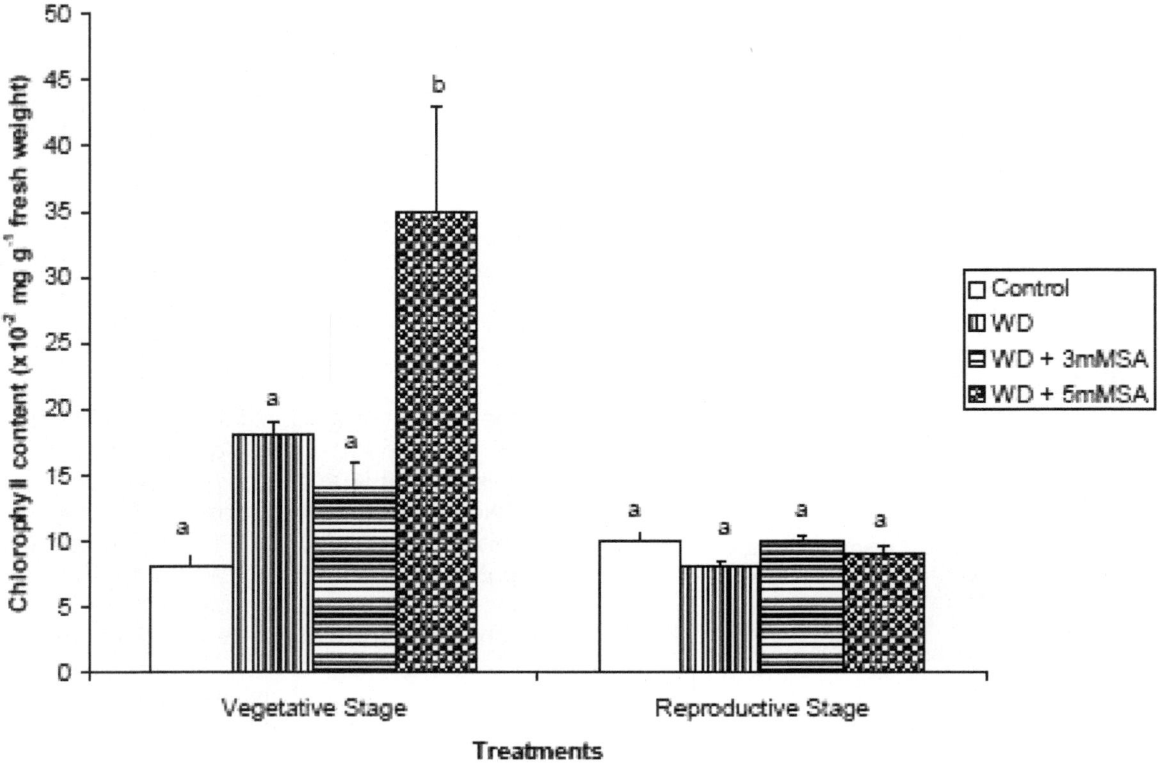

Figure 1. Chlorophyll content of cowpea subjected to water deficit (WD) and salicylic acid (SA) treatment at vegetative and reproductive stages of growth (Bars with same letters at each growth stage are not significantly different at P<0.05 using the Duncan's multiple range test).

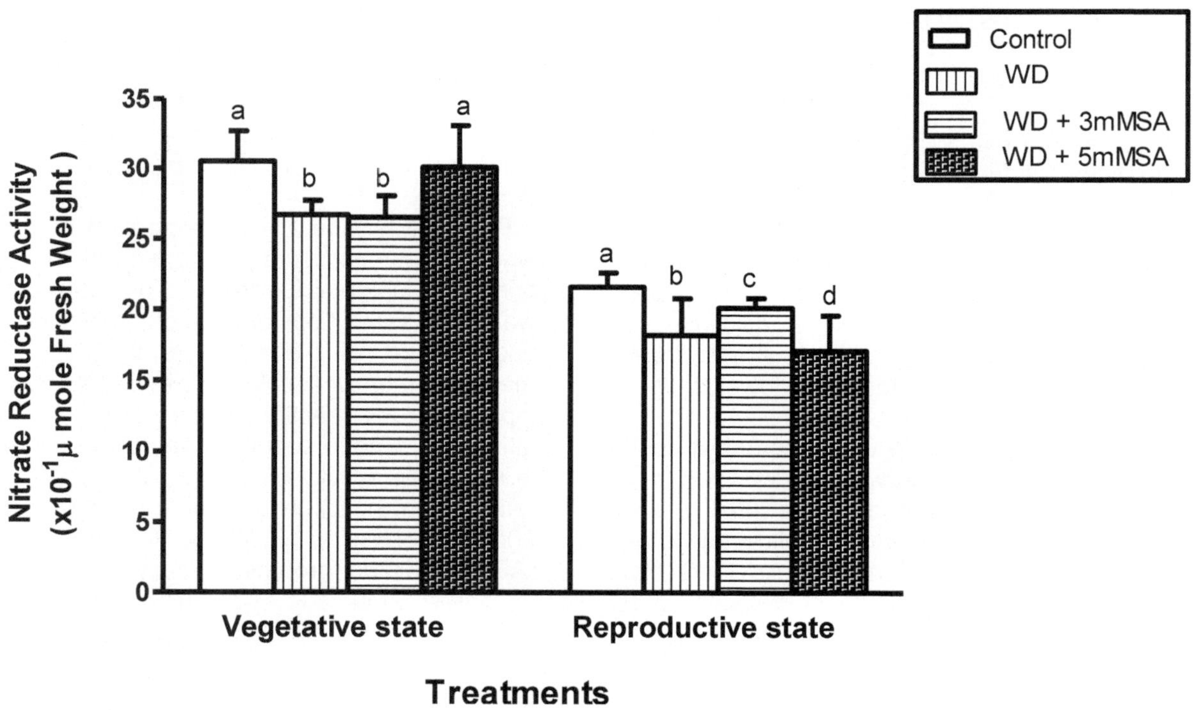

Figure 2. Nitrate reductase activity of cowpea subjected to water deficit (D) and salicylic acid (SA) treatment at vegetative and reproductive stages of growth (Bars with same letters at each growth stage are not significantly different at P<0.05 using the Duncan's multiple range test).

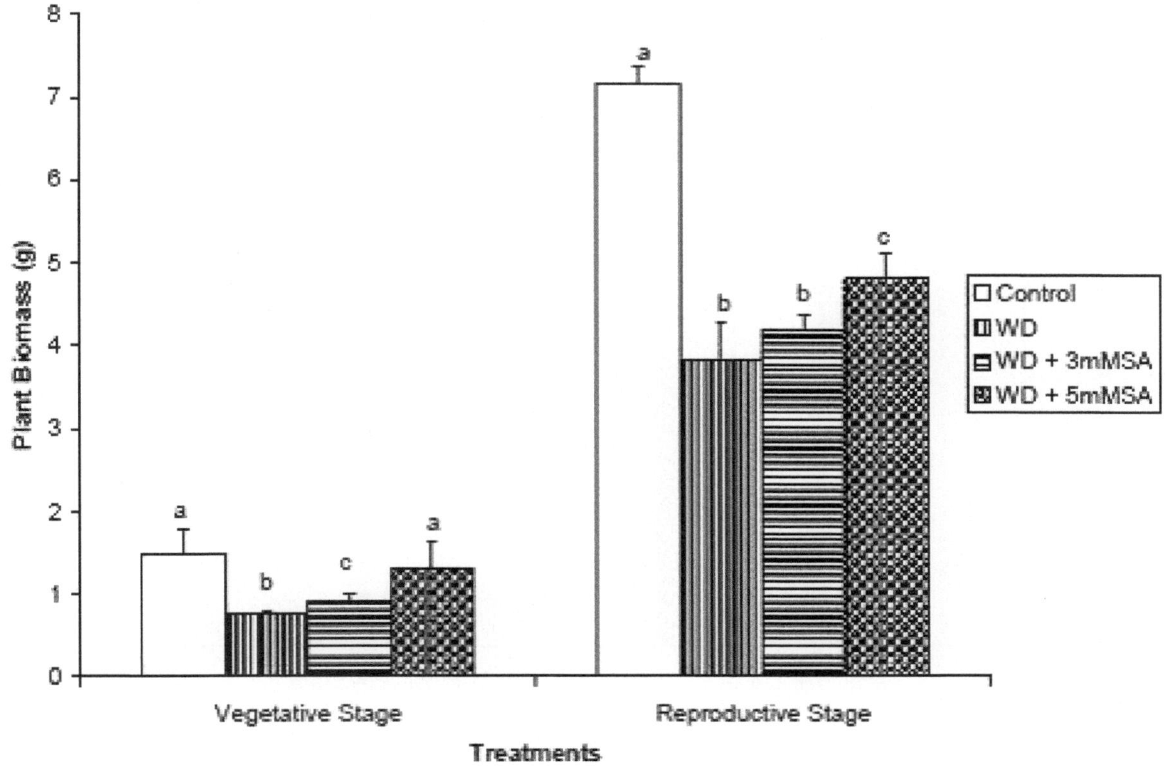

Figure 3. Plant biomass of cowpea subjected to water deficit (D) and salicylic acid (SA) treatment at vegetative and reproductive stages of growth (Bars with same letters at each growth stage are not significantly different at P<0.05 using the Duncan's multiple range test).

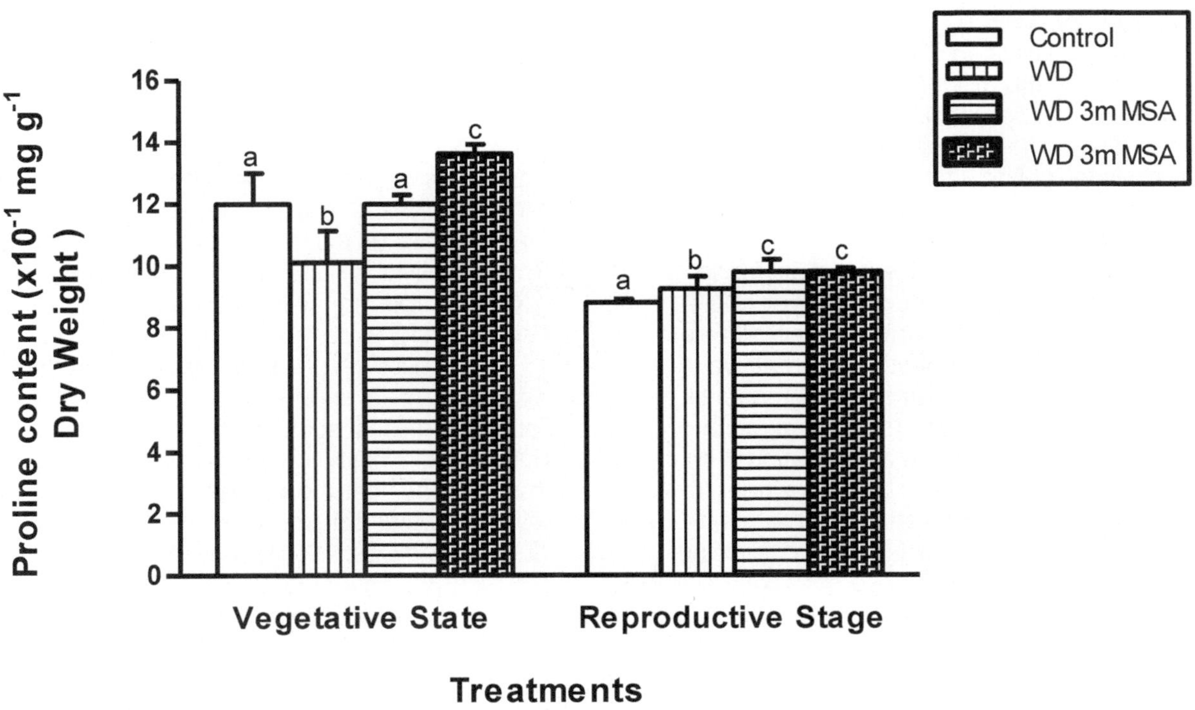

Figure 4. Proline content of cowpea subjected to water deficit (D) and salicylic acid (SA) treatment at vegetative and reproductive stages of growth (Bars with same letters at each growth stage are not significantly different at P<0.05 using the Duncan's multiple range test).

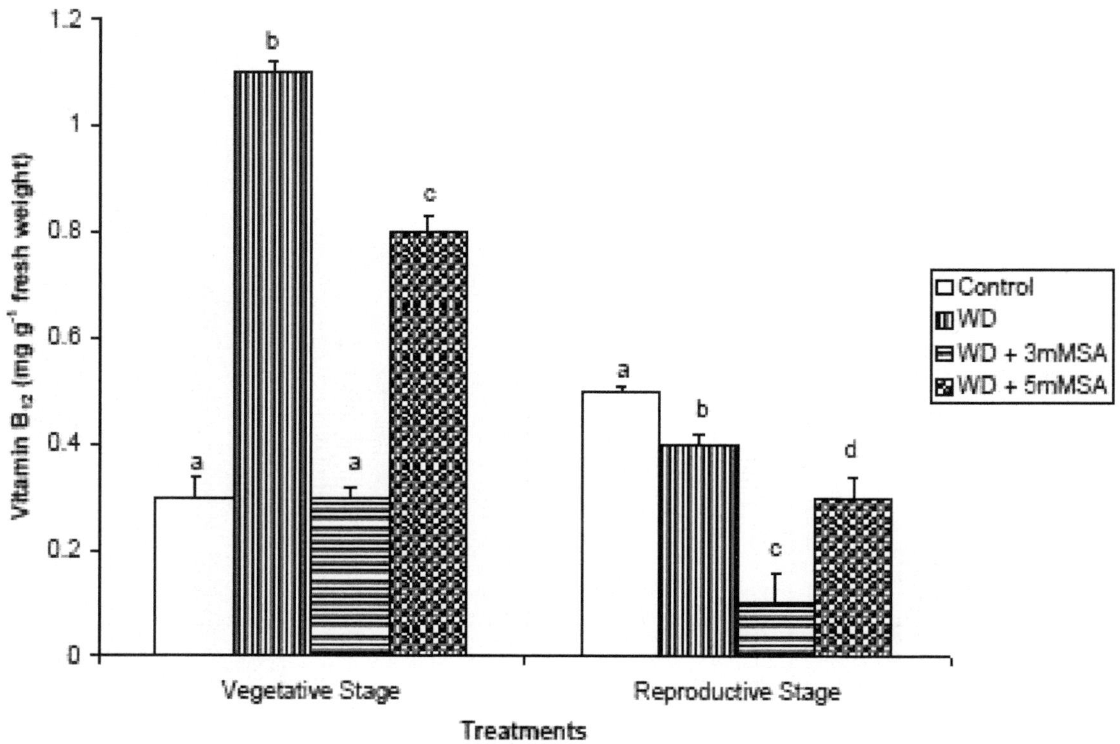

Figure 5. Antioxidant vitamin B_{12} of cowpea subjected to water deficit (D) and salicylic acid (SA) treatment at vegetative and reproductive stages of growth (Bars with same letters at each growth stage are not significantly different at $P<0.05$ using the Duncan's multiple range test).

resulted in the observed enhancement of plant biomass.

The antioxidant vitamin B_{12} was significantly enhanced ($p<0.05$) by water stress in DVS plants but significantly reduced in DRS plants (Figure 5). Vitamin C was also stimulated by water deficit but this was in DRS plants (Figure 6). Thus, the stage at which plants were subjected to water deficit affected the type of antioxidant induced; vitamin B_{12} at the vegetative stage and vitamin C at the reproductive stage. An increase in the level of antioxidants is believed to reflect an increase in the formation of reactive oxygen species (Foyer et al., 1997). All abiotic stresses such as water deficit stimulate the production of significant quantities of reactive oxygen species in sub-cellular compartments or organelles which are highly destructive to their proteins, membrane lipids and nucleic acids (Matsumura et al., 2002). The control of these oxyradicals is achieved by antioxidant systems within the cells. Experiments show that vitamin B is a potent quencher of singlet oxygen with quenching rates comparable to or greater than those of vitamin C and E, two of the most efficient biological antioxidants known (Bilski et al., 2000). Treatment with 3 mM SA resulted in significant reduction in vitamin B content at both stages of growth of stressed plants while 5 mM SA caused a reduction only at the reproductive stage. Thus, SA had a retarding effect on vitamin B_{12} in stressed plants. Both concentrations of SA caused significant reduction in vitamin C content in DVS plants but in DRS plants, 3 mM SA stimulated further increase in this vitamin. This enhancement of vitamin C by 3 mM SA corresponds with the increase in NRA observed in stressed plants subjected to this dose of SA. Vitamin C supplement has been shown to increase NRA in white headed cabbage (Heród-Leszczyńska and Miedzobrodzka, 1993). This shows the ameliorative action of SA when cowpea is subjected to water stress at the reproductive stage of development. The sensitivity of vitamin C to reversible oxidation suggests its role in cellular oxidation-reduction reactions (Verma and Verma, 2007).

Plants must possess efficient antioxidant system to enable them to endure oxidative damage under unfavorable conditions such as high/low temperatures, water deficit and salinity (Sreenivasulu et al., 2000). Various associations between water stress and endogenous levels of water-soluble antioxidants have been described (Zaman and Das, 1991), in which ascorbate (ASC) is also required for the re-conversion of SA to SA yielding monodehydroascorbate, since ASC is highly reactive against phenoxyl radicals generated by peroxidases during oxidative stress (Kawano and Muto, 2000). It can be suggested that the exogenous application of SA is accompanied by increased rates of ROS, which leads to an increased load on the ascorbate-glutathione cycle leading to a depletion of tASC. The results are supported

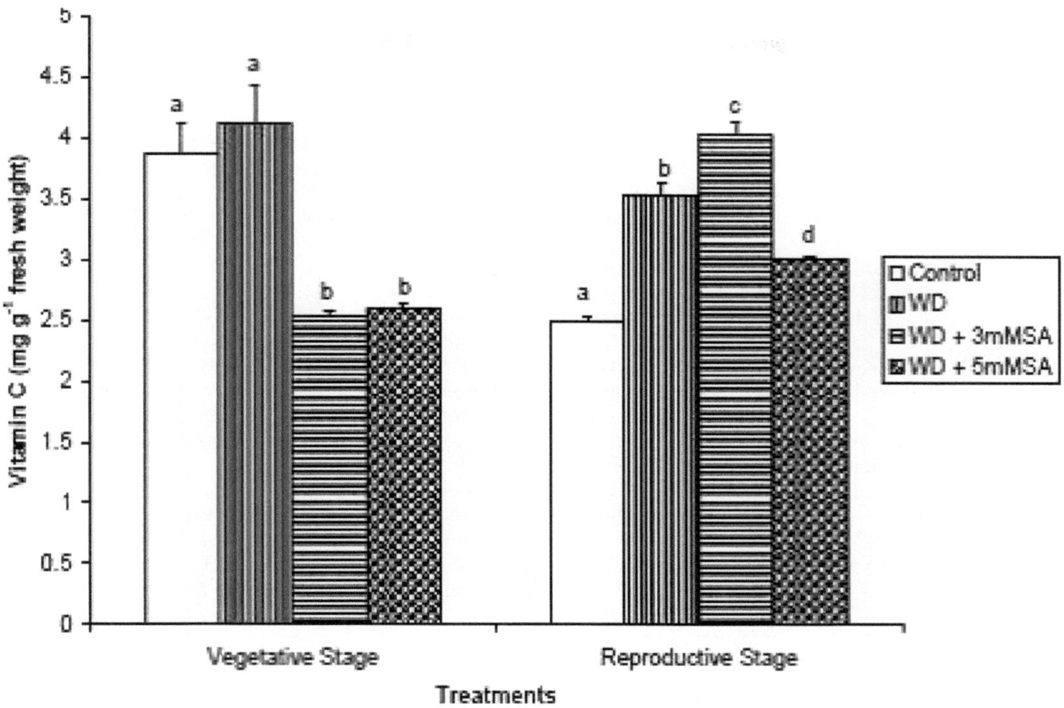

Figure 6. Antioxidant Vitamin C of cowpea subjected to water deficit (D) and salicylic acid (SA) treatment at vegetative and reproductive stages of growth (Bars with same letters at each growth stage are not significantly different at P<0.05 using the Duncan's multiple range test).

by the observation of Shi and Zhu (2008), where SA-treatment enhanced the activities of dehydroascorbate reductase. Mwanamwenge et al. (1999) explained that when ROS increases, chain reactions start, in which superoxide dismutase (SOD) catalyzes the dismutation of O^2 radicals to molecular O_2 and H_2O_2. The H_2O_2 is then detoxified in the ascorbate-glutathione cycle which involves the oxidation and re-reduction of ascorbate and glutathione through the ascorbate peroxidase (APX) and glutathione reductase (GR) action (Ouchi, et al., 1990; Pasternak et al., 2005).

In conclusion, cowpea is more sensitive to water deficit at the reproductive stage than the vegetative stage resulting in reductions in plant biomass and nitrate reductase activity. Treatment of stressed plants with salicylic acid caused significant improvement of those parameters and enhanced proline accumulation, an anti-stress reaction. Plants subjected to water deficit at different stages of growth produced different types of antioxidants: vitamin B_{12} in plants stressed at the vegetative stage and vitamin C in those stressed at the reproductive stage. Salicylic acid enhanced vitamin C content in the latter which may have caused the enhancement of nitrate reductase activity.

REFERENCES

Adejare FB, Umebese CE (2007). Stomatal resistance to low water potential at different growth stages affects plant biomass in *Glycine max* L. Am. J. Agric. Biol. Sci. 3: 136-141.

Amic D, Davidovic D, Beslo D, Trinajstic N (2003). Structure-radical scavenging activity relationships of flavonoids. Croat. Chem. Acta 76: 55-61.

Bates LS, Waldren RP, Tear ID (1973). Rapid determination of free proline for water stress studies. Plant Soil 39: 205-207.

Bezrukova M, Sakhabutdinova VR, Fatkhutdinova RA, Kyldiarova I, Shakirova F (2001). The role of hormonal changes in protective action of salicylic acid on growth of wheat seedlings under water deficits. Agrochemiya (Russ.) 2: 51-54.

Bilski P, Li MY, Ehrenshaft M, Daub ME, Chignell CF (2000). Vitamin B_6 and its derivatives are efficient singlet oxygen quencher and potential fungal antioxidants. Photochem. Photobiol. 71: 129-134.

Borsani O, Valpuesta V, Botella MA (2001). Evidence for a role of salicylic acid in the oxidative damage generated by Nacl and osmotic stress in Arabidopsis seedling. J. Plant Physiol. 126**:** 1024-1030.

Bronzetti G, Cini M, Adrenoli E, Caltavuturo L, Pinunzio M, Croce CD (2001). Protective effects of vitamin and selenium compounds in yeast. Mutat. Res. 496: 105-115.

Delauney AJ, Verma DPS (1993). Proline biosynthesis and osmoregulation in plants. Plant J. 4: 215-223.

Foyer CH, Lopez-Delgado H, Dat JF, Scott IM (1997). Hydrogen peroxide and glutathione-associated mechanisms of acclamatory stress tolerance and signaling. Physiol. Plant 100: 241-254.

Hayat Q, Hayat S, Irfan M, Ahmad A (2010). Effect of exogenous salicylic acid under changing environment: A review. Environ. Exp. Bot. 68: 14–25.

Hayat S, Ali B, Ahmad A (2007). Salicylic acid: biosynthesis, metabolism and physiological role in plants. In: Hayat S, Ahmad A., (Eds.), Salicylic acid – A plant hormone. Springer Verlag, Dordrecht, Boston, London. pp. 1–14.

Heród-Leszczyńska T, Miedzobrodzka A (1993). An attempt to investigate the effect of magnesium ions and vitamin C on activity of nitrate reductase of white headed cabbage. Rocz Panstw Zaki Hig. 44:159-164.

Hopkins WG, Hüner NPA (2004). Introduction to plant physiology. John Wiley and Sons, Inc. USA. pp. 539-540.

Jafar MS, Nourmohammadi G, Meleki A (2004). Effect of water deficit on seedling, plantlets and compatible solutes of forage sorghum cv Speedfield. Proceedings of the 4th International Crop Science Congress Brisbane, Austrialia.

Jaleel C A, Riadh K, Gopi R, Manivannan P, Inès J, Al-Juburi HJ, Hong- Bo ZS, Panneerselvam R (2009). Antioxidant defense responses: physiological plasticity in higher plants under abiotic constraints. Acta Physiol. Plant 31: 427-436.

Janda T, Szalai G, Tari I, Paldi E (1999). Hydroponic treatment with salicylic acid decreases the effect of chilling injury in maize (*Zea mays*). Planta 208: 175 – 180.

Kawano T, Muto S (2000). Mechanism of peroxi-dase actions for salicylic acid-induced generation of active oxygen species and an increase in cytosolic cal-cium in tobacco suspension culture. J. Exper. Bot. 51: 685-693.

Kaya C, Kirnak H, Higgs D, Saltali K (2002). Supplementary calcium enhances plant growth and fruit yield in strawberry cultivars grown at high salinity. Sci. Hortic. 93: 65-74

Khan N, Prithviraj B, Smith DL (2003). Photosynthethic responses of corn and soybean to foliar application of salicylates. J. Plant Physiol. 160: 485-492.

Krause GH (1994). The role of Oxygen in Photoinhibition of Photosynthesis. In: Causes of Photooxidative stress and Amelioration of defense systems in Plants. CH Foyer, PM Mullineaux eds. Boca Raton, Florida, USA. pp. 43-76.

Larsson CM, Whitford PN, Clarkson DT (1989). Influence of osmotic stress on nitrate reductase activity on wheat (*Triticum aestivium*) and the role of abscissic acid. J. Exp. Biol. 40: 1265-1271.

Maclachlam S, Zalik S (1963). Plastid structure, chlorophyll concentration and free amino acid composition of a chlorophyll mutant of barley. Can. J. Bot. 41: 1053-1062.

Matsumura T, Tabayashi N, Kamagata Y, Souma C, Saruyama H (2002). Wheat catalase expressed in transgenic rice can improve tolerance against low temperature stress. Physiol. Plant. 116: 317-327.

McCashin BG (2000). Induction of nitrate reductase in plant shoots, In: Tested studies for laboratory teaching. SJ Karcher ed. volume 21:193-224. Proceedings of the 21st Workshop/Conference of the Association for Biology Laboratory Education (ABLE).

Mwanamwenge J, Loss SP, Siddique KHM, Cocks PS (1999). Effect of water stress during floral initiation, flowering and podding on the growth and yield of faba bean (*Vicia faba* L.). Eur. J. Agron. 11: 1-11.

Ouchi S, Nishikawa A, Kameda E (1990). Soil-improving effect of a super-water-absorbentpolymer II. Evaporation, leaching of salts and growth of vegetables. Jap. J. Soil Sci. Plant Nutr. 61: 606-613.

Pasternak T, Rudas V, Potters G, Jansen MAK (2005). Morphogenic effects of abiotic stress: Reorientation of growth in *Arabidopsis thaliana* seedlings. Environ. Exp. Bot. 53: 299-314.

Rajasekaran LR, Blake TJ (1999). New plant growth regulators protect photosynthesis and enhance Growth under Drought of Jack pine seedlings. J. Plant Growth Regul. 18: 175 – 181.

Senaratna T, Touchell D, Bunn E, Dixon K (2000). Acetyl salicylic acid (Aspirin) and salicylic acid induce multiple stress tolerance on bean and tomato. Plant Growth Regul. 30:157 – 161.

Shi Q, Zhu Z (2008). Effects of exogenous salicylic acid on manganese toxicity, element contents and antioxidative system in cucumber. Environ. Exp. Bot. 63: 317-326.

Singh B, Usha K (2003). Salicylic acid induced physiological biochemical changes in wheat seedlings under water stress. Plant Growth Regul. 39: 137 – 141.

Sreenivasulu N, Grimm B, Wobus U, Weschke W (2000). Differential response of antioxidant compounds to salinity stress in salt tolerant and salt-sensitive seedlings of foxtail millet (*Setaria italica*). Physiol. Plant 109: 435-442.

Umebese CE, Okeola OG (1998). Relationship between nitrate reductase activity, plant water status, growth and yield in four cultivars of *Vigna unguiculata* L. Walp. J. Sci. Res. Dev. 3: 183-190.

Umebese CE, Olatimilehin TO, Ogunsusi TA (2009). Salicylic acid protects nitrate reductase activity, growth and praline in amaranth and tomato plants during water stress. Am. J. Agric. Biol. Sci. 4: 224-229.

Van M (2006). The essential chromatography catalogue. Agilent Technologies Inc., USA.

Vasileva V, Llieva A (2011). Chemical composition, nitrate reductase activity and plastid pigments content in lucerne under the influence of ammonium and nitrate from mineral nitrogen. Agron. Res. 9: 357-364.

Verma SK, Verma M (2007). Plant physiology, biochemistry and biotechnology. S Chand and Company Ltd. pp. 128-129.

Wang LJ, Li SH (2006). Salicylic acid-induced heat or cold tolerance in relation to Ca^{2+} homeostasis and antioxidant systems in young grape plants. Plant Sci. 170: 685-694.

Zaman A, Das PK (1991). Effect of irrigation and nitrogen on yield and quality of safflower. Indian J. Agron. 36: 177-179.

22

Chemical composition and antioxidant activity of essential oil, various organic extracts of *Cistus ladanifer* and *Cistus libanotis* growing in Eastern Morocco

H. Zidane[1] , M. Elmiz[1], F. Aouinti[1], A. Tahani[1], J. Wathelet[2], M. Sindic[2] and A. Elbachiri[1]

[1]Laboratory of Physical Chemistry of the Natural Resources and Environment, University Mohammed Premier, BP 717, 60000 Oujda, Morocco.
[2]Unity of General and Organic Chemistry, Faculty of the Agronomic Sciences of Gembloux, Passage of the Prisoners 2, B-5030 Gembloux, Belgium.

In the present work, we studied the chemical composition of the essential oil of *Cistus ladanifer* and *Cistus libanotis* growing in Eastern Morocco. The essential oils were obtained by hydrodistillation and their chemical composition was analysed using gas chromatography- mass spectrometry (GC–MS). Camphene, borneol, cyclohexanol-2, 2, 6 tremethyl, terpineol-4 and α-pinene were the main constituents of the essential oil of *C. ladanifer*, while in the essential oil of *C. libanotis* we obtained terpineol-4, γ-terpinene, camphene, sabinene, α-terpinene and α-pinene. The antioxidant potential of various extracts (water, ethanol, ethanol: water (50:50), methanol, methanol: water (50:50), acetonitrile) and essential oils of *C. ladanifer* and *C. libanotis* were carried out by the method of 1,1-diphenyl-1-picrylhydrazyl-hydrate (DPPH) free radical scavenging. Total phenolic and flavonoid contents were determined. The result show that *C. ladanifer* of the leaves of methanol: water (50:50) extract had the highest value of total phenolic content and the lowest was present in ethanol: water (50:50) extract of the stem and acetonitrile extract of the flowers of *C. libanotis.* From our experimental results, the extract of flowers, fruit, stem and leaves of those plants showed highest potential as free radical scavengers.

Key words: Antioxidant, phenolics, flavonoids, essential oil, extracts, gas chromatography- mass spectrometry (GC–MS).

INTRODUCTION

Cistaceae comprises eight genera and about 180 species. They are a medium-sized family typically consisting of heliophyte shrubs, subshrubs and herbs occurring in open areas on poor soils (Guzma´n and Vargas, 2009). They are pyrophylic (Alonso et al., 1992) and they are adapted to fires in Mediterranean forests.

The leaves of *Cistus ladanifer* produce a fragrant oleo-resin used in perfumes especially as a fixative. It is used in folk medicine for treatment of various skin diseases, as an anti-diarrheal, and as anti-inflammatory agents (Attaguile et al., 2000). In Eastern Morocco, *Cistus libanotis* and *C. ladanifer* are named by the local population as boubala and touzala, respectively. They are used as an anti-diarrheic, antiacid and antispasmodic by simple decoction of their leaves. In addition, anti-diarrheal effect of aqueous extract of *C. ladanifer* has been proved (Aziz et al., 2011). Furthermore, it produces a number of compounds with pharmacological properties: antihypertensive effect (Belmokhtar et al., 2009), antimicrobial (Guvenc et al., 2005; Greche et al., 2009), antioxidant (Amensour et al., 2010), antifungal (Mrabet et al., 1999), and other more.

The chemical composition of *C. ladanifer* essential oil was characterized by high amount of sesquiterpene (viridiflorol and ledol), and monoterpenoid (bornyl acetate and pinocarveol) (Oller-Lopez et al., 2005), whereas *C. libanotis* oil was dominated by hydrocarbons and oxygenated monoterpenes, monoterpene hydrocarbons, diterpenes, hydrocarbons and oxygenated sesquiterpenes (Ben Jemia et al., 2013).

In the current study, our objective was to evaluate the chemical composition of the essential oil of *C. ladanifer* and *C. libanotis* and to examine the antioxidant potential of oils and various solvent extracts of plant parts (fruits, flowers, stems and leaves) by 1,1-diphenyl-2-picrylhydrazyl-hydrate (DPPH) free radical scavenging assay. Additionally, total flavonoid and phenolic contents were also determined.

MATERIALS AND METHODS

Plant material

C. ladanifer and *C. libanotis* were collected from Tafoughalt (Eastern of Morocco) and they were identified by Dr Haloui of the Department of Biology (University Mohamed Premier Oujda). Collected plants (5 May, 2012) were dried in the laboratory under normal air at room temperature.

Preparation of extract and isolation of essential oil

A weighed portion (2 g) of each dried and powdered sample (fruits, stems, flowers and leaves) was extracted with 100 ml of different solvent (water, ethanol, ethanol: water (50:50), methanol, methanol: water (50:50), acetonitrile) for 8 days at room temperature. The extracts were filtered and evaporated to dryness in a rotary evaporator. The extracts were dissolved in ethanol at 25 mg ml^{-1} concentration and stored until used. The essential oil from leaves of *C. ladanifer* and *C. libanotis* was obtained by hydrodistillation in a Clevenger-type apparatus during 5 and 4 h, respectively. Both oils were dried under anhydrous sulphate and stored at 4°C until used.

Gas chromatography–mass spectrometry (GC–MS) analysis

Analysis of essential oils was carried out by GC–MS using a Hewlett Packard 6890 gas chromatography equipped with a Hewlett Packard 6890 mass selective detector and a HP-5 MS capillary column 30 m x 0.25 mm (cross-linked Phynel-Methyl Siloxane) and film thickness of 0.25 µm. The column was temperature programmed as follows: 40°C for 5 min, and then the temperature was increased to 280°C at a rate of 10°C/min. Helium was used as a carrier gas and the oils (0.1 µl) were injected without any dilution in the splitless mode.

Mass analyzed conditions were ion source of 230°C, ionization energy of 70 eV, electron current of 2 A, and resolution of 1000. Mass unit was monitored from 30 to 450 m/z. Identification of components in the oil was based on retention indices relatives with n-alkanes and computer matching with the WILLEY 275.L library, as well as by comparison of the fragmentation patterns of mass spectra with those reported in the literature (Adams, 1995). The relative proportions (%) of each volatile compound as the percentage ration of the peak area of each compound to the total peak area of all identified compounds were calculated.

DPPH radical scavenging effect

Antioxidant activity was evaluated according to the method cited by Xiao-Juan et al. (2012). 100 µl of different concentration of extract or essential oils or standard (1 to 15 mg ml^{-1}) was added in 1, 4 ml of ethanol and 1 ml of solution of DPPH (0.004 %). The absorbance of the discoloration of the solution was measured at 517 nm. The percentage of inhibition was calculated by:

$$\text{Inhibition (\%)} = [(A_{DPPH} - A_{sample})/A_{DPPH}] \times 100$$

A_{DPPH} is the absorbance of DPPH solution and A_{sample} is the absorbance of the test sample. All tests were performed in triplicate. The lower absorbance of the reaction mixture indicates higher free radical scavenging activity.

Determination of total phenolic content

The concentration of total phenolics in different organic extracts was determined by using Folin Ciocalteu reagent (Singleton and Rossi, 1965); briefly, 20 µl of sample extract (1 mg ml^{-1}) was added to 100 µl of Folin Ciocalteu reagent and 300 µl of Na_2CO_3. 2.5 of distilled water was added after 5 min. The mixture was left 2 h and the supernatant was measured at 576 nm against a blank. The calibration curve was prepared by Ferulic acid at different concentration (0.2 to 2 mg ml^{-1}; Y= 0.0042X + 0.0014; R^2= 0.9991). A test was carried out in triplicate.

Determination of total flavonoid content

Total flavonoid was determined using the method described by Pourmorad et al. (2006); briefly 0.5 ml of each extract (1 mg ml^{-1}) was added to 1.5 ml of ethanol, 0.1 ml of 10% aluminum chloride, 0.1 ml of 1 M potassium acetate and 2.8 ml of distilled water. The resulting solution was left at room temperature for 30 min. The absorbance was measured at 415 nm. The calibration curve was prepared by different concentration of quercetin solutions in ethanol (0.01 to 0.08 mg ml^{-1}; Y= 0.0048X - 0.0128; R^2= 0.9978). A test was carried out in triplicate.

Statistical data

All extractions were conducted in triplicates. Phenolic and flavonoid data are expressed as mean ± standard deviation. The means were compared by using the one-way and multivariate analysis of variance (ANOVA) followed by TuKey's using the SPSS 11.5 for Windows. The differences between individual means were deemed to be significant at $p < 0.05$.

RESULTS AND DISCUSSION

Chemical composition of the essential oils

The essential oil yields in leaf of *C. ladanifer* and *C. libanotis* were respectively 0.14 and 0.13% on dry weight basis (w/w). The essential oil volatile compounds of

Table 1. Chemical composition of the essential oil of *C. ladanifer* and *C. libanotis* .

Compound	Identification	Kovalts Indices	*C. libanotis* (%)	*C. ladanifer* (%)
Tricyclene	IK, MS	923	1.2	2.7
α-Thujene	IK, MS	930	–	1.0
α-Pinene	IK, MS	936	6.9	4.2
Camphene	IK, MS	952	12.2	15.5
Sabinene	IK, MS	977	11.2	0.8
β-Pinene	IK, MS	979	–	0.7
α-Phellandrene	IK, MS	1002	6.2	–
α -Terpinene	IK, MS	1019	7.3	1.8
p-Cymene	IK, MS	1028	4.7	2.3
Cyclohexanone, 2,2,6-trimethyl	IK, MS	1038	–	7.3
γ-Terpinene	IK, MS	1063	12.4	3.8
Terpinolene	IK, MS	1091	3.3	1.3
Sabinaketone	IK, MS	1154	0.8	–
α -Campholene aldehyde	IK, MS	1131	–	1.3
Camphore	IK, MS	1152	–	1.5
Exo-Methyl-Camphenilol	IK, MS	-	–	0.9
Borneol	MS	-	–	11.1
Terpineol-4	IK, MS	1183	17.4	6.3
P-Cymen-8-ol	IK, MS	1190	–	0.6
α-terpineol	IK, MS	1193	0.0	1.2
1-Methyladamantane	MS	1190	0.7	–
Verbenone	IK, MS	1217	–	0.8
l-Bornyl acetate	IK, MS	1293	–	0.9
p-Mentha-1,4-dien-7-ol	IK, MS	1336	–	0.8
(+)-2-Carene	IK, MS	1001	1.1	–
α-Cubebene	IK, MS	1360	–	2.2
6-Amyl-Apha-Pyrone	MS	–	1.0	–
Cyclosativen	IK, MS	1380	–	0.6
1S, Cis-Calamenene	MS	1528	0.7	–
δ-Cadinene	IK, MS	1490	–	6.4
Hexadecanoic acid	MS	1959	0.7	–
Diethyl phthalate	IK, MS	1602	2.2	2.9
Viridiflorol	IK, MS	1610	–	2.8
Cyercene	MS	1619	–	0.5
Bis (2-ethylhexyl) phthalate	MS	2560	–	0.2

%, Relative percentage area; ND, compound not identified in this sample; KI, Kovalts indices calculated using an apolar column (HP-5).

plants, their kovalts indices and their percentage are presented in Table 1.

Twenty eight (28) and 18 compounds were identified in the essential oil of *C. ladanifer* and *C. libanotis*, respectively. As a result of GC-MS analyse, *C. ladanifer* contained, camphene (15.5%), borneol (11.1%), cyclohexanol-2, 2, 6 tremethyl (7.3 %), terpineol-4 (6.3 %) and α-pinene (4.2 %) as the major compounds. The main constituents of the oil of *C. libanotis* were terpineol-4 (17.4%), γ-terpinene (12.4%), camphene (12.2%), sabinene (11:2%), α-terpinene (7.3%), α-pinene (6.9%) and α-phellandrene (6.2%). The comparison of our result with literature shows important qualitative and quantitative differences in compositions. According to Mariotti et al. (1997) report, major compounds of the essential oil of *C. ladanifer* have been determined as α-pinene, viridiflorol, ledol and bornyl acetate.

As can be seen from Table 1, camphene is the major compounds of the *C. ladanifer* oil together with borneol and cyclohexanol-2, 2, 6 tremethyl in the present study. When compared with these compounds, bis (2-ethylhexyl) phthalate and cyercene were detected in a small quantity in the oil. Another study from Portugal has investigated the main constituents in the oil of *C.*

Table 2. Comparison of total phenolic and flavonoid contents of *C. ladanifer* and *C. libanotis*.

Plant		Water extract	Ethanol extract	Methanol extract	Ethanol: water extract (50:50)	Methanol: water extract (50:50)	Acetonitrile extract
Phenolic content (FA mg/g dry weight)							
C. ladanifer	Fl	66.48±0.14^{c}	53.25±0.24^{d}	139.84±0.56^{a}	158.60±0.32^{a}	182.41±0.09^{a}	75.38±0.24^{a}
	F	95.82±0.1^{a}	114.10±0.27^{a}	130.22±0.16^{b}	98.47±0.06^{b}	110.01±0.05^{b}	71.29±0.15^{b}
	S	28.23±0.03^{d}	82.83±0.18^{b}	61.19±0.13^{c}	36.89±0.07^{d}	33.77±0.08^{d}	70.33±0.04^{b}
	L	83.31±0.14^{b}	65.76±0.1^{c}	47.96±0.13^{d}	40.50±0.15^{c}	53.49±0.22^{c}	27.51±0.11^{c}
C. libanotis	Fl	26.07±0.15^{b}	30.88±0.15^{d}	35.93±0.08^{c}	69.36±0.03^{a}	40.98±0.02^{b}	24.14±0.03^{d}
	F	26.08±0.66^{b}	75.62±0.77^{a}	43.87±0.15^{a}	39.30±0.45^{b}	28.96±0.75^{c}	36.17±0.09^{a}
	S	38.58±0.14^{a}	41.70±0.19^{c}	40.50±0.11^{b}	24.14±0.1^{d}	24.39±0.09^{d}	27.51±0.24^{c}
	L	25.11±0.31^{b}	44.59±0.19^{b}	40.02±0.08^{b}	26.55±0.22^{c}	43.15±0.22^{a}	33.53±0.29^{b}
Flavonoid content (QE mg/g dry weight)							
C. ladanifer	Fl	78.13±0.09^{a}	59.64±0.3^{c}	62.76±0.36^{b}	48.43±0.19^{a}	46.14±0.1^{a}	23.71±0.21^{c}
	F	18.93±0.06^{b}	83.94±0.17^{a}	71.27±0.23^{a}	29.32±0.08^{b}	31.19±0.04^{b}	60.26±0.05^{a}
	S	5.85±0.14^{c}	18.10±0.73^{d}	20.59±0.41^{d}	7.93±0.15^{d}	3.98±0.16^{d}	3.56±0.3^{d}
	L	16.44±0.67^{d}	61.10±0.53^{b}	51.13±0.55^{c}	25.99±0.41^{c}	18.10±0.41^{c}	28.69±0.21^{b}
C. libanotis	Fl	10.21±0.19^{c}	63.59±0.15^{c}	35.96±0.23^{a}	41.78±0.17^{a}	32.85±0.17^{a}	22.05±0.11^{a}
	F	8.76±0.15^{d}	14.16±0.28^{d}	14.78±0.24^{d}	11.25±0.12^{d}	3.57±0.14^{d}	6.06±0.11^{d}
	S	16.23±0.09^{b}	31.81±0.14^{b}	26.62±0.14^{b}	12.70±0.11^{c}	14.99±0.04^{c}	11.25±0.07^{b}
	L	20.39±0.1^{a}	15.82±0.55^{c}	24.96±0.18^{c}	18.93±0.74^{b}	17.89±0.12^{b}	10.63±0.89^{c}

Values are mean ± SD (standard mean deviation) (n=3). In each column, different letters mean significant differences between plant parts ($P < 0.001$). Values followed by the same letter under the same column are not significant different ($P \geq 0.05$). FA, Ferulic acid equivalents; QE, quercetin equivalents; Fl, flowers; F, fruit; S, stem; L, leaves.

ladanifer, viridiflorol, glubulol, an unknown sesquiterpene alcohol and 15-nor-labdan-8-ol (Gomes et al., 2005). In some reports, main components have been determined as 1, 8-cineol and viridiflorol (Viuda-Martos et al., 2011). These changes in the essential oil compositions might arise from several factors such as geological, geographical, seasonal and climatic (Perry et al., 1999).

Contents of total phenols and flavonoids

Phenolic compounds are the main class of natural antioxidants and there is a close relationship and positive correlation between the phenolic content and antioxidant activity of plant tissues. Table 2 summarises the results from the quantitative determination of the phenol and flavonoid content of the different extracts. Total phenol content was determined as ferulic acid equivalents in milligrams per gram (FA mg/g dry weight) while total flavonoid contents were calculated as quercetin equivalents in milligrams per gram (mg QE mg/g dry weight). The total phenol contents varied between different plant parts: aqueous methanolic extract from *C. ladanifer* flowers had higher total phenol content (182.41 mg FA/g) than other extract. Significant differences were also found among different plant parts.

As for total flavonoid content, the ethanolic extract of *C. ladanifer* fruit had higher total flavonoid content than other extract, and concerning the total flavonoid content of *C. libanotis* , the highest value was observed in flower from the ethanolic extract. The lowest total flavonoid content was signalled in the acetonitrile extract of stem from *C. ladanifer* (3.56 QE mg/g dry weight) and in the aqueous methanolic extract of fruit from *C. libanotis* (3.57 QE mg/g dry weight). The amount of active compounds like phenolic and flavonoid deposited in each part of the plant is usually different. In similar study, the total phenolic content of ethanolic and acetone extract of *C. ladanifer* (leaf) were 255.19 and 334.46 mg GAE (Gallic acid equivalent)/g of plant extract respectively, which had highest values compared to our results (Andrade et al., 2009). Amensour et al. (2010) reported that total phenolics in methanolic and ethanolic extract of *C. ladanifer* were 18.43 and 11.87 mg GAE/g of extract, respectively; much lower compared to our values. Screening of the polyphenolic composition of aqueous extracts of those plants was studied, given as abundance compounds ellagitannins (Barrajón-catalán et al., 2011).

DPPH radical scavenging activities

Free radical scavenging properties of different extracts from different *C. ladanifer* and *C. libanotis* parts and

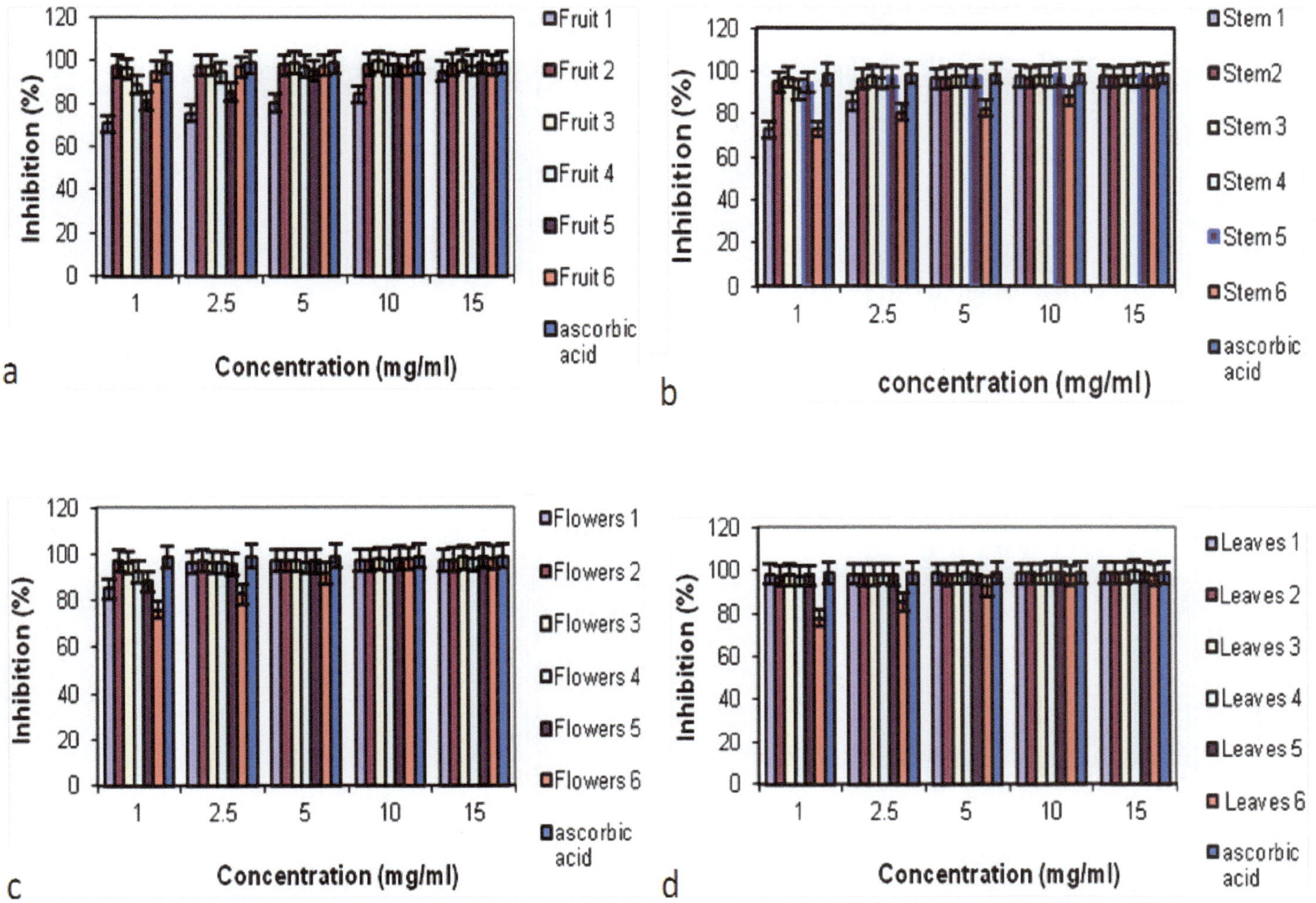

Figure 1. DPPH radical scavenging activities of fruit (a), stem (b), flowers(c) and leaves (d) from *C. ladanifer* of different solvent extracts at various concentrations (1, Water extract; 2, ethanol extract; 3, methanol extract; 4, water : ethanol (50:50) extract; 5, water: methanol (50:50) extract; 6, acetonitrile extract).

essential oils are presented in Figures 1, 2 and 3. Higher percentage inhibition value indicated higher antioxidant activity. All extracts of different plant parts showed higher scavenging ability of DPPH radicals when compared to those reported for essential oils of the leaf of *C. libanotis* and *C. ladanifer* (70.7 and 67.7%, respectively).

For *C. ladanifer*, methanolic extract were observed in this order: 97.8% (leaf) > 97.3% (stem) > 96.1% (flower) > 95.9% (fruit) whereas for *C. libanotis*, it was found as 96.9% (flower and stem) > 95.9% (fruit) > 93.4% (leaf). In addition, DPPH scavenging ability of the extracts of plant parts was similar to ascorbic acid, suggesting that plant extracts have a potent antioxidant activity mainly due to their richness on phenolic compounds. Oils was found less effective than the control. The use of water with alcoholic solvent presents a good advantage to extract polyphenols and can also modulate the polarity of alcohol solvent used (Mohammedi and Atik, 2011). These extract possess a remarkable potential in scavenging of free radicals since they present a number of hydroxyls acting as hydrogen donators.

Conclusion

There is a growing interest of industry to replace synthetic chemicals from natural products with bioactive properties from plant origin. The results indicate a significant qualitative difference found in terms of chemical composition in the oils. The most abundant compounds in *C. ladanifer* were camphene, borneol, cyclohexanol-2, 2, 6 tremethyl, terpineol-4 and α-pinene, while the essential oil of *C. libanotis* were terpineol-4, γ-terpinene, camphene, sabinene, α-terpinene and α-pinene. The present data suggest the importance of comparing and exploring the variety of essential oil and phenolic content from different *C. ladanifer* and *C. libanotis* parts, since this variability of repartition of bioactive substances between leaf, stem, flower and fruit entrained the variability of their potential antioxidant activities.

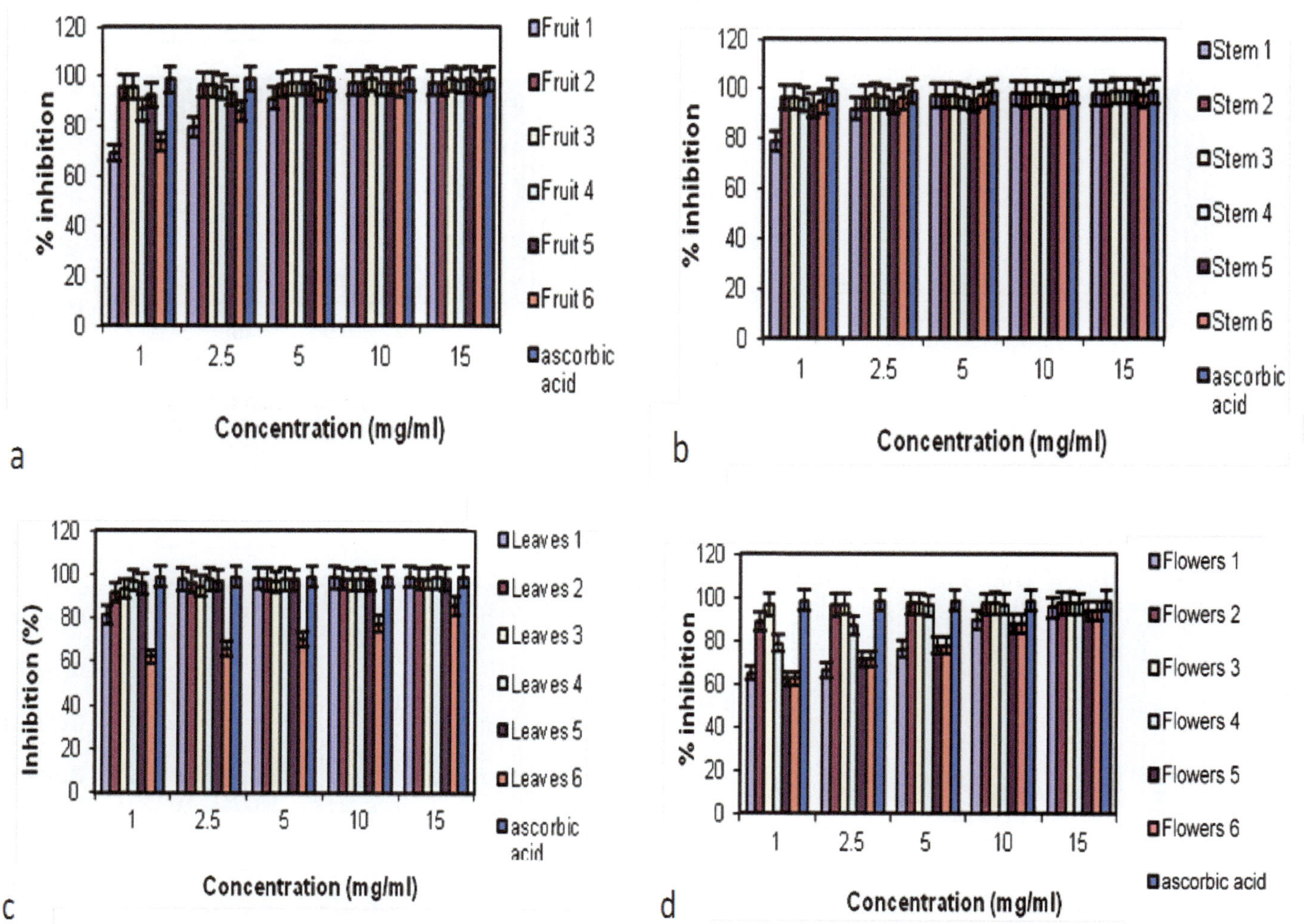

Figure 2. DPPH radical scavenging activities of fruit (a), stem (b), leaves (c) and flowers (d) from *Cistus libanotis* of different solvent extracts at various concentrations (1, Water extract; 2, ethanol extract; 3, methanol extract; 4, water: ethanol (50:50) extract; 5, water: methanol (50:50) extract and 6: acetonitrile extract).

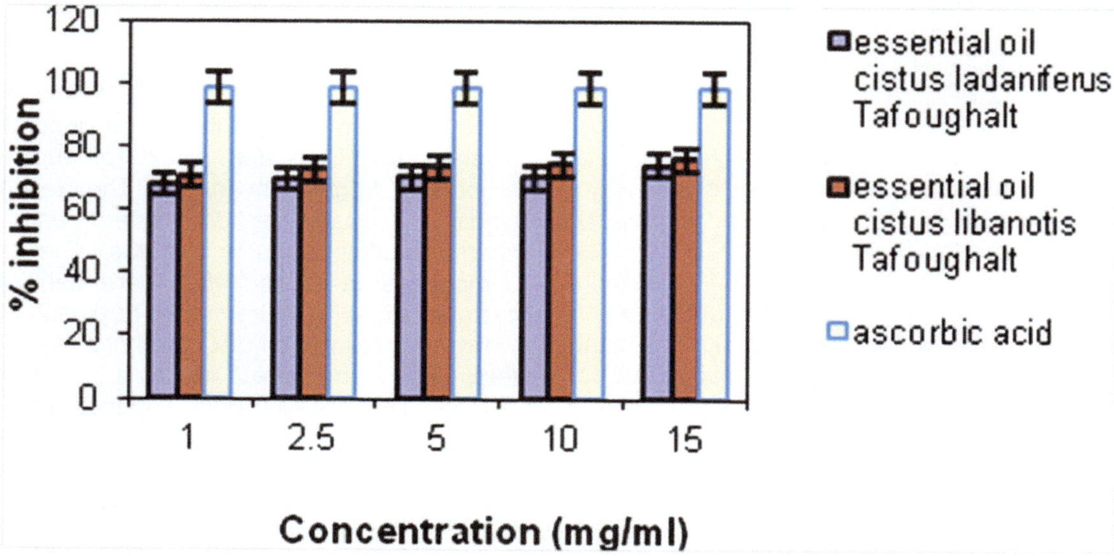

Figure 3. DPPH radical scavenging activities of essential oil of *C. ladanifer* and *C. libanotis*.

REFERENCES

Adams RP (1995). Identification of Essential oil Components by GC-MS (Allured, Carlo stream).

Alonso I, Luis E, Ta´rrega R (1992). First phases of regeneration of Cistus laurifolius and *C. ladanifer* after burning and cutting in experimental plots. Int. J. Wildland Fire 2(1):7–14.

Amensour M, Sendra E, Pérez-Alvarez JA, Skali-Senhaji N, Abrini J, Fernández-López J (2010). Antioxidant activity and chemical content of methanol and ethanol extracts from leaves of rockrose (*C. ladaniferu*s). Plants Foods Hum. Nutr. 65:170-178.

Andrade D, Gil C, Breitenfeld L, Domingues F, Duarte AP (2009). Bioactive extracts from *C. ladanifer* and Arbutus unedo L. Ind. Crops Prod. 30: 165-167.

Attaguile LG, Russo A, Campisi A, Savoca F, Acquaviva R, Ragusa N, Vanella A (2000). Antioxidant activity and protective effect on DNA cleavage of extracts from Cistus incanus L. and Cistus monspeliensis L. Cell Biol. Toxicol. 16: 83-90.

Aziz M, Karim A, Tab N, Mekhfi H, Bnouham M, Ziyyat A, Legssyer A, Atmani F (2011). Antidiarrhoeal activity of *C. ladaniferu*s aqueous extract. Spatulla DD. 13: 175-179.

Barrajón-catalán E, Fernandez-Arroyo S, Roldán C, Guillén E, Saura-Carretero D, Micol V (2011). Asystematic study of the polyphenolic composition of aqueous extracts deriving from several cistus genus species: Evolutionary relationship. Phytochem. Anal. 22(4):303-12.

Belmokhtar M, Bouanani N, Ziyyat A, Mekhfi H, Bnouham M, Aziz M, Mateo P, Fischmeister R, Legssyer A (2009). Antihypertensive and endothelium-dependent vasodilator effects of aqueous extract of *C. ladaniferu*s. Biochem. Biophys. Res. Commun. 389: 145-149.

Ben Jemia M, Kchouk ME, Senatore F, Autore G, Marzocco S, De Feo V, Bruno M (2013). Antiproliferative activity of hexane extract from Tunisian *C. libanotis*, Cistus monspeliensis and Cistus villosus. Chem. Cent. J. 7-47.

Gomes PB, Mata VG, Rodrigues AE (2005). Characterization of the Portuguese-grown *C. ladanifer* essential oil. J. Essent. Oil Res.17.

Greche H, Mrabet N, Zrira S, Ismaili-Alaoui M, Benjilali B, Boukir A (2009). The volatiles of the leaf oil of *C. ladanifer* L. var. albiflorus and Labdanum extracts of Moroccan origin and their antimicrobial activities. J. Essent. Oil Res. 21: 166-173.

Guvenc A, Yildiz S, Ozkan AM, Erdurak CS, Coskun M, Yilmaz G, Okuyama T, Okada Y (2005). Antimicrobiological studies on Turkish Cistus species. Pharm. Biol. 43: 178-183.

Guzma´n B, Vargas P (2009). Historical biogeography and character evolution of Cistaceae (Malvales) based on analysis of plastid rbcL and trnL-trnF sequences. Org. Divers. Evol. 9: 83–99.

Mariotti JP, Tomi1 F, Casanova1 J, Costa J, Bernardini AF (1997). Composition of the Essential Oil of *C. ladaniferu*s L. Cultivated in Corsica (France). Flavour Frag. J. 12: 147.

Mohammedi Z, Atik F (2011). Impact of solvent extraction type on total polyphenols content and biological activity from Tamarix Aphylla (L.) karst. Int. J. Pharma. Bio. Sci. 2:607-615.

Mrabet N, Lahlou H, Benjilali B (1999). Effect of Moroccan *C. ladaniferu*s L. (rockrose) extracts on the growth of four fungi. Cryptogam. Mycol. 20: 23-33.

Oller-Lópeza JL, Rodrígueza R, Cuervaa JM, Oltraa JE, Bazdib B, Dahdouhb A, Lamartib A, Ibn Mansourb A (2005). Composition of the Essential Oils of *C. ladaniferu*s and C. monspeliensis from Morocco. J. Essent. Oil Res. 17:553-555.

Perry NB, Anderson RE, Brennan NJ, Douglas MH, Heaney AJ, McGimpsey JA, Smallfield BM (1999). Essential oils from Dalmation Sage (Salvia officinalis L.): variations among individuals, plant parts, seasons and sites. J. Agric. Food Chem. 47: 2048-2054.

Pourmorad F, Hosseinimehr SJ, Shahabimajd N (2006). Antioxidant activity, phenol and flavonoid contents of some selected Iranian medicinal plants. Afr. J. Biotechnol. 5(11): 1142-1145.

Singleton VL, Rossi JA (1965). Colorimetry of total phenolics with phosphomolybdic-phosphotungstic acid reagents. Am. J. Enol. Vitic. 16: 144-158.

Viuda-Martos M, Sendra E, Pérez-Alvarez JA, Fermández-López J, Amensour M, Abrini J (2011). Identification of flavonoid content and chemical composition of the essential oils of Morocccan herbs: Myrtle (Myrtus communis L.), Rockrose (*C. ladanifer* L.) and Montpellier cistus (Cistus monspeliensis L.). J. Essent. Oil Res. 23.

Xiao-Juan L, Wei W, Meng L, Chun-Ying L, Yuan-Gang Z, Pan-Song M, Yu-Jie F (2012). Solvent-free microwave extraction of essential oil from Dryopteris fragrans and evaluation of antioxidant activity. Food Chem. 133: 437-444.

23

Expression analysis of *OsbZIP* transcription factors in resistance response by the rice blast resistance gene *Pi36*-mediated

Jinglan Han[1,2] , Xiangming Zhang1 , Dan Zhang[1], Xuequn Liu[1], Chuntai Wang[1] and Xinqiong Liu[1]

[1]Hubei Provincial Key Laboratory for Protection and Application of Special Plants in Wuling Area of China, College of Life Science, South-Central University for Nationalities, Wuhan 430074, China.
[2]College of Life Science, South China Agricultural University, Guangzhou 510642, China

Plant basic leucine zipper (bZIP) proteins play an essential role in the genes expression and regulation in higher plants. They have been shown to regulate diverse plant specific phenomena, including germination, floral induction and development, seed maturation, photomorphogenesis, biotic and abiotic stresses. Resistance response mediated by the rice blast resistance gene *Pi36* is a typical signal transduction, in which 12 *OsbZIP* genes were differentially expressed by microarray analyses. To understand the potential function of *OsbZIP* genes during the defense responses against rice blast, the expression analysis of these *OsbZIP* genes, in response to the blast fungus inoculation and the related defense signal molecules induction, were further conducted using real-time fluorescent quantitative polymerase chain reaction (PCR) technique. Our data indicates that among the 12 *OsbZIP* genes, the expression level eight tested genes were differentially regulated and maintained to 96 h points post inoculation in rice resistant and susceptible cultivars during *Magnaporthe oryzae* infection, and all of them were also significantly up-regulated by one or several kinds of exogenous plant hormones stresses. Although, these genes were induced only at early time points (1 to 24 h); it is evident that the *OsbZIP* genes may be involved in different signaling pathway, and play potential important functions in the defense response to rice blast.

Key words: OsbZIP transcription factors, rice blast, resistance response, plant hormones stresses.

INTRODUCTION

The plant defense responses is regulated via a network of signaling pathway involving salicylic acid (SA), abscisic acid (ABA), jasmonate (JA) and ethylene (ET). When the plants are suffering drought, high salt, external hormones, diseases, transcription factors were stimulated to combine with the cis-acting elements of downstream gene promoter through a series of signal transmission (Bari and Jones, 2009; Pieterse et al., 2009), and the target gene was directly regulated to express, or form homologous, heterologous dimers, or interact with other protein becomes a activated form and involved in JA, SA, ABA and other signal transduction pathways, forming a regulation of gene expression networks. More and more evidences show that, in the microbial pathogenic fungus, bacteria and insect resistance, different defense signaling pathways have complicated interactions with each other (Seabolt et al., 2011).

The plant basic leucine-zipper (bZIP) proteins compose a large family of transcriptional regulators exclusively in eukaryotes, which shared two common structures: a DNA-bonding basic region and the Leu zipper dimerization (Nijhawan et al., 2007). bZIP proteins play an essential role in the genes expression and regulation in higher plants, which have been shown to regulate diverse plant-specific phenomena, including photomorphogenesis and light signaling (Jakoby et al., 2002), defense response against pathogens (Zhang et al., 2003; Kesarwani et al., 2007; Thurow et al., 2005) as well as ABA and/or stress signaling (Lu et al., 2009; Xiang et al., 2008). So far, over 120 bZIP transcription factors have been identified in different plants. As a model species of monocotyledonous plants, about 100 bZIP sequences were predicted in rice database, and 89 of them take different roles in rice stress response and plant hormone signal transduction, 16 of those function have been identified (Nijhawan et al., 2007; Lu et al., 2009; Zou et al., 2008).

Plant defense responses are regulated through a network of signaling pathways. To defeat pathogens, the plant has to regulate transcription factors in a timely manner after recognizing the pathogen in order to activate a flood of defense-related genes (Liu et al., 2010). It has been reported that Xa21, an R-protein against *Xanthomonas oryzae pv. oryzae* (*Xoo*) of receptor kinase structure, also interact with a WRKY TF (OsWRKY62) in yeast two hybrid assay (Peng et al., 2008). Barley CC-NBS-LRR protein MLA10 interacts by CC domain with HvWRKY1/2 transcription factor (Shen et al., 2007). Rice blast, caused by the filamentous ascomycete *Magnaporthe oryzae*, remains the most serious pathogen in most rice producing (*Oryza sativa L.*) regions. Genetic resistance has been and undoubtedly will continue to be the major means of disease control for blast. To date, more than 80 major blast resistance genes have been identified (Zhai et al., 2011). Among them, 17 blast *R* genes have been characterized. With the exceptions of *pi21* and *Pid-2*, all of them are predicted to encode intracellular proteins with a coiled-coil (CC) domain at the N terminus and a nucleotide binding site (NBS) and leucine-rich repeat (LRR) domains at the C terminus. CC-NBS-LRR genes might be the most important genes to respond to the rice blast pathogens. Although so many resistance genes have been cloned, the mechanism of resistance remains unknown.

The blast resistance gene *Pi36* encodes a CC-NBS-LRR disease resistance protein which confers a stable and high level of resistance against isolates of the rice blast pathogen *M. oryzae* in Northern China (Liu et al., 2010; Liu et al., 2007). In this study, the expression profiles of 12 *OsbZIP* genes were reported, which were differentially expressed by microarray analyses, in various stages before and after rice blast pathogen *M. oryzae* inoculated in resistance cultivar (cv.) Kasalath with *Pi36* and susceptible cv. Lijiangxintuanheigu (LTH), which carries no known major resistance genes, as the susceptible one. The cv. Kasalath was also detected the treatments with three exogenous plant hormones ABA, methyl jasmonate (MeJA) and benzothiadiazole (BTH) which is fuctional analogues of SA, by real-time PCR analysis using gene-specific primers. A better understanding of the gene network associated with *Pi36* may help in the elaboration of strategies to promote durable resistance against rice blast, and the *OsbZIP* genes at transcriptional level are important for resistance to the rice blast.

MATERIALS AND METHODS

Plant materials, growth condition and stress treatments

The rice *indica* cv. Kasalath was used for this study because this culture was the donor of resistance gene *Pi36* (Liu et al., 2010; Liu et al., 2007), and the cv. Lijiangxintuanheigu (LTH), which carries no known major resistance genes, as the susceptible one. Seeds were sterilized in 1.5% NaClO and germinated at 28°C for three days and then transferred to plant growth chamber to grow for 14 days before treatments. For the hormone treatments, seedlings were fully sprayed with 0.05 mM ABA, 0.1 mM MeJA, 0.05 mM BTH, respectively, and covered with a plastic film over 48 h to keep a humidity environment more than 95%. And leaves were harvested at 0, 1, 3, 6, 12, 24, 48 and 72 h after treatment. For *M. oryzae* isolate treatment, three week-old seedlings were infected with a suspension of (1 to 2) × 10^5 spores per mL in 0.02% (v/v) Tween 20. Negative control leaves were sprayed with 0.02% (v/v) Tween 20. After inoculation, the seedlings were maintained at 24°C in darkness for 24 h, and then under a 14/10 h (light/dark) regime at 95% humidity. Leaf samples were taken at 0, 1, 2, 3, 6, 12, 24, 48, 72 and 96 hpi.

RNA extration and reverse transcription

Total RNA was extracted from a sample of leaf tissue at each time point using the TRIzol reagent (Invitrogen). The RNA was reverse transcribed using an avian myeloblastosis virus (AMV) reverse transcriptase kit (Promega Inc., Madison, WI, USA), following the manufacturer's instructions. For each sample, two independent reverse transcription (RT) reactions were pooled. All quantitative reverse transcriptions polymerase chain reaction (qRT-PCR) were based on the same pooled cDNA, and three technical replicates were run for each cultivar/time point combination. Three biological replicates (independent experiments) were included in the analysis.

Quantitative reverse transcriptions polymerase chain reaction (qRT-PCR)

Gene-specific primer sets were designed from nucleotide sequences available in the public domain (Table 1). All the primers were also designed to span intron-splicing sites in order to reduce

Table 1. The primer sets sequences used in this study.

Gene	Forward (5'-3')	Reverse (5'-3')	Product size/bp	Tm/°C
OsNIF1	CCTGAGTTCTCTCTGGCTTGC	ACTGAGCCACCAATCTTCTGC	141	62
OsNIF2	GTGGTCCGTCAGGGAGATT	TGCTCTCTCCTTTCTGCCA	131	62
OsNIF3	CAATGACGGAGTTCTTACACCC	CTGTCCTAGGGCTCATATCTGC	139	61
OsNIF4	TCAGTAGTTGACCCTCAGACAC	TCCTTGGACTTGTATCAGCC	121	56
OsbZIP5	CATCCAGCCCTTAGTCCATACC	CAAGATCCCTGGGCCGTAAT	65/160*	62
OsbZIP10	CGGTGGTAGCCTGGAATTCTT	ACGCAACTTTGGTTCCATGAA	76	63
OsbZIP36	GCTGCAGATGCACTAAATGAC	GCATTAACCTGACCTGTTGCAA	73	60
OsbZIP45	GGAACCAGCCAAGTCTGCCAA	CAGCCCCATCAACTTCCTTAAG	70	61
OsbZIP47	CAGGGAATCTTCATTGCAACTG	GCGCGTACTCAAGGTCGAA	90/187*	64
OsbZIP58	TGCTTGGCATGAGCATCTG	AAGTCCTGGCCCAAATGGA	65	60
OsbZIP62	AACAAGATCTCGCGGTTGGA	TCAGGAAATCGACGCAGCTA	97/743*	63
OsbZIP76	CTCGTGTTGAGTGATCCGAAGA	TTGGACCTCGCAGCAGATCT	67/403*	61
Actin1	TGCTATGTACGTCGCCATCCAG	AATGAGTAACCACGCTCCGTCA	209/458*	60

*The intron size if genomic DNA is used as the template.

the effect of potential genomic DNA contamination. Each PCR reaction was prepared as follows: 1 μL cDNA as template, 0.4 μL primer pair, 5 μL SYBR® Green Real-time PCR Master Mix (TOYOBO, Japan) and 3.6 μL ddH_2O. The reaction was initiated by pre-denaturation 5 min at 94°C, then followed by 40 cycles of 30 s at 94°C, 30 s at 55°C, and 30 s at 72°C. At the end of each extension step, the fluorescence signal was detected, and used to generate an amplification profile.

The rice *Actin1* gene, a constitutively expressed housekeeping gene, was used to normalize samples. The comparative C_t method, as described by Reyna and Yang (2006), was used to describe all genes within a sample relative to the time 0 sample. Briefly, the relative C_t method consists of the following formula

$$2^{-(\Delta Ct\ treatment - \Delta Ct\ control)}$$

Where, (C_t $^{target\ gene}$ to C_t ^{normalizer}) and (C_t $^{time\ zero}$ to C_t ^{normalizer}) represent ΔC_t ^{treatment} and $\Delta C_t^{control}$, respectively.

Real-time PCR analysis was based on at least three technical replicates and a repetitive average C_t value was taken no more than 0.5. In this study, only genes that showed five-fold increase compared with the 0 h control were considered to have been induced.

RESULTS

Expression profiles of OsbZIPs challenged with *M. oryzae*

The expression profiles of 12 OsbZIP genes in cv. Kasalath and LTH challenged with blast isolate RB11, which is avirulent on Kasalath and virulent on LTH, were examined by qRT-PCR at a series of time points postinoculation (p.i.). The TGA transcription factors were documented to involve in basal resistance and PR gene expression (Zhang et al., 2003; Kesarwani et al., 2007). Of the four TGA type OsbZIP genes (*OsNIF1*, *OsNIF2*, *OsNIF3* and *OsNIF4*) assayed, *OsNIF1* was rapidly induced in cv. Kasalath, and reached a peak of expression at 2 hpi. Thereafter, its expression was temporarily reduced, before increasing again to a second maximum between 24 and 48 hpi. In cv. LTH, this same gene was down-regulated throughout the measurement period. *OsNIF2* was strongly up-regulated in cv. Kasalath by 1 hpi, and reached a peak of expression at 12 hpi, followed by a gentle decline, but maintaining a level above background throughout. In cv. LTH, this same gene was down-regulated throughout; while *OsNIF3* was rapidly up-regulated (peaking at 1 hpi) in cv. Kasalath, its expression level fell rapidly thereafter; in cv. LTH, the same gene was induced gradually, reaching 500-fold of the background level by 12 hpi. *OsNIF4* was down-regulated in both cv. Kasalath and cv. LTH at most of the measurement times, with the exception of 1, 3 and 12 hpi. in cv. LTH, when its expression level was somewhat raised (Figure 1). Therefore, our results reveal that OsNIF1 and OsNIF2 may be involved in the rice blast resistance response.

Of the eight other structural type *OsbZIPs* genes, which were reported to be involved in light regulation and abiotic stress signaling (Kesarwani et al., 2007; Lu et al., 2009), *OsbZIP62* displayed the similar expression profiles between cv. Kasalath and LTH. *OsbZIP45* was weakly up-regulated in cv. Kasalath at late period, whereas the gene was down-regulate throughout the measurement period. As for the other six genes, *OsbZIP5*, *OsbZIP10*, *OsbZIP36*, *OsbZIP47*, *OsbZIP58* and *OsbZIP76*, the expression of level strongly up-regulated in resistant cv. Kasalath, with a peaking of, respectively, 320-, 500-, 45-, 450-, 650- and 250-fold of the background level at 24 hpi, respectively. Notably, the expression level of these genes in cv. Kasalath was maintained to 96 hpi, but the six genes were down- regulated in cv. LTH throughout the measurement period (Figure 1).

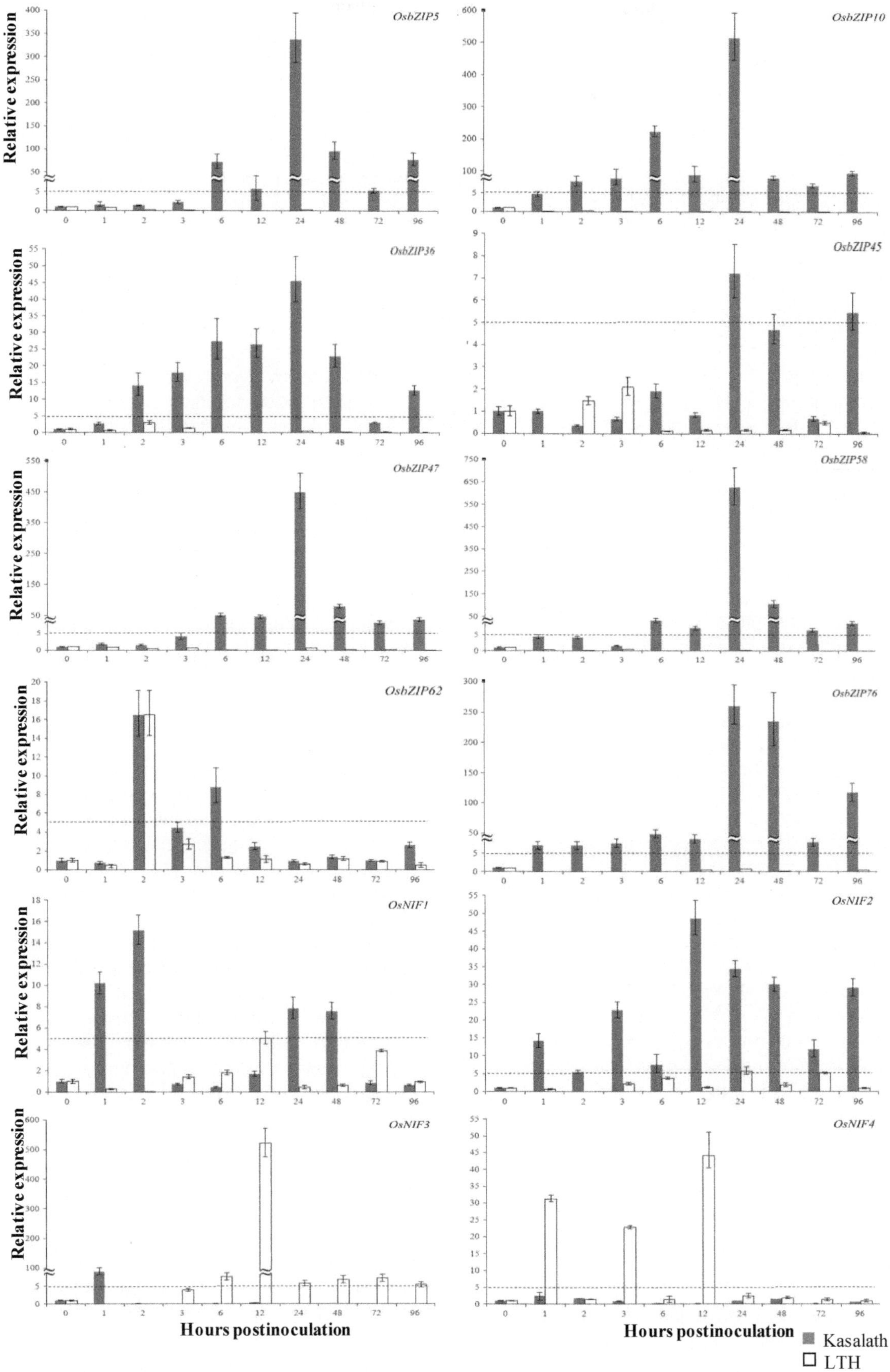

Figure 1. *OsbZIP* genes induced by rice blast infection

To our knowledge, this is the first report of the involvement of these bZIP transcriptional factors in the blast-induced defence response.

Expression analysis of OsbZIP gene after plant hormone treatments

To investigate expression profiles of the 12 *OsbZIP* genes in cv. Kasalath response to plant hormones treatments, we further treated the rice cv. Kasalath with ABA, MeJA and BTH (Figure 1). When the seedling were treated with ABA, most of the *OsbZIP* genes which displayed the similar expression pattern, *OsbZIP5*, *OsbZIP10*, *OsbZIP45*, *OsbZIP47*, *OsbZIP58* and *OsbZIP76* were up-regulated rapidly to reach the expression peaks at 6 or 24 h, but not so remarkable, only *OsbZIP58* and *OsbZIP76* got to 45- and 55-fold, respectively. *OsbZIP36* and *OsNIF2* were weakly induced at 6 and 24 h, respectively, but *OsbZIP62*, *OsNIF1*, *OsNIF3* and *OsNIF4* were not induced (Figure 2).

When the rice cv. Kasalath was treated by MeJA, the expression of *OsNIF3* and *OsNIF4* showed no significant changes, but *OsNIF1* and *OsNIF2* were up-regulated 5.5-fold at 3 and 6 h, respectively, while *OsbZIP76* was up-regulated 22-fold only at 6 h. The other OsbZIPs displayed different expression profiles. *OsbZIP5* and *OsbZIP45* were rapidly induced at 1 hpi and the expression level increased up to maximum by 6 and 12 hpi, respectively; whereas *OsbZIP10*, *OsbZIP36*, *OsbZIP47* and *OsbZIP62* were induced at 3 hpi and reached a peak of expression by 6 hpi and then followed by a gently decline by 24 hpi. *OsbZIP58* was strongly induced at 6 hpi reaching 175-fold of the background level and then slightly decreased by 24 hpi (Figure 3).

To investigate the expression pattern of OsbZIPs in cv. Kasalath under BTH stress, the results showed that the expression level of eight genes increased up to maximum by 12 hpi at different degree, and decreased slightly by 48 hpi. Among these genes, *OsbZIP62* and *OsbZIP76* were up-regulated only at 12 h and reached 6 and 7-fold, respectively. Whereas *OsbZIP5* and *OsbZIP45* were rapidly induced at 1 hpi and reached a peak by 12 hpi with 175 and 27-fold, decreased slight by 24 hpi. And, *OsbZIP10*, *OsbZIP36* and *OsbZIP47* were induced at 6 h and increased up to maximum by 12 hpi and declined slight by 24 hpi. As for *OsbZIP58*, the expression was induced at 3 hpi and reached a peak by 12 hpi, finally declined by 48 hpi. While the four genes *OsNIF1*, *OsNIF2*, *OsNIF3* and *OsNIF4* were not induced (Figure 4).

DISCUSSION

In many higher plants, bZIP proteins play crucial roles in the genes expression and regulation, which have been shown to regulate diverse plant-specific phenomena, as well as biotic and abiotic stresses (Nijhawan et al., 2007; Jakoby et al., 2002; Kesarwani et al., 2007). So far, most of the *OsbZIP* genes identified are related to drought, low temperature, high salt and other abiotic stress (Fujita et al., 2005; Achuo et al., 2006; Asselbergh et al., 2008). In the present study, 12 OsbZIP possible differentially expressed by microarray analyses involved in resistance response *Pi36*-mediated were analyzed by two induction patterns, such as *M. oryzae* infection and exogenous hormones induction. Our results reveal that, except *OsbZIP45*, *OsbZIP62*, *OsNIF3* and *OsNIF4* which were not induced among 12 *OsbZIP* tested genes, the other genes were up-regulated in the incompatible *Pi36/AvrPi36* interaction, indicating that these genes are active in the response of *Pi36* to the infection with an avirulent race of blast. To our knowledge, this is the first report of the involvement of bZIP transcriptional factors in the blast-induced defence response.

Plant defense responses are regulated through a network of signaling pathways that involve endogenous plant signaling molecules such as SA, JA, ET, and ABA (Mauch-Mani and Mauch , 2005; An and Mou, 2011). In this study, among 12 *OsbZIP* genes, they were up-regulated by blast fungus and could also be induced by one or more hormones resulting to expression. Interestingly, it was observed that the expression was induced by blast fungus in early time points (1 to 24 h) and lasted to late period (24 to 96 h) in resistance response, whereas all expression was up-regulated by the early time points. Apparently, our data is beneficial to dissect the roles of plant hormones in resistance response.

It has become evident that SA and JA signaling pathways are mutually antagonistic in *Arabidopsis* (Kunke and Brooks, 2002). Recently, ABA has been reported to have an antagonistic interaction with SA signaling pathways in the rice-*M. oryzae* interaction (Achuo et al., 2006; Asselbergh et al., 2008; Fan et al., 2009; Jiang et al., 2010). In this study, our results show that in rice, some of the OsbZIP members were differentially regulated in rice plants in different hormones signaling pathway, implicating that OsbZIPs may be involved in SA, JA and ABA-dependent pathway, consistent with the suggestion that SA, JA and ABA-dependent defence signals interact with one another, either synergistically or antagonistically (Kunke and Brooks, 2002; Fujita et al., 2006).

The redundant function of *OsbZIP* genes may be beneficial in protecting the cell or organism under various stress conditions and in eliciting the multiple pathways that lead to the wide array of physiological responses that occur following infection with pathogens. Recently, we have produced transgenic rice plants with overexpression or RNAi construct for each differentially regulated OsbZIP. These lines provide the opportunity to study the contribution of each OsbZIP under normal or defined stress condition, and should eventually lead to the determination of the functions of *OsbZIP* genes.

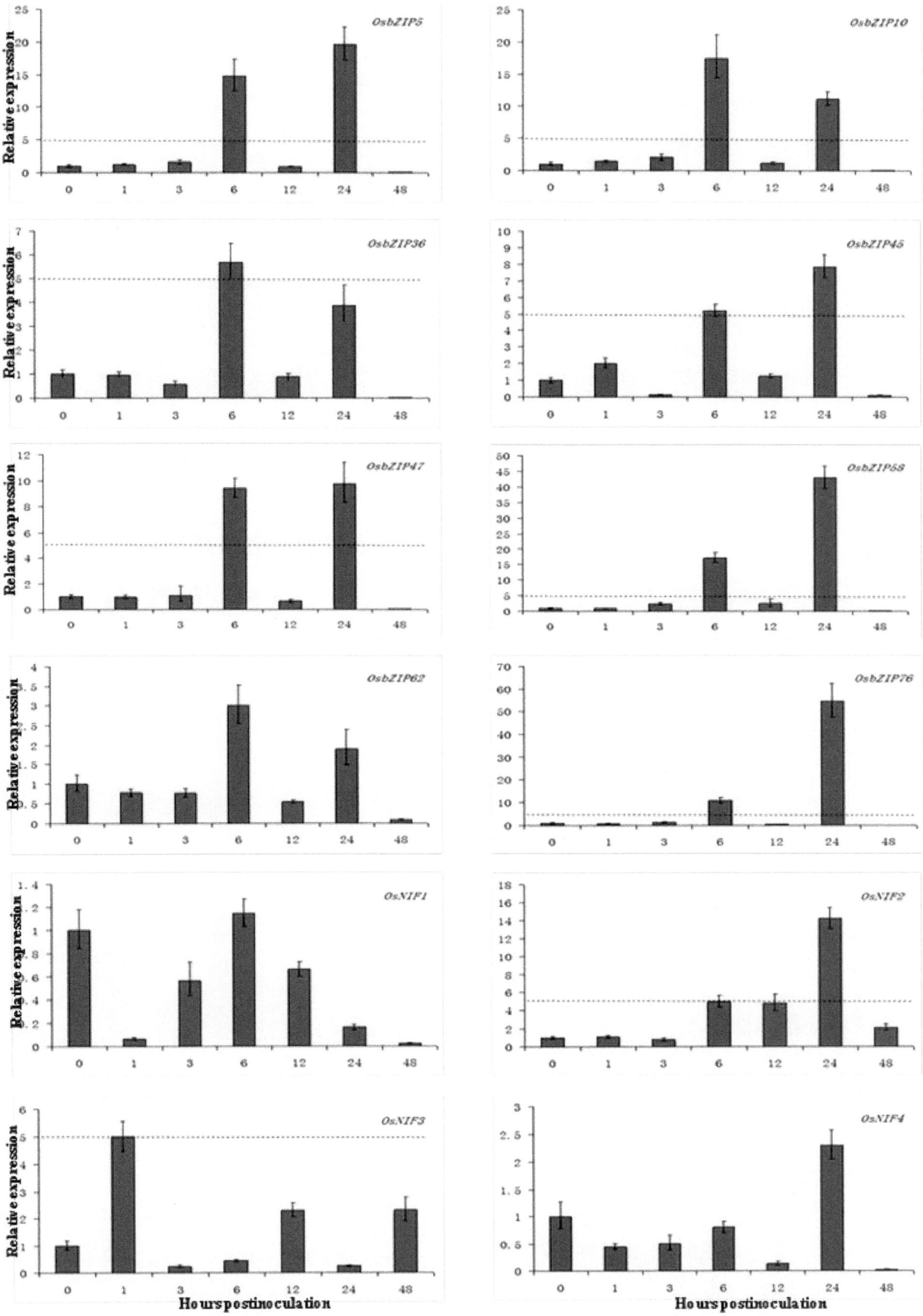

Figure 2. *OsbZIP* genes induced by abscisic acid (ABA) treatment.

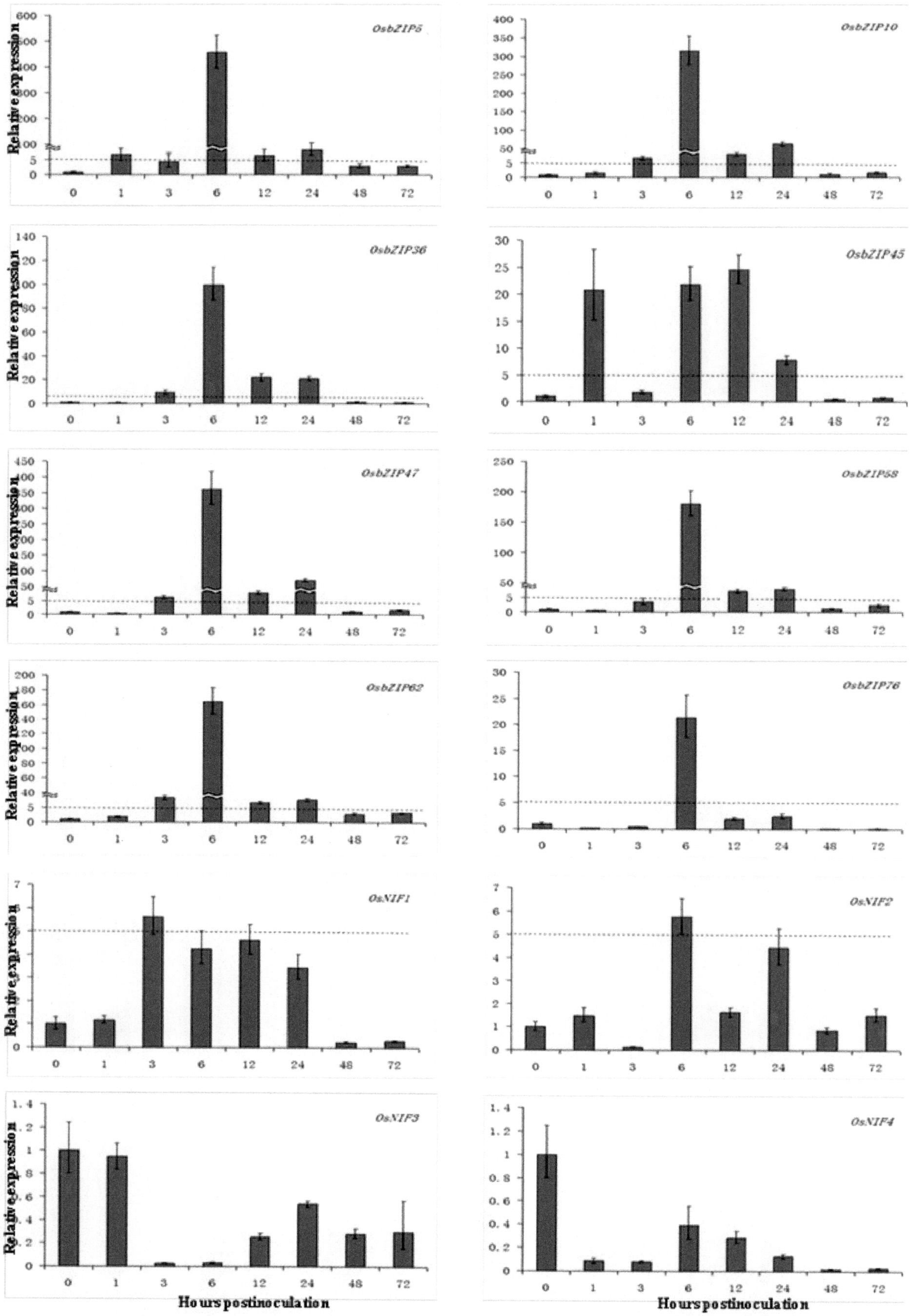

Figure 3. *OsbZIP* genes induced by methyl jasmonate (MeJA) treatment.

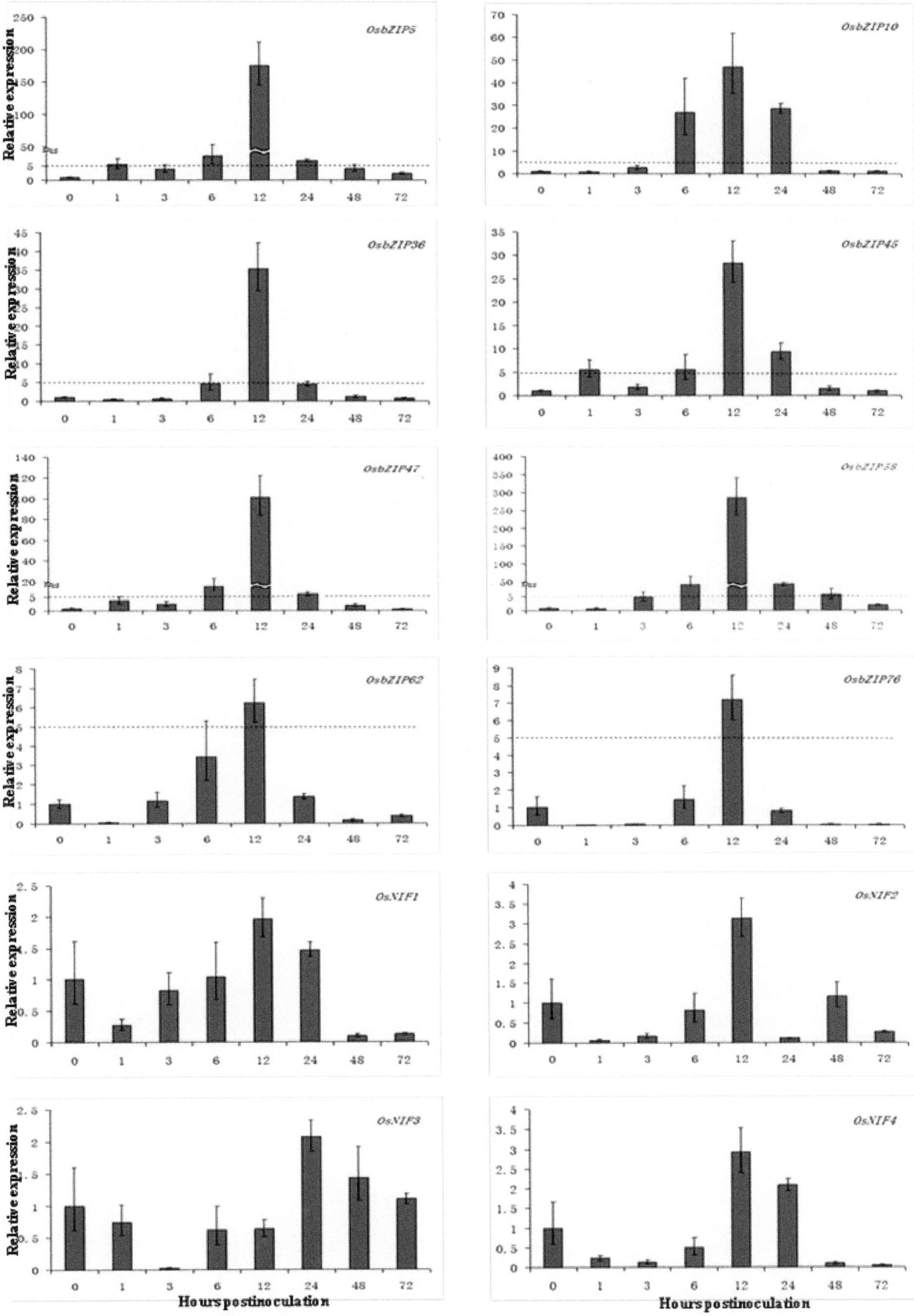

Figure 4. *OsbZIP* genes induced by benzothiadiazole (BTH) treatment.

ACKNOWLEDGMENTS

This work was supported by the National Natural Science Foundation of China (31370306), the Key Project of Chinese Ministry of Education (No. 212202), the Hubei Provincial Natural Science Foundation (2010CDB02006) and Hubei Provincial Creative Group Project (2011CDA125).

REFERENCES

Achuo EA, Prinsen E, Höfte M (2006). Influence of drought, salt stress and abscisic acid on the resistance of tomato to Botrytis cinerea and Oidium neolycopersici. Plant Pathol. 55:178-186.

An CF, Mou ZL (2011). Salicylic acid and its function in plant immunity. J. Integr. Plant Biol. 53:412-428.

Asselbergh B, De Vleesschauwer D, Hofte M (2008). Global switches and fine-tuning-ABA modulates plant pathogen defense. Mol. Plant-Microbe Interact. 21:709-719.

Bari R, Jones JDG (2009). Role of plant hormones in plant defense response. Plant Mol. Biol. 69:473-488.

Fan J, Hill L, Crooks C, Doerner P, Lamb C (2009). Abscisic acid has a key role in modulating diverse plant-pathogen interactions. Plant Physiol. 150:1750-1761.

Fujita M, Fujita Y, Noutoshi Y, Takahashi F, Narusaka Y, Yamaguchi-Shinozaki K, Shinozaki K (2006). Crosstalk between abiotic and biotic stress responses: a current view from the points of convergence in the stress signaling networks. Curr. Opin. Plant Biol. 9:436-442.

Fujita Y, Fujita M, Satoh R, Maruyama K, Parvez MM, Seki M, Hiratsu K, Ohme-Takagi M, Shinozaki K, Yamaguchi-Shinozaki K (2005). AREB1 is a transcription activator of novel ABRE-dependent ABA signaling that enhances drought stress tolerance in Arabidopsis. Plant Cell. 17:3470-3488.

Jakoby M, Weisshaar B, Droge-Laser W, Vicente-Carbajosa J, Tiedemann J, Kroj T, Parcy F (2002). bZIP transcription factors in Arabidopsis. Trends Plant Sci. 7(3):106-111.

Jiang CJ, Shimono M, Sugano S, Kojima M, Yazawa K, Yoshida R, Inoue H, Hayashi N, Sakakibara H, Takatsuji H (2010). Abscisic acid interacts antagonistically with salicylic acid signaling pathway in rice-Magnaporthe grisea interaction. Mol. Plant Microbe Interact. 23:791-798.

Kesarwani M, Yoo J, Dong XN (2007). Genetic interaction of TGA transcription factors in the regulation of pathogenesis-related genes and disease resistance in Arabidopsis. Plant Physiol. 144:336-346.

Kunke B N, Brooks DM (2002). Cross talk between signaling pathways in pathogen defence. Curr. Opin. Plant Biol. 5:325-331.

Liu XQ, Li YY, Wang L Y, Liu XQ, Wang C T, Wang L, Pan QH (2010). The effect of the rice blast resistance gene Pi36 on the expression of disease resistance-related genes. Chin. Sci. Bull. 55:1881-1888.

Liu XQ, Lin F, Wang L, Pan QH (2007). The in Silico map-based cloning of Pi36, a rice coiled-coil-nucleotide-binding site-leucine-rich repeat gene that confers race-specific resistance to the blast fungus. Genetics 176:2541-2549.

Lu GJ, Gao CX, Zheng XN, Han B (2009). Identification of OsbZIP72 as a positive regulator of ABA response and drought tolerance in rice. Planta. 229:605-615.

Mauch-Mani B, Mauch F (2005). The role of abscisic acid in plant pathogen interactions. Curr. Opin Plant Biol. 8:409-414.

Nijhawan A, Jain M, Tyagi AK, Khurana JP (2007). Genomic survey and gene expression analysis of the basic leucine zipper transcription factor family in rice. Am. Soc. Plant Biol. 146:333-350.

Peng Y, Bartley LE, Chen X, Dardick C, Chern M, Ruan R, Canlas PE, Ronald PC (2008). OsWRKY62 is a negative regulator of basal and Xa21-mediated defense against Xanthomonas oryzae pv. oryzae in rice. Mol. Plant 1:446-458.

Pieterse CMJ, Leon-Reyes A, Van Der Ent S, Van Wees SCM (2009). Networking by small-molecule hormones in plant immunity. Nat. Chem. Biol. 5:308-316.

Reyna NS, Yang YN (2006). Molecular analysis of the rice MAP kinase gene family in relation to Magnaporthe oryzae infection. Mol. Plant Microbe Interact. 19:530-540.

Seabolt G, Ng S, Zhang C, Salimian S, Watkins TA, Lu H (2011). Genetic dissection of salicylic acid-mediated defense signaling networks in Arabidopsis. Genetics 189:851-859.

Shen QH, Saijo Y, Mauch S, Biskup C, Bieri S, Keller B, Seki H, Ulker B, Somssich IE, Schulze-Lefert P (2007). Nuclear activity of MLA immune receptors links isolate-specific and basal disease-resistance responses. Science 315:1098-1103.

Thurow C, Schiermeyer A, Krawczyk S, Butterbrodt T, Nickolov K, Gatz C (2005). Tobacco bZIP transcription factor TGA2.2 and related factor TGA2.1 have distinct roles in plant defense responses and plant development. Plant J. 44:100-113.

Xiang Y, Tang N, Du H, Ye HY, Xiong LZ (2008). Characterization of OsbZIP23 as a key player of the basic leucine zipper transcription factor family for conferring abscisic acid sensitivity and salinity and drought tolerance in rice. Plant Physiol. 148:1938-1952.

Zhai C, Lin F, Dong Z, He X, Yuan B, Zeng X, Wang L, Pan QH (2011). 2011 The isolation and characterization of Pik, a rice blast resistance gene which emerged after rice domestication. New Phytol. 189:321-334.

Zhang Y, Tessaro MJ, Lassner M, Li X (2003). Knockout analysis of Arabidopsis transcription factors TGA2, TGA5, and TGA6 reveals their redundant and essential roles in systemic acquired resistance. Plant Cell. 15:2647-2653.

Zou M, Guan Y, Ren H, Zhang F, Chen F (2008). A bZIP transcription factor, OsABI5, is involved in rice fertility and stress tolerance. Plant Mol. Biol. 66:675-683.

Effect of propolis consumption on hepatotoxicity and brain damage in male rats exposed to chlorpyrifos

Newairy, A. A.[1] and Abdou, H. M.[2]

[1]Department of Biochemistry, Faculty of Science, Moharam Bik. Alexandria University, Egypt.
[2]Department of Zoology, Faulty of Science, Moharam Bik. Alexandria University, Egypt.

This study was undertaken to determine the protective effect of propolis against the hepatotoxicity and brain damage of chlorpyrifos (CPF) in male rats. Animals were assigned to one of four groups. The first group was used as control. Groups 2, 3 and 4 were treated with 6.8 mg CPF /kg BW (1/20 LD50); 50 mg propolis/kg BW; CPF (6.8 mg CPF/kg BW) plus propolis (50 mg propolis/kg BW). Rats were orally administered their respective doses daily for 28 days. Serum transaminases, glucose, lactate dehydrogenase, total proteins, albumin, reduced glutathione, thiobarbituric acid-reactive substances, superoxide dismutase and catalase were assessed. Also, total lipid, cholesterol, triglyceride and LDL-c were assayed. Oral treatment with CPF was found to elicit significant deterioration in all the tested parameters confirming its toxicity. The injury of liver tissues after CPF-treatment was confirmed by the histological changes. Also, CPF caused significant decrease in the activity of serum and brain cholinesterase but increased glial fibrillary acidic protein-expression and cause some histological changes in the brain tissues. While, oral treatment with propolis plus CPF could antagonize CPF-toxicity. These results suggest that propolis may become a promising tool for wide use in reducing the liver and brain damage during CPF-exposure.

Key words: Chlorpyrifos, propolis, rats, lipid peroxidation, antioxidant enzymes, biochemical parameters.

INTRODUCTION

Chlorpyrifos [O, O-diethyl-O-(3, 5, 6-trichloro-2-pyridyl) phosphorothioate] is an organophosphate (OPIs) pesticide used to control household, public health and stored product insects. It is effective against mushroom flies, aphids, spider mites, caterpillars, thrips and white flies in greenhouse, outdoor fruit and vegetable crops (Ambali, 2009). Chloropyrifos (CPF) induces toxicity through inhibition of acetyl cholinesterase (AChE) in addition to multiple other mechanisms (Slotkin et al., 2005). Pesticide chemicals can induce oxidative stress by generating free radicals and altering antioxidant levels of the free radical scavenging enzyme activity (Sharma et al., 2005). Exposure to CPF can differentially modify endogenous antioxidants like superoxide dismutase (SOD), glutathione peroxidase (GPX) and reduced glutathione (GSH), which can lead to the development of oxidative stress in some tissues (Bebe and Panemangalore, 2003). Also, it has been reported that CPF-intoxication causes a significant decrease in the reduced glutathione (GSH) and the activities of some antioxidant enzymes (Goel et al., 2005).

The mammalian cells reduced the adverse effect

of lipid peroxidation via the utilization of both enzymatic and non-enzymatic antioxidants, which scavenge the free radicals in the living system. To control the level of reactive oxygen species (ROS) and to protect cells under stress conditions, mammalian tissues contain several enzymes that scavenge ROS such as catalase (CAT) and glutathione-S-transferase (GST). Therefore, during oxidative stress, an increase in the exogenous supply of antioxidants improves the capacity of the tissue to cope with high antioxidant demands such as propolis (Goel et al., 2009). Propolis or bee glue is a resinous hive product collected by honey bees from plant exudates and contains more than 160 constituents. Historically, it has been used for various purposes such as a medicine (Ghisalberti, 1979). Flavonoids are thought to be responsible for many of its biological and pharmacological activities including anticancer (Valente et al., 2011), anti-inflammatory (Nirala and Bhadauria, 2008) and antioxidant effects (Yousef et al., 2004a).

Flavonoids and various phenolics are the most important pharmacologically active constituents in propolis capable of scavenging free radicals and thereby protecting lipids from being oxidized or destroyed during oxidative damage (Nieva Moreno et al., 2000). Propolis has gained popularity and used extensively in healthy drinks and foods to improve health and prevent diseases such as inflammation, heart disease, diabetes and even cancer (Matsuno, 1995). The role of propolis against CPF-induced toxicity in rats has not so far been studied. Therefore, the present study was carried out to investigate: 1) the alterations in biochemical parameters, free radicals and antioxidant enzymes induced by CPF in the serum of male rats; 2) the histological alterations induced by CPF in the liver and brain tissues and 3) the protective effect of propolis against the possible toxicity caused by CPF-exposure.

MATERIALS AND METHODS

Chemicals

Chlorpyrifos (CPF) was applied as a commercial emulsified concentrate formulation containing 48% emulsifiable concentrate chlorpyrifos (EC) (Trade name pestban) active ingredient from Egychem a chemicals company (El-Watania Co), Egypt. The EC was emulsified in water just before use and orally administrated to animals by esophageal intubation, and propolis was obtained from Superior Nutrition and Formulation by Jarrow Formulas, Los Angeles, USA. All other used chemicals in our experiment were of analytical grade. The dose of chloropyrifos (CPF) was 6.8 mg CPF /kg BW (1/20 LD_{50}). The acute oral LD_{50} of CPF was 136 mg /kg BW for rats (Suleiman et al., 2010). The dose of propolis was 50 mg/kg BW. This dose was used according to the previous study of Newairy et al. (2009).

Experimental design

Forty male Wistar rats were obtained from the animal house, Faculty of Medicine, Alexandria University, Egypt. The weights of rats were about 180 to 225 g. The local committee approved the design of the experiments, and the protocol conforms to the guidelines of the National Institutes of Health (NIH). Animals were caged in groups, each of five rats, and were given feed and water *ad libitum*. After one week of acclimatization, animals were divided into four equal groups. The first group was used as control. Groups 2, 3 and 4 were orally treated with CPF (6.8 mg/kg BW), the combination of CPF (6.8 mg/kg BW) and propolis (50 mg/kg BW) and propolis (50 mg/kg BW), respectively. Rats were orally administered their respective doses daily, for 28 days. Rats of each group were euthanized at the end of treatment period. Trunk blood samples were collected from the sacrificed animals and placed immediately on ice. Serum samples were obtained by centrifugation at 860 *g* for 20 min and stored at -60°C till measurements. Specimens of the liver and brain were fixed immediately in 10% buffered formalin for histological study.

Biochemical parameters

The level of glucose was determined with kits from Biosystems, S.A. Costa Brava, 30-Barcelona (Spain). The extent of lipid peroxidation in terms of thiobarbituric acid reactive substances (TBARS) formation was measured according to the method of Esterbauer and Cheeseman (1990). Tissue homogenate was mixed with 1 ml TCA (20%), 2 ml TBA (0.67%) and heated for 1 h at 100°C. After cooling, the precipitate was removed by centrifugation. The absorbance of the sample was measured at 535 nm using a blank containing all the reagents except the sample. Glutathione content (GSH) was determined using commercial glutathione reduced kits (Biodiagnostic for diagnostic reagents: Dokki, Giza, Egypt) according to the method of Beutler et al. (1963). The activities of liver aspartate aminotransferase (AST; EC 2.6.1.1) and alanine aminotransferase (ALT; EC 2.6.1.2) were assayed by the kinetic methods of Bergmeyer et al. (1986). Determination of lactate dehydrogenase (LDH; EC 1.1.1.27) activity was carried out using kits from SENTINEL CH. (via principle Eugenio 5-20155 MILAN-ITALY). The catalase enzyme (CAT; EC 1.11.1.6) converts H_2O_2 into water. The CAT activity was measured spectrophotometrically at 240 nm by calculating the rate of degradation of H2O2; the substrate of the enzyme (Xu et al., 1997).

Super oxide dismutase (SOD; EC 1.15.1.1) was assayed according to Misra and Fridovich (1972). The assay procedure involves the inhibition of epinephrine auto-oxidation in an alkaline medium (pH 10.2) to adrenochrome, which is markedly inhibited by the presence of SOD. Epinephrine was added to the assay mixture that contains the tissue supernatant. The change in extinction coefficient was followed at 480 nm in a Spectrophotometer. Plasma concentrations of total lipids (TL) and cholesterol were assayed by the method of Knight et al. (1972) and Carr et al. (1993), respectively. High density lipoprotein (HDL) and low density lipoprotein (LDL) were determined according to the methods of Warnick et al. (1983) and Bergmenyer (1985), respectively. The protein content was determined by the method of Lowry et al. (1951) using bovine serum albumin as a standard. Acetyl cholinesterase (AChE; EC 3.1.1.7) activity was measured according to the method of Blawen et al. (1983).

Immunohistological examination

Brain samples were fixed in 10% neutral-buffered formalin and embedded in the paraffin wax. Six micrometer thick sections were cut, placed on glass slides and incubated at 37°C overnight. Dewaxed sections were incubated with protein block for 30 min and then incubated overnight at 4°C with the polyclonal anti-glial fibrillary acidic protein (GFAP) antibody (1:1000). Sections were then incubated with biotinylated antirabbit IgG antibody (1:100) and

Table 1. The effect of CPF and/or propolis administration on the levels of glucose, total proteins and albumin and the specific activities of AST, ALT and LDH in the serum samples of male rats.

Parameter	Control	CPF	CPF + Propolis	Propolis
Glucose (mg/dL)	80.86±1.40	136.31± 1.26**	81.10 ± 0.97	78.68 ± 1.06
AST (U/L)	37.05 ± 0.46	46.23± 1.09**	39.13 ± 0.41	31.56 ± 0.45**
ALT (U/L)	26.08 ± 0.33	54.35± 0.81**	29.54± 0.28**	22.82± 0.49**
LDH (U/L)	737.02 ± 13.64	1153.45±21.39**	775.37 ± 12.81	716.12 ± 15.07
Total Proteins (g/L)	8.25 ± 0.16	4.55± 0.18**	8.0 ± 0.15	8.35 ± 0.23
Albumin (g/L)	3.75 ± 0.23	2.82± 0.27*	3.20 ± 0.23	4.12 ± 0.24

Values are expressed as means ± SE; *significantly different at P <0.05 compared to the control group; **, significantly different at P <0.01 compared to the control group; CPF, chlorpyrifos; AST, aspartate transaminase; ALT, alanine transaminase; LDH, lactate dehydrogenase.

Table 2. The effect of CPF and/or propolis administration on the levels of GSH and TBARS and the specific activities of SOD and CAT in the serum samples of male rats.

Parameter	Control	CPF	CPF + Propolis	Propolis
GSH (mg/dL)	30.07 ± 0.57	16.40± 0.32**	29.45 ± 0.57	30.97 ± 0.58
TBARS (mg/dL)	0.66 ± 0.13	1.36± 0.06**	0.70 ± 0.01	0.62 ± 0.04
SOD (U/mL)	2.021 ± 0.07	1.23± 0.10**	1.87 ± 0.12	3.01± 0.05**
CAT (U/mL)	48.98 ± 1.0	30.42± 0.70**	47.71 ± 1.04	49.12 ± 1.0

Values are expressed as means ± SE; *significantly different at P <0.05 compared to the control group; **significantly different at P <0.01 compared to the control group; CPF, chlorpyrifos, GSH: reduced glutathione; TBARS, thiobarbituric acid-reactive substances; SOD, superoxide dismutase; CAT: Catalase.

finally with the avidin-biotin-peroxidase reagent for 30 min at room temperature. The peroxidase reaction was visualized with DAB. Samples were counterstained with haematoxylin. To visualize neurons, sections were stained with Nissl staining according to the method of Malkiewicz et al. (2006).

Histopathology

Specimens of the liver and brain tissues were fixed immediately in 10% formalin for histological studies. Then the tissues were treated with conventional grades of alcohol and xylol, embedded in paraffin and sectioned at 4 to 6 µm thickness. The sections were stained with Hematoxylin and Eosin (H and E) stain for studying the histopathological changes (Drury and Wallington, 1980).

Statistical analysis

Data were analyzed according to Steel and Torrie (1981). Statistical significance of the difference in values of control and treated animals was calculated by F test at 5% significance level. Data of the present study were statistically analyzed by using Duncan's Multiple Range Test (SAS, 1986).

RESULTS

The effect of CPF and/or propolis on the levels of serum biochemical parameters is summarized in Table 1. Results from this study shows that the oral ingestion of CPF caused functional and structural damage of liver tissue. This damage was indicated in part through highly significant increases in the levels of serum transaminases (ALT and AST). They increased from 26.08 ± 0.33 and 37.05 ± 0.46 to 54.35 ± 0.81 and 46.23 ± 1.09, respectively as shown in Table 1. Concurrent with this result, significant elevations in the activity of LDH from 737.02 ± 13.64 to 1153.45 ±21.39 and the level of glucose from 80.86±1.40 to 136.31 ± 1.26, respectively were recorded in the same group when compared to the control one. The functional damage caused by CPF-treatment was indicated by the significant decrease in the serum total protein (4.55 ± 0.18) and albumin (2.82 ± 0.27) concentrations compared to the control levels; 8.25 ± 0.16 and 3.75 ± 0.23, respectively (Table 1). This indicates impaired synthetic and excretory functions of the liver, respectively. The treatment with CPF plus propolis ameliorated the levels of the tested serum biochemical parameters compared to the control group as shown in Table 1.

The results of the reduced glutathione content are represented in Table 2. It is clear that, the level of reduced GSH in the serum (16.40 ± 0.32, $P < 0.05$) was greatly depleted by CPF-treatment compared to the control group (30.07 ± 0.57). In contrast, the level of TBARS was greatly elevated in CPF-treated group (1.36 ± 0.06, $P < 0.05$) compared to the control group (0.66 ± 0.13). The treatment with the CPF plus propolis could normalize the levels of reduced GSH (29.45 ± 0.57) and

Table 3. The changes in the serum lipids profile after oral administration of male rats with chlorpyrifos and/or propolis.

Parameters (mg/dl)	Control	CPF	CPF +Propolis	Propolis
Total lipids	460.39 ± 0.80	561.24** ±3.44	519.13** ±0.71	395.26** ±1.81
Total cholesterol	109.98 ± 0.57	126.85** ±1.57	110.10 ± 0.47	105.66* ±0.81
LDL-c	97.10 ± 2.28	125.43** ±1.25	101.05 ± 2.61	85.92 ± 1.80
HDL-c	31.42 ± 0.41	20.92** ± 0.44	31.16 ± 0.48	41.03** ± 0.66
TG	81.46 ± 0.97	92.10** ± 0.95	82.20 ± 0.81	75.18** ± 1.32

The data are presented as mean ± SE, (n = 10); *significantly different at P <0.05 compared to the control group; **significantly different at P <0.01 compared to the control group; CPF, chlorpyrifos; LDL-c, low density lipoprotein-cholesterol; HDL-c, high density lipoprotein-cholesterol Lipoprotein; TG, triglycerides.

TBARS (0.70 ± 0.01) compared to the control group (30.07 ± 0.57 and 0.66 ± 0.13, respectively). Table 2 shows also that CPF-treatment caused significant (P≤ 0.05) inhibition in the specific activities of catalase (30.42 ± 0.70) and super oxide dismutase (1.23 ± 0.10) compared to the control values; 48.98 ± 1.0 and 2.021 ± 0.07, respectively. On contrast, the specific activities of serum catalase and super oxide dismutase could be normalized after treatment with CPF plus propolis (47.71 ± 1.04 and 1.87 ± 0.12, respectively).

Table 3 shows the lipids profile of all the studied groups in this study. The results shows that the treatment of rats with CPF induced significant ($p < 0.05$) increases in the serum total lipids from 460.39 ± 0.80 to 561.24 ±3.44, total cholesterol from 109.98 ± 0.57 to 126.85 ±1.57, triglycerides from 81.46 ± 0.97 to 92.10 ± 0.95 and LDL-c from 97.10 ± 2.28 to 125.43 ±1.25 while HDL-c was significantly decreased compared to the control group from 31.42 ± 0.41 to 20.92 ± 0.44. On the other hand, treatment with CPF plus propolis showed amelioration in the levels of serum total lipids, total cholesterol, triglycerides, LDL-c and HDL-c (519.13 ±0.71, 110.10 ± 0.47, 82.20 ± 0.81, 101.05 ± 2.61 and 31.16 ± 0.48, respectively).

Light micrographs of the liver showed degenerated hepatocytes, dilatation and vascular congestion in sinusoids, cytoplasmic vacuolation and degenerated hepatocytes with pyknotic nuclei are shown in the CPF-treated rats, compared to central vein with normal hepatocytes architecture in the control group and propolis-treated group. Histological changes induced after CPF-treatments were markedly reduced in the combination group showing moderate improvement in the hepatic cells (Figure 1).

Furthermore, CPF-treatment caused significant ($p < 0.05$) decrease in the activity of serum and brain AChE (64.80 ± 0.55 and 5.49 ± 0.10, respectively) compared to the control group (71.22 ± 0.34 and 20.23 ± 0.42, respectively) as shown in Table 4. On contrast, the specific activity of AChE in the serum and brain extract could be normalized after treatment with CPF plus propolis (70.77 ± 0.38 and 19.74 ± 0.42, respectively). Histopathological examination of the brain showed mild neuronal degeneration in purkinje cells, neuronal shrinkage and nuclear pyknosis after CPF-treatment (Figure 2), characterizing CPF-induced neuronal necrosis as compared to the controls. The changes observed in GFAP immunostaining revealed longer and more numerous astrocytic processes in the brain tissues of CPF-treated group as compared to controls. Our study shows that the astrocytic density was raised in the CPF-treated group (Figure 3).

DISCUSSION

Chlorpyrifos

Chlorpyrifos is an extensively used organophosphorus pesticide having many urban and agricultural crop pest control uses. The present study is an attempt to protect against CPF-induced toxicity by propolis-treatment in male rats. The present data showed that the activities of serum AST, ALT and LDH were significantly ($p<0.05$) increased after CPF-treatment compared to the control group (Table 1). This fact is a conventional indicator of liver injury (Rao, 2006). Elevation of serum AST and ALT indicates liver damage (Akhtar et al., 2009). This agrees with the low ALT and AST activities in the liver extract of rats exposed to CPF in the previous findings of Zama et al., (2007). When the liver cell membrane is damaged, varieties of enzymes normally located in the cytosol are released into the blood stream. Also, the elevation in the activity of LDH suggests an increase in lysosomal mobilization, cell necrosis and liver damage due to pesticide toxicity (Etim et al., 2006). Chen et al. (2000) observed a significant rise in serum LDH activity after carbon tetrachloride-induced hepatotoxicity.

On the other hand, Table 1 indicates that the serum total proteins and albumin were significantly ($p<0.05$) decreased in the CPF-treated rats compared to the control group, which agrees with the report of Peeples et al. (2009). Normally, the reduction of serum proteins and albumin levels as shown in Table 1 indicates a liver disease. This reduction could be attributed to the changes in the metabolism and the synthesis of the

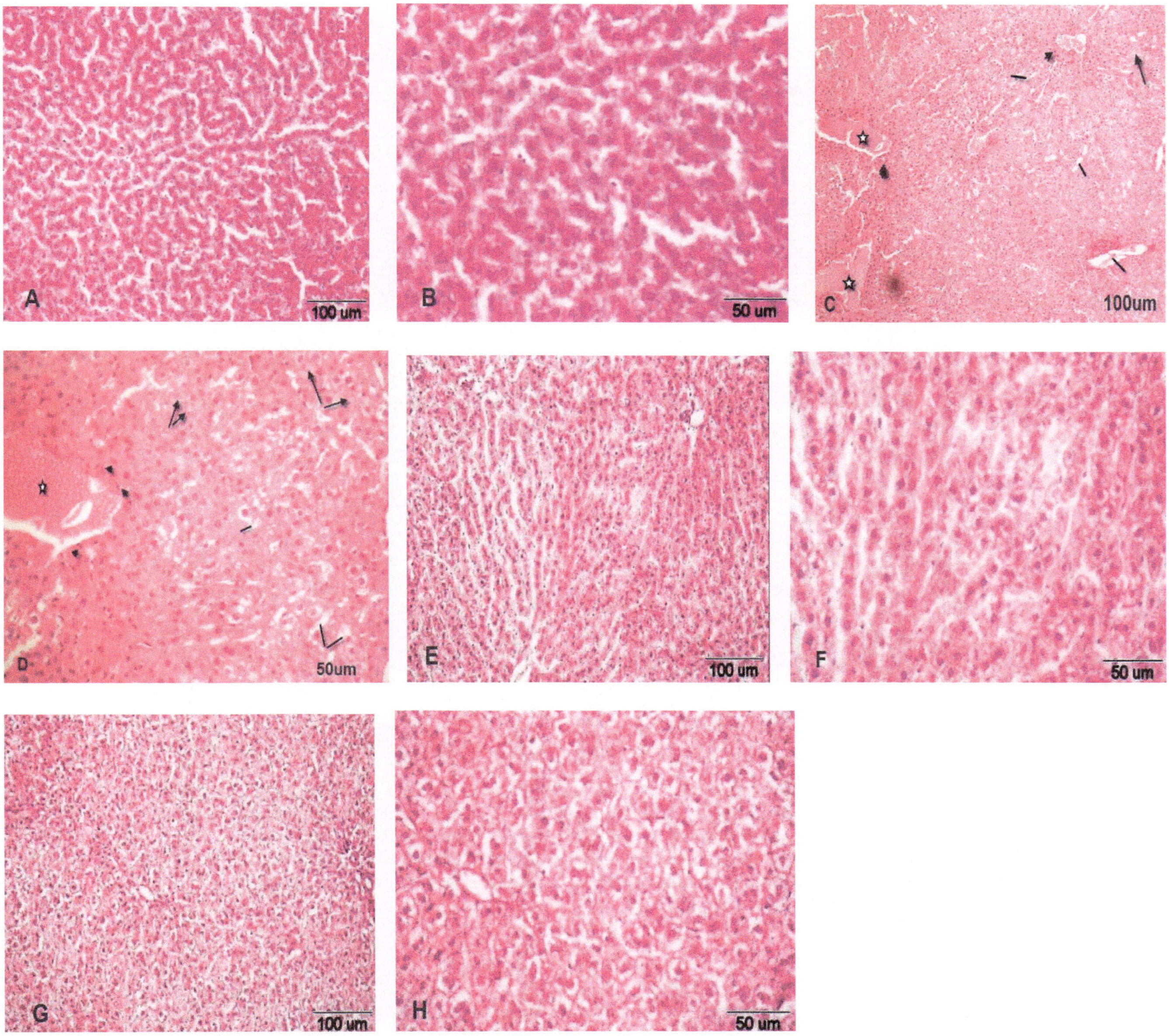

Figure 1. Light micrographs of the liver of rats treated: A and B, control; C and D, chlorpyrifos-treated rats; E and F, chlorpyrifos plus propolis-treated rats and G and H, propolis-treated group.

Table 4. The effect of CPF and/or propolis administration on the specific activity of AchE in the serum samples and brain extracts of male rats.

Parameter	Control	CPF	CPF + Propolis	Propolis
AchE- serum (1)	71.22 ± 0.34	64.80**± 0.55	70.77 ± 0.38	71.35 ± 0.37
AchE- brain (2)	20.23 ± 0.42	5.49**± 0.10	19.74 ± 0.42	20.98 ± 0.44

Values are expressed as means ± SE; **significantly different at $P < 0.01$ compared to the control group; 1, U/l; 2, n mol/min/mg protein; CPF, chlorpyrifos; AchE, acetylcholine esterase.

protein and free amino acid in the liver (Li et al., 2007). They suggested that albumin could be used as a biomarker of CPF-toxicity. Generation of oxidative stress and consequent lipid peroxidation by pesticides was reported in many species. Due to the high concentration of polyunsaturated fatty acids in cells, lipid peroxidation is

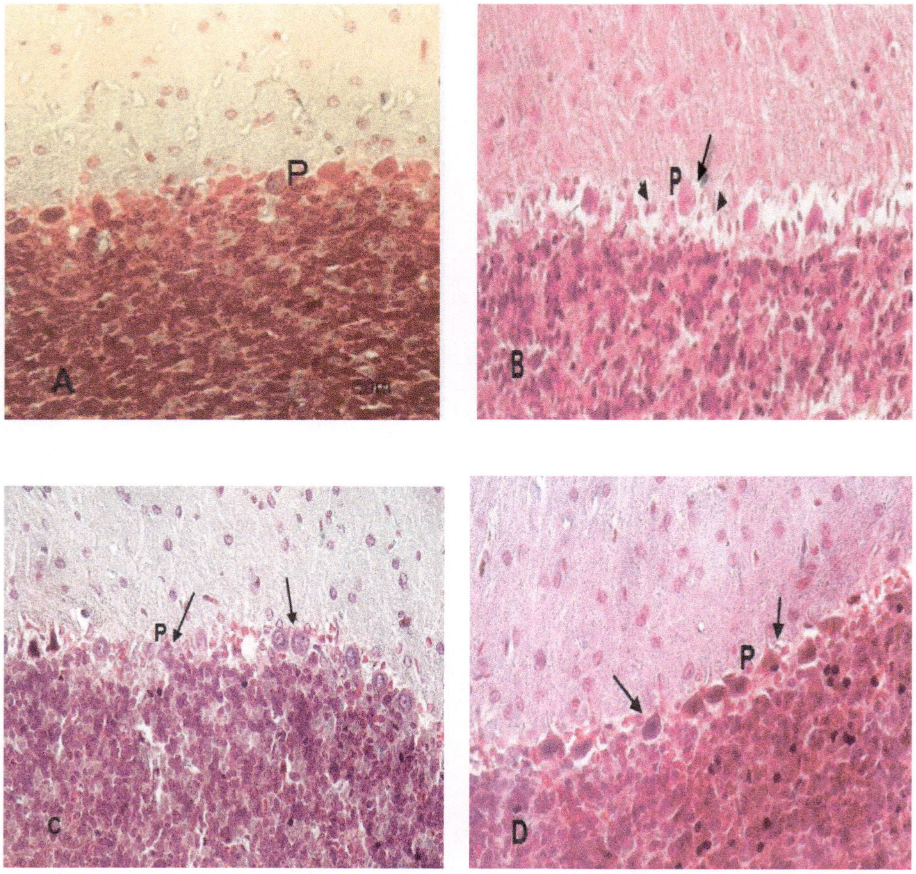

Figure 2. Light micrographs of cerebellar cortex in rats: Control group (A) and propolis-treated rats (D) showed normal histological structure of brain with dark-staining Purkinje cell (arrow). Chlorpyrifos-treated rats (B) showed contracted, dark-staining cell bodies described as "degenerating" Purkinje neurons (arrow) with loss of dendrites and loss of granules in granular layer. Chloropyrifos plus propolis-treated rats (C) revealed some amelioration of the neuron cells.

a major outcome of the free radical-mediated injury (Verma et al., 2007).

Results of the present study show that CPF-exposure caused significant decreases in the activities of serum SOD and CAT (Table 2). CPF has been shown to impair antioxidant enzyme activities either directly or through the induction of free radicals resulting in oxidative stress (El-Banna et al., 2009). Also, the decrease of antioxidant enzyme activities was observed after CPF-exposure in rats (El-Kashoury and Tag El-Din, 2010). Our results show also that CPF-exposure to male rats caused an increase in the lipid peroxidation, as evidenced by elevated level of TBARS and the decreased level of GSH (Table 2). These findings agree with the results obtained in a previous study (Saxena and Garg, 2010). They reported that tissue lipid peroxidation is a degradative phenomenon as a consequence of free radical chain production and propagation which affects mainly polyunsaturated fatty acids. Lipid peroxidation has been used as a measure of this xenobiotic-induced oxidative stress, which was originally defined as the disequilibrium between prooxidants and antioxidants in the biological systems (Bebe and Panemangalore, 2003). Once this imbalance appears, cellular macromolecules may be damaged by the predominant free radicals. Lipid peroxidation was reported to be increased after exposure to various xenobiotics (Bebe and Panemangalore, 2003).

Thus, it is an important and necessary consequence of oxidative stress and is involved in the pathophysiology of the number of diseases and other natural degenerative conditions. The role of oxygen derived species in causing cell injury or death is increasingly recognized: superoxide and hydroxyl radicals are involved in a large number of degenerative changes often associated with an increase in peroxidative process and linked to low antioxidant concentration (Slotkin et al., 2005). Mammalian cells are equipped with both enzymatic and non-enzymatic antioxidant defense mechanisms to minimize the cellular damage resulting from the interaction between cellular constituents and ROS (Goel et al., 2009). The enzymatic antioxidant defense mechanism contains various forms of superoxide dismutases, catalase and glutathione peroxi-

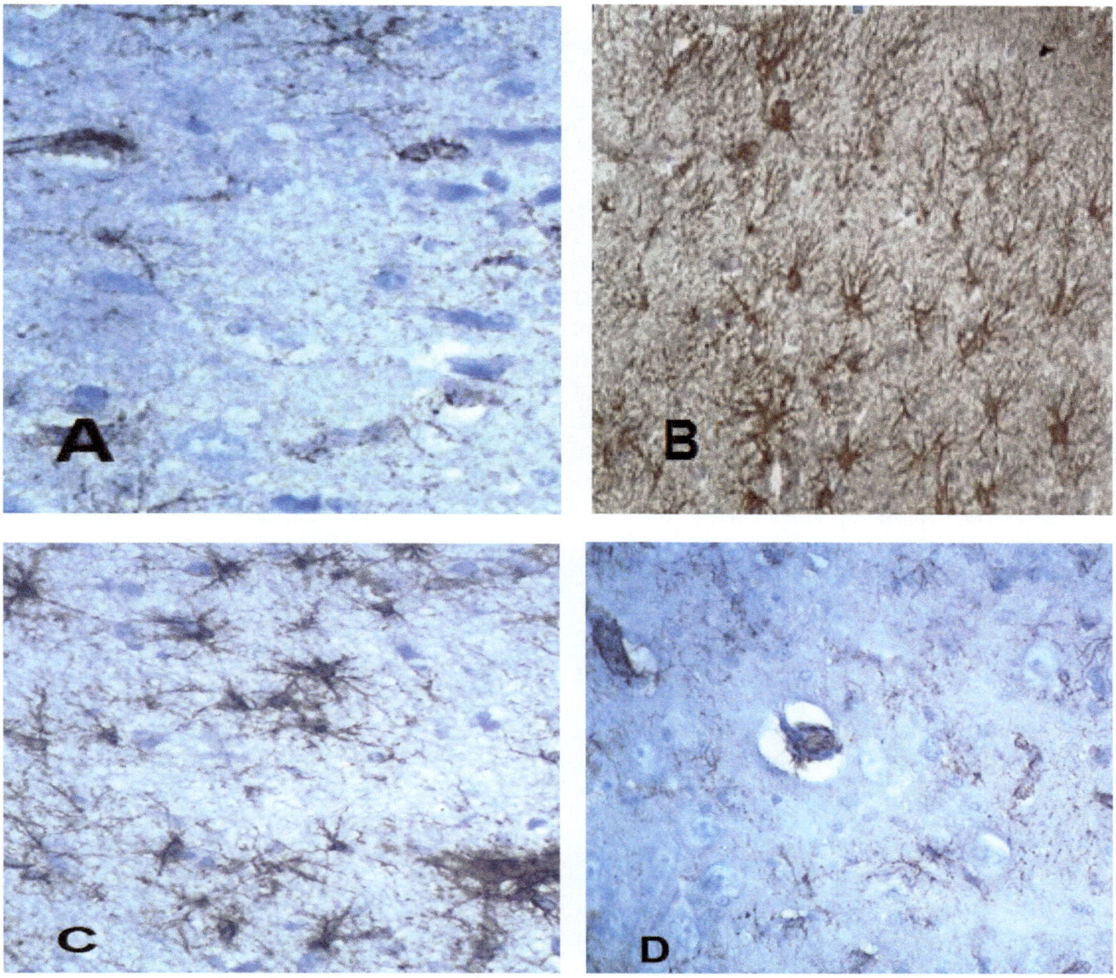

Figure 3. Photomicrograph showing the immunohistochemical staining of anti-glial fibrillary acidic protein (GFAP) expression in stratum moleculare-lacunosum of the cerebellar cortex in groups of rat at the end of experiment. Brown colour rounded cells with processes are the astrocytes. A, Control group; B, chlorpyrifos group; C, chlorpyrifos plus propolis group; D, propolis group (GFAP, 400×).

dase. Despite of the presence of these delicate cellular antioxidant systems, an overproduction of ROS in both intra- and extracellular spaces often occurs upon exposure of cells or individuals to certain chemicals (Goel et al., 2005).

An unbalanced production of ROS in localized compartments has been reported to play a role in the pathogenesis of a number of clinical disorders which illustrates the importance of antioxidant defense system in maintaining normal cellular physiology (El-sharaky et al., 2009). Also, Suleiman et al., (2010) have reported that lipid peroxidation was increased and the activities of antioxidant enzymes were decreased in rat erythrocytes on CPF-treatment which were restored to the normal values by vitamine C-treatment. GSH acts as reducing agent and a vital substance in detoxification and provides antioxidant protection in the aqueous phase of cellular systems (El-sharaky et al., 2009); its antioxidant activity is through the thiol group of its cysteine residue. Like ascorbic acid, another important water soluble antioxidant, GSH can directly reduce a number of ROS and is oxidized to GSSG in this process. Intrahepatic glutathione was reported to afford protection against liver dysfunction by at least two ways: 1) as a substrate of GPx, GSH serves to reduce large variety of hydroperoxides before they attack unsaturated lipids or convert the formed lipid hydroperoxides to the corresponding hydroxyl compounds and 2) as a substrate of glutathione-S-transferase, it enables the liver to detoxify foreign compounds or their metabolites and to excrete the products, preferably into the bile.

Relatively high ratios of GSI-I/GSSG are maintained intracellularly through the action of glutathione reductase (GR) and NADPH dependent reaction (El-Banna et al., 2009). Also, a similar decrease in the level of GSH and corresponding increase in GSSG has also been reported in different parts of brain on CPF-exposure (Suleiman et al., 2010). The present study clearly shows that CPF-exposure generates oxidative stress in male rats.

Table 3 also shows that treatment of rats with CPF induced significant ($p < 0.05$) increases in serum total lipids, total cholesterol, triglycerides and LDL-c compared to the control group. The same trend was seen with other organophosphorus pesticide (Kalender et al., 2005). Previous studies demonstrated increases in the concentration of serum triglycerides in the experimental animals that were treated with different insecticides, including the organophosphate, dichlorvos (Ranjbar et al. 2002) and carbamate furadan (Ibrahim and El-Gamal, 2003). This elevation of serum or plasma triglycerides has been attributed to an inhibition of the lipase enzyme activity of both the hepatic triglycerides and plasma lipoproteins (Goldberg et al., 1982). Also, our study showed that oral treatment with CPF induced significant decrease in serum HDL-c. HDL-c is mainly synthesized in the liver and intestinal cells. It plays an important role in cholesterol efflux from tissues and carries it back to the liver for removal as bile acids (Shakoori et al. 1988). It has been established that the reduced levels of HDL-c are associated with an increased risk for coronary artery disease (Stain, 1987).

In addition, the elevated cholesterol level in the present study due to CPF-treatment is consistent with the finding of Ashgar et al. (1994), who reported that total cholesterol was elevated in male rabbits treated by methyl parathion. The increase in the level of serum cholesterol may be due to an increased synthesis of cholesterol in the liver (Enan et al., 1987). The present study is in agreement with other investigations which reported that organophosphorus pesticide cause increases in the total cholesterol and the total lipid levels (Adham et al., 1997). The induced increase in serum cholesterol can be attributed to the effect of the pesticide on the permeability of liver cell membrane (Kalender et al., 2005). Also, the increase in serum total cholesterol level may be attributed to the blockage of liver bile ducts causing reduction or cessation of its secretion to the duodenum (Aldana et al., 1998).

Also, Figure 1 demonstrates the degenerative changes in liver after CPF-treatment comprise fatty change, individual cell necrosis, cellular infiltration and areas of necrosis that are suggestive of acute hepatitis. Also, cellular infiltration and the presence of newly forming bile ducts suggest a persistent sub acute or chronic liver injury. The damage of the liver tissues in CPF-treated rats in the present study could be due to injured mitochondria (Goel et al., 2005). The present study suggests that the administration of CPF in rats induced pathological changes in the liver tissue.

Results of this study shows that CPF-treatment caused significant ($p < 0.05$) decrease in the activity of serum and brain AChE compared to the control group (Table 4). This is consistent with the study of Celik and Isik (2009). They reported that in animals intoxicated with methyl parathione and dichlorvos at sub lethal concentrations, AChE and butyl cholinesterase (BChE) activities are inhibited. They also reported that inhibition of AChE may be a better biomarker for the assessment of neurotoxic effects in the living.

Organophosphates toxicants generally elicit their effects by inhibition of acetyl cholinesterase, which lead to accumulation of the neurotransmitter acetylcholine in synapses; in the neuromuscular junction, over stimulation of postsynaptic cholinergic receptors leads to muscle fasciculation and eventual paralysis (Mileson et al., 1998)

Furthermore, our study showsthat the astrocytic density was raised in the CPF-treated group (Figure 3). One of the proteins associated with neuronal damage is glial fibrillary acidic protein (GFAP). GFAP is a cytoplasmic intermediate filament protein found in astrocytes.

They maintain the structural integrity of astrocytes, especially when these cells undergo hypertrophy and hyperplasia in response to a non-invasive CNS injury (Norenberg, 1994) whereby, expression of GFAP is upregulated (Abou-Donia et al., 2004). GFAP expression in the astrocytes was more prominent compared to the control groups. The astrocytic density of the groups receiving CPF was greater as compared to controls. This may be attributed to the neuroprotective effect of astrocytes limiting neuronal damage.

It has been suggested that the metabolites of CPF; trichloropyridinol (TCP) exert strong toxic effects on astrocytes, compromising their neuroprotective effects and thus increasing the neurotoxicity of CPF (Zurich et al., 2004). A characteristic feature of gliosis, GFAP upregulation often occurs in response to injury in the brain (Abou-Donia et al., 2004).

Ho et al. (2007) reported that GFAP is a sensitive and early biomarker of neurotoxicity, it's up-regulation preceding anatomically perceptible damages in the brain. Findings of Lim et al. (2011) suggest that GFAP expression is up regulated with dermal exposure to low dose of CPF.

Propolis

Although certain compound has been tested for the detoxification of CPF (Suleiman et al., 2010), there are no previous studies that carried out to evaluate the curative effect of propolis against CPF-toxicity that cause dysfunction of the liver and brain. Treatment with chlorpyrifos plus propolis could normalize the serum glucose level compared to the control group (Table 1). This suggests that propolis can control blood glucose and modulate the metabolism of glucose (Fuliang et al. 2005). The present data indicated that treatment with propolis alone decreased the activities of AST, ALT and LDH in the serum samples (Table 1).

Also, the present data showed that the treatment with propolis alone caused reduction in TBARS level and increased the activities of SOD and CAT and the level of GSH in the serum (Table 2). These data are in agreement with the results obtained by Jasprica et al., (2007) who reported that propolis caused reduction in the malondialdehyd (MDA) level and increased the activities of the antioxidant enzymes (SOD, GPx and CAT). Moreover, the present results show that propolis could normalize the levels of AST, ALT and LDH in the serum (Table 1), indicating that propolis tended to prevent the damage and suppressed the leakage of enzymes through cellular membranes. This result is in accordance with the findings that propolis induced reduction of the increased activity of AST and ALT concentrations in the plasma of galactoseamine-treated rats (Gonzalez et al., 1994). Our results indicate an increased TBARS levels in the serum of CPF-treated rats, implying the increased oxidative damage to the tissues. Propolis treatments returned the increased TBARS levels back to the control levels (Table 2). This result is in accordance with the findings that propolis extract induced reduction of the increased TBARS concentrations in serum of galactoseamine-treated rats (Gonzalez et al. 1994).

In the present work, CPF caused oxidative stress and consequently decreased the activities of the antioxidant enzymes (SOD and CAT). After treatment of rats with chlorpyrifos plus propolis, the activities of these antioxidant enzymes were normalized to their control values (Table 2). Jasprica et al. (2007) showed that propolis and related flavonoids exercise their activity through the scavenging of hydroxyl, superoxide free radicals and lipid peroxides. The antioxidant activities of propolis and its polyphenolic/flavonoid components are related to their ability to chelate metal ions and scavenge singlet oxygen, superoxide anions, peroxyl radicals, hydroxyl radicals and peroxynitrite (Fonseca, 2011). Also, other previous studies of Yousef et al. (2004a) showed that the treatment with isoflavones as antioxidant in combination with cypermethrin minimized its hazardous effects.

Moreover, the results of Yousef et al. (2004a) concluded that isoflavone dosages (2.5 or 5 mg/kg body weight), which is more than two or four times to the amounts consumed (40 mg) in many Eastern nations, have beneficial effect on the level of plasma lipids and lipoproteins in rabbits. Fuliang et al. (2005) showed that propolis elevates GPX, SOD, GST and catalase activities. Therefore, the flavonoids of the propolis can increase the activities of the antioxidant enzymes and reduce the levels of the ROS. Kanbura et al. (2009) found decreases in the plasma and the tissues (liver, kidney and brain) malondialdehyde (MDA) levels and normalization in the antioxidant enzyme parameters (SOD, CAT and GSH-Px) of animals that were administered propolis in association with propetamphos, in comparison to the group that was administered propetamphos alone.

Moreover, Newairy et al. (2009) found decreased levels of the plasma and tissue (liver, kidney and brain) malondialdehyde and ameliorations in the antioxidant enzymes (SOD, CAT and GSH-Px) of animals that were administered propolis in association with aluminium chloride, in comparison to the group that was administered aluminium chloride alone. The primary mechanism of the effect of propolis may involve the scavenging of free radicals that cause lipid peroxidation. The other mechanism may comprise the inhibiting effect of propolis on the activity of xanthine oxidase, which is known to cause free radicals to be generated (Harris et al. 2000). Among other studies that demonstrate the mechanisms responsible for the antiradical and antioxidant activities of propolis, in a trial conducted by Matsushige et al. (1995), propolis has been found to exhibit antioxidant activity against 1, 1-diphenyl-2-picrylhydrazyl (DPPH) and the superoxide radical by means of xanthine/xanthine oxidase and NADH/ phenazine reactions. Also, propolis was able to induce hepatoprotective effects on paracetamol induced liver damage in mice (Gonzalez et al.1994). Taken together, these findings constitute evidence that the antioxidative properties of the propolis contribute to the prevention of damage induced by CPF in rats.

This study shows that treatment of rats with CPF plus propolis normalized the total lipids, cholesterol, triglycerides and low-density lipoprotein-cholesterol (LDL-c) and increased high-density lipoprotein cholesterol (HDL-c) level compared to the control group (Table 3). These results are in agreement with those of Fuliang et al. (2005), who found that oral administration of propolis significantly lowered total cholesterol, triglycerides, LDL-c and very low-density lipoprotein cholesterol (VLDL-c) in the serum of rats while increased the serum level of HDL-c. Also, Kolankaya et al. (2002) found that propolis significantly decreased cholesterol and triglycerides. Some studies suggested that propolis can act in several ways to lower plasma LDL-bound cholesterol. First, uptake of cholesterol in the gastrointestinal tract could be inhibited. Second, LDL-c could be eliminated from the blood via LDL receptor and finally, the activity of cholesterol-degrading enzymes, namely cholesterol-7-hydroxylase could be increased. It has been suggested that propolis decreased total cholesterol and LDL-c while increased HDL-c due to absorption, degradation or elimination of cholesterol.

Other studies also show that propolis reduced cholesterol and increased HDL-c, indicating that it may mobilize cholesterol from extrahepatic tissues to the liver where it is catabolised (Fuliang et al.2005). Alves et al. (2008) reported that the hypocholesterolemic effect of propolis is the result of a direct effect on the liver or an indirect effect through the thyroid hormones, since the thyroid hormones affect reactions in almost all the pathways of lipid metabolism. Furthermore, the specific activity of AChE in the serum and brain extract could be normalized after treatment with CPF plus propolis.

This study shows also that the astrocytic density was raised in the CPF-treated group (Figure 3). El-Masry et al. (2011) reported that propolis increased brain vitamin C, vitamin E and P-SH levels in rat's brain that was treated with propolis alone/or in combination with Pb. They concluded that propolis has beneficial effects and could be able to antagonize Pb-induced neurotoxicity. The biological effects exhibited by propolis could be related to an overall effect of the phenolic compounds present in propolis. Caffeic acid phenethyl ester (CAPE) is an active component of propolis and has been used in traditional medicine to treat a number of diseases, CAPE treatment have been shown to protect tissues from ROS mediated oxidative stress and reduce lipid peroxidation in ischemia and toxic injuries. The antioxidant activity of CAPE is due to the presence of two hydroxyl groups in its structure (Sud'ina et al., 1993).

Twelve (12) different flavonoids, pinocembrin, acacetin, chrysin, rutin, catechin, naringenin, galangin, luteolin, kaemferol, a pigenin, myricetin and quercetin, two phenolic acids, cinnamic and caffeic acid are present in propolis (El-Masry et al., 2011). Propolis contains acid derivatives such as benzoic-4- hydroxy benzoic which improves the digestive utilization of calcium, phosphorus and magnesium (Haro et al., 2000). Furthermore, it has been reported that Chinese propolis contains many biologically active constituents expected to be useful for improvement of the neuropathological conditions in the injured spinal cord (Usia et al., 2002) and to protect the brain from ischemia-reperfusion injury (Khan et al., 2007). Therefore, CAPE is one of the most probable candidates that may act beneficially after traumatic spinal cord injury (Song et al., 2002). Chrysin, a flavonoid of highest concentration in Chinese propolis, potentially represents anti-oxidative capacity in neuronal cell death and exhibits an anti-inflammatory effect by inhibiting iNOS mRNA and NO production (Izuta et al., 2008).

Conclusion

CPF has adverse effects on human health. Our results report that CPF induced oxidative damage and caused damage to the liver and brain of the male rats while, propolis administered in combination with chloropyrifos minimized its hazardous effect. In addition, propolis alone was proved to be beneficial in decreasing the levels of free radicals and lipids and increasing the activities of the antioxidant enzymes. Consequently, the exposure to CPF should be reduced. Also, the intake of propolis as supplement could be a beneficial way to overcome the toxicity of CPF-exposure.

ACKNOWLEDGEMENT

The authors gratefully acknowledge critical Immunohistological examinations by Prof. Dr. Sheiref Bayomy, Prof of Pathology, Faculty of Science, Alexandria University, Egypt.

REFERENCES

Abou-Donia MB, Khan WA, Suliman HB, Abdel-Rahman AA, Jensen KF (2004). Stress and combined exposure to low doses of pyridostigmine bromide, DEET, and ermethrin produce neurochemical and neuropathological alterations in cerebral cortex, hippocampus, and cerebellum. J. Toxicol. Environ. Health 67(2):163-192.

Adham KG, Khairalla A, Abu-Shabana M, Abdel-Maguid N, Abd El-Moneim A (1997). Environmental stress in Lake Maryut and physiological response of Tilipia Zilli Gerv. J. Environ. Sci. Health Part A. 32:2585-2598.

Akhtar N, Srivastava MK, Raizada RB (2009). Assessment of chlorpyrifos toxicity on certain organs in rat, Rattus norvegicus. J. Environ. Biol. 30(6):1047-1053.

Aldana L, De Mejía EG, Craigmill A, Tsutsumi V, Armendariz-Borunda J, Panduro A, Rincón AR (1998). Cypermethrin increases apo A-1 and apo BmRNA but not hyperlipidemia in rats. Toxicol. Lett. 95:31-39.

Alves MJQF, Mesquita FF, Sakaguti M, Tardivo AC (2008). Hypocholesterolemic effect of propolis caffeic acids. Revista Brasileira de Plantas Medicinais. 10(1):100-105.

Ambali SF (2009). Ameliorative effects of vitamins C and E on neurotoxicological, haematological and biochemical changes induced by chronic chlorpyrifos in Wistar rats. PhD Dissertation. Ahmadu Bello University, Zaria, Nigeria.

Ashgar M, Sheikh MA, Hashmi A (1994). Effects of orally fed methyl parathion on some hematochemical parameters of rabbits. Pakistan Vet. J. 14:34-36.

Bebe FN, Panemangalore M (2003). Exposure to low doses of endosulfan and chlorpyrifos modifies endogenous antioxidants in tissues of rats. J. Environ. Sci. Health Part B. 38:349-163.

Bergmenyer HU (1985). Methods of Enzymatic Analysis. 3rd ed., vol. VIII:154-160.

Bergmeyer HU, Herder M, Red R (1986). Approved recommendation on IFCC methods for the measurement of catalytic concentration of enzymes. Part 3. FCC Method for alanine aminotransferase. J. Clin. Chem. Clin. Biochem. 24:481-489.

Beutler E, Duron O, Kelly BM (1963). Improved method for the detection of blood glutathione. J. Lab. Clin. Med. 61:882-888.

Blawen DH, Poppe WA, Trischler W (1983). The effect of subacute doses of organophosphorus pesticide, nuvacron, on the biochemical and cytogenetic parameters of mice and their embryos. J. Clin. Chem. Biochem. 21:381-386.

Carr TP, Andressen CJ, Rudel LL (1993). Enzymatic determination of triglyceride, free cholesterol and total cholesterol in tissue lipid extracts. Clin. Biochem. 26:39-42.

Celik I, Isik I (2009). Neurotoxic effects of subacute exposure of dichlorvos and methyl parathion at sublethal dosages in rats. Pestic. Biochem. Physiol. 94:1-4.

Chen W, Kennedy DO, Kojma A, Matsui-Yuasa I (2000). Poly-amines and thiols in the cytoprotective effect of L-cystein and L-methionine on carbon tetrachloride-induced hepatotoxicity. Amino acids. 18:319-327.

Drury RAB, Wallington EA (1980). Carleton's Histological Technique. 5th ed. Oxford University Press, Oxford, New York, Toronto.

El-Banna SG, Attia AM, Hafez AM, El-Kazaz SM (2009). Effect of garlic consumption on blood lipid and oxidant/antioxidant parameters in rat males exposed to chlorpyrifos. Slovak J. Anim. Sci. 42(3):111 - 117.

El-Kashoury AA, Tag El-Din HA (2010). Chlorpyrifos (from different sources):Effect on Testicular Biochemistry of Male Albino Rats. J. Am. Sci. 6(7):252-261.

El-Masry TA, Emara AM, El-Shitany NA (2011). Possible protective effect of propolis against lead induced neurotoxicity in animal model. J. Evol. Biol. Res. 3(1):4-11

El-sharaky AS, Newairy AA, Kamel MA, Eweda SM (2009). Protective effect of ginger extract against bromobenzene-induced hepatotoxicity in male rat. Food Chem. Toxicol. 47:1584-1590.

Enan E, Berberian IG, El-fiky S, El-Masry M, Enan OH (1987). Effects of two organophosphorus insecticides on some biochemical constituent in the nervous system and liver of rabbits. J. Environ. Sci. Health Part B. 22:149-170.

Esterbauer H, Cheeseman KH (1990). Determination of aldehydic lipid peroxidation products:malonaldehyde and 4-hydroxynonenal. Methods Enzymol. 186:407-421.
Etim OE, Farombi EO, Usoh IF, Akpan EJ (2006). The protective effect of aloe vera juice on lindane induced hepatotoxcity and genotoxicity Pakistan. J. Pharmaceut. Sci. 19:337-340.
Fonseca YM, Marquaele-Oliveira, Vicentini FT, Furtado NAJC, Sousa JP, Lucisano-Valim YM, Fonsecno F (2011). Evaluation of the potential of Bazlian propolis against UV-induced oxidative stress. Evidence-Based Complement. Alter. Med. 2011:1-8.
Fuliang HU, Hepburn HR, Xuan H, Chen M, Daya S, Radloff SE (2005). Effects of propolis on blood glucose, blood lipid and free radicals in rats with diabetes mellitus. Pharmacol. Res. 51(2):147-152.
Ghisalberti EL (1979). Propolis:A review. Bee World. 60:59-84.
Goel A, Danni V, Dhawan DK (2005). Protective effects of zinc on lipid peroxidation, antioxidant enzymes and hepatic histoarchitecture in chlorpyrifos-induced toxicity. Chem. Biol. Interact. 156:131-140.
Goel M, Eraslan G, Silici S (2009). Antioxidant effect of propolis against exposure to propetamphos in rats. Ecotoxicol. Environ. Safety 72:900-915.
Goldberg IJ, Le NA, Paterniti JR, Ginsberg HN, Lindgren FT, Brown WV (1982). Lipoprotein metabolism during acute inhibition of hepatic triglyceride lipase in the Cynomolgus monkey. J. Clin. Invest. 70 (6):1184-1192.
Gonzalez R, Rernirez D, Rodriguez S (1994). Hepatoprotective effects of propolis extract on paracetarnol induced liver damage in mice. Phytother. Res. 8:229-232.
Haro A, Aliaga LI, Lisbona F, Barrionuevo M, Alferez M, Campos M (2000). Beneficial effect of pollen and/or propolis on the metabolism of iron, calcium, phosphorus and magnesium in rats with nutritional ferropenic anemia. J. Agric. Food Chem. 48:5715-5722.
Harris SR, Panaro NJ, Thorgeirsson UP (2000). Oxidative stress contributes to the anti-proliferative effects of flavone acetic acid on endothelial cells. Anticancer Res. 20(4):2249-2254.
Ho G, Zhang C, Zhuo L (2007). Non-invasive fluorescent imaging of gliosis in transgenic mice for profiling developmental neurotoxicity. Toxicol. Appl. Pharmacol. 221(1):76-85.
Ibrahim NA, El-Gamal BA (2003). Effect of Diazinon, an organophosphate Insecticide, on plasma Lipid Costituents in Experimental animals. J. Biochem. Mol. Biol. 36(5):499-504.
Izuta H, Shimazawa M, Tazawa S, Araki Y, Mishima S, Hara H (2008). "Protective effects of Chinese propolis and its component, chrysin, against neuronal cell death via inhibition of mitochondrial apoptosis pathway in SH-SY5Y cells." J. Agr. Food Chem. 56(19):8944- 8953.
Jasprica D, Mornar A, Debelijak Z, Smolcic-Bubalo A, Medic- Saric M, Mayer L, Romic Z, Bucan K, Balog T, Sobocanec S, Sverko V (2007). In vivo study of propolis supplementation effects on antioxidative status and red blood cells. J. Ethnopharmacol. 110:548-554.
Kalender S, Ogutcu A, Uzunhisarcikli M, Acikgoz F, Durak D, Ulusoy Y, Kalender Y (2005). Diazinon-induced hepatotoxicity and protective effect of vitamin E on some biochemical indices and ultrastructural changes. Toxicology 211:197-206.
Kanbura M, Eraslan G, Silici S (2009). Antioxidant effect of propolis against exposure to propetamphos in rats. Ecotoxicol. Environ. Safety 72:900-915.
Khan M, Elango C, Ansari MA, Singh I, Singh AK (2007). Caffeic acid phenethyl ester reduces neurovascular inflammation and protects rat brain following transient focal cerebral ischemia. J. Neurochem. 102(2):365-377.
Knight JA, Anderson S, Rawle JM (1972). Chemical basis of the sulfo-phosphovanillin reaction estimating total serum lipids. Clin. Chem. 18:199-202.
Kolankaya D, Selmanoglu G, Sorkun K, Salih B (2002). Protective effects of Turkish propolis on alcohol-induced serum lipid changes and liver injury in male rats. Food Chem. 78:213-217.
Li B, Lawrence MS, Steven HH, Masson B, Lockridge O (2007). Matrix-assisted laser desorption/ionization time-of-flight mass spectrometry assay for organophosphorus toxicants bound to human albumin at Tyr411. Anal. Biochem. 361:263-272.
Lim KL, Tay A, Nadarajah VD, Mitra NK (2011). The effect of consequent exposure of stress and dermal application of low doses of chlorpyrifos on the expression of glial fibrillary acidic protein in the hippocampus of adult mice. J. Occupational Med. Toxicol. 6:4-13.
Lowry OH, Rosebrough NJ, Farr AL, Randall RJ (1951). Proteinmeasurement with the Folin phenol reagent. J. Biol. Chem. 193:265-275.
Malkiewicz K, Koteras M, Folkesson R, Brzezinski J, Winblad B, Szutowski M, Benedikz E (2006). Environ. Toxicol. Pharmacol. 21:51-55.
Matsuno T (1995). A new clerodane diterpenoid isolated from propolis. Zeitschrift. Für. Naturforschung. Tiel. 50C 93-97.
Matsushige K, Kusumoto IT, Yamomoto Y, Kadota S, Namba T (1995). Quality evaluation of propolis. A comparative study on radical scavenging effects of propolis and vespae nidus. J. Trad. Med. 12:45-53.
Mileson BE, Chambers JE, Chen WL, Dettbran W, Ehrich M (1998). Common mechanism of toxicity:a case study of organophosphorus pesticides. Toxicol. Sci. 41:8-20.
Misra HP, Fridovich I (1972). The role of superoxide anion in the autoxidation of epinephrine and a simple assay for superoxide dismutase. J. Biol. Chem. 247:3170-3175.
Newairy AA, Salama AF, Hussien HM, Yousef MI (2009). Aluminium induced alterations in biochemical parameters and lipid peroxidation of male rats: protective role of propolis. Food Chem. Toxicol. 47:1093-1098.
Nieva Moreno M I, Isla M I, Sampietro AR, Vattuone MA (2000). Comparison of the free radical - scavenging activity of propolis from several regions of Argentine. J. Ethnopharmacol. 71:109-114.
Nirala S K, Bhadauria M (2008). Propolis reverses acetaminophen induced acute hepatorenal alterations: a biochemical and histopathological approach. Arch. Pharmacol. Res. 31:451-461.
Norenberg M (1994). Astrocyte responses to CNS injury. J. Neuropathol. Exp. Neurol. 53:213-220.
Peeples ES, Schopfer L M, Duysen EG, Spaulding R, Voelker T, Thompson CM (2009). Original paper Slovak. J. Anim. Sci. 42 (3):111-117.
Ranjbar A, Pasalar P, Abdollahi M (2002). Induction of oxidative stress and acetyl cholin esterase inhibition in organophosphorus pesticide manufacturing workers. Hum. Exp. Toxicol. 21(4):179-182.
Rao JV (2006). Biochemical alterations in euryhaline fish, Oreochromis mossambicus exposed to sub-lethal concentrations of an organophosphorus insecticide, monocrotophos. Chemosphere 65(10):1814-1820.
SAS (Statistical Analysis System) (1986). SAS User's Guide:Statistics, Version 5, SAS Inst., Inc., Cary, NC, USA.
Saxena R, Garg P (2010). Vitamin E provides protection against in vitro oxidative stress due to pesticide (Chlorpyrifos and Endosulfan) in goat RBC. GERF Bull. Biosci. 1(1):1-6.
Shakoori AR, Ali SS, Saleem MA (1988). Effects of six months feeding of cypermethrin on the blood and liver of albino rats. J. Biochem. Toxicol. 3:59-72.
Sharma Y, Bashir S, Irshad M, Gupta SD, Dogra TD (2005). Effects of acute dimethoate administration on antioxidant status of liver and brain of experimental rats. Toxicol. 206:49-57.
Slotkin TA, Olivier CA, Seidler FJ (2005). Critical periods for the role of oxidative stress in the developmental neurotoxicity of chlorpyrifos and terbutaline, alone or in combination. Brain Res. Dev. Brain Res. 157:172-80.
Song YS, Park EH, Hur GM (2002). Caffeic acid phenethyl ester inhibits nitric oxide synthase gene expression and enzyme activity. Cancer Letters. 175(1):53-61.
Steel RGD, Torrie JH (1981). Principle and procedure of statistics. A biometrical approach, second ed. Mc Gvaus-Hill Book Company, New York, USA.
Stein EA (1987). Lipids and lipoproteins and apolipoproteins, in: N.W. Tietz (ed.), Fundamentals of Clinical Chemistry, 3rd ed., W.B. Saunders Company, Philadelphia, USA.
Sud'ina GF, Mirzoeva OK, Puskareva MA, Korshunova GA,Sumbatyan NV, Varfolomeev SD (1993). Caffeic acid phenethyl ester as a lipooxygenase inhibitor with antioxidant properties. FEBS Lett. 329:21.
Suleiman FA, Dayo OA, Mufta S, Abdul Ganiyu G, Olushola OO, Joseph OA (2010). Chlorpyrifos-Induced Clinical, Hematological and Biochemical Changes in Swiss Albino Mice- Mitigating effect by co-

administration of vitamins C and E. Life Sci. J. 7(3):37- 44.
Usia T, Banskota AH, Tezuka Y, Midorikawa K, Matsushige K, Kadota S (2002). Constituents of Chinese propolis and their antiproliferative activities. J. Natural Products. 65(5):673-676
Valente MJ, Baltazar AF, Henrque R, Estevinho L, Carvalho M (2011). Biological activities of Portuguese Propolis:Protection against free radical-induced erythrocyte damage and inhibition of human renal cancer cell growth in vitro. Food Chem. Toxicol. 49:86-92.
Verma RS, Mehta A, Srivastava N (2007). In vivo Chlorpyrifos induced oxidative stress:attenuation by antioxidant vitamins. Pesticide Biochem. Physiol. 88:191-196.
Warnick GR, Benderson JM, Albers JJ (1983). Selected methods Clin. Chem. 10:91-99.
Xu JB, Yuan XF, Lang PZ (1997). Determination of catalase activity and catalase inhibition by ultraviolet spectrophotometry. Chin. Environ. Chem. 16:73-76.
Yousef MI, Esmail AM, Baghdadi HH (2004a). Effect of isoflavones on reproductive performance, testosterone levels, lipid peroxidation and seminal plasma biochemistry of male rabbits. J. Environ. Sci. Health part B. 39:819-833.
Zama D, Meraihi Z, Tebibel S, Benayssa W, Benayache F, Benayache S, Vlietinck AJ (2007). Chlorpyrifos-induced oxidative stress and tissue damage in the liver, kidney, brain and fetus in pregnant rats:The protective role of the butanolic extract of Paronychia argentea L. Indian J. Pharmacol. 39:145-150.
Zurich MG, Honegger P, Schilter B, Costa LG, Monnet-Tschudi F (2004). Involvement of glial cells in the neurotoxicity of parathion and chlorpyrifos. Toxicol. Appl. Pharmacol. 201(2):97-104.

25

Purification and properties of N-acetyl-β-D-glucosaminidase from bovine testicle

Xiao-Hong Huang, Yan-Dong Jin, Yi-Fan Huang, Wei-Chao Han and Wen Zhang

Institute of Animal Health, College of Animal Science, Fujian Agriculture and Forestry University, Fuzhou 350002, China.

N-Acetyl-β-D-glucosaminidase (NAGase) from bovine testicle was purified by ammonium sulfate fractionation followed by diethylaminoethyl (DEAE)-cellulose (DEAE-32) and Sephacryl S-300 chromatography. The enzyme was purified to homogeneity as analyzed by polyacrylamide gel electrophoresis (PAGE) and isoelectric focusing gel electrophoresis (IFGE). The specific activity of the purified enzyme was 658.21 U/mg. The enzyme was a single subunit with molecular weight of 68.3 kDa and contained 3.03% sugar. The pI value was calculated to be 5.54 using IFGE. The optimal pH and temperature of the enzyme for hydrolysis of *p*-Nitrophenyl-N-acetyl-*β*-D-glucosaminide (pNP-NAG) were found to be pH 5.6 and 50°C, respectively. The kinetics results showed that the enzyme hydrolyzed pNP-NAG following Michaelis-Menten with K_m of 0.71 mM and V_m of 16.72 μM/min at pH 5.6 and 37°C. The enzyme was stable at pH values ranging from 2 to 6.5 and at temperatures below 60°C. The activation energy was determined to be 64.19 kJ/mol.

Key words: Bovine testicle, N-Acetyl-β-D-glucosaminidase, purification, characteristic.

INTRODUCTION

N-Acetyl-*β*-D-glucosaminidase (NAGase, EC 3.2.1.52), an intracellular lysosomal hydrolase, is widely distributed in nature (Liu et al., 2009; Keyhani et al., 1996). The enzyme not only is involved in glycosylation of glycol-protein and glycolipid biosynthesis but also plays an important role in oligosaccharide and polysaccharide chain formation and degradation. Therefore, an in-depth study of NAGase can provide a basis for life science research. Research reports have shown that NAGase has been investigated more in microbial systems and has been the focus of a large number of gene cloning and expression studies (Mamarabadi et al., 2009). NAGase has been purified from the human spleen (Robinson et al., 1968), bovine liver (Bernard, 1964) and spleen (Vehpoorte et al., 1972), and cortical granules of Xenopus eggs (Prody et al., 1985). The enzymatic characteristic of NAGase, which was obtained from rat spleen cytoplasm, has been described in detail (Dong and Hart, 1994). Using muscovy duck testicles as a source, NAGase was extracted, and some of its enzymatic properties and the effect of metal ions on enzyme activity have been determined (Wang et al., 2007). Moreover, high NAGase activity is present in rabbit semen (Farooqui and Srivastava, 1980), human sperm (Perez et al., 2007), and ascidian eggs (Koyanagi and Honegger, 2003). The activity of NAGase changes regularly in female buffalo saliva in the telocinesia of estrus, which can be used to determine ovulation time and the optimal time for artificial insemination to improve calving rates (Feng et al., 1997). Mouse egg and sperm lost combination capacity after the eggs were treated with NAGase (Miller et al., 1993). Additionally, NAGase can help the sperm penetrate the eggs' zona pellucida (Shur, 1982). Therefore, NAGase is relevant to the reproductive function of animals. However, there are very few studies of animal NAGase at the molecular level. In the present research, we purified NAGase from bovine testicle and studied the enzymatic characteristics to provide further details on the physiological functions of NAGase in bovines.

MATERIALS AND METHODS

Bovine testicle was purchased from Fuzhou in China. p-Nitrophenyl (pNP) was made in England. p-Nitrophenyl-N-acetyl-β-D-glucosaminide (pNP-NAG) was purchased from the Biochemistry Lab of Shanghai Medicine Industry Academy (Shanghai China). Diethylaminoethyl (DEAE)-cellulose (DE-32) was purchased from Whatman. Sephadex G-200 was purchased from Pharmacia. Sephacryl S-300, Con A-Sepharose and standard marker proteins were purchased from Amersham. All other reagents were local products of analytical grade. The water used was re-distillated and ion-free.

Enzyme activity and protein concentration assays

Enzyme activity was determined at 37°C by following the increase in absorbance at 405 nm accompanying the hydrolysis of pNP-NAG as described by Xie et al. (2004). A 10 μl aliquot of the enzyme solution was added to the reaction media (2 ml) containing 0.5 mM pNP-NAG in 0.15 M NaAc-HAc buffer (pH 5.2). After incubation for 10 min at 37°C, 2 ml of 0.5 M NaOH was added to the reaction mixture to stop the reaction. Absorption was recorded using a Beckman UV-650 spectrophotometer. The molar absorption coefficient of the product pNP is 1.73×10^4 $M^{-1}cm^{-1}$ (Xie et al., 2004). One unit (U) of enzymatic activity was defined as the amount of enzyme catalyzing the formation of 1 μM pNP-NAG at the conditions denoted above. The protein concentration was measured by the method of Follin phenol with bovine serum albumin as the standard or by measuring the absorbance at 280 nm.

Enzyme purification and determination of homogeneity

Unless otherwise noted, all operations were performed at 4°C. One hundred grams of frozen bovine testicle tissue was homogenized in 250 ml of 50 mM Tris-HCl buffer (pH 7.5) containing 0.2 M NaCl and incubated for 2 h. The homogenate was centrifuged at 10, 000 *g* for 30 min. The supernatant was fractionated with ammonium sulfate. The fraction between 30 and 50% saturation ammonium sulfate was collected by centrifugation, dissolved in a small volume of 50 mM Tris-HCl buffer (pH 7.5), and then dialyzed against this buffer until sulfate could not be detected.

The crude preparation was concentrated by ultrafiltration and loaded onto a DEAE-cellulose column (2.5 × 20 cm) pre-equilibrated with the same buffer. The elution was carried out using a linear gradient of 0 to 1.5 M NaCl at a flow rate of 20 ml/h, and the eluate was collected at 2.5 ml per tube. The fractions with enzyme activity were pooled. The collected enzyme solution was further chromatographed on a Sephacryl S-300 gel filtration column (2.5 × 60 cm) pre-equilibrated with the same buffer containing 0.2 M NaCl. The elution was carried out at a flow rate of 20 ml/h, and eluate was collected at 2 ml per tube. The fractions with enzyme activity were pooled. The enzyme homogeneity was assessed using the methods of polyacrylamide gel electrophoresis (PAGE) and sodium dodecyl sulfate polyacrylamide gel electrophoresis (SDS-PAGE).

Determination of enzyme molecular weight and subunit number

The molecular weight was determined using gel filtration and SDS-PAGE. The gel filtration was performed on a Sephadex G-200 (2.5 × 70 cm) column with 0.01 M Tris-HCl buffer, pH 7.5, containing 0.2 M NaCl. Pepsin (PEP, 35 kDa), ovalbumin (OA, 45 kDa), bovine serum albumin (BSA, 68 kDa), rabbit phosphorylase B (RPR, 97.2 kDa), and bovine gamma globulin (BGG, 140 kDa) were used as protein standards. SDS-PAGE was performed on a discontinuous horizontal thick-layer with a stacking gel of 3% acrylamide in Tris-HCl buffer (pH 6.7) at 10 mA current and a separating gel of 12% acrylamide in Tris-HCl buffer (pH 8.9) 0.1% SDS at 20 mA current. Both the NAGase sample and marker were combined with one-fold solubilizing mixture containing 1% SDS, 1% 2-mercaptoethanol, 40% sucrose, 0.02% bromophenol blue (tracking dye), and 0.01 M Tris-HCl buffer (pH 8.0). The mixtures were heated in a boiling water bath for 3 min. After electrophoresis, the gel was stained with 0.05% Coomassie brilliant blue R-250.

Determination of isoelectric point (pI)

The pI value of the enzyme was determined using isoelectric focusing electrophoresis. The electrophoresis was run on a continuous disc (5 × 100 mm) with 7% acrylamide containing ampholyte (pH 3 to 10) at 150 V for 4 h. After electrophoresis, the gel column was immersed into a 10% trichloroacetic acid (TCA) solution, and the enzyme protein band was visualized. The other gel column was sliced into 5 mm pieces. Each gel piece was immersed into 1 ml distilled water for 8 h, and the pH value was measured.

Glycoprotein assay

Concanavalin A (con A)-sepharose column was used to assess whether the enzyme was a glycoprotein. The sulfate-phenol methods introduced by Xu et al. (2005) were used to test the saccharide contents of the purified enzyme preparation.

Determination of the kinetic parameters of NAGase

The enzyme hydrolysis of pNP-NAG was studied using various concentrations of pNP-NAG. The Lineweaver-Burk plot was used to determine the kinetic parameters of the K_m and V_m values.

Optimal pH and stability assays

The optimal pH was determined by measuring the enzyme activities as described above at different pH values at 37°C. The pH-stability of the enzyme was monitored by incubating the enzyme in buffers with different pH values for 20 h at 4°C. A 10 μl aliquot of the mixture was used for the activity assay.

Optimal temperature and thermal stability assays

The optimal temperature of the enzyme was determined by measuring the activity as described above at various temperatures at pH 5.6. The thermal stability of the enzyme was monitored by incubating the enzyme at different temperatures for 30 min. A 10 μl aliquot of the mixture was used for the activity assay as described

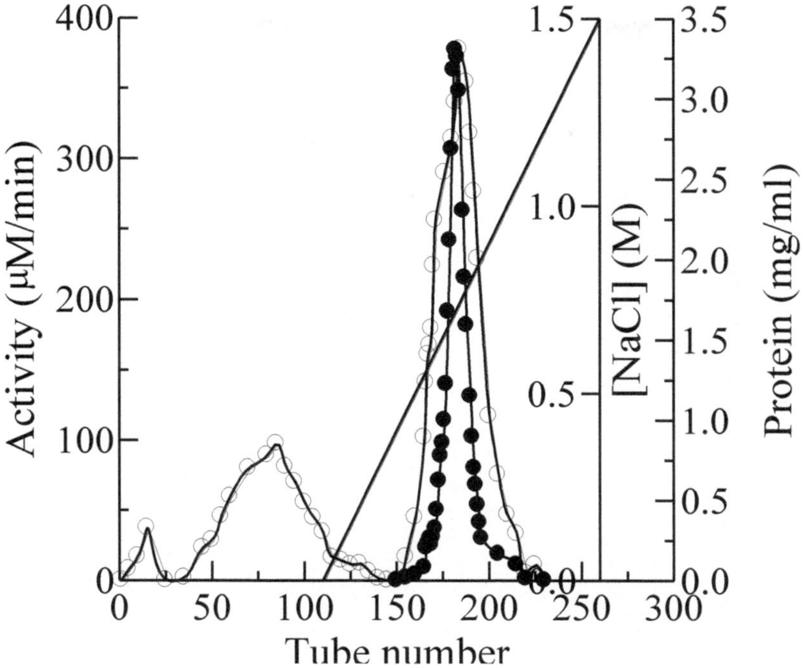

Figure 1. Column chromatography of bovine testicle NAGase on DEAE-cellulose (DE-32). (-○-) Protein concentration; (-●-) activity of NAGase. NaCl concentration in a linear gradient: 0 to 1.5 M.

above.

RUSULTS

Purification of NAGase from bovine testicle

NAGase (EC3.2.1.52) was first prepared from bovine testicle according to the method of Xie et al. (2004) until the ammonium sulfate fractionation step; the crude preparation was chromatographed by ion-exchange with DEAE-cellulose (DE-32). The elution profile is shown in Figure 1. Most of the enzymes eluted at above 0.5 M of NaCl. The activity peak (from tubes 175 to 190) overlapped with the protein peak. The active fractions were pooled and chromatographed by gel filtration with a Sephacryl S-300 column for further purification. The active fractions were pooled. The final preparation was homogeneous as determined by PAGE (Figure 2A) and SDS-PAGE (Figure 2B). The specific activity of the purified enzyme was determined to be 658 U/mg.

Determination of enzyme molecular weight and subunit number

The molecular weight of bovine testicle NAGase was determined by gel filtration on a Sephadex G-200 column, and the result is shown in Figure 3. The molecular weight of the enzyme was estimated to be 68.5 kDa. In addition, the enzyme showed a single band on SDS-PAGE (Figure 2B). The molecular mass of the enzyme subunit was estimated to be 68.3 kDa. From these results, we concluded that the enzyme is composed of a single polypeptide chain.

Determination of the isoelectric point of the enzyme

The isoelectric point of the enzyme was determined by IFGE. The result shows that the pI value of the enzyme was approximately 5.54, which is close to that of porcine semen acidic NAGase (Huang et al., 2009), and same as papaya NAGase (Chen et al., 2011). However, this value was different from that of the *Bombyx mori* alimentary canal (pI 4.86) (Koga et al., 1986) and *Penaeus vannamei* (pI 4.8) NAGases.

Glycoprotein assay

A concanavalin A (Con A) column was used to determine whether the enzyme is a glycoprotein. In this study, we found that the purified NAGase is bound to the Con A column. These results show that this enzyme contains oligosaccharides and is a glycoprotein. Using the sulfate-phenol method introduced by Xu et al. (2005), we tested the saccharide contents of the purified enzyme. The results show that bovine testicle NAGase contained approximately 3.03% saccharides.

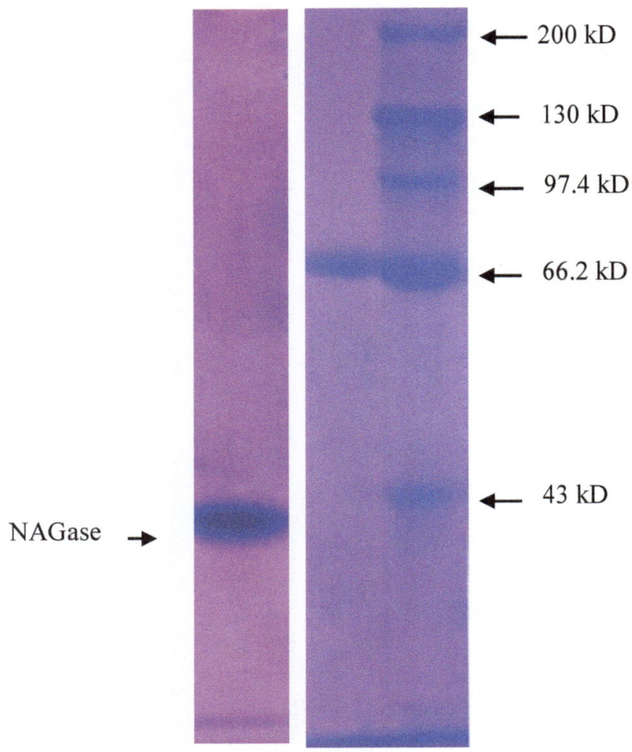

Figure 2. PAGE (A) and SDS-PAGE (B) of purified NAGase.

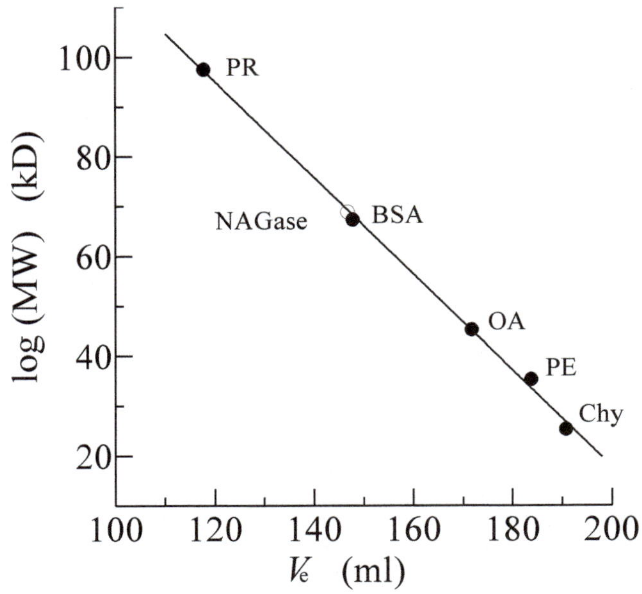

Figure 3. Measurement of the molecular weight of the enzyme on a Sephadex G-200 gel filtration. •, Standard proteins; ○, purified NAGase.

Determination of the kinetic parameters of NAGase

The kinetic behavior of the enzyme during hydrolysis of pNP-NAG was studied. Under the conditions employed in the present investigation, the hydrolysis of pNP-NAG by the enzyme followed Michaelis-Menten kinetics (Figure 4). The kinetic parameters for the enzyme obtained from a Lineweaver-Burk plot showed that K_m and V_m were 0.71 mM and 16.72 μM/min, respectively.

Effects of pH on the enzyme activity and stability

The effect of pH on the enzyme activity for the hydrolysis of pNP-NAG was determined at 37°C. The results demonstrated that the optimal catalysis reaction occurred at pH 5.6. The enzyme activity decreased rapidly at pH values either lower or higher than the optimal pH (Figure 5a). In addition, when the enzyme was incubated at different pH values for 20 h at 4°C (Figure 5b), the enzyme was stable in the pH range between 2 and 6.5. The optimal pH of bovine testicle NAGase was at pH 5.6, and this enzyme was stable at pH 3.0 to 6.0. The optimal pH of hetian chicken testicle NAGase was 5.6, and the enzyme was active at pH 3.0 to 9.2. The optimal pH of avian broilers chicken testicle NAGase was pH 5.7, and the enzyme maintained its activity at pH 4.0 to 9.2 (Jin et al., 2009). The optimal pH of muscovy duck testicle NAGase was 5.6, and the enzyme maintained its activity at pH 3.0 to 8.0 (Wang et al., 2007). The results described above indicate that the optimal pH of NAGase from the same tissue of different varieties is similar, although the tolerance to acidic and basic conditions is different. The optimal pH of human liver NAGase was 4.5, and it was stable at pH 6.5 to 8.5 (Sasaki et al., 1991). The optimal pH of crustacean NAGase was at pH 5.0 to 6.0 (Zhang et al., 2006; Huang et al., 2005).

Effects of temperature on the enzyme activity and thermal stability

The enzyme activities were measured at various temperatures to determine the effect of temperature on the enzyme activity. The results indicate that the optimal temperature was 45°C, while the activity decreased rapidly at temperatures higher or lower than the optimal temperature (Figure 6a). The enzyme was stable below 60°C and unstable above 60°C after incubation for 30 min (Figure 6a). The optimal temperature of bovine testicle NAGase was 45°C. The enzyme maintained activity at a temperature of 60°C when preincubated for 30 min. The results show that bovine testicle NAGase can be stable under the normal temperature of a bull, but it did not have full activity. It has been reported that the optimal temperature of boar semen alkaline NAGase is 50°C and that the enzyme activity remains stable at 10 to 45°C (Huang et al., 2009). The optimal temperature of hetian chicken testicle NAGase is 55°C, and it is stable at 10 to 60°C (Jin et al., 2009). The optimal temperature of papaya NAGase is 50°C, and it is stable at 30 to 45°C; however, its enzyme activity rapidly decreases at temperatures higher than 50°C (Chen et al., 2011).

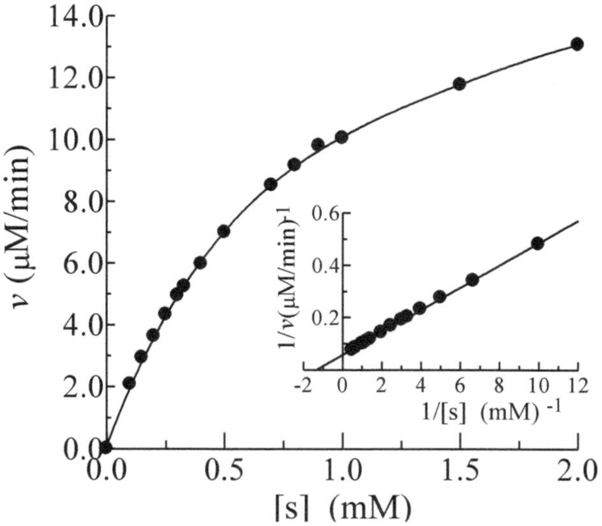

Figure 4. Michaelis-Menten kinetics for the hydrolysis of pNP-NAG by bovine testicle NAGase. Conditions were 0.15 M NaAc-HAc buffer (pH 5.2) and different concentrations of pNP-NAG at 37°C. The enzyme concentration was 15.6 μg/ml. The inset is the Lineweaver-Burk plot used to determine the kinetic parameters of K_m and V_m values.

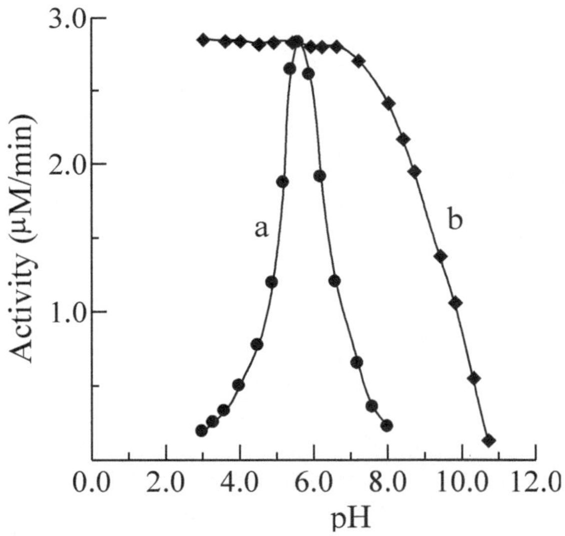

Figure 5. Effects of pH on bovine testicle NAGase activity. (a) Optimal pH: enzyme activity was assayed in 0.15 M NaAc-HAc buffer (1) or in 0.15 M Na_2HPO_4-NaH_2PO_4 buffer (2), the concentration of substrate was 0.5 mM. (b) pH stability: the enzyme (15.6 μg/ml) was incubated at different pH buffers of Gly-HCl (pH 2.5 to 4.0), NaAc-HAc (pH 4.0 to 5.8), Na_2HPO_4-NaH_2PO_4 (pH 5.8 to 8.0) and Gly-NaOH (pH 8.0 to 10.6) for 20 h at 4°C. A 10-μl aliquot of the mixture was used for the activity assay.

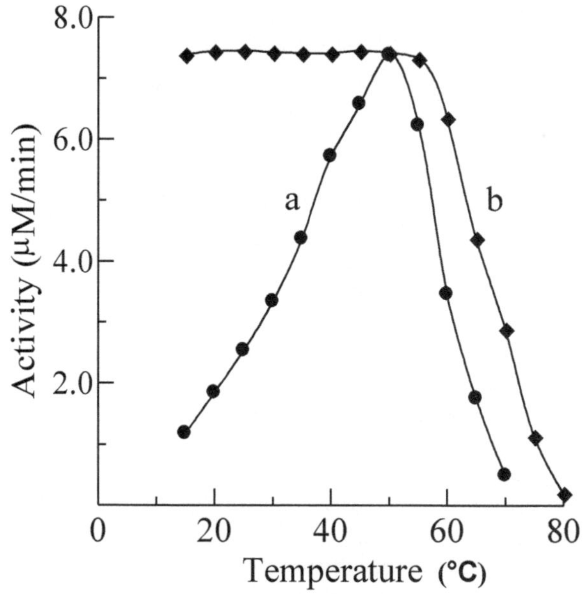

Figure 6. Effects of temperature on bovine testicle NAGase activity. (a) Optimal temperature, (b) thermal stability of the enzyme. The enzyme (15.6 μg/ml) was incubated in 0.15 M NaAc-HAc buffer (pH 5.2) at different temperature for 30 min. A 10 μl-aliquot of the mixture was used for the activity assay at 37°C.

Enzyme activation energy

The activation energy (E_a), which is the energy produced when reactants (initial state) change into intermediate products (transition state), can reflect the degree of an enzymatic reaction. The E_a was assayed by measuring the maximum velocity of the enzymatic hydrolysis of pNP-NAG, and the value of E_a was calculated from the plot of log V_m versus 1/T to be 64.19 kJ/mol, which is different for a variety of organisms. It has been reported that the E_a values of NAGase from pig semen alkaline (Huang et al., 2008), pig semen acid (Huang et al., 2009), muscovy duck testicle (Wang et al., 2007), hetian chicken testicle and avian broilers testicle (Jin et al., 2009) are 88.73, 41.70, 69.05, 85.74 and 73.47 kJ/mol, respectively. The degree of catalytic reaction was as follows: pig semen alkaline > hetian chicken testicle > avian broilers testicle > muscovy duck testicle > bovine bull testicle > pig semen acid.

DISCUSSION

N-Acetyl-*β*-D-glucosaminidase, belonging to the glycolside hydrolase family, has the function of degrading N-Acetyl-*β*-D-glucosamine, which is connected with *β*-1,4-glycosidic non-restored glycoconjugates in organisms. We found that bovine testicle has a high activity of NAGase, which is directly related to the reproductive activity of male cattle.

Therefore, it is of significant value in theory and application to study the enzyme. Bovine testicle NAGase is a sugar protein composed of a single polypeptide chain that weighs 68.3 kD. Most bacterial NAGases are monomeric enzymes with molecular weights greater than 150 kD. However, most of the fungal NAGases are homodidimers with molecular weights between 110 and 150 kD. Most animal NAGases are less than 150 kD, such as the krill epidermal NAGase, which is a single subunit enzyme with a molecular weight of 120 kD, and the krill visceral NAGase is a 150 kD single subunit enzyme (Peters et al., 1998). *Scylla serrata* NAGase is composed of two homotypic subunits, and its molecular weight is 132 kD (Zhang et al., 2006). *Bombyx mori* digestive tract NAGase is composed of two special subunits with molecular weights of 67.5 and 57.5 kD (Koga et al., 1986). *Manduca sexta* NAGase has a molecular weight of 62 kD (Zen et al., 1996). *Helicoverpa armigera* pupal NAGase has a molecular weight of 89.3 kD (Huang et al., 2005). It was recently found that there are three types of lysosomal NAGases (HexA, HexB, and HexS). All three are dimers composed of α- and β-subunits. HexA is a heterotypic dimer, while HexB is homodimer (Mark et al., 2003). The molecular weight of bovine liver NAGase is 50 kD (Bernard, 1964). NAGase A and B are purified from bovine spleen, with NAGase B having only one subunit with a molecular weight of 140 kD (Vehpoorte, 1972). Pig semen alkaline NAGase has only one subunit, and its molecular weight is 58.03 kD (Huang et al., 2008). NAGase is a glycoprotein in bovine testicle. It has been shown that *Antarctic krill* NAGase has two types; one is a glycoprotein and the other is not (Peters et al., 1998).

NAGases have different biological functions in different organisms. The lysosomes of vertebrate contain NAGase, which participates in the hydrolysis of glycoprotein, ganglioside, and osamine. It is also related to the differentiation, recognition, and adhesion of somatic and germ cells (Slamova et al., 2010). Bacterial and insect NAGases are mainly involved in chitin degradation (Liu et al., 2009; Keyhani et al., 1996; Mamarabadi et al., 2009). Crustacean NAGase is related to chitin metabolism, and it participates in the molting process (Xie et al., 2004). NAGase in plants is related to glycan metabolism and defense functions (Chen et al., 2011). NAGase in mammals is mainly related to the synthesis and degradation of glycan in glycoproteins and is involved in the physiological process of sperm and egg binding (Miller et al., 1993; Shur and Hall, 1982). Through the study of NAGase characteristics from bovine testicle, we provided a theoretical basis to further study the enzyme's physiological function in bull reproduction. This study also provided a theoretical foundation and technical reference for related enzyme research on mammalian fertilization processes.

ACKNOWLEDGEMENT

The present investigation was supported by grant 2010J01059 from the Natural Science Foundation of the Fujian Province.

REFERENCES

Bernard W (1964). Oligosaccharase activity of β-N-Acetyl-D-glucosaminidase of beef liver. Biol. Chem. 239 (1): 59-63.

Chen LC, Chung YC, Chang YM, Chang CT (2011). Characterisation of a β-N-acetylhexosaminidase from a commercial papaya latex preparation. Food Chem. 124: 1404-1410.

Dong DLY, Hart GW (1994). Purification and characterization of an O-GlcNAc Selective N-Acetyl-β-glucosaminidase from rat spleen cytosol. J. Biol. Chem. 269(30): 19321-19330.

Farooqui AA, Srivastava PN(1980). Isolation of N-acetylhexosaminidase from rabbit semen and its role in fertilization. Biochem. J. 191: 827.

Feng JC, Yuan HJ, Pan HY (1997). Relationship between salivary activity of N-acetyl-β-D-glucosaminidase and ovulation time in swamp buffalo cows. Huazhong (Central China) Agric Univ. 16 (4): 357-360 (In Chinese).

Huang XH, Chen QX, Wan J, Guan X (2005). Isolation, purification and partial characterization of the β-N-Acetyl-D-glucosaminidase from the pupae of *Helicoverpa armigera*. Acta Entomologica Sinica. 48 (4): 498-502 (in chinese).

Huang XH, Luo ZB, Li MW, Huang YF (2008). Study on purification and properties of the alkaline N-Acrtyl-*β*-D-glucosaminidase from procine semen. Scientia Agric. Sinica. 41 (9): 2843-2849.

Huang XH, Luo ZB, Huang YF, Chen XY, Huang JP (2009). Purification and properties of the acidic N-acetyl-β-D-glucosaminidase from porcine semen. Chin. J. Appl. Environ. Biol. 15 (6): 819-823 (In Chinese).

Jin YD, Huang XH, Li WJ (2009). Comparison of the properties of β-N-Acetyl-D-glucosamibidase from Hetian Chicken and Avian broiler. J. Fujian Agric. Forest. Univ. (Nat Sci). 38 (3): 265-270 (In Chinese).

Keyhani NO, Roseman S (1996). The chitin catabolic cascade in the marinebacterium Vibrio furnissii. identification, molecular cloning, and characterization of a N,N'-diacetylchitobiose phosphorylase. J. Biol. Chem. 271: 33425-33432.

Koga D, Nakashima M, Matsukura T, Kimura S, Ide A (1986). Purifications and properties of β-N-acetyl-D-glucosaminidase from alimentary canal of the silkworm, *Bombyx mori*. Agric. Biol. Chem. 50 (9): 2357-2368.

Koyanagi R, Honegger TG (2003). Molecular cloning and sequence analysis of an ascidian egg β-N-acetylhexosaminidase with a potential role in fertilization. Develop. Growth Differ. 45 (3): 209-218.

Liu T, Liu F, Yang Q, Yang J (2009). Expression, purification and characterization of the chitinolytic β-N-acetyl-D-hexosaminidase from the insect Ostrinia furnacalis. Protein Expression and Purif. 68 (1): 99-103.

Mamarabadi M, Jensen DF, Lubeck M (2009). An N-acetyl-β-D-glucosaminidase gene, cr-nag-1, from the biocontrol agent clonostachys rosea is up-regulated in antagonistic interactions with Fusarium culmorum. Mycol. Res. 113: 33-43.

Mark BL, Mahuran DJ, Cherney MM, Zhao DL, Knapp S, James MN (2003). Crystal structure of human β-hexosaminidase B: understanding the molecular basis of Sandhoff and tay-Sachs disease. J. Mol. Biol. 327 (5): 1093-1109.

Miller DJ, Gong XH, Shur BD (1993). Sperm require β-N-acetylglucosaminidase to penetrate through the egg zona pellucida. Development. 118: 1279-1289.

Perez MS, Menendez HR, Zitta KS, Brandelli A, Miranda PV (2007). Characterization of human sperm N-acetylglucosaminidase. International J. Androl. 31: 315-324.

Peters G, Saborowski R, Mentlein R (1998). Isoforms of an N-acetyl-β-D-glucosaminidase from the Antarctic krill, Euphausia superba: purification and antibody production. Comp Biochem Physiol. 120B: 743-751.

Prody GA, Greve LC, Hedrick JL (1985). Purification and characterization of an N-Acetyl-β-D-Glucosaminidase from cortical granules of Xenopus laevis eggs. J. exp zool. 235: 335-340.

Robinson D, Stirling JL (1968). N-Acetyl-β-Glucosaminidases in human spleen. Biochem. J. 107: 321-327.

Vehpoorte J. A (1972). Purification of two β-N-Acetyl-D-glucosaminidase from beef spleen. Biol. Chem. 247 (15): 4781-4793.

Sasaki T, Sukegawa K, Masue M, Fukuda S, Tomatsu S, Orii T (1991). Purification and partial characterization of α-N-acetylglucosaminidase from human liver. J. Biochem. 110 (5): 842-846.

Shur BD, Hall NG (1982). A role for mouse sperm surface galactosyl transferases in sperm binding to the egg zona pellucida. J. Cell Biol. 95: 574-579.

Slamova K, Bojarova, P, Petraskova L, Kren V (2010). β-N-acetyl hexosaminidase: what's in a name ...?. Biotechnol Adv. 28 (6): 682-693.

Wang CK, Huang YF, Huang XH, Chen HH (2007). Priliminary purification and properties of the β-N-Acetyl-D-glucosaminidase from Cairina moschata. Acta Veterinaria et Zootechnica Sinica. 38 (6): 552-557 (In Chinese).

Xie XL, Chen QX, Lin JC (2004). Wang, Y. Purification and some properties of b-N-acetyl-D-glucosaminidase from prawn (*Penaeus vannamei*). Mar. Biol. 146: 143-148.

Xu GY, Yan J, Guo XQ, Liu W, Li XG, Gou XJ (2004). The betterment and apply of phenol-sulphate acid method. Sci. Food. 26(8):342-346. (In Chinese).

Zen KC, Choi HK, Krishnamachary N, Muthukrishnan S, Kramer KJ (1996). Cloning, expression, and hormonal regulation of an insect β-N-acetylglucosaminidase gene. Insect Biochem. Molbiol. 26 (5): 435 - 444.

Zhang JP, Chen QX, Wang Q, Xie JJ (2006). Purification and Some Properties of β-N-Acetyl-D-glucosaminidase from Viscera of Green Crab (*Scylla serrata*). Biochemistry (Moscow). 71 (Suppl. 1): S55-S59.

Relationship between soil contents and plasma levels of selenium, chromium and manganese in healthy adult Nigerians

Kolawole Sunday E.[1]* and Obueh Henrietta O.[2]

[1]National Open University of Nigeria, Benin Study Centre, Benin City, Nigeria.
[2]Department of Biology, College of Education, P.M.B 1144, Benin City, Nigeria.

The soil contents of trace elements selenium, chromium and manganese were measured to determine their impact on the plasma levels of 160 healthy adult Nigerians in five different experimental locations in Cross River and Akwa Ibom States, South - South Nigeria. The mean (±SD) soil selenium, chromium and manganese concentrations were 2.86 ± 3.97, 15.93 ± 3.35 and 90.33± 42.65 mg/kg, respectively. Manganese soil concentration was the highest and selenium soil concentration was the lowest in all the locations. The mean (±SD) plasma selenium, chromium and manganese concentrations were 0.006, 0.111 ± 0.031 and 0.028 ± 0.024 mg /1, respectively. There were no significant differences between the concentrations of the trace elements in the soil and the plasma levels as determined by t-test. Plasma levels of chromium and manganese according to the study were adequate but that of selenium was low.

Key words: Healthy adult Nigerians, trace elements, plasma level, soil content.

INTRODUCTION

Trace elements like selenium, chromium and manganese are inorganic nutrients that originate in the earth and cannot be made in the body. They play important roles in the various bodily functions and are necessary to sustain life and maintain optimal health, and thus, are essential nutrients (Tourmma, 1995). Most of the minerals in the human diet come directly from plants and water, or indirectly from animal foods. However, the mineral content of water and plant foods varies according to different geographical zones because of variations in the mineral content of soil from region to region (Combs, 2001). For many people with access to an adequate energy intake, an extensive freedom of choice exists in selection of food. However, the existence of widespread poverty in Nigeria precludes the opportunity to consume adequate energy let alone a diet balanced with micronutrients. Micronutrient deficiency is caused by inadequate access to micronutrient-rich food, high burden of infection and parasites, and detrimental feeding and dietary practices (Tuormaa, 1995). Micronutrients deficiency adversely affects the health and function of individuals and the economic and social development of communities and nations. Micronutrients usually function as essential cofactors in the numerous enzymes catalyzed reactions. Serious disease conditions due to the impairment of metabolic functions can occur in the absence of micronutrients. For example, obesity and diet-related chronic diseases such as diabetes and cardiovascular diseases have been reported to be on the increase in many developing countries affected by rapid industrialization and urbanization (Eckhardt, 2006). Selenium may have an anti-carcinogenic effect. Observational studies indicate that death from cancer, including lung, colorectal, and prostrate cancers is lower among people with higher blood levels or intake of selenium (Combs et al., 1997).

The role of selenium for the proper functioning of the thyroid gland has also been documented. Selenium status has been suggested to be a significant predictor of survival for persons infected with HIV (Fairweather-Tait, 1997). Chromium has been reported to play a role in a number of metabolic systems. The role of chromium in insulin function, diabetes mellitus, metabolic syndrome, polycystic ovarian syndrome and depression has been revealed by current research (Vincent, 2000). Manganese, a nutritionally essential antioxidant nutrient is reported to play an important role in the breakdown of amino acids and the production of energy. Manganese is a component of manganese superoxide dismutase (MnSoD). The role of this antioxidant enzyme in the dismutation of superoxide radicals has also been documented. There are also reports associating low dietary levels of manganese with abnormal bone development and abnormal wound healing in a number of animal species (Keen and Zindenberg - Cherr, 1996).

Many minerals and trace elements are inefficiently and variably absorbed from the diet. Determination of adequacy of dietary intake is complicated by the numerous interactions which exist among trace elements among these and other dietary components and variations in the bio-availability of trace elements in different foods. Interactions of trace metals are potent and could be either positive or negative (Tuormaa, 1995). For the healthy population, consumption of a nutritionally adequate diet balanced with regard to major nutrients generally is relied upon to provide an adequate and safe intake of trace element. Selenium, chromium and manganese unlike macronutrients are ingested in very small amounts (microgrammes and milligrams per day). Despite their relative paucity in the diet and body, they perform important roles in regulating whole body metabolism. This study was designed to investigate the relationship between soil contents and the plasma levels of selenium, chromium and manganese in healthy adult Nigerians. The study also determines the effect of age and sex of the subjects on their selenium, chromium and manganese plasma levels. Few studies have been reported on the plasma levels of selenium, chromium and manganese in Nigeria, in spite of the accumulating evidence that these trace elements are of immense benefit in the health and wellbeing of man.

MATERIALS AND METHODS

Sample collection

Soil samples were collected from Uyo, Calabar, Oron, Ikom and Ikot Okpora in Cross River State and Akwa Ibom State (Figures 1, 2 and 3), which were the five different experimental location for the determination of selenium, chromium and manganese. Blood was drawn in the morning into heparinised disposable plastic syringes from 160 healthy adult Nigerians from rural and urban communities of Akwa Ibom and Cross River States in the South-south Zone of Nigeria. Plasma was immediately separated out by centrifuging the blood samples at 3000 rpm for 10 min. The samples were stored at -20°C until required for analysis. The population was made up of 24 elderly subjects (16 men and eight women) aged between 51 to 90 years and 126 younger adults made up of 42 men and 54 women aged 18 to 50 years. All the subjects ate ordinary foods and diets purchased from local restaurants and markets in the five experiment areas.

Trace metals analysis

Trace metals analysis was done using a Unicam 939/959 atomic absorption spectrophotometer. The assessment of the trace metals nutritional status on the basis of plasma concentration must be used with several important caveats. Assay methods and reference standards are at an early stage. Taking all the studies available into account, it is possible to state with some confidence that at blood concentration of approximately 1.0 U mol/l, there is good correlation between plasma selenium concentration and the activity of Glutathione Peroxidase (GPx) in erythrocytes and therefore plasma selenium concentration may be taken as good index of nutritional status with respect to selenium.

Statistical analysis

The statistical analysis of the data was done with the statistical package for the social science (SPSS) software. The values obtained from this study were expressed as mean and standard deviation when compared using t-test. Pearson correlation was used to determine relationships between variables.

RESULTS

The mean values ± SD for selenium, chromium and manganese concentrations in soil samples are shown in Table 1. The mean ± SD soil selenium concentration was 2.86 ± 3.97 mg/kg and values ranged from 0.002 to 8.08 mg/kg. The mean soil chromium concentration was 15.93 ± 3.35 mg/kg and values ranged from 6.86 to 26.61 mg/kg. The mean soil manganese concentration was 90.33 ± 42.65 mg/kg and values ranged from 23.95 to 130.00 mg/kg. The concentration of manganese was the highest and selenium concentration was the lowest in the locations. The plasma selenium, chromium and manganese of the subjects categorized by age and sex are shown in Table 2. The mean ± SD plasma selenium concentration for 59 healthy subjects, females (n = 29) and males (n = 30) was 0.004 ± 0.006 mg/l. Selenium was not detected in other subjects. Statistically, there was no significant difference ($p > 0.05$) between values for males and females as determined by t-test. Since there was no significant effect of sex on selenium values obtained, data from males and females were pooled. No significant correlation between age and plasma selenium levels was obtained ($p > 0.05$). However, there was negative correlation for plasma selenium concentration in males aged 18 to 30 years ($r = -0.0757$, $p = 0.05$). There was precipitous fall in plasma selenium concentration from ages 25 to 35 years (Figure 4). There was further

Figure 1. Map showing Uyo, Calabar and Oron, Nigeria.

decline until age 65. Plasma selenium concentration rose very slightly from age 75 to 85 years. There was no significant difference between selenium values in plasma and soil as determined by t-test.

The mean ± SD plasma chromium concentration was 0.111 ± 0.03 mg/l for 73 healthy subjects. No chromium was detected in the other subjects. The mean plasma chromium concentration for females (n = 38) was 0.108 ± 0.03 mg/l and 0.114 ± 0.32 mg/l for males (n = 35). There was no significant difference between the values for males and females. No statistically significant correlation between age and plasma chromium levels was observed ($p>0.05$). Plasma chromium levels tended to be stable (0.126 mg/l) between ages 25 and 45. There was a sharp decline between ages 45 and 55 from 0.126 to 0.096 mg/l (Figure 4). Plasma chromium levels stabilized again between ages 55 and 65 (0.063 mg/l). It rose sharply thereafter and tended to be stable between the ages of 75 and 85. There were no significant correlation between chromium values in plasma and soil. The mean plasma manganese concentration was 0.028 ± 0.02 mg/l for 89 adults. No manganese was detected in other healthy subjects. The mean ± SD plasma manganese concentration for females (n = 50) was 0.023 ± 0.01 mg/l and 0.035 ± 0.04 mg/l for males (n = 39). Manganese was not detected in the other healthy subjects.

There were no differences between manganese values in plasma and soil as determined by t-test. Statistically, there was no significant correlation between age and plasma manganese levels ($p>0.05$). However, there was a highly significant correlation ($r = 1.000$, $p = 0.01$) for plasma manganese concentration in females aged 51 and 60 years. There was a sharp decline from age 25 (0.041 mg/l) to 45 years (0.018 mg/l). It rose thereafter

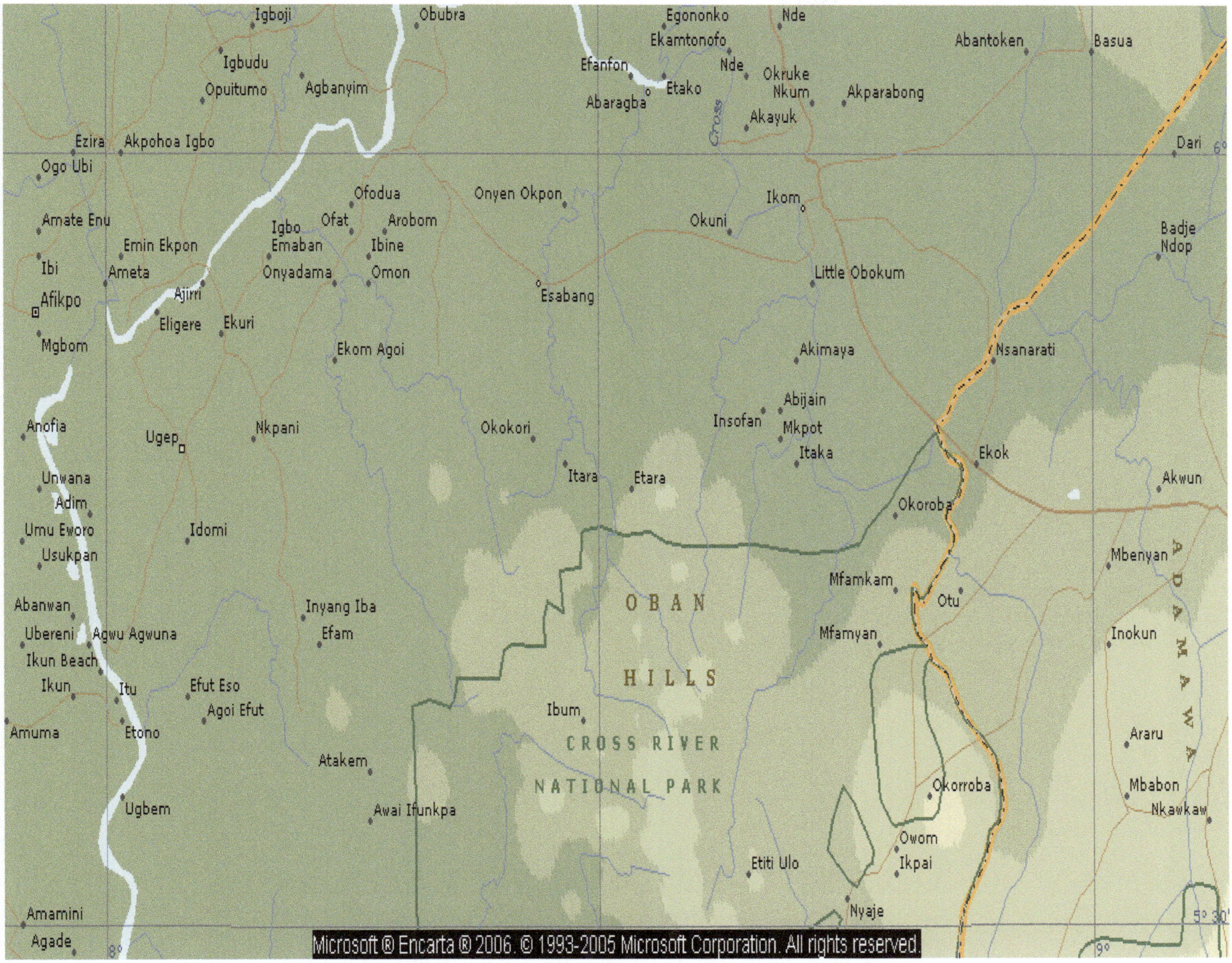

Figure 2. Map showing Ikom, Nigeria.

from age 45 to 60 years (Figure 4) and then declined sharply at age 75 years. No significant correlation was observed between plasma manganese concentration and soil manganese (p>0.05).

DISCUSSION

There is an increasing evidence for the important role which micronutrients play in the prevention of disease and promotion of overall health. These nutrients can be obtained from a diet containing a variety of foods from a variety of sources. Environmental changes and agricultural practices may be reducing nutrient concentration in the soil (Combs, 2001). The mean value of selenium (0.004 ± 0.006 mg/l, obtained in this study for apparently healthy adults was lower than that reported for most countries for example, New Zealand (0.056 mg/l) (Robinson et al., 1997), France (0.087 mg/l) (Coudray et al., 1997), Lagos, Nigeria, (57.05 ± 2.50 µg/l) (Arinola and Charles - Davies, 2008), Greece (108.86 ± 33.88 µg/l) (Sotiropoulos et al., 2011), UK (88.8 ± 19.2 ng/g) (Rayman et al., 2011), England (88.5 ± 19.1 ng/g) (Rayman et al., 2012) and Taiwan (106.4 ± 12.7 ng/l) (Guo., 2013). The low content of selenium in the soils could be attributed to a decline in the availability of selenium through the food chain and in human selenium status (Daniels, 1996). Also, the low plasma selenium concentration observed in this study may reflect the intrinsically low selenium content of the soil in which the foods were grown and the low protein intake of the participants, who took mostly starchy foods. Selenium has been shown to be more available from diet of animal origin than those of plant sources (Combs, 2001). With more than 80% of the dietary selenium intake coming directly from plant sources, it is likely that availability is

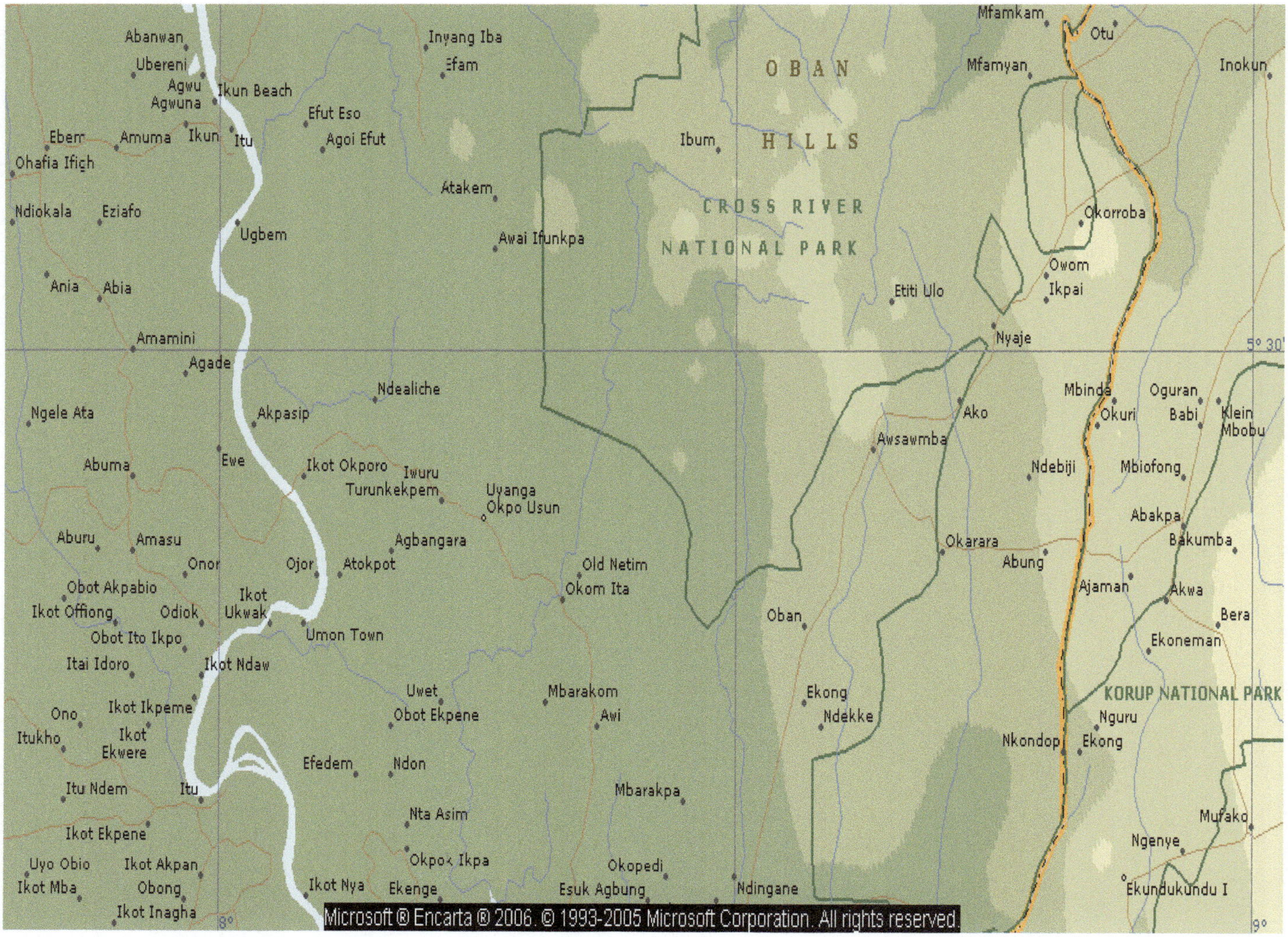

Figure 3. Map showing Ikot Ikpora, Nigeria.

Table 1. The selenium, chromium and manganese content in soil samples.

Location	Selenium (mg/kg)	Chromium (mg/kg)	Manganese (mg/kg)
Calabar	0.005	14.73	23.95
Ikot Okpora	0.009	26.61	102.58
Ikom	0.002	8.53	74.72
Uyo	8.08	6.86	130.00
Oron	6.22	12.93	120.4.
Mean±SD	2.86±3.97	15.93±3.35	90.33±42.65

low. The correlation between the local availability of selenium and the selenium status of the population might often be significant. Thus, when selenium availability is low, the local soil and plant content of the element gives an indication of the likely occurrence of diseases that may involve selenium in its etiology (Diplock, 1993).

Low levels of selenium status have been linked with senility and cognitive decline in the elderly and with Alzheimer's disease (DC Nutrition, 2003). Plasma concentration is the most commonly used indicator of selenium status. From the results, there was no significant effect of sex on the selenium values obtained. Surveys carried out by Lyons et al. (2004) showed that plasma selenium was affected by age ($p = 0.008$) and

Table 2. The plasma selenium, chromium and manganese categorized by age and sex.

Age group / sex	Age (years)	Selenium (mg/l)	Chromium (mg/l)	Manganese (mg/l)
18-90 (pooled)				
M + F	n = 106	n = 59	n = 73	n = 89
Mean ± SD	22.5 ± 3.15	0.004 ± 0.006	0.111 ± 0.031	0.028 ± 0.024
18-30				
M	n = 22	n = 10	n = 13	n = 14
Mean ± SD	25 ± 4.17	0.003 ± 0.002	0.115 ± 0.023	0.046 ± 0.056
F	n = 52	n = 5	n = 5	n = 9
Mean ± SD	23 ± 2.76	0.005 ± 0.009	0.109 ± 0.032	0.023 ± 0.006
31-40				
M	n = 16	n = 9	n = 11	n = 10
Mean ± SD	36.58 ± 2.84	0.003 ± 0.001	0.113 ± 0.047	0.026 ± 0.008
F	n = 12	N = 5	n = 5	n = 9
Mean ± SD	35 ± 3.09	0.002 ± 0.001	0.100 ± 0.034	0.022 ± 0.003
41-50				
M	n = 4	n = 2	n = 2	n = 3
Mean ± SD	46.50 ± 0.71	0.002	0.125 ± 0	0.019 ± 0.006
F	n = 11	n = 3	n = 4	n = 6
Mean ± SD	46.27 ± 3.20	0.002 ± 0	0.125 ± 0	0.023 ± 0
51-60				
M	n = 6	n = 3	n = 3	n = 4
Mean ± SD	56.5 ± 2.74	0.002 ± 0	0.104 ± 0.036	0.020 ± 0.005
F	n = 4	n = 1	n = 1	n = 3
Mean ± SD	55.7 ± 2.87	0.004 ± 0	0.125 ± 0	0.031 ± 0.014
61-70				
M	n = 5	n = 1	n = 3	n = 3
Mean ± SD	63 ± 0.00	0.004 ± 0	0.104 ± 0.009	0.036 ± 0.026
F	n = 4	n = 1	n = 1	n = 2
Mean ± SD	66 ± 2.00	0.002 ± 0	0.063 ± 0	0.035 ± 0.026
71-80				
M	n = 4	n = 4	n = 2	n = 4
Mean ± SD	75.5 ± 3.3	0.003 ± 0.0012	0.125 ± 0	0.035 ± 0.014
F	n = 1	n = 1	n = 1	n = 1
Mean ± SD	85	0.004	0.125	0.047

M, Male; f, female.

sex ($p = 0.01$). Plasma concentration of selenium from this study showed that it increased with age. Women had lower selenium levels than men (Kaifa et al., 2003). In the study by Akbaraly et al. (2010), the selenium level was 1.08 ± 0.21 µmol/l in men and 1.10 ± 0.20 µmol/l in women. Most studies have found little relationship between sex and blood selenium concentration in adults (Bates et al., 2002) unless in late term pregnancy or those who are lactating (Reyes et al., 2000). But Arinola and Charles - Davies (2008) in a study on females with breast cancer had selenium value of 52.25 ± 2.17 µg/l. The mean value of chromium (0.111 ± 0.031 mg/l) obtained in this study for 73 apparently healthy subjects was higher than that reported by Arinola and Akinbinu (2006) with 0.047 ± 0.019 mg/l and Arinola and Charles - Davies (2008) with 52.25 ± 2.17 µg/l mean plasma

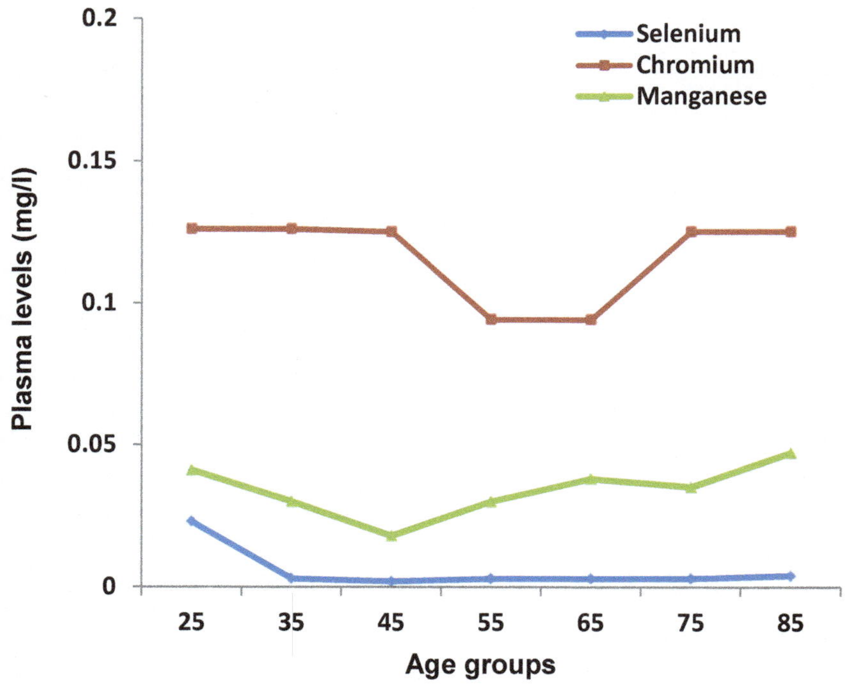

Figure 4. Plasma distribution with age of healthy adult Nigerians.

chromium concentrations. The mean value of chromium in plasma in healthy adults in studies by Chernecky and Berger (1997) was between 0.18 and 0.47 mg/l, Van Buynder (2010) was 5.17 ± 1.53 nmol/l and Newton et al. (2012) was 53.8 nmol/l.

Diet is the main source of chromium in humans; it is found in minute amounts in foods. Processed meats, whole grain products, ready-to-eat bran cereals, broccoli and spices are relatively high in chromium. But only 1 to 2% of the chromium delivered by the diet is absorbed. From the study, there was no significant effect of sex on the plasma values obtained. Plasma chromium level tended to be stable between the ages of 25 to 45 years and declined sharply thereafter. Plasma chromium level stabilized again between the ages of 55 to 65 and 75 to 85 years and rose sharply thereafter. This agreed with the study by Nourmohammadi et al. (2005), who reported that chromium levels stabilize at certain ages and then decline. Several disease states have been associated with low levels of serum manganese (Freeland-Graves and Llanes, 1994). The mean ± SD plasma manganese value obtained in this study for healthy adults 0.028 ± 0.02 mg/l was higher than the normal range (0.003 to 0.010 mg/l) (Nourmahammadi et al., 2005). This corroborates with studies by Arinola and Charles - Davies, (2008) with manganese value of 65.35 ± 2.96 µg/l in Lagos, Nigeria, Chan et al., (2009) with manganese value of 121.4 ng/ml in Boston, Omobola et al.,(2011) with manganese value of 63.20±3,9 µg/dl in Abeokuta, South-west Nigeria, Roos et al.,(2012) with manganese value of 0.78µg/l and Yuyan et al. (2012) with manganese value of 8.2 µg/l in China. The high plasma manganese value obtained in this study may reflect the high soil content in the region.

Values obtained in this study show that there was no significant correlation between age and plasma manganese. However, there was a significant correlation for plasma concentration in females aged 51 to 60 years. There was no significant effect of sex on plasma values although females had lower plasma manganese levels than the males.

Conclusion

Soil contents of micronutrients have significant impact on their plasma concentrations. Plasma levels of chromium and manganese seem to be adequate in healthy adult Nigerians. The plasma levels of selenium of healthy adults Nigerians appeared to be low. Supplementation may therefore be of benefit. Statistically, age and sex appear to have no significant effects on the plasma levels of selenium, chromium and manganese. Interactions between selenium, chromium and manganese in soil and subsequently in foods and diet showed the influence of the plasma levels of some of these trace nutrients in varying degrees.

REFERENCES

Akbaraly TN, Arnaud J, Rayman MP, Hininger-Favier I, Roussel AM, Berr C, Fontbonne A (2010). Plasma selenium and risk of dysglycemia in an elderly French population: results from the prospective epidemiology of vascular ageing study. Nutr. Metab. 7:21.

Arinola OG, Akinbinu MO (2006). The levels of antioxidants and some trace metals in Nigerians that are occupationally exposed to chemicals. Official Publication of Indian Association of Occupational Health 10(2):65-68.

Arinola OG, Charles-Davies (2008). Micronutrient levels in the plasma of Nigerian females with breast cancer. Afr. J. Biotechnol. 7(11):1620.

Bates CJ, Thane CW, Prentice A, Delves HT (2002). Selenium status and its correlates in a British national diet and nutrition survey: people ages 65 years and over. J. Trace Elem. Med. Biol. 16:1-8

Chan JM, Oh WK, Xie W, Regan MM, Stampfer MJ, King IB, Abe M, Kantoff PW (2009). Plasma selenium, manganese superoxide dismutase,and intermediate-or high-risk prostate cancer. J. Clin. Oncol. 27(22):3577-3583.

Chernecky CC, Berger BJ (1997). Laboratory tests and diagnostic procedures 2nd ed. Philadelphia, Saunders. pp. 367-368.

Combs GF (2001). Selenium in global food systems. Br. J. Nutr. 85(5):517-547.

Combs GF, Clark LC, Turnbull BW (1997). Reduction of cancer risk with an oral supplement of selenium. Biomed. Environ. Sci. 10:227-234.

Coudray C, Roussel AM, Mainard F, Arnaud J, Favier A (1997). Lipid peroxidation level and antioxidant micronutrient status in a pre-aging population; correlation with chronic disease prevalence in a French epidemiological study (Nantes, France) J. Am. Coll. Nutr. 16:584-591.

Daniels LA (1996). Selenium metabolism and bioavailability. Biol. Trace Elem. Res. 54:185-199

DC Nutrition (2003). Selenium - general discussion. Selenium deficiency diseases pl file: //A:/DC Nutrition (selenium).htm .

Diplock AT (1993). Indexes of selenium states in human populations. Am. J. Clin. Nutr. Suppl. 57:256S-258S

Eckhardt CL (2006). Micronutrient malnutrition, obesity and chronic disease in countries undergoing the nutrition transition: potential links and program/policy implication: International Food Policy Research Institute.

Fairweather - Tait SJ (1997). Minerals and trace elements - summary. Europ. J. Clin. Nutr. 51: supplement 1, S3.

Freeland - Graves J, Llanes C (1994). Models to study manganese deficiency in Klinmics - Tavantzis DL, ed. Manganese in health and disease. Boca Raton CRC Press, Inc. pp. 59-86.

Guo CH, Chan PC, Hsu GS, Wang CL (2013). Zinc supplementation alters plasma aluminium and selenium status of patients under - going dialysis: a pilot study. Nutrients 5:1456-1470.

Kafai MR, Gangi V (2003). Sex, age, geographical location, smoking and alcohol consumption influence serum selenium concentration in the USA, third health and nutrition survey, 1993 - 1994. J. Trace Elem. Med. Biol. 17:13-18.

Keen CL, Zindenberg - Cherr S (1996). Manganese. In: Ziegler E.E., and Filer L. J. (eds). Present Knowledge in Nutrition. 7th ed. Washington DC ILSI Press. pp. 334-343.

Lyons GH, Judson GT, Stangoulis JCR, Palmer LT, Jones JA, Graham RD (2004). Trends in selenium status of South Australians. Med. J. Aust. 180(8): 383-386.

Newton AW, Ranganath L, Armstrong C, Peter V, Roberts NB (2012). Differential distribution of cobalt, chromium, and nickel between whole blood, plasma and urine in patients after metal-on-metal (MOM) hip arthroplasty. J. Orthop. Res. 30(10):1640-1646.

Nourmohammadi I, Kocheki-Shalmani, Shaabani M, Gohari L, Nazari H (2005). Zinc, copper, mangnese and magnesium levels in serum and hair of insulin-dependent diabetics: internet available from http://www.ams.ac.ir /AIM0033/nour 00333.html.

Omobola A. Ogundahunsi, Olaniyi O. Duduyemi and Oluwatosin O. Soyinka (2011). Effect of antipsychotic drugs on the levels of selected trace elements in schizophrenia: African Scientist 12(2):77-82.

Paul Van Buynder (2010). Evaluation of chromium in blood. A report prepared by ChemRisk Canada, a division of ChemRisk LLP, Guelph, Ontario N1H7L6.

Rayman MP, Stranges S, Griffin BA, Pastor - Barriuso R, Guallar E (2011). Effect of supplementation with high - selenium yeast on plasma lipids: a randomized trial. Ann. Intern. Med. 154(10): 656 - w 240. ISSN 0003 - 4819.

Rayman, M. P, Blundell - Pound, G, Pastor - Barriuso, R, Guallar, E, Steinbrenner, H and Stranges, S (2012). A randomized trial of selemium supplementation and risk of type - 2 diabetes as assessed by plasma adiponectin. PLOS ONE (7) 9: e 45269.

Reyes H, Baez ME, Gonzalez C (2002). Selenium, zinc, copper plasma levels in intrahepatic cholestatis of pregnancy, in normal pregnancies and in healthy individuals in Chile. J. Hepatol. 32:542-549

Robinson MF, Thomson CD, Jenkins P (1997). Long term supplementation with selenate and selenomethionine urinary excretion by New Zealand women. Br. J. Nutr. 77(4): 551-563.

Roos PM, Lierhagen S, Flaten TP, Syversen T, Vesterberg O, Nordberg M (2012): Mangenese in cerebrospinal fluid and blood plasma of patients with amytrophic lateral lateral sclerosis. Experimental Biology and Medicine: 237:803-810.

Sotiropoulos A, Papadodina SA, Papazafiropoulou AK, Ioannidis A, Kokkinari A, Apostolou O, Spiliopoulou CA, Athanaselis S (2011). Serum selenium levels do not differ in type 2 diabetic subjects with and without coronary artery disease. BMC Res. Notes 4:270.

Tuormaa TE (1995). The role of chromium, selenium and copper in human and animal metabolism. J. Orthomol. Med. 10(384):149-164.

Vincent JB (2000). Elucidation of a biological role for chromium at a molecular level. Acc. Chem. Res. 33(7):503 - 510.

Yuyan, Li, Junqing Wu, Weijin Zhou and Ersheng Gao (2012): Effects of managanese on routine semen quality parameters: results from a population - based study in China: BMC Public Health12:919.

Metabolic response of *Cyclotella meneghiniana* Kutzing to toxicity of the antifouling agent Irgarol 1051

Hanan A.Said[1], Wafaa M. Abdel-Aziz[2] and Hala M.Taha[2*]

[1]Botany Department, Faculty of Science, Fayoum University, El Fayoum, Egypt.
[2]Botany Department, Faculty of Science, Alexandria University, Alexandria, Egypt.

Soluble toxic compounds that are leached out from antifouling paints cause unpalatable taste and disagreeable odor. This in turn will directly or indirectly affect the metabolic activities of the aquatic biota. In this work, reliable studies about the expected effects of these toxins on living organisms were conducted. Microorganisms especially algae were used as powerful tools to assess *in vitro* metabolic response on several environmental toxins. The obtained results reveal that the antifouling Irgarol 1051 was a strong inhibitor to the studied microalga *Cyclotella meneghiniana*. It became apparent from the laboratory results that water polluted by these leached toxins caused a substantial metabolic inhibition and the degree of inhibition is selectively concentration dependent.

Key words: Antifouling agent, *Cyclotella meneghiniana*, Irgarol 1051, metabolic activities.

INTRODUCTION

Water purity can have a great influence on the ability of plants and animals to exist and grow, which in turn affects human health since it can carry toxins and bacteria, as well as, having unpalatable taste and disagreeable odor (Palmer, 1980). The main cause of pollution is the discharge of solid or liquid waste products containing pollutants onto the land surface or into the aquatic habitats. The wastes that contribute towards water pollution may be broadly grouped into domestic, industrial and agricultural types (Seiki et al., 1991). An additional environmental problem which has increased during recent years is due to the soluble active ingredients (biocides) in the paints of ships that have been leached out from the antifouling paints which contained usually very toxic compounds (Hoch, 2001). These in turn cause damage to the metabolism of living organisms in aquatic and land environments. However, some microorganisms have the ability to resist and accumulate these toxins, which in turn will be more serious to consumers whether fishes, animals or humans.

Irgarol 1051 (2-methylthio -4-terbulylamino -6-cyclopropylamino -s-triazine) is used now in aquatic antifouling paints after the restriction of using tributyltin (TBT) as antifouling agent (Zamora-Ley et al., 2006). The active ingredients in the paints biocides -mainly copper and organostanous compounds- leach from the coating of ships and thereby contaminate the aquatic environment (Konstantinou and Albanis, 2004). Thus, in areas with high boating activities many of these leached substances (biocides) occur as mixtures (Biselli et al., 2000; Lamoree et al., 2002; Thomas et al., 2003). Occurrence of Irgarol 1051 has been widely reported in coastal waters of many countries; Red Sea, Mediterranean Sea, Japan, Australia, Europe, United States and many others (Biselli et al., 2000; Konstantinou and Albanis, 2004; Key et al., 2008). Increasing concentration of Irgarol in coated waters up to 0.048 to 0.068 µg/l have been reported by Bascher et al. (2002). The toxic effects of Irgarol 1051 against algae, crustaceous and fish have been reported by Van Wezel and Van Vlaardingen (2004). Concentrations of Irgarol 1051 have been found to be highest in and adjacent to marines and fishing harbours range between 0.02 and 0.70 µg/l. Readmann et al. (2004) reported toxic effects

*Corresponding author. E-mail: taha_hm@yahoo.com.

of Irgarol 1051 at low concentrations of 0.07 ng/l. Irgarol 1051 was firstly reported as an aquatic contaminant in the Mediterranean since 1993 (Readmann et al., 1993). Knowledge about the fate of these toxic chemicals in aquatic environments (fresh or marine) is essential for understanding and predicting the possibility of ecotoxic ecology effects.

This work is concentrated on the response of the microalga *Cyclotella meneghiniana* to differences of the ambient antifouling agent Irgarol 1051 on its growth and content of some important metabolites viz; chlorophyll fractions, protein, protein profile, photosynthesis, respiration and the destructive effects of the total antioxidants.

MATERIALS AND METHODS

The selective biological material is the unicellular microalga *Cyclotella meneghiniana* Kutzing -usually used for fish feeding; obtained from the culture collection of Prof. Khaleafa, Faculty of Science, Alexandria University. The tested microalga was grown in brackish water medium as described by Schlösser (1994). The pH of the culture media was adjusted to 7.5 using pH-meter then sterilized by using the bacterial filter. The tested alga was grown in Erlenmeyer pyres-glass flasks (capacity 250 ml) containing 50 ml culture medium. The cultures were maintained under controlled laboratory conditions (temperature of 22 ± 2°C and light intensity of 80 □mol m^{-2} s^{-1}) in controlled culture chamber for 16 days and under a photoperiod of 16/8 h light/dark cycles. Three independent experiments were performed for each treatment.

The herbicide triazine (Irgarol 1051) was purchased from Fluka Company. The stock solution 1000 mg of the standard Irgarol 1051 was prepared in acetone and stored in dark at 4°C. 0.1 ml of stock solution was mixed with 100 ml of sterilized distilled water. The mixtures were stirred by a magnetic stirrer for half an hour. Dilution of this stock solution was done by the selected culture medium for preparing the different concentrations of Irgarol. These concentrations were nearly within the range of those found in marines.

Growth rate was estimated by cell number using haemocytometer slide. The growth rate was calculated by using the formula proposed by Stein (1979). Growth rate (number of cells / day) = $(\log N_2/N_1) \times (3.322/T_2\text{-}T_1)$, where N_1 and N_2= number of cells; T_1 and T_2= times in days.

The chlorophylls content were estimated in the acetone extracts by using the dichromatic equations of Jeffrey and Humphrey (1975): Chl. a (µgml^{-1}) = $11.47\ E_{664} - 0.40\ E_{630}$; Chl. c_1+c_2 (□gml^{-1}) = $24.36\ E_{630} - 3.73\ E_{664}$.

Protein was extracted from the algal cells by the method described by Rausch (1981). Protein content and its fractions were estimated by the method described by Hartree (1972). The protein content was calculated from a standard curve made from working standards of similarly treated bovine serum albumin as mgl^{-1} cultures.

Protein profile was performed according to the method described by Laemmli (1970). The freshly cell pellets obtained by centrifugation of 10 ml culture were ground in a mortar in the presence of small volume of 0.5 M Tris-HCl buffer (pH 7.2). The suspension was centrifuged and the supernatant was concentrated in pre-activated dialysis tubing over a sucrose bed. Few drops of glycerol were added to increase viscosity and 2 to 3 drops of bromophenol blue as an electrophoresis marker. Electrophoresis was carried out in 12% acrylamide running gel and a 4% stacking gel, with 0.025 M tris 0.19 M glycine buffer pH 8.3. Further details of the methods used in gel preparation and electrophoresis were described previously by Moore et al. (1980). Electrophoresis was performed at 25°C in a vertical tank apparatus maintained at 100 V using a constant-voltage power supply, until bromophenol blue tracking dye reached the bottom of the gel. After electrophoresis, gels were stained with 2% Coomassie brilliant blue G 250 in water-methanol-acetic acid (4.5:4.5:1) for 4 h at 37°C, after that, they were destained by agitation in the same solvent without dye, and stored in a fixing solution (7% acetic acid) for at least 30 min at 30°C and rinsed with deionized distilled water. The gels were then photographed.

Photosynthesis and respiration

The photosynthetic activity was measured polarographically as oxygen evolution using a Clark type electrode (YSI, model 53). The actinic white light was obtained from a 150 W. tungsten lamp. The test organism was harvested every 4 days intervals till the end of the experiment. Photosynthetic activity was measured as (µmolO_2 evolution/mg chlorophyll/hour). Respiration was measured in the dark and expressed as (O_2 uptake µmolO_2h^{-1}).

Total antioxidants activity

Determination of the total antioxidants activity was performed by the reaction of antioxidants in the sample with a definite amount of exogenously provided hydrogen peroxide as described by Koracevic et al. (2001). The antioxidants in the sample eliminated a certain amount of the provided H_2O_2. The residual H_2O_2 was determined colorimetrically by using the standard package (R1, R2 and R3) supplied from the manufacture's instruction (Biodiagnostic and research reagents Egypt); an enzymatic reaction which involves the conversion of 3, 5, dichloro-2-hydroxy benzensulphonate to a colored product. The absorbance of the blank (AB) and sample (ASA) against distilled water was read immediately at 505 nm. The overall antioxidants capacity was then calculated as mM l^{-1} by using the following formula; mM l^{-1} = $A_B - A_{SA} \times 3.33$ (Comba et al., 2004).

Statistical analysis

The data were the means ± SD of at least three independent experiments. The obtained data were analyzed statistically using the least significant difference (LSD) at 0.05.

RESULTS AND DISCUSSION

The data recorded in Figures 1 and 2 show that suppression in growth parameters (growth rate and chlorophylls content) of *Cyclotella meneghiniana* under the effect of the tested different concentrations of Irgarol 1051 may be due to the increasing toxicity of the biocide. The same results were also obtained by Munnas (2003) and Singla and Garg (2005). The rate of growth depended mainly on the level of irgarol 1051 concentration and length of culture period. This idea is in harmony with those obtained by Gatidou et al. (2003). The stress effect of Irgarol 1051 on growth of *C. meneghiniana* may be due to the toxic metals (biocides) that leached out from the antifouling paint. These cause inhibition of normal cell division due to the binding of the

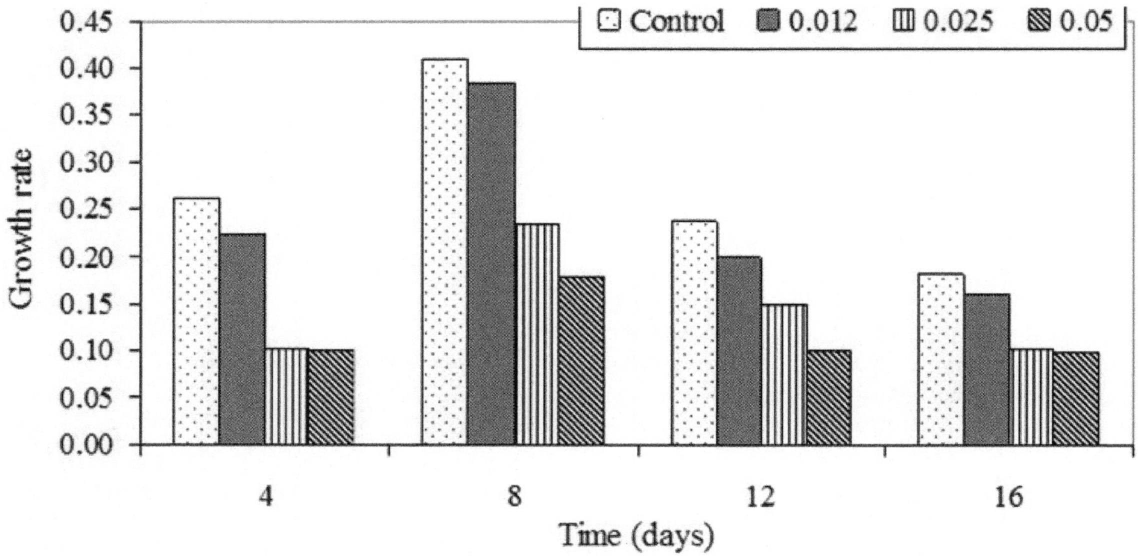

Figure 1. Effect of different concentrations of Irgarol 1051 (μg/l) on growth rate (number of cells/day) of *Cyclotella meneghiniana*.

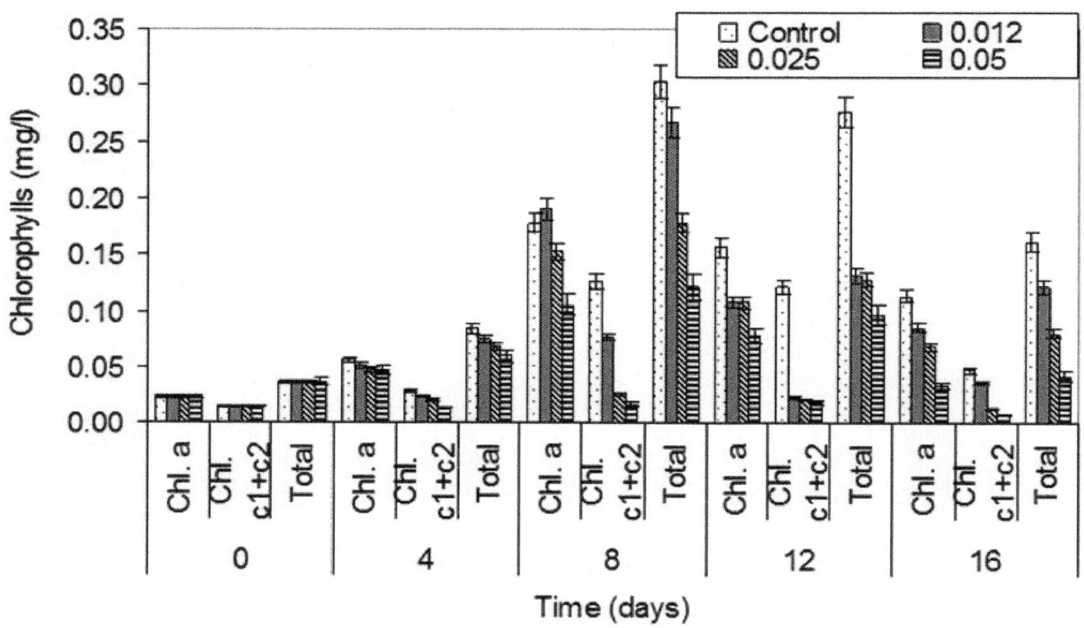

Figure 2. Effect of different concentrations of Irgarol 1051 (μg/l) on chlorophylls content of *Cyclotella meneghiniana*. (Data are mean of three replicates ± SD).

biocides to sulfhydryle groups which are important in the regulation of cell division (Visviki and Rachlin, 1992; Sidharthan et al., 2002). Moradi and Ismail (2007) and Ghezelbash et al. (2008), reported that reduction in chlorophyll content with reduced growth rate was due to decrease in photosynthetic rate. Also, Jones (2005) and Zamora-Ley et al. (2006) reported that the antifouling boostering agent Irgarol 1051 was a strong inhibitor of photosystem II (PSII).

The changes of biochemical composition of some microorganisms under the stress effect of toxins may be due to the block conversion of excitation energy into chemical energy (Jones, 2005; Sheikh et al., 2009). The data recorded in Table 1 shows that protein content increased gradually by increasing the period of culturing. At the end of the experiment (16 days), the content of total proteins decreased. This may be due to decrease in insoluble protein fractions and deficiency of nutrients

Table 1. Effect of different concentrations of Irgarol 1051 (μg/l) on protein fractions content (mg/l) of *Cyclotella meneghiniana*.

Time (days)	Parameter	Control	0.012	0.025	0.050
	Insoluble	14.221±0.002[a]	14.456±0.001[a]	13.752±0.002[b]	14.135±0.004[a]
4	Soluble	15.226±0.002[a]	15.516±0.002[q]	12.234±0.002[b]	12.012±0.001[b]
	Total	29.447	29.972	25.956	22.147
	Insoluble	22.720±0.200[a]	15.752±0.100[b]	17.214±0.012[c]	16.321±0.201[d]
8	Soluble	43.310±0.150[a]	43.195±0.104[a]	35.310±0.130[b]	21.214±0.420[c]
	Total	66.030	58.947	52.524	37.445
	Insoluble	30.641±0.143[a]	20.197±0.013[b]	25.701±0.205[c]	18.502±0.104[d]
12	Soluble	48.123±0.031[a]	40.161±0.022[b]	30.214±0.001[c]	20.370±0.108[d]
	Total	78.764	60.358	55.915	38.872
	Insoluble	34.210±0.200[a]	25.241±0.022[b]	27.520±0.103[c]	20.210±0.040[d]
16	Soluble	42.420±0.220[a]	40.021±0.001[b]	30.120±0.024[c]	22.341±0.051[d]
	Total	76.630	65.262	57.640	42.551

Data are mean of three replicates ± SD. Different superscripts are significant ($P \leq 0.05$).

Table 2. Effect of different concentrations of Irgarol 1051 (μg/l) on the activity of photosynthesis (μmolO_2 evolution / mg chlorophyll / h) and respiration (O_2 uptake μmolO_2 h^{-1}) of *Cyclotella meneghiniana*.

Time (days)	Parameter	Control	0.012	0.025	0.050
4	Photosynthesis	1.67±0.24[a]	1.22±0.17[b]	0.82±0.10[c]	0.51±0.12[d]
	Respiration	4.73±0.14[a]	3.02±0.11[b]	1.80±0.21[c]	1.42±0.19[d]
8	Photosynthesis	1.93±0.22[a]	1.21±0.15[b]	0.51±0.18[c]	0.32±0.21[d]
	Respiration	3.52±0.18[a]	2.30±0.10[b]	1.34±0.18[c]	1.22±0.27[d]
12	Photosynthesis	1.71±0.19[a]	1.20±0.13[b]	0.42±0.25[c]	0.22±0.24[d]
	Respiration	3.16±0.31[a]	1.64±0.22 [b]	1.24±0.20[c]	1.04±0.18[d]
16	Photosynthesis	1.60±0.10[a]	1.02±0.18[b]	0.42±0.11[c]	0.19±0.19[d]
	Respiration	1.70±0.17[a]	1.31±0.21[b]	1.02±0.13[c]	1.01±0.27[c]

Data are mean of three replicates ± SD. Different superscripts are significant ($P \leq 0.05$).

which increase proteolysis (Cooke et al., 1979, 1980) and/or due to decrease in the rate of protein synthesis (Vaadia and Waisel, 1967; Tam and Wong, 1995). The the toxicity effects of the different levels of Irgarol 1051. The greater the toxic effects of the antifouling agent and the length of culturing period, the greater was the content of soluble proteins and the less was the content of insoluble ones. The decrease in content of insoluble protein may be due to its degradation and/or weak synthesis of the soluble ones. This in turn will cause accumulation of the soluble ones (Salah El-Din, 1994; Ahmed, 2010; Kaamoush 2011).

Data represented in Table 2 shows that the quantity of O_2 evolution in photosynthesis and O_2 uptake in data recorded also cleared that although total protein content decreased under the stress effect of Irgarol 1051, yet protein groups (soluble and insoluble) differed under respiration of *C. meneghiniana* depended mainly on the concentration and length of culturing period. Under these conditions, both photosynthesis and respiration decreased gradually till the end of the experiment. The rate of the stress effect on the activity of these two parameters increased gradually by increasing the concentration of Irgarol 1051. At the 8th day of culturing, the percent of decrease in respiration reached 34.86, 62.78 and 65.34% at the different concentrations of the tested Irgarol 1051, respectively. In the case of photosynthesis, it decreased by 37.31, 73.58 and 83.42%

Table 3. Effect of different concentrations of the antifouling agent Irgarol (1051) on protein profile bands in *Cyclotella meneghiniana*.

Rf	KDa	Control (un-treated)	0.012 µg/l			0.025 µg/l			0.050 µg/l		
			a	b	c	a	b	c	a	B	c
0.0 - 0.1	222 - 201	1	-	-	1	-	-	1	-	-	1
0.1 - 0.2	174 - 139	-	-	-	-	-	-	-	-	-	-
0.2 - 0.3	118 - 92	2	2	-	-	1	-	1	1	-	1
0.3 - 0.4	89 - 64	-	1	1	-	-	-	-	1	1	-
0.4 - 0.5	63 - 48	1	1	-	-	1	-	-	1	-	-
0.5 - 0.6	30 - 27	-	-	-	-	1	1	-	-	-	-
0.6 - 0.7	24 - 22	2	1	-	1	1	-	1	-	-	2
0.7 - 0.8	17 - 14	2	1	-	1	1	-	1	1	-	1
0.8 - 0.9	12 - 9	1	1	-	-	1	-	-	1	-	-
Total		9	7	1	3	6	1	4	5	1	5

a = Present bands; b = newly appeared bands; c = disappeared bands.

respectively. This means that the process of photosynthesis was affected to a larger extent under the toxicity stress of Irgarol 1051 than respiration. Moradi and Ismail (2007) and Ghezelbash et al. (2008) represented that reduction in chlorophyll content with reduced growth rate was due to decrease in photosynthetic rate. Also, Jones (2005), Zamora-Ley et al. (2006) and Kaamoush (2011) reported that the antifouling boostering agent Irgarol 1051 is a strong inhibitor of photosystem II (PSII) with high efficiency (toxicity) to photosynthetic micro-organisms.

Most of the living microorganism during their adaptation to the external stresses changes in the macromolecule content and in the enzyme activities usually occurs (Schiewer and Jonas, 1977; Hagmann and Wittenburg, 1989; Hagemann et al., 1989). Therefore, changes in the protein composition of these organisms may also be involved in adaptation processes. During these processes of adaptations, disappearance and synthesis of stress, induced proteins has been demonstrated in living organism (Hecker et al., 1988; Apte and Bhagwat, 1989; Hagemann et al., 1990). In our work, the gel of the total soluble protein profile of the tested organism at the different concentrations of Irgarol 1051 showed bands distributed through the gels. The number of bands that disappeared, were newly formed and remained compared to the control which depended mainly on the concentration of Irgarol 1051 (Table 3 and Figure 3). The results of Hoyos and Zhang (2000) are in agreement with these results. The results of Salah El-D (1994) confirmed that most of the living organism species have similar physiological fractions which are related to biosynthesis or biodegradation of some macromolecules. This conclusion seems to explain the different changes of the amount of total soluble protein bands in the stressed microorganisms. Exss-Sonne et al. (2000), Hifney (2002) and Ahmed (2010) reported that tolerance of an organism to stress conditions could be achieved through the synthesis or accumulation of new types of protein. The above results were nearly in harmony with our results for *C. meneghiniana*.

From the results obtained concerning the antioxidative defense systems (Figure 4), it is clear that the activity of total antioxidants in *C. meneghiniana* amended with different concentrations of Irgarol 1051 showed a significant progress, if there was any increase through the entire period of the experiment. Careful examination of the results obtained reveal that by increasing concentration of Irgarol 1051, a progressively greater increase in the activity of the total antioxidants was obtained under low and mild stress (Comba et al., 2004). However, Jungklang (2005) reported that in living organisms subjected to high stress, the balance between the production of ROS and quenching activity of the antioxidants in upset, results in antioxidant stress. Our results lend a strong support to the results obtained by Bor et al. (2003), Chaparzadeh et al. (2004) and Jungklang (2005).

Based on the fact that the different environmental stress conditions are likely to affect the same physiological parameters, it is not really surprising to find that the present state of Irgarol 1051 stress triggers the expression of the antioxidative defense systems. This is evident from the fact that total antioxidants which included non-enzymatic and enzymatic antioxidants, showed in general an increase with increasing stress indicating the cell capacity to overcome the stress. The conclusion can be substantiated if the degree of stress is not so high to concentration limit that may cause damage and death of the organism. In this work, the same conclusion was also observed for *C. meneghiniana* where the total antioxidant activity began to decrease at the higher concentration of Irgarol 1051.

In conclusion, this work clears that waters polluted by these booster biocides of Irgarol 1051 which leached out

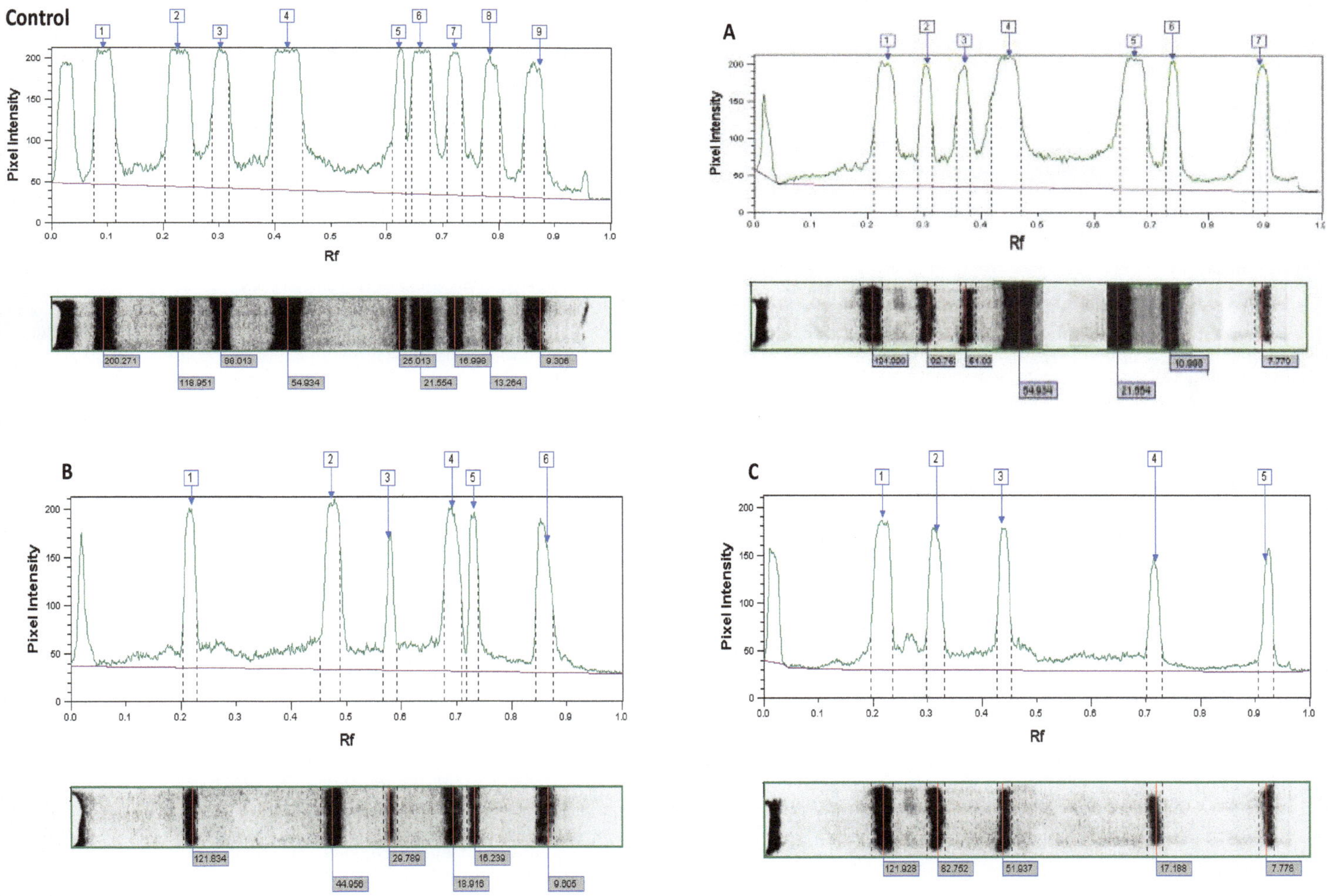

Figure 3. Electropherogram showing the scanning results of protein profile bands in *C. meneghiniana* cells asthe control and at the effect of different concentrations of Irgarol 1051. **(A)** 0.012 µg/l. **(B)** 0.025 µg/l. **(C)** 0.050 µg/l.

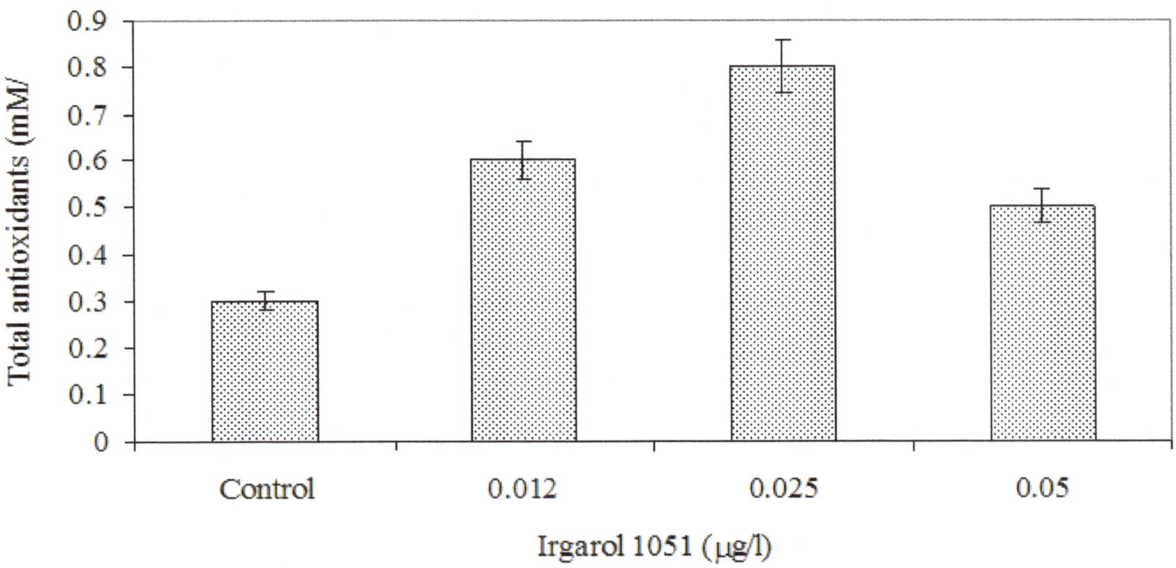

Figure 4. Total antioxidant capacity (mM/l) of *C. meneghiniana* at control and at the three tested Irgarol 1051 concentrations. (Data are mean of three replicates ± SD).

from the antifouling paints will cause weak growth, damage or death of the periphyton communities.

REFERENCES

Ahmed EAM (2010). Impact of Tributyltin (TBT) on metabolism of some marine algae. Ph.D. Thesis . Faculty of science, Alazhar University. Egypt.

Apte SK, Bhagwat AA (1989). Salinity-stress-induced proteins in two nitrogen-fixing *Anabaena* strains differentially tolerant to salt. J. of Bacteriol. 171: 909 - 9 1 5.

Bascher C, Tan KS, Lee HK (2002). Organotin and Irgarol-1051 contamination in Singapore coastal waters. Mar. Poll. Bull. 44:697-703.

Biselli S, Bester K, Huhnerfuss H, Fent K (2000). Concentrations of the antifouling compound Irgarol 1051 and of organotins in water and sediments of German North and Baltic Sea marinas. Mar. Poll. Bull. 40:233-243.

Bor M, Özdemir F, Türkan I (2003). The effect of salt stress on lipid peroxidation and antioxidants in leaves of sugar beet, *Beta vulgaris L.* and wild beet- *Beta maritime L.* Plant. Sci.164:77-84.

Chaparzadeh ND, Amico ML, Khavari-Nejad RA, Izzo R, Navari-Izzo F (2004). Antioxidative response of *Calendula afficinales* under salinity stress. Plant Physiol. Biochem. 42:695-701.

Comba ME, Benavides MP, Tomaro LT (2004). Effect of salt stress on antioxidant defense system in Soy bean root nodules. Aust. J.Plant Physiol. 25: 665-671.

Cooke RJ, Oliver J, Davis DD (1979). Stress and protein turnover in *Lemna minor*. Plant Physiol. 64: 1109-1113.

Cooke RJ, Roberts K, Davis DD (1980). Model for stress induced protein degradation in *Lemna minor*. Plant Physiol. 66: 1119-1122.

Exss-Sonne P, Tölle J, Bader KP, Pistorius E K, Michel K P (2000). The Idil A protein of *Synechococcus* sp. PCC7942 functions in protecting the acceptor side of photosystem II under oxidative stress. Photosynth. Res. 63: 145-157.

Gatidou G, Kotrikla A, Rontogianni V, Thomaidis NS (2003). Toxic effect of the antifouling biocide Irgarol 1051 and its principal metabolites on the green alga *Dunaliella tertiolecta*. 8th international conference on environmental science and technology.

Ghezelbash F, Farboodnia T, Heidari R, Agh N (2008). Biochemical effects of different salinities and luminance on green microalgae *Tetraselmis chuii*. Research J. Biology. Sci. 3(2): 217-221.

Hagemann M, Wolfel L, Krl Ulger B (1990). Alterations of protein synthesis in the cyanobacterium *Synechocystis* sp. PCC 6803 after salt shock. J. Gen. Microbiol. 136: 1393-1399.

Hagemann M, Erdmann N, Schiewer U. (1989). Salt adaptation of the cyanobacterium *Microcystis firma* and *Synechocystis aquatilis* in turbidostate cultures, I. Steady state values. Arch. Hydrobiol. Suppl.82:425-435.

Hagmann M, Wittenburg E (1989). Salt-induces changes in the RNA and DNA content of cyanobacteria (blue-green algae) *Synechocystis aquatilis* and *Microcystis firma* in batch and turbidostate cultures. Arch. Hydrobiol. Suppl. 82, Algological Studies. 56: 381-391.

Hartree EF (1972). A modification of lowry method that gives a linear photometric response. Analyt. Biochem. 48: 422.

Hecker M, Heim C, Volker U, Wolfel L (1988). Induction of stress proteins by sodium chloride treatment in *Bacillus subtilis*. Archives of Microbiology. 150: 564-566.

Hifney AF (2002). Metabolic response of *Synechococcus* sp. Strain PCC 7942 some nutrient deficiencies with special reference to Iron. Ph. D. Thesis. Faculty of Science. Assiut University.

Hoch M (2001). Organotin compounds in the environment-an overview. Appl. Geochem. 16: 719-743.

Hoyos ME, Zhang S (2000). Calcium independent activation of salicylic acid - induced protein Kinase and 40- Kilodalton protein kinase by hyperosmotic stress. Plant physiol. 122: 1355-1363.

Jeffrey SW, Humphrey GF (1975). New spectrophotometric equations for determing chlorophylls a, b, c_1 and c_2 in higher plants, algae and natural phycoplankton. Biochem. Physiol. Pflanzen. 167: 191-194.

Jones R (2005). The exotoxicological effects of photosystem II herbicides on corals. Mar. Pollut. Bull. 51(5-7):495-506.

Jungklang J (2005). Physiological and biochemical mechanisms of salt tolerance in *Sesbania rostrata* Berm and Obem and *Phaseolus vulgaris* L. Ph D Thesis E C Agric Univ Teckuba, Japan.

Kaamoush M A (2011). Metabolic response of the two marine algae *Dunaliella bardawil* and *Chlorella salina* to toxicity of the antifouling agent Irgarol 1051. Ph.D. Thesis. Botany Department, Faculty of Science, Alex. Univ.

Key PB, Chung KW, Hoguet J, Sapozhnikova Y, Fulton MH (2008). Effects of the anti-fouling herbicide Irgarol 1051 on two life stages of the grass shrimp, *Palaemonetes pugio*. J. Environ. Sci. Health B. 43(1):50 - 55.

Konstantinou IK, Albanis TA (2004). Worldwide occurrence and effects of antifouling paint booster biocides in the aquatic environment: A review. Environ. Intern. 30: 235-248.

Koracevic D, Koracevic G, Djordjevic V, Andrejevic S, Cosic V (2001). Method for the measurement of antioxidant activity in human fluids. J. Clin. Pathol. 54:356-361.

Laemmli UK (1970). Cleavage of structural proteins during assembly of the head of bacteriophage T4. - Nature. 227(5259): 680 - 685.

Lamoree M, van der Horst A, Swart CP and van Hattum B (2002). Determination of diuron and antifouling paint biocide Irgarol 1051 in Dutch marinas and coastal waters. J. Chromatogr. A. 970: 183-190.

Moore WEC, Hash DE, Holdeman LV, Cato EP (I980). Polyacrylamide slab gel electrophoresis of soluble proteins for studies of bacterial floras. Appl Environ Microbiol. 39: 900- 907.

Moradi M, Ismail AM (2007). Response of photosynthesis, chlorophyll flouresence and ROS-Scavaning system to salt stress during speedling and reproductive stages in rice. Ann. Botany.99: 1161-1173.

Munnas R (2003). Comparative physiology of salt and water stress. Plant Cell Environ. 25: 239-250.

Palmer CM (1980). Algae and water pollution. Castle House publication Ltd. England. 123 p.

Rausch T (1981). The estimation of microalgal protein content and its meaning to the evolution of algal biomass. I., Comparison method for extracting protein. Hydrobiologia. 78: 237-251.

Readman JW, Liong LL W, Grondin D, Bartocci J, Villeneuve JP, Mee LD (1993). Coastal water contamination of a triazine herbicide used in antifouling paints. Environ.Sci. Technol. 27(9): 1940-1942.

Readman JW, Devilla RA, Tarran G, Llewellyn CA, Fileman TW, Easton A, Buckhill PH, Mantoura RFC (2004). Flow cytometry and pigment analyses as tools to investigate the toxicity of herbicides to natural phytoplankton communities. Mar. Environ. Res. 58:353.

Salah El-Din RA (1994). Contribution to the biological and phytochemical studies of marine algal vegetation on the coast of Red-Sea and Suez-Canal (Egypt). Ph.D. Thesis. Botany Department. Faculty of Science. Al-Azhar University, Cairo. Egypt.

Schiewer U, Jonas L (1977). Die Wirkung unterschielicher NaCl-Konzentrationen auf die Ultrastruktur von Blaualgen. I. Microcystis firma. Archive fur Protistenkunde. 119: 127-145.

Schlösser (1994). SAG: Samlung von Algenkulturen at the University of Gottingen, catalogue of strains. Stuttgart; New York: Tieme Med. Publ. Botanica Acta. 107(3).

Seiki T, Date E, Lzawa H. (1991). Eutrophication in Hiroshima Bay, Mar. Poll. Bull. 23: 95.

Sheikh MA, Higuchi T, Fujimura H, Imo TS, Miyagi T and Oomori T (2009). Contamination and impact of new antifouling biocides Irgarol-1051 on subtropical coral reef waters. Environ. Sci. Tech. 6(3): 353-358.

Sidharthan M, Kim SY, Lee HW, Park K S, Hyun WS (2002). TBT toxicity on the marine microalga *Nannochloropsis oculata*. Mar. Pollut. Bull. 45:177-180.

Singla R, Garg N (2005). Influence of salinity on growth and yield Attributes in Chickpea Cultivars. Turk. J.Agric. For. 29: 231-235.

Stein JR (1979). Hand book of phycological methods. Culture method and growth measurements. Cambridge University press, London.

Tam NFY, Wong YS (1995). Wastewater treatment with microorganisms. The commercial press (H.K.) Ltd. 2D Finnie St. Quarry Bay, Hong Kong.215 pages.

Thomas KV, McHugh M, Hilton M, Waldock M (2003). Increased persistence of antifouling paint biocides when associated with paint particles. Environ. Pollu. 123:153-161.

Vaadia Y, Waisel Y (1967). Physiological processes as affected by water balance. In: "Irrigation of Agricultural lands" (R.M.Hagen, H.R.Haise, T.W.Edminster, eds).

Van Wezel AP, Van Vlaardingen P (2004). Environmental Risk Limits for Antifouling Substances. Aquat. Toxicol. 66: 427- 444.

Visviki I, Rachlin JW (1992). Ultra structural changes in *Dunaliella minuta* following acute and chronic exposure to copper and cadmium. Arch. Environ. Contam. Toxicol. 23: 420-425.

Zamora-Ley I M, Gardinali PR, Jochem FJ (2006). Assessing the effect of Irgarol 1051 on marine phytoplankton populations in Key Largo Harbor, Florida. Marine Pollution Bulletin. 52: 935-941.

Anti-diabetic effect of Cyclo-His-Pro (CHP)-enriched yeast hydrolysate in streptozotocin-induced diabetic mice

Yooheon Park[1], Hyun Jung Lee[1], Jang Won Choi[2], Song Hwan Bae[3] and Hyung Joo Suh[1]

[1]Department of Food and Nutrition, Korea University, Seoul 136-703, Republic of Korea.
[2]Department of Bioindustry, College of Life and Environment, Daegu University, Gyeongsan 712-714, Republic of Korea.
[3]Department of Food and Biotechnology, Hankyong National University Anseong, Anseong 456-749, Republic of Korea.

The present study was designed to investigate the hypoglycemic effects of the daily oral dose of 0.50 to 0.75 g/kg of yeast hydrolysate (YH) containing high Cyclo-His-Pro (51.0 mg CHP/g YH) on normal and streptozotocin (STZ)-induced diabetic rats for 14 days. In STZ-induced diabetic rats, after administrations of the YH for 14 days, the body weight gain was significantly increased in dose dependent manner, and the plasma glucose levels were decreased approximately (60%) as compared to the STZ induced diabetic control group. Glucose level showed significant differences between the diabetic control (DC) and the YH administered groups in oral glucose tolerance test (OGTT) ($P<0.05$). Results of the OGTT showed a significant decrease in the area under curve (AUC) value of YH supplemented groups as compared to the DC group. The present data suggests that the CHP-enriched YH has potential anti-diabetic effect, which can help in the cure and management of diabetes.

Key words: Yeast hydrolysate, Cyclo-His-Pro (CHP), diabetes, streptozotocin.

INTRODUCTION

Diabetes mellitus (DM) is one of the major endocrine disorder and global public health problems, affecting nearly 10% of the world's population. DM consists of a group of syndromes characterized by hyperglycemia; altered metabolism of lipids, carbohydrates and proteins; and the increased risk of complications from vascular disease (Davis and Granner, 2001). Conventional drugs used in its treatment are associated with drawbacks such as rigid and multiple dosing regimen, high-cost, inaccessibility and untoward effects. Scientific investigations of alternatives for diabetes have provided valuable clues for the development of therapeutic strategies. In the last few decades, there has been exponential growth in the field of alternatives by natural supplements, owing to their natural origin and lesser side effects (Yeh et al., 2003), since natural sources are usually considered to be less toxic with fewer side effects than synthetic sources (Pari and Umamaheswari, 2000). They have the potential to impart therapeutic effect in complicated disorders such as diabetes and its complications (Tiwari and Rao, 2002).

With more than 1000 strains of *Saccharomyces cerevisiae* categorized as generally recognized as safe, yeast is utilized in biochemical and medical applications (Carver, 1994) as well as many food industries such as brewing (Casey and Ingledew, 1983), winemaking (Shinohara et al., 2000) and baking (Chell, 1997). Several studies have demonstrated that yeast hydrolysate (YH) displays antiobesitic (Jung et al., 2009) and antidiabetic activity (Morley et al., 1981; Steiner et al., 1989). For these reasons, YH is receiving remarkable attention as a functional substance in the diet food market. The continuing development of functional foods is likely to entail increased use of different protein sources known to contain bioactive components.

Cyclo-His-Pro (CHP) is a naturally occurring cyclic dipeptide consisting of histidyl and proline and is a metabolite of thyrotropin-releasing hormone. As plasma levels of CHP in humans is increased after ingestion of glucose, CHP activity has been suggested to be related to glycemic control in patients with DM. Dietary feeding of CHP with zinc supplementation significantly improves insulin sensitivity and glucose clearance in diabetic animals and humans (Hwang et al., 2003; Song et al., 2001; Rosenthal et al., 2001). Furthermore, several studies have demonstrated that CHP decreases food intake, consequently, mimicking the action of leptin, the appetite control hormone (Morley et al., 1981; Steiner et al., 1989). Hence, CHP plays an important role in regulating insulin and leptin sensitivity, while no evidence of toxicity or side effects associated with its oral administration has been reported (Song et al., 2005, 2009).

We developed CHP containing high content of YH for possible applications of this cyclic dipeptide in metabolic disorder therapy. In the previous study (Jung et al., 2011), we increased the CHP content in the YH that was obtained from enzymatic hydrolysis and briefly assessed the antidiabetic effect by the oral glucose tolerance test (OGTT) in a type 1 diabetic animal model. The purpose of this study was to assess the anti-diabetic effect of YH with a high CHP content in normal and streptozotocin (STZ)-induced rats.

MATERIALS AND METHODS

Animals and diets

The animal protocol was approved by the Korea University Animal Care Committee. Male Sprague-Dawley (SD) rats, obtained at 6 weeks of age (200 ± 10 g), from Central Lab. Animal Inc. (Seoul, Korea) were housed individually in plastic cages with grated stainless steel floor. The colony was maintained at 24 ± 1°C with 60% relative humidity and a 12 h light/12 h dark cycle. The rats had *ad libitum* access to water and the rodent chow (Samyang Co., Seoul, Korea). Composition (g/100 kg of diet) of the chow was: moisture, 80; protein, 230; fat, 35; fiber, 50; carbohydrate, 600; and water.

Preparation of the diabetic rats

STZ (Sigma-Aldrich, St. Louis, MO, USA) was used to induce diabetes. The animals were given intraperitoneal injections of freshly prepared STZ (40 mg/kg in 0.01 M citrate buffer, pH 4.5) for 5 days, and the normal control (NC) group was injected with 0.01 M citrate buffer only. Five days after STZ treatment, blood was collec-ted from the tip of the tail vein, and fasting glucose level was measured (Takamura et al., 1999). The STZ injection destroyed a sufficient number of islet beta cells to be a model of type 1 diabetes with the glucose level of over 300 mg/dl. After the STZ-induction period, the rats were separated into four groups (six rats/group) based on their blood glucose level and body weight. Treatment was started 8 days after the STZ injection.

YH preparation

An 8% yeast suspension was hydrolyzed for 48 h using Flavourzyme (endoprotease and exopeptidase from *Aspergillus oryzae*). The hydrolysis temperature was 50°C for the crude enzymes, and the enzyme : yeast substrate ratio was 1:100 for the enzymes. The hydrolysis of yeast with enzymes was performed in 0.01 M phosphate buffer. pH was adjusted to the optimal values for the Flavourzyme specific proteases (pH 7.0) before hydrolysis was initiated. Hydrolysis was inactivated by heating at 90°C for 5 min. The YH obtained from enzymatic hydrolysis was first passed through a 0.2 µm membrane filter (Satocon cassette, Sartorius, Goettingen, Germany). A portion of the solution was removed immediately, and the filtrate was then pumped through a 10 kDa molecular weight cut-off membrane (Satocon cassette, Sartorius). The YH obtained from ultrafiltration was adsorbed with active carbon. The YH obtained from ultrafiltration, and the resulting substance were dried and used as the YH (51.0 mg/g CHP).

YH administration

YH was suspended in distilled water and administered orally through an intragastric tube at doses of 0.50 or 0.75 g/kg body weight. The volume of administrated extract was 1 ml for each animal. The rats were divided into four groups (eight rats/group): NC: 1 ml water for normal control; DC: 1 ml water for diabetic rats; YH-1: 0.50 g/kg body weight YH; or YH-2: 0.75g/kg body weight YH for diabetic rats, and each group of rats was treated daily for 14 days using an intragastric tube.

At the end of the experimental period, the rats were anesthetized with ethyl ether, and blood was collected from the abdominal artery into a heparinized sterile tube. Plasma was obtained by centrifugation at 1,800 × *g* for 30 min and stored at -80°C until further analysis. The liver, spleen and kidneys were excised and weighed after sacrificing the animals.

Fasting blood glucose level

Blood glucose levels were monitored every week after a 12 h fast in venous blood taken from the tail vein using a glucose analyzer. Fasting blood glucose was measured after 14 days of YH treatment during which the animals were fed a normal diet. Rats were fasted for 12 h, and blood was collected from the tip of the tail vein. Blood glucose level was measured using a blood glucose analyzer (Superglucocard II, Arkray Inc., Kyoto, Japan) based on the glucose oxidase method (Barham and Trinder, 1972). Results are expressed as glucose mg/dl blood.

Oral glucose tolerance test

On the day of animal sacrifice, after an overnight fast, 0 min blood was taken from the tip of the tail vein from all the rats. Rats that had been administered YH (0.75 g/kg body weight), received oral load of 30% glucose solution (2 g glucose/kg body weight). Blood samples were collected from the tail vein at 30, 60, 90 and 120 min after the oral glucose load and treated as described previously for the plasma glucose analysis (Han et al., 2007). Blood glucose level was expressed as increments from baseline. Incremental areas under the response curves (AUC) were calculated using the trapezoidal rule, with fasting levels considered as baseline.

Statistical analysis

All statistical analyses were performed using the Statistical Package

Table 1. Effect of yeast hydrolysate on body weight in normal and STZ-induced diabetic rats.

Group[1]	Initial body weight (g)	Final body weight (g)	Body weight gain (g/14 days)
Normal control	211.20 ± 1.02[2a]	236.50 ± 3.96[a]	25.30 ± 2.94[a]
Diabetic control	195.80 ± 1.39[b]	169.30 ± 2.69[c]	-26.50 ± 1.31[d]
Diabetic YH-1	192.10 ± 0.65[b]	179.80 ± 0.94[c]	-12.30 ± 0.29[c]
Diabetic YH-2	195.80 ± 1.22[b]	197.00 ± 5.88[b]	8.40 ± 2.56[b]

[1]Diabetic YH-1; STZ-induced diabetic rats given 0.50 g/kg of yeast hydrolysate; Diabetic YH-2, STZ-induced diabetic rats given 0.75 g/kg of yeast hydrolysate. [2]Mean ± SEM; values with different superscripts letters within the same column are significantly different at $P<0.05$ by Duncan's multiple range test.

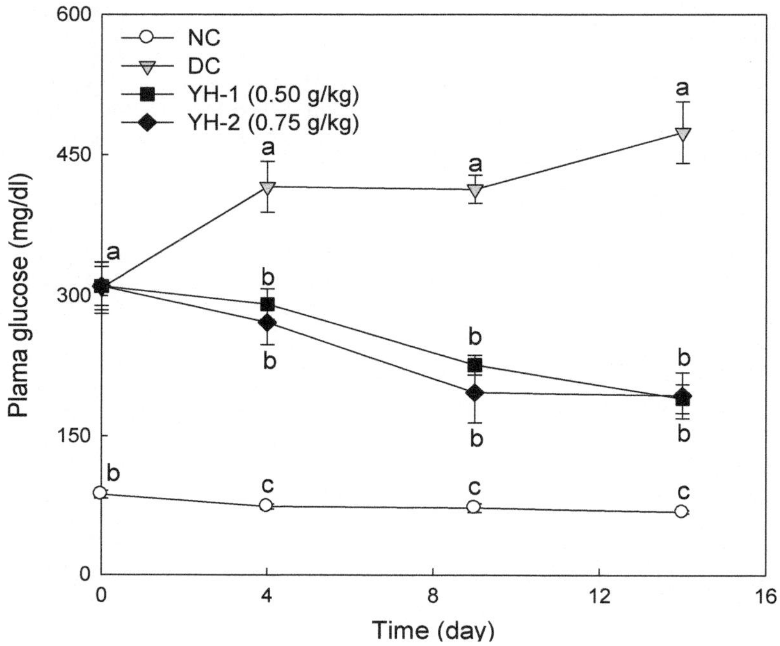

Figure 1. Effect of yeast hydrolysate on plasma glucose in normal and STZ-induced diabetic rats. Values with different superscripts letters within the column are significantly different at $P<0.05$ by Duncan's multiple range test; NC, 1 ml water for normal control; DC, 1 ml water for diabetic rats; YH-1, 0.50 g/kg body weight YH; YH-2, 0.75 g/kg body weight YH for diabetic rats.

for Social Sciences (SPSS) version 12.0 (SPSS Inc., USA). The differences among groups were evaluated by one-way analysis of variance (ANOVA) and Duncan's multiple range tests. All data were reported as the mean and standard error. A level of $P<0.05$ was used as the criterion for statistical significance.

RESULTS

Body weight gain in STZ-induced mice during the 14 day period

YH or vehicle was administered orally to STZ-induced mice for 14 days, and body weight was measured. As shown in Table 1, no significant differences in initial body weight were observed among the STZ-diabetic groups. The NC group continued to gain weight until the end of the study. STZ produced significant body weight losses as compared to that in the NC group. After 14 days of testing, final body weights and mean body weight gains were significantly lower ($P<0.05$) in all STZ-diabetic groups as compared to those of the NC group. Mean weight gains in the YH administered groups were significantly higher ($P<0.05$) than those in the DC group, and significant improvements ($P<0.05$) in body weight were observed in the YH-2 group.

Fasting blood glucose levels in STZ-induced mice during the 14 day period

The effect of repeated oral administration of YH or vehicle on fasting blood glucose levels in the STZ-diabetic groups during 14 days of treatment are presented in Figure 1. Intraperitoneal administration of STZ to rats

Table 2. Effect of yeast hydrolysate on organ weights in normal and STZ-induced diabetic rats.

Group[1]	Liver (g/100 g of body weight)	Spleen (g/100 g of body weight)	Kidney (g/100 g of body weight)
Normal control	4.42 ± 0.10[2b]	0.57 ± 0.02[NS]	0.48 ± 0.04[NS]
Diabetic control	4.92 ± 0.10[a]	0.60 ± 0.03	0.60 ± 0.06
Diabetic YH-1	4.52 ± 0.10[ab]	0.58 ± 0.03	0.54 ± 0.04
Diabetic YH-2	4.42 ± 0.01[b]	0.59 ± 0.05	0.51 ± 0.05

[1]Diabetic YH-1, STZ-induced diabetic rats given 0.50 g/kg of yeast hydrolysate; Diabetic YH-2, STZ-induced diabetic rats given 0.75 g/kg of yeast hydrolysate. [2]Mean ± SEM; values with different superscripts letters within the same column are significantly different at *P*<0.05 by Duncan's multiple range test; NS, not significant.

caused a significant diabetogenic response with significant increases (*P*<0.05) in fasting blood glucose levels as compared to those in the NC group. Plasma glucose level increased continuously in the DC group during the experimental period, reaching a final level of approximately 500 mg/dl. YH administered at two different doses of 0.50 and 0.75 g/kg body weight to the STZ-treated diabetic groups resulted in a significant reduction (*P*<0.05) of fasting blood glucose levels, which was associated with treatment duration. After 14 days of treatment, the fasting blood glucose levels in the STZ-induced diabetic mice treated with YH at 0.75 g/kg (193.80 ± 10.12 mg/dl) were significantly lower (*P*<0.05) as com-pared to those of the DC group (474.00 ± 13.35 mg/dl).

Organ weights in the STZ-induced mice

We investigated the influence of YH on the liver, spleen and kidney weight (Table 2). The DC group (4.92 ± 0.10 g/100 g body weight) had significantly greater (*P*<0.05) liver weights than the NC group (4.42 ± 0.10 g/100 g body weight). The liver weight was significantly changed by YH administration in a dose dependent manner. Spleen weight was not significantly different (*P*<0.05) regardless of diabetes induction. However, the DC group (0.60 ± 0.03 g/100 g body weight) exhibited greater spleen weight than the NC (0.57 ± 0.2 g/100 g body weight) and YH administered groups (0.58 ± 0.03 and 0.59 ± 0.05 g/100 g body weight, respectively). Diabetes induced by STZ injection resulted in increased kidney weight; the YH administered groups had decreased kidney weight when compared with the DC group.

Oral glucose tolerance test

The OGTT and the calculated AUC were used to determine the glucose response to YH administration. As a result, the DC, YH-1, YH-2 and NC groups showed significant increases (*P*<0.05) in blood glucose levels at 30 min after approximately 522.00 ± 6.21, 443.00 ± 3.59, 431.50 ± 6.29 and 209.30 ± 3.10 mg/dl glucose gavages, respectively. However, the DC group (248.80 ± 7.68 mg/dl) did not recover baseline glucose level even after 120 min. The OGTT results show that blood glucose decreased significantly (*P*<0.05) in the YH administered groups when compared with that in the DC group (Figure 2A). Furthermore, the AUC was significantly smaller in the YH administered groups than in the DC group (Figure 2B). The AUC of the STZ-induced diabetic mice treated with YH-2 was significantly smaller (*P*<0.05) than that in the YH-1 group.

DISCUSSION

DM is a chronic metabolic disorder affecting a major proportion of the population worldwide. Conventional therapies for DM have many disadvantages such as side effects and a high rate of secondary failure. In contrast, natural sources are expected to have similar efficacy without side effects as those of conventional drugs. CHP exists in several common nutritional supplements, and it is known to have antiobesity effects (Jung et al., 2009) as well as antidiabetic activity (Morley et al., 1981; Steiner et al., 1989). This study was undertaken to evaluate the hypoglycemic activity of YH with a high content of CHP in normal and STZ-induced diabetic rats.

STZ is used to induce DM by selectively affecting pancreatic β-cells. Thus, STZ affects endogenous insulin release and increases blood glucose levels (So et al., 2011). Final body weights and mean body weight gains were significantly lower (*P*<0.05) in all STZ-diabetic groups after 14 days of testing, as compared to those of the NC group (Table 1). The induction of DM with STZ is associated with a characteristic loss of body weight due to increased muscle loss (Chatterjee and Shinde, 1994). However, hypoglycemic treatment with medicinal herbs may help STZ-induced diabetic rats to recover from hyperglycemia (Hwang et al., 2005; Prasad et al., 2009). Treatment with YH significantly improved body weight, indicating that it prevented muscle loss due to hyperglycemic conditions.

The liver weights were significantly greater (*P*<0.05) in the DC group as compared to those of the NC group, and the YH administered groups showed significant changes in liver weight in a dose-dependent manner as compared to those of of the DC group. The lipid metabolic pathway is activated by acetyl-CoA in STZ-induced diabetic rats, rather than normal glucose metabolism, due to reduced

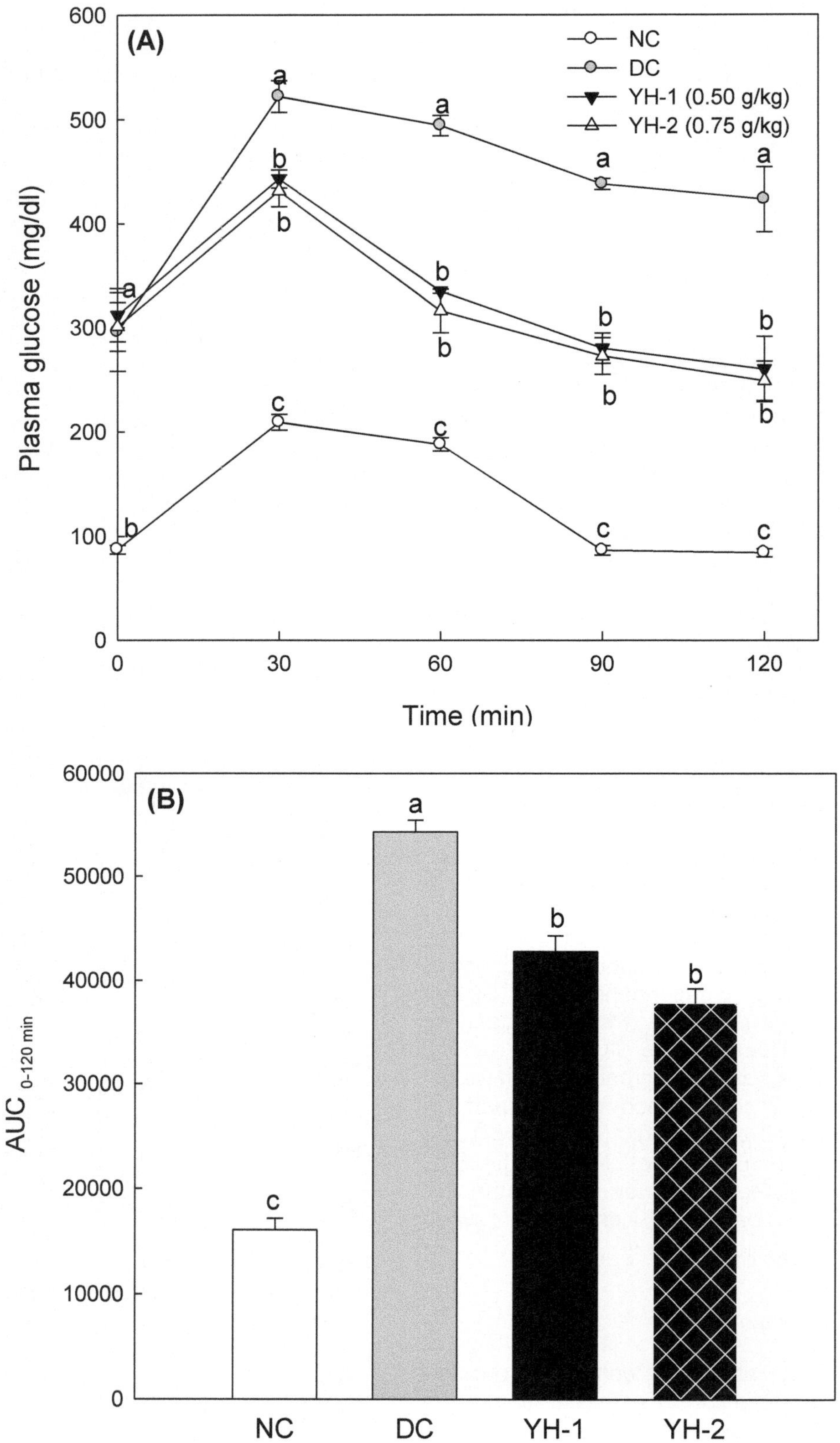

Figure 2. Effect of yeast hydrolysate on oral glucose tolerance test (A) OGTT and integrated area under the curves (AUCs) over a period of 0 to 120 min (B) in normal and STZ-induced diabetic rats). Values with different superscripts letters within the column are significantly different at *P*<0.05 by Duncan's multiple range test. NC, 1 ml water for normal control; DC, 1 ml water for diabetic rats; YH-1, 0.50 g/kg body weight YH; YH-2, 0.75 g/kg body weight YH for diabetic rats.

insulin levels (Goldstein and Brown, 1977). Based on this hypothesis, increased synthesis and accumulation of heaptic triglycerides can be well explained by the liver hypertrophy in STZ-induced diabetic rats. DM induced by STZ resulted in increased kidney weight. The DC group exhibited greater kidney weight gains than the NC group. However, the YH administrated groups showed decreased kidney weights as compared to the DC group, indicating that renal enlargement commonly occurs in DM patients. The increase in kidney volume and size may be due to the increased glomerular filtration rate in early stage DM. McNeill (1999) reported that such diabetic animals develop glomerular hypertrophy, renal hyper filtration and enlargement, increased urinary albumin excretion, and glomerular extracellular matrix accumu- lation within weeks of DM onset. The results of this study support the possibility that CHP in YH might be effective in preventing renal failure in diabetic animals.

The intraperitoneal administration of STZ to SD rats caused a significant diabetogenic response with significant increases in the levels of fasting blood glucose as compared to those of the NC group. After 14 days of treatment, the fasting blood glucose levels of the STZ-induced diabetic mice treated with YH at 0.75 g/kg were significantly lower ($P<0.05$) than those of the DC group. As a result, *in vivo* treatment with YH resulted in a significant reduction in blood glucose level due to an increased rate of glucose elimination. The OGTT results reveal that blood glucose level was significantly decreased in the YH administered groups as compared to that of the DC group. Furthermore, the AUC was significantly lower in a dose-dependent manner ($P<0.05$) in the YH administered groups as compared to the DC group. Hyperglycemia was induced by intraperitoneal STZ. Maintaining blood glucose at levels close to normal and preventing diabetic complications are major goals in the treatment of DM (Control and Group, 1994). YH may enhance glucose utilization, because it significantly decreases blood glucose levels in glucose-loaded rats, although the exact mechanisms are unclear.

Similar results were reported by Song et al. (2001), in which a prostate extract containing zinc and CHP significantly decreased blood glucose and improved glucose tolerance and insulin sensitivity in STZ-induced diabetic rats. These results indicate that CHP has a strong ability to stimulate intestinal zinc absorption, cellular zinc uptake and promoted glucose utilization (Rosenthal et al., 2001; Song et al., 2001, 2003). Zinc deficiency critically affects DM, because zinc activates insulin receptor β-subunits, thereby exerting an influence on glucose metabolism (Song et al., 2001). Although, the exact mechanism of the antidiabetic effect of CHP has not been clearly established, it is likely that CHP is involved in regulating insulin and leptin sensitivity by stimulating zinc metabolism because of the following arguments: oral intake of CHP plus zinc significantly improves glucose tolerance in diabetic and overweight animals in the state of decreased or unchanged insulin levels (Hwang et al., 2003; Song et al., 2001); high CHP concentrations inhibit insulin and glucagon secretion from islet cells *in vitro* (Wilber et al., 1984); and CHP stimulates zinc transport mechanisms across the small intestine and muscle cell membrane to increase zinc use (Rosenthal et al., 2001). Accordingly, it is assumed that the YH-containing CHP might affect zinc metabolism.

In conclusion, YH containing CHP was useful as a therapeutic agent to improve glucose tolerance in diabetic rats. Further investigations are needed to elucidate the molecular mechanisms by which the antidiabetic effects of CHP are mediated.

ACKNOWLEDGEMENT

This study was supported by the Technology Development Program for Food, Ministry for Food, Agriculture, Forestry and Fisheries, Republic of Korea (Project No.: 109144-03-3).

REFERENCES

Barham D, Trinder P (1972). An improved colour reagent for the determination of blood glucose by the oxidase system. Analyst 97(151):142-145.

Chatterjee MN, Shinde K (1994). Text Book of Medical Biochemistry. Metabolism of Carbohydrate. Jaypee Brothers Medical Publishers Private Ltd., New Delhi, India. p 284 - 322.

Chell M (1997). New developments in breadmaking. Food Manu. 72:21–22

Control D, Group CTR (1994). Effect of intensive diabetes treatment on the development and progression of long-term complications in adolescents with insulin-dependent diabetes mellitus: Diabetes Control and Complications Trial. J. Pediatr. 125(2):177-188.

Carver JD (1994). Dietary nucleotides: cellular immune, intestinal and hepatic system effects. Journal of Nutrition 124:144S-148S.

Casey GP, Ingledew WM (1983). High gravity brewing: influence of pitching rate and wort gravity on early yeast viability. J. Am. Soc. Brew. Chem. 41:481-488.

Davis SN, Granner DK (2001). Insulin, oral hypoglycemic agents, and the pharmacology of the endocrine pancreas. Goodman & Gilman's: The Pharmacological Basis of Therapeutics 10 th. McGraw Hill, New York, USA. p. 1679-1714.

Goldstein L, Brown S (1977). The low-density lipoprotein pathway and its relation to atherosclerosis. Annu. Rev. Biochem. 46(1):897-930.

Han GC, Ko SK, Sung JH, Chung SH (2007). Compound K enhances insulin secretion with beneficial metabolic effects in *db/db* mice. J. Agric. Food Chem. 55(26):10641-10648.

Hwang HJ, Kim SW, Lim JM, Joo JH, Kim HO, Kim HM, Yun JW (2005). Hypoglycemic effect of crude exopolysaccharides produced by a medicinal mushroom *Phellinus baumii* in streptozotocin-induced diabetic rats. Life Sci. 76(26):3069-3080.

Hwang I, Go V, Harris D, Yip I, Kang K, Song M (2003). Effects of cyclo (his–pro) plus zinc on glucose metabolism in genetically diabetic obese mice. Diabetes Obes. Metab. 5(5):317-324.

Jung EY, Kang DH, Suh HJ, Chang UJ (2009). Effects of yeast hydrolysate on neuropeptide Y (NPY) and tryptophan hydroxylase (TPH) immunoreactivity in rats. Phytother. Res. 23(5):619-623.

Jung EY, Lee HS, Choi JW, Ra KS, Kim MR, Suh HJ (2011). Glucose tolerance and antioxidant activity of Spent Brewer's yeast hydrolysate with a high content of cyclo-his-pro (CHP). J. Food Sci. 76(2):272-278.

McNeill JH (1999). Experimental models of diabetes. Boca Raton, CRC Press LLC, Florida, USA. p. 3-17.

Morley JE, Levine AS, Prasad C (1981). Histidyl-proline diketopiperazine decreases food intake in rats. Brain Res. 210:475-478.

Pari L, Umamaheswari J (2000). Antihyperglycaemic activity of *Musa sapientum* flowers: effect on lipid peroxidation in alloxan diabetic rats. Phytother. Res. 14(2):136-138.

Prasad S, Kulshreshtha A, Qureshi TN (2009). Antidiabetic activity of some herbal plants in streptozotocin induced diabetic albino rats. Pak. J. Nutr. 8(5):551-557.

Rosenthal M, Hwang I, Song M (2001). Effects of arachidonic acid and cyclo (his-pro) on zinc transport across small intestine and muscle tissues. Life Sci. 70(3):337-348.

So O, Eai AOA, Akinpelu D (2011). Antidiabetic and haematological effect of aqueous extract of stem bark of *Afzelia africana* (Smith) on streptozotocin-induced diabetic Wistar rats. Asian Pac. J. Trop. Biomed. 1(5):353-358.

Shinohara T, Kubodera S, Yanagida F (2000). Distribution of phenolic yeasts and production of phenolic off-flavors in wine fermentation. J. Biosci. Bioeng. 90:90-97.

Song M, Rosenthal M, Song A, Uyemura K, Yang H, Ament M, Yamaguchi D, Cornford E (2009). Body weight reduction in rats by oral treatment with zinc plus cyclo-(his-pro). Br. J. Pharmacol. 158(2):442-450.

Song MK, Hwang IK, Rosenthal MJ, Harris DM, Yamaguchi DT, Yip I, Go VLW (2003). Anti-hyperglycemic activity of zinc plus cyclo (his-pro) in genetically diabetic Goto-Kakizaki and aged rats. Exp. Biol. Med. 228(11):1338-1345.

Song MK, Rosenthal MJ, Hong S, Harris DM, Hwang I, Yip I, Golub MS, Ament ME, Go VLW (2001). Synergistic antidiabetic activities of zinc, cyclo (his-pro), and arachidonic acid. Metabolism 50(1):53-59.

Song MK, Rosenthal MJ, Song AM, Yang H, Ao Y, Yamaguchi DT (2005). Raw vegetable food containing high cyclo (his-pro) improved insulin sensitivity and body weight control. Metabolism 54(11):1480-1489.

Steiner H, Wilber JF, Prasad C, Rogers D, Rosenkranz RT (1989). Histidyl proline diketopiperazine (cyclo [his-pro]) in eating disorders. Neuropeptides 14(3):185-189.

Takamura T, Ando H, Nagai Y, Yamashita H, Nohara E, Kobayashi K (1999). Pioglitazone prevents mice from multiple low-dose streptozotocin-induced insulitis and diabetes. Diabetes Res. Clin. Pract. 44(2):107-114.

Tiwari AK, Rao JM (2002). Diabetes mellitus and multiple therapeutic approaches of phytochemicals: Present status and future prospects. Curr. Sci. 83(1):30-38.

Wilber JF, Mori M, Kandarakis ED, Iriuchijima T, Sacks H, Kitabchi AE (1984). Histidyl-proline diketopiperazine [cyclo(his-pro)]: a new neuropeptide modulator of insulin and glucagon secretion. Trans. Assoc. Am. Physicians 97:88-94.

Yeh GY, Eisenberg DM, Kaptchuk TJ, Phillips RS (2003). Systematic review of herbs and dietary supplements for glycemic control in diabetes. Diabetes Care 26(4):1277-1294.

Study of root *para*-nodules formation in wheat (*Triticum durum*) inoculated with *Frankia* strain Ccl3 and treated with 2, 4-dichlorophenoxyacetate (2, 4-D)

Djemel Amina N., Benmati Mahbouba, Ykhlef Nadia, Belbekri Nadir and Djekoun Abdelhamid

Laboratoire de Génétique, Biochimie et Biotechnologies Végétales University Mentouri of Contantine, Algeria.

***Frankia* strains can induce N_2-fixing root nodules on certain non-leguminous plants. It is known that exogenous application of 2,4-dichlorophenoxyacetate (2,4-D) affects root morphology. In this work, wheat roots were treated with 2,4-D and inoculated with the actinomycete *Frankia*. Wheat plants grew in a growth chamber with hydroponic medium. Binocular observation revealed that para-nodules were formed when wheat roots were inoculated with *Frankia* and the root length was enhanced. When the inoculation with *Frankia* was combined to 2,4-D treatment, the para-nodules formed were bigger and more numerous, while the root length was shortened.**

Key words: *Frankia*, wheat, roots, *para*-nodules, 2,4-dichlorophenoxyacetate.

INTRODUCTION

Actinomycetes are found in soil, rhizosphere, ponds and lake sediments. They are a large group of bacteria that can produce secondary metabolites which have important applications in pharmacy, medicine and agriculture (Suttiviriya et al., 2008). Actinorhizal symbioses are mutual interactions between plants, such as *Casuarina glauca* and *Alnus glutinosa*, and the soil bacteria *Frankia* that lead to the formation of root nodules (Perrine-Walker et al., 2010), which are specialized organs for assimilation of the N derived from fixation by the symbiotic bacteria (Berry et al., 2004). Little is known about the signals exchanged between the two partners during the establishment of actinorhizal symbioses (Perrine-Walker et al., 2010).

Synthetic auxins exogenously applied stimulated lateral root primordium initiation on *Alnus* roots, an important step in nodule formation, and thus auxins may play a role in the actinorhizal nodulation process (Berry et al., 1989). It is known that nodule like structures, could be induced by 2,4-D on plant roots and that they could be colonized by microorganisms (Zeman et al., 1992). Little studies were realized about the interaction between *Frankia* and wheat. Soil with *Frankia* had significant effects on canola shoot length, plant dry weight, number of leaves, and total N, but not on wheat shoot length, fresh and dry weight of shoots and ears, dry weight of roots and total N (Cusato and Tortosa, 2000). In this work, para-nodules were induced using 2,4-D and inoculation with *Frankia* and an eventual growth promoting on root system was elucidated.

MATERIALS AND METHODS

Plant host

Seeds of wheat (Triticum durum) were surface-sterilized in 70% ethanol for 20 s, followed by 30 min in a 30% solution of hypochlorite. After various washings with sterile distilled water, the

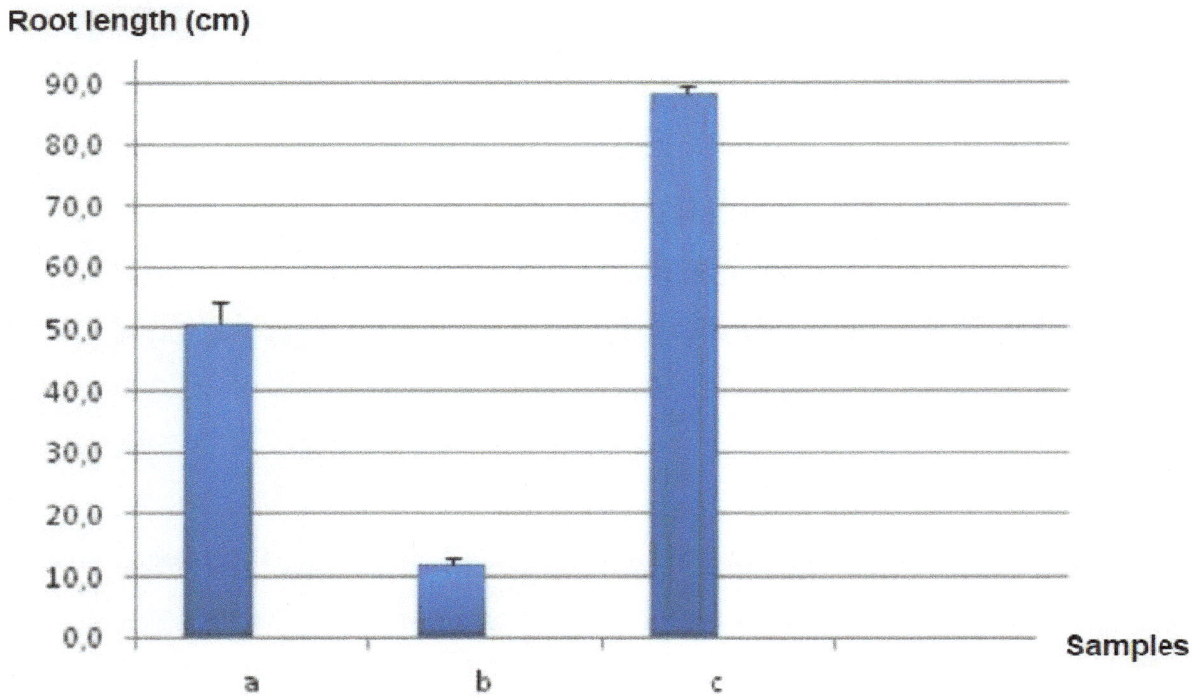

Figure 1. Root length difference (cm). **a.** control. **b.** inoculated root with Frankia and treatment by 2,4-D. **c.** inoculated root with Frankia.

seeds were germinated in the dark under sterile conditions on moist filter paper in Petri dishes (Saatovich, 2006). The germinating seeds were then placed in 1 L bottles (sterilized by UV) containing sterile hydroponic solution and grown until one week old before inoculation with Frankia. 0.4 mg/l of 2,4-D was added to hydroponic solution for para-nodule induction.

Bacterial material

20 ml of Frankia strain Ccl3 culture (isolated from Casuarina glauca) were centrifuged at 5000 rpm for 15 min. The supernatant was eliminated; 5 ml of sterile distilled water were added to the residuum to reduce Frankia hyphae, using sterile syringe. 15 ml of sterile distilled water were added to the suspension and used for inoculation. Seedlings of wheat were grown for three months.

RESULTS AND DISCUSSION

Effects of 2, 4-D and *Frankia* on root system

Inoculation with *Frankia* alone increased the root length. When combined with 2,4-D treatment, the root length was considerably shortened (Figure 1), and the lateral ones were hardly absent. These results are in accordance with those of El-Shahed and Abdel-Wahab (2006), who established that different 2,4-D concentrations in combination with *Nostoc rivulare* decrease root length in wheat, maize and rice, while the simple application of *N. rivulare* significantly enhances shoot and root weight as well as plant height as we observed in this work with *Frankia*. Fischer et al. (2000), showed that wheat plants are treated with 2,4-D (0.4 µg /ml) and inoculated with *Azospirillum brasilense* present shorter roots.

Seedlings of wheat (*Triticum durum*) grew, during 3 months, in bottles containing hydroponic solution which is composed of 0.4 mg/l of 2,4-D and 20 ml of suspended *Frankia* strain Ccl3 (isolated from *Casuarina glauca*). A represent control, B as inoculated root with *Frankia* and treatment by 2,4-D and C as inoculated root with *Frankia*. The significant elongation of wheat root only inoculated with *Frankia* due to the root oxygenation in bottles which was reduced, in comparison with some actinorhizal genera where the length of nodule roots is inversely correlated with the external oxygen tension (Pawlowski and Demchenko, 2012).

A pink coloration was observed on certain inoculated root parts. These pink zones were probably colonized by *Frankia* and the coloration was related to hemoglobin protein. Pawlowski and Demchenko (2012) noted that the promoters of soybean leghemoglobin and of the symbiotic hemoglobin from *C. glauca* are both specifically active in cells that stably accommodate the microsymbiont.

Effects of 2,4-D and *Frankia* on *para*-nodulation

Para-nodulation was more pronounced when inoculation was combined with 2,4-D, in terms of number and size, and the distribution of *para*-nodules was irregular but observed along the root length. The *para*-nodules resulting from the simple inoculation with *Frankia* were smaller, less numerous, had an irregular distribution and presented

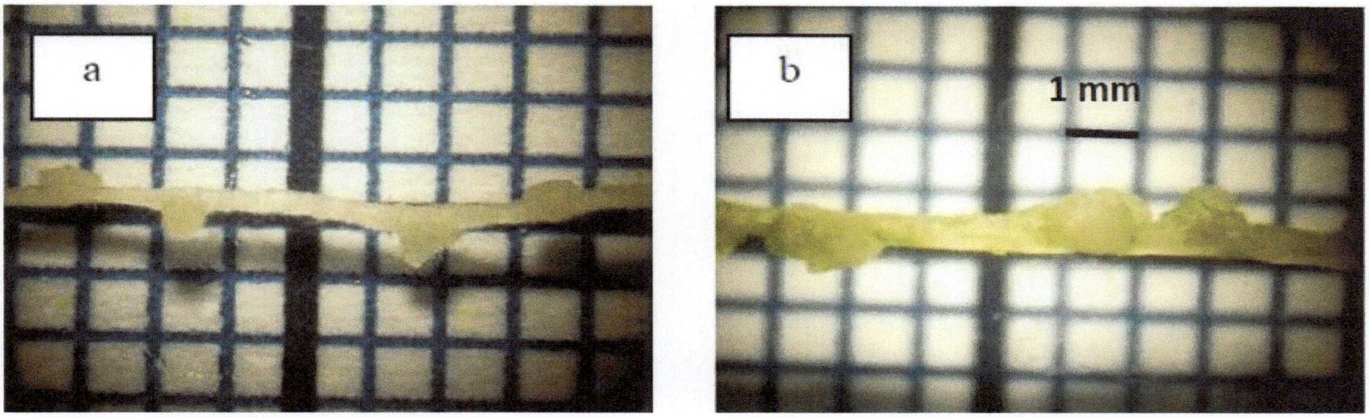

Figure 2. Binocular observation of para-nodules. **a.** Para-nodules induced by the inoculation with Frankia. **b.** Para-nodules induced by the combination between Frankia and 2,4-D.

several shapes (Figure 2). *Para*-nodule formation through 2,4-D treatment is a physiological process independent from bacterial action (Francisco and Akao, 1993). It is strongly possible that *para*-nodules obtained in our study without adding 2,4-D were resulting from the *Frankia* colonization, comparing to Saatovich (2006) who obtained *para*-nodules without the addition of phytohormones on the roots of wheat plants inoculated with *Azospirillum*. Biabani et al. (2012) reported that the colonization of *para*-nodules is more extensive than that of lateral root emergence sites in plants which are not treated with 2,4-D as we found in the present study.

Actinorhizal nodules have an indeterminate growth pattern, which generally consist of numerous conical-shaped lobes and may or may not exhibit nodular roots. The *Rhizobium*-induced nodule can be spherical, cylindrical or collar-shaped according to the patterns of plant cell division and growth of cortical cells as reported by Diouf et al. (2003), while in this study other shapes were obtained in addition to those described by Diouf et al. (2003) and some shapes exhibited root that emerged from the nodules and 2,4-D induced *para*-nodules.

Isolation of *Frankia* from wheat root and nodules

After an isolation experiment, by disinfecting the nodules with 30% of H_2O_2 (Mansour et al., 1990), pleasuring them, then incubating these ones in liquid DPM medium at 30°C (Gtari et al., 2004) for 5 to 9 weeks, it was determined that *Frankia* colonized root and both *para*-nodules obtained by the combination of 2,4-D and *Frankia*, and those obtained only by inoculation with *Frankia*. The quantity of *Frankia* filaments was higher in *para*-nodules obtained when 2,4-D was combined with *Frankia*.

In this study we found that, for the first time, that the actinomycete *Frankia* stimulated the root growth and induced the *para*-nodulation of wheat root. We suggest that *Frankia* may be regarded as Plant Growth Promoting Bacteria and may be used as biofertilizer. These results indicate that *Frankia* has the potential to be applied in natural conditions to increase the root length of wheat plants.

ACKNOWLEDGEMENT

The authors are grateful to the engineers of the laboratory and to Mrs Bouldjadj Ryma for providing materials of this work.

REFERENCES

Berry A, Kahn R, Booth M (1989). Identification of indole compounds secreted by *Frankia* HFPArl3 in defined culture medium. Plant Soil 118: 205-209.

Berry A, Murphy T, Okubara P, Jacobsen KR, Swensen SM, and Pawlowski K (2004). Novel expression pattern of cytosolic Gln synthetase in nitrogen-fixing root nodules of the actinorhizal host, *Datisca glomerata*. Plant Physiol. 135(3): 1849-1862.

Biabani A, Boggs LC, Kovalskaya N, Umarov M (2012). Microscopic morphology of nitrogen fixing paranodules on wheat roots. Afr.J.Biotechnol. 11(12): 2971-2976.

Cusato MS, Tortosa RD (2000). Frankia and crops interactions. Phyton 68: 47- 53.

Diouf D, Diop TA, Ndoye I (2003). Actinorhizal, mycorhizal and rhizobial symbioses: how much do we know? Afr.J.Biotechnol. 2 (1): 1–7.

El-Shahed AM, Abdel-Wahab AM (2006). Para-nodule induction in wheat, maize and rice with 2,4-D and its infection with *Nostoc rivulare* kutzing. Pak.J.Biol.Sci. 9(9): 1693-1699.

Fischer S, Rivarola V, Mori G (2000). Colonization of wheat by *Azospirillum brasilense* Cd is impaired by saline stress. Plant Soil 225:187-191.

Francisco P, Akao S (1993). The 2, 4-D-induced wheat para-nodules are modified lateral roots with structure enhanced by rhizobial inoculation. Plant Soil 157:121-129.

Gtari M, Brusetti L, Gharbi S, Mora D, Boudabous A, Daffonchio D (2004). Isolation of *Elaeagnus*-compatible *Frankia* from soils collected in Tunisia. FEMS Microbiol.Lett. 234:349-355.

Mansour SR, Dewedar A, Torrey JG (1990). Isolation, Culture, and Behavior of *Frankia* Strain HFPCgl4 from Root Nodules of *Casuarina* glauca. Bot.Gaz. 151: 490-496.

Pawlowski K, Demchenko KN (2012). The diversity of actinorhizal symbiosis. Protoplasma 249(4):967-79.

Perrine-Walker F, Doumas P, Lucas M, Vaissayre V, Beauchemin NJ, Band LR, Chopard J, Crabos A, Conejero G, Péret B, King JR, Verdeil JL, Hocher V, Franche C, Bennett MJ, Tisa LS, Laplaze L (2010). Auxin carriers localization drives auxin accumulation in plant

cells infected by *Frankia* in *Casuarina glauca* actinorhizal nodules. Plant Physiol.154:1372–1380.
Saatovich SZ (2006). Azospirilli of Uzbekistan soils and their influence on growth and development of wheat plants. Plant Soil 283:137-145.
Suttiviriya P, Vajrodaya S, Thamchaipenet A (2008). Production of plant growth promoting agents from endophytic actinomycetes. 34th Congress on Science and Technology of Thailand.
Zeman AM, Tchan YT, Elmerich C, Kennedy IR (1992). Nitrogenase activity in wheat seedlings bearing *para*-nodules induced by 2,4-dichlorophenoxyacetic acid (2,4-D) and inoculated with *Azospirillum*. Res. Microbiol.143:847-855.

Fluctuating levels of reprogramming factor expression in cultured human undifferentiated keratinocytes

Mingjun Wu[1,2], Qiang Chen[1], Wei Wu[1], Zhongwei Zheng[1], Bisong Yue[2], Zhiqiang Wu[3], Yali Zhao[3], Weidong Han[3] and Xiaobing Fu[3]

[1]Institute of Blood Transfusion, Chinese Academy of Medical Science, Chengdu 610081, Sichuan Province, China.
[2]Key Laboratory of Bio-resources and Eco-environment (Ministry of Education, College of Life Sciences, Sichuan University, Chengdu 610064, Sichuan Province, China.
[3]Laboratory of Molecular Biology, Institute of Basic Medicine, Chinese People's Liberation Army General Hospital, Beijing 100853, China.

Although human undifferentiated keratinocytes (HUKs) can be reprogrammed to become induced pluripotent stem cells (iPSCs) with high efficiency and rapid kinetics by transducing reprogramming factors (RFs), the endogenous expression of reprogramming factors in cultured HUKs is not clear at different stages. In this study, keratinocytes were isolated from foreskin of adult subject (18 years old) and cultured on collagen type IV-coated culture dishes in a low-calcium, serum-free medium (Epilife, Invitrogen). In order to clarify the expression patterns of RFs and other stem cell markers in cultured human keratinocytes, total RNA was extracted using Trizol reagent, and polymerase chain reaction (PCR) was performed using established GenBank sequences to design primers. The subsequent PCR analysis was carried out by agarose gel electrophoresis. The expression levels of RFs and other stem cell markers in human HUKs clearly fluctuated during culturing, which supports the hypothesis that HUKs might be reprogrammed into a pluripotent state when the maximum levels of RFs expression are maintained by appropriate culture conditions.

Key words: Human undifferentiated keratinocytes, reprogramming factors, expression fluctuation.

INTRODUCTION

Human somatic cells can be directly reprogrammed to become induced pluripotent stem cells (iPSCs) by ectopic expression of reprogramming factors (RFs). Results from many independent groups have suggested that human iPSCs, once established, generally exhibit a normal karyotype, are transcriptionally and epigenetically similar to embryonic stem cells (ESCs) and maintain the potential to differentiate into derivatives of all germ layers (Takahashi and Yamanaka, 2006; Takahashi et al., 2007; Yu et al., 2007). This strategy, therefore, represents a significant breakthrough toward the practical use of iPSCs in regenerative medicine. For instance, iPSCs could serve as disease models and aid in the discovery of drugs and genes; furthermore, this approach to generating patient-specific iPSCs would undoubtedly transform regenerative medicine in numerous ways. Current reprogramming protocols, however, predominantly involve the expression of exogenous RFs using virus-mediated transduction. The potentially harmful genomic insertion of these viruses has been the major barrier to the clinical use of iPSCs generated using this approach. Therefore, recruiting endogenous Oct4 (octamer-binding transcription factor 4), Sox2 (SRY <sex determining region Y>-box 2), Klf4 (Krüeppel-like factor 4) and c-Myc (v-myc myelocytomatosis viral oncogene homolog) genes for reprogramming, eliminating the need for transgenes and other factors, would be a more practical approach with regard to the therapeutic use of iPSCs. Such an approach would make it possible for safe, viral-free human iPSCs to be derived routinely in the near future. The high efficiency and rapid kinetics of the generation of iPSCs from human undifferentiated keratinocytes (HUKs) involve retrovirus-mediated expression of three (Oct4, Sox2 and Klf4) or four (Oct4, Sox2, Klf4 and c-Myc) RFs (Aasen et al., 2008), suggesting that an accessible source of more easily reprogrammable cells from epidermal biopsies should facilitate progress toward the goal of safe, efficient iPSCs generation using non-transgenic approaches. Until recently, it was shown that mouse somatic cells-derived iPSCs can be generated using a combination of small-molecule compounds(Hou, Li et al. 2013).

In addition, the endogenous expression of RFs in HUKs should be characterized before non-transgenic reprogramming approaches are developed. In this study, therefore, we clarified the expression patterns of RFs and other stem cell markers in cultured human keratinocytes.

MATERIALS AND METHODS

Cell resources and culture medium

Foreskin removed from adult subject (18 years old) who underwent a routine circumcision was provided by the Chinese People's Liberation Army General Hospital. The patient and his family members signed an informed consent form. The fresh foreskin was collected, and the subcutaneous tissue was removed. Keratinocytes were separated from the foreskin using dispase and trypsin and cultured on collagen type IV-coated culture dishes in low-calcium, serum-free defined medium (Epilife, Invitrogen) as described below.

Isolation and culture of keratinocytes

Human adult foreskin from routine circumcisions was processed within 2 h of collection. Epithelial sheets were obtained after overnight incubation with 3 mg/ml dispase at 4°C, and the separated epidermal sheet was placed in 0.25% trypsin for 30 min at 37°C. Medium was then added to inactivate the trypsin, and the tubes were shaken violently to dissociate individual basal keratinocytes. The keratinocytes were plated on collagen type IV-coated culture dishes in low-calcium, serum-free defined medium (Epilife, Invitrogen).

RNA preparation and RT-PCR

The expression of selected genes (seven stem cell markers: Oct4, Sox2, c-Myc, Klf4, Nanog, teratocarcinoma-derived growth factor 1 (CRIPTO) and RNA exonuclease 1 (REX1), and six keratinocyte-specific markers: keratin (KRT) 1, KRT 5, KRT 10, KRT 14, KRT 15, and KRT 19) was analyzed by reverse transcription polymerase chain reaction (RT-PCR). Three wells of human keratinocytes (in 6-well plates) were collected on days 0, 1, 5, 18 and 33 of culture. Total cellular RNA was isolated and reversely transcribed using conventional protocols. PCR amplification was performed using the primer sets shown in Table 1. All primer sequences were determined using established GenBank sequences. Duplicate PCR reactions were amplified using primers designed against β-actin as a control to assess PCR efficiency and for subsequent analysis by agarose gel electrophoresis.

RESULTS

Morphological inspection of human keratinocytes during differentiation

Figure 1 shows the morphological changes observed in human keratinocytes after cultivation in low-calcium, serum-free defined medium for 33 days. The undifferentiated monolayer cultures showed small, uniform, polygonal cells attached to the bottom of collagen type IV-coated plastic dishes soon after plating. From about the fourth or fifth day in culture, the keratinocytes grew and formed colonies at random over the small, uniform, polygonal cells (Figure 1A and B). In a few days, some keratinocytes developed into a stratified epithelium, and intact contaminating fibroblasts were seen among disaggregated keratinocytes via cell proliferation and differentiation (Figure 1C, D, and E).

Expression patterns of genes involved in stem cell defferentiation in human keratinocytes

In order to examine the expression patterns of stem cell markers, human keratinocytes were cultured in low-calcium, serum-free defined medium. The results of the RT-PCR analysis are summarized in Figure 2. The expression profiles of selected genes (seven stem cell markers: Oct4, Sox2, c-Myc, Klf4, Nanog, CRIPTO and REX1, and six keratinocyte-specific markers: KRT 1, KRT 5, KRT 10, KRT 14, KRT 15, and KRT 19) involved in the stem cell development pathway of keratinocytes were analyzed. The expression of each gene was analyzed during the development of keratinocyte stem cells from day 0 to day 33 in order to show the stability and reproducibility of the selected genes in culture.

The stem cell markers for undifferentiated cells, including Oct4, Sox2 and Nanog, and the keratinocyte-

Table 1. PCR primers used in this study.

Gene	Primer sequence	Annealing temperature (°C)	PCR product size (bp)
Oct4	FP:CGTGAAGCTGGAGAAGGAGAAGCTG RP:GAACATGTGTAAGCTGCGGCCCTTG	62	247
Sox2	FP:GCTGCACATGAAGGAGCACCC RP: CGGACTTGACCACCGAACCCA	62	443
c-Myc	FP:ACTCTGAGGAGGAACAAGAA RP:TGGAGACGTGGCACCTCTT	58	159
Klf4	FP:CCAGAGGAGCCCAAGCCAAAG RP:CGAGGTGGTCCGACCTGGAAA	58	289
Nanog	FP:CCCAAAGGCAAACAACCCACT RP:ATTGCTATTCTTCGGCCAGTT	58	276
CRIPTO	FP:TGCCCAAGAAGTGTTCCCTGT RP: GCAGCAGCCTTTACTGGTCAT	60	269
REX1	FP: CGCTGACACCATCCTCATCGG RP: GGCGTCATCGCTTGGTCTTGG	62	269
KRT1	FP: AGGATGTGGATGGTGCTTAT RP: GCTTTGCTCTTCTGGGCTAT	58	235
KRT5	FP: CTGGACACCAAGTGGACCCT RP: GCTCCGCATCAAAGAACATC	60	346
KRT10	FP: TGATAATGCCAACATCCTGC RP: CCTCCTCGTGGTTCTTCTTC	60	224
KRT14	FP: GGAGATGATTGGCAGCGTGGA RP: GGACCTGCTCGTGGGTGGACA	68	281
KRT15	FP: AGCCTACCTGAAGAAGAACCACG RP: TGGCATAGCGGCACTCTGTCT	62	365
KRT19	FP: GCGACTACAGCCACTACTACACGAC RP: CGACCTCCCGGTTCAATTCTT	58	474
β-actin	FP: AAAGACCTGTACGCCAACAC RP: GTCATACTCCTGCTTGCTGAT	62	219

specific marker KRT 19 were first detectable in vivo (day 0) and were expressed at maximum levels on day 5 of cultivation before becoming undetectable. Klf4 and the keratinocyte-specific marker KRT 1 were undetectable in vivo (day 0), and their maximum expression levels were observed on day 5 before becoming undetectable at day 10. The stem cell markers for undifferentiated cells, including c-Myc and CRIPTO, and keratinocyte-specific markers, including KRT 5, KRT 14 and KRT 15, were detectable *in vivo* (day 0) and were expressed at maximum levels on day 5 before showing reduced gene expression after day 10. In addition, the stem cell marker REX1 and keratinocyte-specific marker KRT 10 were undetectable *in vivo* (day 0) and were expressed at maximum levels on day 5 before showing reduced gene expression after day 10.

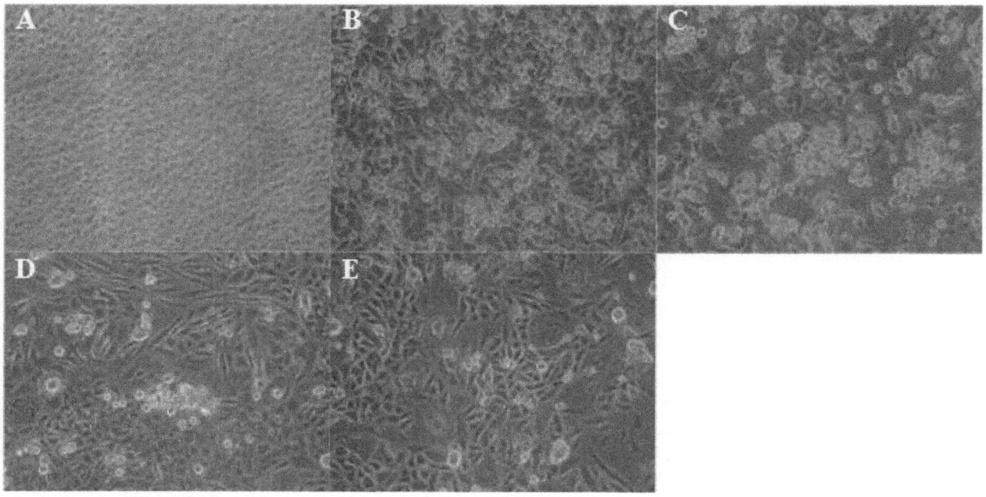

Figure 1. Morphological changes observed in human keratinocytes during cultivation in low-calcium, serum-free defined medium. (A) Morphology of human keratinocytes on day 1 in culture. (B) From the fifth day, the keratinocytes grew and formed colonies at random over the small, uniform, polygonal cells. (C-E) In a few days, some keratinocytes developed into a stratified epithelium (C: day 10), and intact contaminating fibroblasts were seen among disaggregated keratinocytes via cell proliferation and differentiation (D: day 18; E: day 33). Scale bar 100 µm.

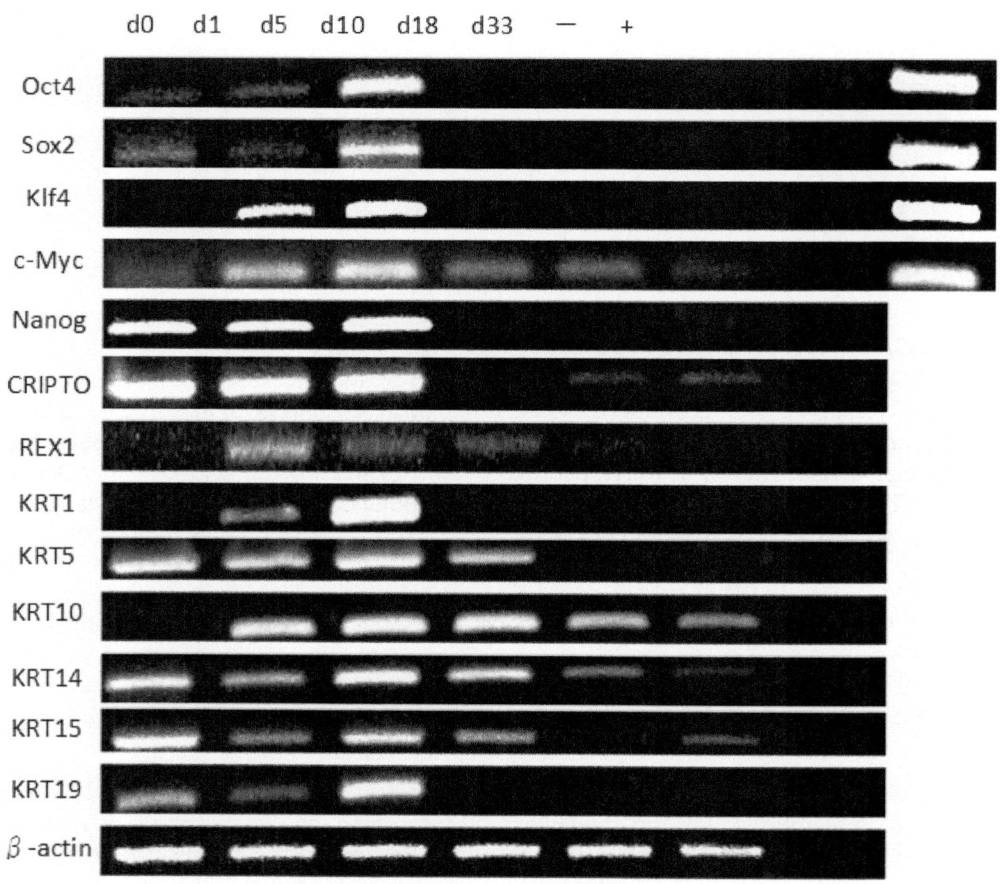

Figure 2. RT-PCR analysis of the expression of selected genes (seven stem cell markers: Oct4, Sox2, c-Myc, Klf4, Nanog, CRIPTO and REX1, and six keratinocyte-specific markers: KRT 1, KRT 5, KRT 10, KRT 14, KRT 15 and KRT 19) in human keratinocytes during cultivation in vitro. We used the pMIG-hOCT4, pMIG-hSOX2, pMIG-hKLF4 and pBABE-c-myc-zeo vectors as PCR templates for positive controls.

DISCUSSION

In this study, we evaluated the expression patterns of RFs and other stem cell markers in cultured human keratinocytes and found that the expression levels of almost all of the selected genes, including Oct4, Sox2, c-Myc and Klf4, were significantly increased in the early phase of culturing and decreased upon cellular differentiation. Our results strongly suggest that the fluctuation of stem cell marker expression is ubiquitous in cultured human keratinocytes. The culture characteristics of human keratinocytes support the hypothesis that the reprogramming process could be more easily achieved when pluripotent markers are expressed at maximum levels during cultivation *in vitro*. Oct4 expression is essential for the development of the inner cell mass (ICM) *in vivo*, the derivation of ESCs and the maintenance of a pluripotent state (Nichols et al., 1998), and the precise levels of Oct4 govern three distinct fates of ESCs (Niwa et al., 2000). A less than twofold increase in expression causes differentiation into primitive endoderm and mesoderm, whereas inhibition of Oct4 expression induces a loss of pluripotency and differentiation into trophectoderm (Niwa et al., 2000).

Sox2 is a transcription factor involved in the self-renewal of ESCs. It plays an important role in maintaining ES-cell pluripotency and heterodimerizes in a complex with Oct4 (Yuan et al., 1995). Sox2 expression is restricted to cells with stem cell characteristics, supporting their self-renewal capability, and is no longer expressed in cells with more restricted developmental potential (Avilion et al., 2003). In addition, forced expression of Oct4 can compensate for the loss of Sox2 in ESCs (Masui et al., 2007).

c-Myc is a pleiotropic transcription factor and has been linked to several cellular functions, including cell-cycle regulation, proliferation, growth, differentiation and metabolism (Schmidt, 1999). It tends to be highly expressed in the majority of rapidly proliferating cells and is generally expressed at low levels or absent during quiescence (Murphy et al., 2005). The role of c-Myc in reprogramming is not yet clear. It is dispensable for the generation of iPSCs in mice and humans (Eminli et al., 2008; Huangfu et al., 2008; Kim et al., 2008; Nakagawa et al., 2008; Wernig et al., 2008), but the efficiency of reprogramming decreases dramatically without c-Myc.

Klf4 is a transcription factor expressed in a variety of tissues, including the epithelium of the intestine, kidney and skin (Segre et al., 1999). Depending on its interaction partner and the target gene, Klf4 can both activate and repress transcription (Rowland and Peeper, 2006), and a growing body of evidence suggests that Klf4 can function as both an oncoprotein and a tumor suppressor (Zhao et al., 2004). Constitutive expression of Klf4 suppresses cell proliferation by blocking G1–S progression through the cell cycle (Zhao et al., 2004). Recently, it has been demonstrated that forced over expression of Klf4 in ESCs inhibits differentiation in erythroid progenitors, suggesting a role for this factor in ES cell function (Li et al., 2005). Its exact role in the reprogramming process is still not fully understood.

Transcripts of human keratinocyte stem cell markers have been detected in vitro (Aasen et al., 2008), and keratinocytes treated with ESC-conditioned medium (CM) change their morphology and express the pluripotency regulator Oct4 and its target transcripts Sox2, Nanog and REX1 (Grinnell and Bickenbach, 2007). In our gene expression profile, no transcripts of Klf4 or REX1 were detected *in vivo* (day 0). The different expression levels observed for Klf4 and REX1, however, suggest that these two markers may be involved in the differentiation of epithelial cells.

It has been speculated that the expression of Oct4 fluctuates during the development of keratinocytes from cultured human ESCs (Green et al., 2003) as it is expressed at high levels in ESCs and shows reduced expression at day 6 of cellular differentiation (Pellizzer et al., 2004). Furthermore, key genes that control pluripotency, including Oct4, Sox2 and Nanog, undergo dynamic changes in transcript abundance during porcine embryo cleavage development (Magnani and Cabot 2008). In the course of keratinocyte differentiation, the expression levels of KRT 5 and KRT 14 decline, whereas the expression levels of KRT 1 and KRT 10 are augmented (Kartasova et al., 1992). It has also been reported that CaCl2 induces the differentiation of epidermal keratinocytes in culture (Hennings et al., 1980), with decreased expression of KRT 5 and 14, markers of the basal cell layers, and increased expression of KRT 1 and 10, markers of the prickle cell layers (Vellucci et al., 1995; Tennenbaum et al., 1996; Amoh et al., 2005; Yano et al., 2005). The expression of KRT 15, which has been reported to be a bulge stem cell marker, is significantly induced in both NHEKs and BDKs after CaCl2 treatment (Sasahara et al., 2009). Previous studies have evaluated the expression pattern of KRT 19 in the phase following the degeneraion of the lowest part of the follicle. When epithelial cell populations were cultured, the percentage of KRT 19-positive keratinocytes significantly increased in primary culture, remained high for the first subcultures and then decreased prior to senescence (Michel et al., 1996). A similar expression pattern was observed in our study, except for the fact that no transcripts of KRT 1 or KRT 19 were detected after day 10.

Fluctuation in the expression levels of stem cell markers occurs during increased proliferation, rather than terminal differentiation, in their initial developmental stage, while the converse situation is observed in later stages. Our results show that the fluctuation in the expression levels of stem cell markers could provide an opportunity for reprogramming using non integrating viruses or transient episomal gene expression or, more favourably, for the generation of iPSCs by biochemical means alone.

In conclusion, this study highlights the expression pat-

patterns of stem cell markers that are present during early cellular development in cultured human keratinocytes, guiding future induction protocols for pluripotent stem cells and bioengineering research in this field.

REFERENCES

Aasen T, Raya A, Barrero MJ, Garreta E, Consiglio A, Gonzalez F, Vassena R, Bilic J, Pekarik V, Tiscornia G, Edel M, Boue S, Izpisua Belmonte JC (2008). Efficient and rapid generation of induced pluripotent stem cells from human keratinocytes. Nat. Biotechnol. 26:1276-1284.

Amoh Y, Li L, Katsuoka K, Penman S, Hoffman RM (2005). Multipotent nestin-positive, keratin-negative hair-follicle bulge stem cells can form neurons. Proc. Natl. Acad. Sci. USA102:5530-5534.

Avilion AA, Nicolis SK, Pevny LH, Perez L, Vivian N, Lovell-Badge R (2003). Multipotent cell lineages in early mouse development depend on SOX2 function. Genes Dev. 17:126-140.

Eminli S, Utikal J, Arnold K, Jaenisch R, Hochedlinger K (2008). Reprogramming of neural progenitor cells into induced pluripotent stem cells in the absence of exogenous Sox2 expression. Stem Cells. 26:2467-2474.

Green H, Easley K, Iuchi S (2003). Marker succession during the development of keratinocytes from cultured human ESCs. Proc. Natl. Acad. Sci. USA100:15625-15630.

Grinnell KL, Bickenbach JR (2007). Skin keratinocytes pre-treated with embryonic stem cell-conditioned medium or BMP4 can be directed to an alternative cell lineage. Cell Prolif. 40:685-705.

Hennings H, Michael D, Cheng C, Steinert P, Holbrook K, Yuspa SH (1980). Calcium regulation of growth and differentiation of mouse epidermal cells in culture. Cell 19:245-254.

Huangfu D, Osafune K, Maehr R, Guo W, Eijkelenboom A, Chen S, Muhlestein W, Melton DA (2008). Induction of pluripotent stem cells from primary human fibroblasts with only Oct4 and Sox2. Nat. Biotechnol. 26:1269-1275.

Hou P, Li Y, Zhang X, Liu C, Guan J, Li H, Zhao T, Ye J, Yang W, Liu K, Ge J, Xu J, Zhang Q, Zhao Y, Deng H (2013). Pluripotent stem cells induced from mouse somatic cells by small-molecule compounds. Science 341(6146): 651-654.

Kartasova T, Roop DR, Yuspa SH (1992). Relationship between the expression of differentiation-specific keratins 1 and 10 and cell proliferation in epidermal tumors. Mol. Carcinogenesis. 6:18-25.

Kim JB, Zaehres H, Wu G, Gentile L, Ko K, Sebastiano V, Arauzo-Bravo MJ, Ruau D, Han DW, Zenke M, Scholer HR (2008). Pluripotent stem cells induced from adult neural stem cells by reprogramming with two factors. Nature 454:646-650.

Li Y, McClintick J, Zhong L, Edenberg HJ, Yoder MC, Chan RJ (2005). Murine embryonic stem cell differentiation is promoted by SOCS-3 and inhibited by the zinc finger transcription factor Klf4. Blood 105:635-637.

Magnani L, Cabot RA (2008). In vitro and in vivo derived porcine embryos possess similar, but not identical, patterns of Oct4, Nanog, and Sox2 mRNA expression during cleavage development. Mol. Reprod. Dev. 75:1726-1735.

Masui S, Nakatake Y, Toyooka Y, Shimosato D, Yagi R, Takahashi K, Okochi H, Okuda A, Matoba R, Sharov AA, Ko MS, Niwa H (2007). Pluripotency governed by Sox2 via regulation of Oct3/4 expression in mouse ESCs. Nat. Cell Biol. 9:625-635.

Michel M, Torok N, Godbout MJ, Lussier M, Gaudreau P, Royal A, Germain L (1996). Keratin 19 as a biochemical marker of skin stem cells in vivo and in vitro: keratin 19 expressing cells are differentially localized in function of anatomic sites, and their number varies with donor age and culture stage. J. Cell Sci. 109 (Pt 5):1017-1028.

Murphy MJ, Wilson A, Trumpp A (2005). More than just proliferation: Myc function in stem cells. Trends Cell Biol. 15:128-137.

Nakagawa M, Koyanagi M, Tanabe K, Takahashi K, Ichisaka T, Aoi T, Okita K, Mochiduki Y, Takizawa N, Yamanaka S (2008). Generation of induced pluripotent stem cells without Myc from mouse and human fibroblasts. Nat. Biotechnol. 26:101-106.

Nichols J, Zevnik B, Anastassiadis K, Niwa H, Klewe-Nebenius D, Chambers I, Scholer H, Smith A (1998). Formation of pluripotent stem cells in the mammalian embryo depends on the POU transcription factor Oct4. Cell 95:379-391.

Niwa H, Miyazaki J, Smith AG (2000). Quantitative expression of Oct-3/4 defines differentiation, dedifferentiation or self-renewal of ESCs. Nat. Genet. 24:372-376.

Pellizzer C, Adler S, Corvi R, Hartung T, Bremer S (2004). Monitoring of teratogenic effects in vitro by analysing a selected gene expression pattern. Toxicol. In Vitro 18:325-335.

Rowland BD, Peeper DS (2006). KLF4, p21 and context-dependent opposing forces in cancer. Nat. Rev. Cancer 6:11-23.

Sasahara Y, Yoshikawa Y, Morinaga T, Nakano Y, Kanazawa N, Kotani J, Kawamata S, Murakami Y, Takeuchi K, Inoue C, Kitano Y, Hashimoto-Tamaoki T (2009). Human keratinocytes derived from the bulge region of hair follicles are refractory to differentiation. Int. J. Oncol. 34:1191-1199.

Schmidt EV (1999). The role of c-myc in cellular growth control. Oncogene 18:2988-2996.

Segre JA, Bauer C, Fuchs E (1999). Klf4 is a transcription factor required for establishing the barrier function of the skin. Nat. Genet. 22:356-360.

Takahashi K, Tanabe K, Ohnuki M, Narita M, Ichisaka T, Tomoda K, Yamanaka S (2007). Induction of pluripotent stem cells from adult human fibroblasts by defined factors. Cell 131:861-872.

Takahashi K, Yamanaka S (2006). Induction of pluripotent stem cells from mouse embryonic and adult fibroblast cultures by defined factors. Cell 126:663-676.

Tennenbaum T, Belanger AJ, Quaranta V, Yuspa SH (1996). Differential regulation of integrins and extracellular matrix binding in epidermal differentiation and squamous tumor progression. J. Investig. Dermatol. Symp. Proc. 1: 157-161.

Vellucci VF, Germino FJ, Reiss M (1995). Cloning of putative growth regulatory genes from primary human keratinocytes by subtractive hybridization. Gene 166:213-220.

Wernig M, Meissner A, Cassady JP, Jaenisch R (2008). c-Myc is dispensable for direct reprogramming of mouse fibroblasts. Cell Stem Cell 2:10-12.

Yano S, Ito Y, Fujimoto M, Hamazaki TS, Tamaki K, Okochi H (2005). Characterization and localization of side population cells in mouse skin. Stem Cells 23:834-841.

Yu J, Vodyanik MA, Smuga-Otto K, Antosiewicz-Bourget J, Frane JL, Tian S, Nie J, Jonsdottir GA, Ruotti V, Stewart R, Slukvin II, Thomson JA (2007). Induced pluripotent stem cell lines derived from human somatic cells. Science 318:1917-1920.

Yuan H, Corbi N, Basilico C, Dailey L (1995). Developmental-specific activity of the FGF-4 enhancer requires the synergistic action of Sox2 and Oct-3. Genes Dev. 9:2635-2645.

Zhao W, Hisamuddin IM, Nandan MO, Babbin BA, Lamb NE, Yang VW (2004). Identification of Kruppel-like factor 4 as a potential tumor suppressor gene in colorectal cancer. Oncogene 23:395-402.

Impact of physico-chemical parameters on the physiological growth of *Arthrospira* (*Spirulina platensis*) exogenous strain UTEXLB2340

Yahia Mustafa A. Fagiri[1], Aisha Salleh[1] and Saifeldin Ahmed F. El-Nagerabi[2]

[1]Institute of Biological Sciences, Faculty of Science, University of Malaya, 50603 Kuala Lumpur, Malaysia.
[2]Department of Biological Sciences and Chemistry, College of Arts and Sciences, University of Nizwa, P.O. Box 33, Postal Code 616, Birkat Al Mouz, Nizwa, Oman.

Spirulina platensis **is an attractive human food supplement, animals feed and of numerous medicinal uses. The effect of physico-chemical factors on the physiological growth rates as optical densities (O.D.) of *S. platensis* strain (UTEXLB2340) were investigated. Experiments were conducted at different pH levels, light intensities, and temperature regimes, and monitored for 20 days. The half concentration of Zarrouk media was found to be suitable for cultivation of this strain under indoor (1.14 O.D.) and outdoor conditions (0.99 O.D.). The nitrogen source sodium nitrate (2.5 gl^{-1}) can be replaced by inexpensive and low concentration of urea fertilizer (0.12 gl^{-1}). The growth rates of this strain were optimum at light intensities of 1500 to 2500 lux of continuous light (0.99 to 1.71 O.D.) or 2500 lux of intermittent light (1.32 O.D.). The optimal growth was recorded at lower pH 7 to 8 (1.19 to 1.76 O.D.), and temperature of 25 to 35°C (1.13 to 1.71 O.D.). Therefore, this microalga can be cultivated for commercial uses in the available natural ponds where the average temperature reached 30°C and light intensities fluctuated between 1000 to 4000 lux in the morning, 12,000 to 22,000 lux midday, and 7000 to 10,000 lux in the noon.**

Key words: *Arthrospira* (*Spirulina*) *platensis*, indoor and outdoor cultures, light intensity, pH, sodium nitrate, temperature, urea.

INTRODUCTION

Arthrospira (*Spirulina*) *platensis* is a planktonic, filamentous cyanobacterium that grows in tropical, subtropical and temperate water habitats (Vonshak, 1997; Colla et al., 2007; Rodrigues et al., 2011; Madkour et al., 2012). This microalga deserves special attention due to its nutritional properties as high single cell protein for human food supplement, animal and fish feed. It produces numerous bioactive compounds such as vitamins, essential amino acids, minerals, polysaturated fatty acids, biopigments, and antioxidents (Kamat, 1995; Abu et al., 2007; Pandey and Tiwari, 2010; Pandey et al., 2010; Madkour et al., 2012; Thirumala, 2012). It has been used as an alternative healthy food for malnutrition, growth stimulation, anticancer, and enhancing milk secretion (Richmond, 1986a; Ogbonda et al., 2007). The microalgae industries have been receiving increasing interest and utilized in both traditional and biotechnological processes during the last decades (Celekli et al., 2009; Thirumala, 2012). Increase in human population stimulating the search for alternative food and feed sources through exploiting biotechnological techniques for cultivation of useful microalgae (Venkataraman et al., 1981; Celekli et al., 2009; Jitendra et al., 2012). Many algal genera such as *Spirulina*, *Chlorella*, *Dunaliella*, and *Scenedesmus* were used to fulfill these purposes (Pulz and Gross, 2004). The physiological growth and large-scale production of *Spirulina* depends on many physical, environmental and nutritional factors. The most important are nutrients availability and composition, pH, light and temperature under both indoor and outdoor conditions (Abu et al., 2007; Colla et al., 2007; Madkour et al., 2012). The challenging cultivation of *S. platensis* depends on nutrients availability (Paoletti et al., 1975; Richmond, 1986; Rodrigues et al., 2010; 2011). Although nitrates are commonly used as growth media, many studies supported the benefit of using alternative cheaper nutritional sources such as ammonia and urea (Bezerra et al., 2008; Rodrigues et al., 2010). This will help to avoid difficulties encountered in cultivation with only one cultivation source (Piorreck et al., 1983; Stanca and Popovici, 1996; Danesi et al., 2002, 2011; Volkman et al., 2008; Rodrigues et al., 2010). The inhibition of the growth was less marked with urea comparable to ammonia (Converti et al., 2006; Celekli and Dönmez, 2006; 2009). Different low cost media were used for cultivation of *Spirulina* such as Zarrouk media (Cola et al., 2007; Jitendra et al., 2012; Madkour et al., 2012; Thirumala, 2012).

The pH plays apparent role in the metabolic activities of microalgae (Richmond, 1986b; Rafiqual et al., 2005; Ogbonda et al., 2007). It strongly affects biomass production (Celekli and Dönmez, 2006; 2009), chemicals dissociation and cell physiology (Kim et al., 2007; Ogbonda et al., 2007; Celekli et al., 2009). Therefore, the effect of different pH levels on the growth of microalgae was continuously evaluated under different environmental conditions (Kim et al., 2007; Ogbonda et al., 2007; Celekli et al., 2009). On the other hand, light is the most important factor representing the main source of energy for *S. platensis* (Soletto et al., 2008). The effect of light and temperature under both indoor and outdoor conditions on the algal growth was under continuous investigation (Richmond, 1986; Richmond and Grobbelaar, 1986; Torzillo et al., 1986, 1991; Torzillo and Vonshak, 1994; Boussiba et al, 1988; Pandey et al., 2010). At lower temperature, the photo inhibition effect results in both low final cell concentration and productivity of *S. platensis* (Richmond and Grobbelaar, 1986; Boussiba et al, 1988; Pandey et al., 2010). Photoinhibition is a reduction of the photosynthetic activities with high light intensity (Soletto et al., 2008). This is associated with climatic variances and the photoperiods (Tamiya et al., 1953; Samuelsson et al., 1985; Jensen and Knutsen, 1993; Danesi et al., 2011).

The studies of the major limitations on the growth of *S. platensis* were carried in open-race way (Richmond et al., 1990). Economically, the open-air system is useful in the large-scale production compared to expensive closed systems (Borowitzka, 1999). Indoor controlled cultivation of *Spirulina* sp. facilitates checking of the simultaneous effect of light and temperature (Danesi et al., 2011). Therefore, the present study was undertaken to evaluate the physiological growth as optical density of exogenous strain of *S. platensis* (UTEXLB2340) under indoor and outdoor conditions of Malaysia. Different concentrations of Zarrouk media, urea fertilizer, sodium nitrate, pH levels, light intensities, and temperature regimes were investigated. Hopefully these findings will help in introducing micro-algae industry in Malaysia and exploiting several hundred natural ponds available.

MATERIALS AND METHODS

Microorganism strain and cultivation media

In the present study, *S. platensis* strain namely UTEXLB2340 was obtained from the culture collection centre of the University of Texas, Austin, USA. The strain was inoculated in 2 l Erlenmeyer flask with 1 l of sterilized Zarrouk media containing (gl^{-1}): $NaHCO_3$, 8.0; $NaNO_3$, 2.5; NaCl, 1.6; KCl, 1.0;K_2HPO_4, 0.5; $MgSO_4.7H_2O$, 0.2; $(NH_2)_2CO$, 0.1; EDTA-Na_2, 0.08; $CaCl_2.2H_2O$, 0.04 and $FeSO_4.7H_2O$, 0.01. The media was sterilized at 12°C under pressure of 15 Lb/in^2 for 20 min. The inoculated flasks were incubated at 30°C with cool white florescent lamps of continuous light intensity of 1500 lux and continuous aerator (10 vvm), and kept as stock cultures in the algal culture room of the Institute of Biological Sciences, University of Malaya.

Inoculum preparation

For inoculum preparation, an aliquot from the stock culture was added to the sterilized conical flask (500 ml) and diluted with Zarrouk media until the algal biomass of optical density 0.2 and 0.5 O.D. were maintained using spectrophotometer at 560 nm. The algal biomass of 0.5 O.D. was used as initial inoculum to study the effect of different concentration of Zarrouk media, light intensity and urea on the physiological growth of this strain under both outdoor and indoor conditions, whereas the concentration of 0.2 O.D. was used as inoculum for testing the indoor impacts of pH and temperature on the growth of the alga (Canizares et al., 1995).

Effect of outdoor, indoor conditions and media on *Spirulina* growth

The outdoor experiment was carried out in 45 cm Plexiglass tanks of 30 cm width and depth. The tanks were located in area of 40 m^2 in the Institute of Biological Sciences, University of Malaya. The top of the shelter was sealed by cement roof with the western side covered by perforated folding cover to maintain light intensity of 100,000 lux in a sunny day which is normally reduced by cloud. The light intensity in the outdoor area was fluctuating between 1000-4000 lux in the morning, 12,000-22,000 lux midday, and 7000-10,000 lux in the noon, with the average temperature of between 24-35°C. The tanks were filled with 15 L of Zarrouk media of full, half, and quarter concentrations of the ingredients. The media were inoculated with 1.5 L of algal inoculum of 0.05 optical densities with continuous aeration (10 vvm) throughout the duration of the experiment.

For the indoor culture experiment, the algal culture room of the Institute of Biological Sciences, University of Malaya was used. The room was supplied with continuous light illumination from cool white florescent lamps of 1500 lux and average temperature of 30°C. For inoculation, 150 ml of sterilized Zarrouk media of full, half, and quarter concentrations of the ingredients in 250 ml conical flasks was adjusted to pH 9 using sodium hydroxide (40% w/v). Four replicates from each concentration were inoculated with 15 ml of the inoculum (0.05 O.D.), stopped by cotton wool, covered by aluminum foil, sealed with sealing film and were then shaken on rotatory shaker of 140 rpm.

Effect of urea and pH on *Spirulina* growth

For testing the efficiency of urea fertilizer as feasible and cheaper nitrogen source, 0.12 gl^{-1} urea were used as alternative to 2.5 gl^{-1} sodium nitrate ($NaNO_3$) in Zarrouk media. The other culture conditions were similar to that of indoor experiment by inoculating 150 ml of Zarrouk media with 15 ml algal inoculum of 0.05 O.D. in 250 ml conical flasks. The algal growth was measured as optical density and compared with the growth in Zarrouk media containing 2.5 gl^{-1} sodium nitrate. To determine the effect of pH on the growth and biomass production of this strain, 150 ml Zarrouk media in 250 Erlenmeyer conical flasks with different pH (7, 8, 9, 10, and 11) were adjusted using 40% (w/v) sodium hydroxide and 1.0 N hydrochloric acid. Four replicates were inoculated with 15 ml of 0.2 O.D. pure inoculum of algal strain. The inoculated flasks were incubated at 30°C in growth chamber of lighting provided through continuous light from white cool florescent lamp of 2500 lux and were then shaken on rotatory shaker at 140 rpm in addition to hand shaken three times per day at morning, midday and night.

Effect of light and temperature on *Spirulina* growth

To investigate the effect of light intensity on the growth of *S. platensis* strain UTEXLB2340, the algal culture room was adjusted at 30°C with various light intensity of 1500 and 2500 lux of intermittent light and dark cycle of 12 h each. Similarly, other three sets of growth chambers were set at 30°C and adjusted to 500, 1500, and 2500 lux with continuous white cool light intensity. For this test, 150 ml of Zarrouk media were added to 250 ml conical flasks, inoculated with 15 ml of algal inoculum (0.05 O.D.) and the pH was maintained at 9 in all cultures. The inoculated flasks were covered with cotton wool and aluminum foil and were then shaken using rotary shaker at 140 rpm for 20 days. The algal growth rate as an optical density was measured every 4 days using UV spectrophotometer at 560 nm. To evaluate the impact of temperature on the growth of this microalga strain, thermostatic controlled growth chambers were set at different temperature regimes (20, 25, 30, 35, and 40°C). For this, 150 ml of Zarrouk media were added to 250 ml conical flasks and inoculated with 15 ml from 0.2 O.D. algal inoculum with pH adjusted to 9 and incubated under continuous light illumination of 2500 lux. The inoculated flasks were then vigorously shaken manually at three times a day in the morning, midday, and night for 20 days.

Measurement of the physiological growth rates

The effect of growth conditions, media concentrations, urea and sodium nitrate ingredients, different pH levels, light intensities and temperature regimes on *S. platensis* strain on physiological growth rates and biomass production were measured as optical densities using UV spectrophotometer at wavelength of 560 nm.

Statistical analyses

For comparison between different concentrations of Zarrouk media, urea and sodium nitrate, different pH levels, light intensities, and temperature regimes, Duncan's multiple range test and one way ANOVA were used with $p<0.05$. The analysis was conducted with statistical package software SPSS of version 11.0.

RESULTS AND DISCUSSION

Influence of indoor, outdoor conditions and media on growth

The growth and biomass production of *Spirulina* depends on many physico-chemical factors such as nutrients availability, pH, light, and temperature. For economic reasons, the open-air system is predominating in the large-scale commercial cultivation of these types of organisms (Celekli et al., 2009). On the other hand, closed system is being very expensive and often difficult to scale up (Borowitzka, 1999). Cost and composition of media are challenging factors for viable and mass production of cyanobacteria (Jitendra et al., 2012). Different media were used for cultivation of *Spirulina* such as Zarrouk media, CFTIR media, OFERR media, and Bangladesh media (Belay et al., 1993; Jitendra et al., 2012). Zarrouk media have been successfully served as standard media (SM) for cultivation of *Spirulina* (Madkour et al., 2012). Higher growth rates and lipids content of *Spirulina* grown on Zarrouk media compared to cultivation under nitrogen starvation (Olguin et al., 2001; Colla et al., 2007). In the present study, the growth rates of *S. platensis* strain UTEXLB2340 on full, half, and quarter concentrations of Zarrouk media significantly ($p<0.05$) increased with incubation time and the maximum growth was obtained on 20th day and 16th day of cultivation under outdoor and indoor conditions, respectively (Table 1). The full and half concentrations of Zarrouk media displayed significantly ($p<0.05$) higher and optimum growth of *Spirulina* compared with quarter concentration. Therefore, from economic point of view, the half concentration of this media is evidently suitable for cultivation of this *Spirulina* strain in both indoor and outdoor conditions of Malaysia as suggested in our earlier reports (Fagiri et al., 2013). This supports the basic aim of providing simple and inexpensive media to decrease the cost of the large-scale production under open-air system as suggested by many researchers (Madkour et al., 2012; Jitendra et al., 2012). These intensions were implemented by substituting all the ingredients of Zarrouk media with cheaper and locally available commercial fertilizers and chemicals (Madkour et al., 2012).

Effect of urea and sodium nitrate on *Spirulina* growth

Various studies focused on the use of a cheaper nitrogen sources and locally available fertilizers for cultivation of microalgae (Rodrigues et al., 2011; Madkour et al., 2012). In the biotechnological process, the cultivation media components are responsible for higher costs (Danesi et al., 2011). The cost of the nutrients is considered the second to labour as major factors influencing the cost of *Spirulina* biomass production (Vonshak, 1997). Therefore, production of *Spirulina* with reduced cost is necessary when considering large-scale cultivation for industrial purposes. The growth of *Spirulina* was best when using urea compared to potassium nitrate (KNO_3) (Danesi et al., 2011). Therefore, urea is cheap effective alternative which significantly provides higher cell growth (Piorreck et al., 1983; Stanca and Popovici, 1996; Volkmann et al., 2008). In contrast, cultures supplemented with urea showed slow growth rates and higher rates when compared to

Table 1. Effect of Zarrouk media concentrations on the physiological growth rates of *S. platensis* strain UTEXLB2340 under outdoor and indoor conditions.

Time (day)	Outdoor conditions			Indoor conditions		
	Concentrations of Zarrouk medium			Concentrations of Zarrouk medium		
	Full	Half	Quarter	Full	Half	Quarter
0	0.03Fa	0.04Fa	0.05Fa	0.03Fa	0.05Ea	0.04Ea
4	0.17Ea	0.20Ea	0.16Ea	0.32Ea	0.26Da	0.26Da
8	0.32Db	0.42Da	0.40Da	0.65Da	0.60Ca	0.45Cb
12	0.59Cb	0.69Ca	0.61Cab	0.86Cb	1.12Aa	0.68Ac
16	0.78Bb	0.86Ba	0.70Bc	1.06Ab	1.14Aa	0.60Bc
20	1.04Aa	0.99Aa	0.89Ab	0.99Ba	0.94Ba	0.56Bb

Within rows and columns, means followed by different lower case and upper case letters differ significantly ($p<0.05$) according to the Duncan's multiple range test, respectively.

Table 2. Effect of urea and sodium nitrate as ingredients in Zarrouk media on the growth physiology of *S. platensis* strain UTEXLB2340.

	Growth rate (optical density)	
Time (day)	Urea (0.12g/L)	Sodium nitrate (2.5 g/L)
0	0.03Fa	0.03Ea
4	0.18Eb	0.32Da
8	0.45Db	0.65Ca
12	0.81Ca	0.80Ba
16	1.15Ba	1.01Ab
20	1.56Aa	1.06Ab

Within rows and columns, means followed by different lower case and upper case letters differ significantly ($p<0.05$) according to the Duncan's multiple range test, respectively.

Zarrouk media, and ammonium nitrate media, respectively (Madkour et al., 2012).

The growth parameters in urea containing media showed a significant increase urea concentration. Although, urea has been know as an excellent nitrogen source and successfully metabolized by algae, *Spirulina* could most efficiently utilize ammonia nitrate compared to urea (Bezerra et al., 2008; Rodrigues et al., 2010). However, the inhibition effect was less marked with urea due to enzymatic hydrolysis of this compound by urease enzyme (Converti et al., 2006). On the other hand, the concentration of sodium nitrate in Zarrouk media (2.5 gl^{-1}) can be reduced without loss of productivity, as an important cost-saving factor in large-scale cultivation of *S. platensis* (Colla et al., 2007).

In the present results, it is evident that the use of low concentration of urea (0.12 gl^{-1}) and high concentration of sodium nitrate (2.5 gl^{-1}) significantly ($p<0.05$) increased the growth rates of *Spirulina* strain with time and maximum growth was obtained with urea (1.56 O.D.) compared to sodium nitrate (1.06 O.D.) after 20 days of cultivation (Table 2). These findings support the effective use of inexpensive urea as alternative to sodium nitrate in large-scale commercial cultivation of this microalga as concluded by many authors (Converti et al., 2006; Colla et al., 2007; Vonshak, 1997; Volkman et al., 2008). Therefore, in similar studies, KNO3 was replaced by urea (Piorreck et al., 1983; Danesi et al., 2011) whereas the use of higher concentrations of sodium nitrate (1.875 and 2.500 gl^{-1}) showed no increase in the algal growth and level of protein (Colla et al., 2007) and increase lipids (Manabe et al., 1992). Therefore, in this study the high concentration of sodium nitrate can be reduced or replaced by urea in Zarrouk media as an important cost-saving in large-scale cultivation (Colla et al., 2007).

Effect of pH on *Spirulina* growth

pH is one of the limiting factors which affect the metabolic activities of the microalgae (Richmond, 1986b; Rafiqual et al., 2005; Ogbonda et al., 2007). It affects the physiological growth and biomass production (Celekli and Dönmez, 2006; 2009). The cyanobacterium massively grows in tropical and subtropical bodies of water which have pH of up to 11 (Kim et al., 2007; Ogbonda et al., 2007; Celekli et al., 2009). The optimization of the growth of *S. platensis* culture was recorded at pH 9-10 (Soundarapandian and Vasanthi, 2008; Pandey et al.,

Table 3. Effect of different pH levels on the growth physiology of *S. platensis* strain UTEXLB2340.

	Growth rate (optical density)				
Time (day)	**pH 7**	**pH 8**	**pH 9**	**pH 10**	**pH 11**
0	0.04Fa	0.06Ea	0.06Fa	0.05Fa	0.05Ca
4	0.19Ec	0.28Db	0.41Ea	0.39Ea	0.39ABa
8	0.59Dd	0.71Cc	1.34Aa	0.81Ab	0.37Ae
12	1.19Ab	1.76Aa	1.10Bc	0.67Bd	0.34Ae
16	0.79Bc	1.52Ba	0.92Cb	0.53Cd	0.30Ae
20	0.68Ca	0.74Ca	0.70Da	0.42DEb	0.28ABc

Within rows and columns, means followed by different lower case and upper case letters differ significantly ($p<0.05$) according to the Duncan's multiple range test, respectively.

2010; Thirumala, 2012). In these investigations, inoculation of *S. platensis* at different levels of pH (7, 8, 9, 10, and 11) for 20 days showed that the physiological growth rates were significantly ($p<0.05$) increased up to 12th day and eventually declined towards the end of the cultivation time (Table 3). The highest growth rates were recorded at pH 8 (1.76 O.D.), followed by pH 7 (1.19 O.D.) and evidently reduced pH 9 (1.10 O.D.) and pH 10 (0.42 O.D.) and pH 11 (0.28 O.D.). Similar high biomass production was obtained at pH 8.5 and temperature of 35°C, whereas at pH values of 9, 9.5, and 10 the highest biomass production occurred at 30°C (Ogbonda et al., 2007) and at moderate pH 7-9 (Fagiri et al., 2013). It was also reported that pH 9 and temperature of 32°C as optimal conditions for biomass production by this microalga (Rafiqul et al., 2005). The same author reported the pH 10 and 37°C as optimal conditions for *Spirulina fusiformis*. Here the optimum conditions for our strained ranged between 7-8 at ambient condition (30°C) as concluded in similar studies under controlled temperature of 30-35°C (Rafiqul et al., 2005; Ogbonda et al., 2007; Soundarapandian and Vasanthi, 2008; Pandey et al., 2010; Thirumala, 2012). Thus, low alkalinity is required for optimal growth of this strain where solubility of CO_2 and other mineral compounds was affected by pH (Ogbonda et al., 2007). Therefore, pH conditions will help in avoiding the auto inhibition effect with increased pH on the cell growth (Richmond, 1986).

Influence of light and temperature on *Spirulina* growth

In literature, the interaction between illumination and temperature on the growth of *S. platensis* was extensively studied showing that at low temperature, the photo inhibition effect is more accentuated and results in low cells concentrations and productivities (Danesi et al., 2011). However, these studies covered the cultivation under natural illumination where the temperature effect and light intensity associated with seasonal climatic changes and the photoperiod of the year (Samuelsson et al., 1985; Jensen and Knutsen, 1993). Various studies were investigating the optimal light intensity and temperature for the growth of *S. platensis* under different climatic conditions. For almost all algal species, growth increases nearly proportional with light intensity when it is lower than the lower saturation point, while above saturation point, significant light inhibition may occur (Ogbonda and Tanaka, 2000; Yuan et al., 2011). The algal cultures produce small amounts of biomass when grown in darkness or light intensity below 300 μmolm^{-2}s^{-1} (Wang et al., 2007). On the contrary, higher light intensities could harvest more algal biomass. The growth of *S. maxima* was inhibited at light intensity higher than 25 to 30 Klux (300-360 μmolm^{-2}s^{-1}). The growth of *S. platensis* reached the maximum at light intensity exceeding 150-200 μmolm^{-2}s^{-1} (Vonshak, 1997). At 5 Klux light intensity, the dry weight of *S. maxima* was 0.72 g/500 ml (Pandey and Tiwari, 20109), and the dry weight of *S. platensis* was 0.85 g/500 ml (Pandey et al., 2010). The best was achieved at light intensity of 60 μmolm^{-2}s^{-1} (Danesi et al., 2011). It is also inclined to drop with light intensity below 5 Klux. The dry weight of *S. platensis* is 0.91 g/500 ml at 5 Klux light intensity (Pandey et al., 2009). In the present investigations, incubation of *S. platensis* strain on Zarrouk media under different light intensities regimes of 500, 1500, and 2500 lux continuous light and 1500 and 2500 lux intermittent light (12 light/12 dark) for 20 days, showed that the growth rates were significantly ($p<0.05$) increased with time and the maximum growth rates achieved at the end of cultivation time (Table 4). The best growth rates were obtained with continuous light of 2500 lux (1.71 O.D.), and 1500 lux (0.99 O.D.), followed by intermittent light of 2500 lux (1.33 O.D.), whereas continuous light of 500 lux and intermittent light of 1500 lux displayed the lowest growth rates of 0.71 and 0.58 O.D., respectively. Therefore, the optimal light intensity for this algal strain is continuous light of 1500-2500 lux and 2500 lux of intermittent light as suggested in similar studies on different strain of *Spirulina* (Fagiri et al., 2013). It is noticeable in this study that the growth rates of this strain increases with increase in light intensities as suggested by Pandey et al. (2010). However, in similar studies growth was inhibited at light intensity higher than 25-30 Klux (Vonshak, 1997), or in darkness below 300 μmolm^{-2}s^{-1} (Wang et al., 2007), or light intensity above 5 Klux (Pandey et al., 2010). Nonetheless, the best growth rates were recorded at light of 60 μmolm^{-2}s^{-1} (Danesi et

Table 4. Effects of continuous and intermittent light intensities on the growth rate of *S. platensis* strain UTEXLB2340.

	Growth rate (optical density)				
Time (day)	500 Cont.[a]	1500 Cont.[a]	2500 Cont.[a]	1500 Int.[b]	2500 Int.[b]
0	0.02Ea	0.02Ea	0.02Fa	0.02Ea	0.02Fa
4	0.11Dd	0.32Da	0.15Ec	0.18Dbc	0.21Eb
8	0.17De	0.65Ca	0.50Db	0.31Cd	0.39Dc
12	0.29Ce	0.86Bb	1.00Ca	0.52Bd	0.60Cc
16	0.49Be	1.06Ab	1.50Ba	0.66Ad	0.91Bc
20	0.71Ae	0.99Ac	1.71Aa	0.58Bd	1.32Ab

Within rows and columns, means followed by different lower case and upper case letters differ significantly ($p<0.05$) according to the Duncan's multiple range test, respectively. Light intensity (lux): Cont.[a], Continuous light; Int.[b], intermittent light.

Table 5. Effect of different temperature regimes on the growth rate of *S. platensis* strain UTEXLB2340.

Time (day)	Growth rate (Optical density)				
	20°C	25°C	30°C	35°C	40°C
0	0.04Ca	0.06Fa	0.06Fa	0.06Ea	0.05Ca
4	0.25Ab	0.15Ec	0.41Ea	0.44Da	0.15Bc
8	0.17Bc	0.50Db	1.34Aa	1.36Ba	0.22Bc
12	0.15BDe	1.00Cc	1.13Bb	1.98Aa	0.42Ad
16	0.12BDe	1.50Bb	0.92Cc	2.01Aa	0.36Ad
20	0.10Bd	1.71Aa	0.70Db	0.71Cb	0.18Bc

Within rows and columns, means followed by different lower case and upper case letters differ significantly ($p<0.05$) according to the Duncan's multiple range test, respectively.

al., 2011).

The optimization of the growth of *S. platensis* was noticed at 35°C with maximum biomass production (Soundarapandian and Vasanthi, 2008; Thirumala, 2012). The highest biomass concentration of 4.4 mg/ml was obtained at 30°C without limiting effect of lighting (Ogbonda et al., 2007). Under laboratory controlled conditions, the optimum growth temperature for *Spirulina* was reported at 35 to 37°C, and 39°C for outdoor cultivation (Richmond, 1986b), and 25-30°C for indoor cultivation (Fagiri et al., 2013). Thermotolerant *Spirulina* grows at 35 to 40°C (Vonshak et al., 1982). In the present study (Table 5), cultivation of *Spirulina* strain at different levels of temperature (20, 25, 30, 35, and 40°C) showed significantly ($p<0.05$) higher growth rates at 35°C (2.01 O.D.) after 16 days of cultivation, followed by 25°C (1.71 O.D.) at the end of the cultivation period (20 days) and at 30°C (1.13 O.D.) after 12 days. However, the growth rates and biomass production were significantly reduced ($p<0.05$) above 35°C (40°C) and below 25°C (20°C). This is may be attributed to photoinhibition effect of temperature and light (Jensen and Knusten, 1993). Therefore, the optimum temperature for the cultivation of this cyanobacteria strain in the range of 25 to 35°C as concluded by many researchers (Richmond, 1986b; Ogbonda et al., 2007; Soundarapandian and Vasanthi, 2008; Thirumala, 2012). This temperature is the average level in the outdoor conditions of Malaysia (30°C).

Conclusions

S. platensis strain UTEXLB2340 cultivations were carried out in different concentrations of Zarrouk media under outdoor and indoor conditions and at 0.12 gl^{-1} of urea and 2.5 g/l sodium nitrate, different pH levels, light intensities, and temperature regimes. The growth of this microalga strain was best using half concentration of Zarrouk media under both indoor and outdoor conditions. The maximum growth rate was obtained at low concentration of urea (0.12 gl^{-1}) compared to 2.5 gl^{-1} of sodium nitrate commonly used as nitrogen source in Zarrouk media. The pH of between 7-8 and low alkalinity were optimum ranges for achieving highest growth rates as well as light intensities of 1500-2500 lux and temperature in the range of 25 to 35°C. Therefore, from economic point of view, it is feasible to cultivate this strain of cyanobacteria under the outdoor conditions of Malaysia (24-35°C) using half concentration of Zarrouk media and replacing sodium nitrate with effective, inexpensive and available urea fertilizer. This will help in introducing the biotechnology of microalgae industry and exploiting the available natural ponds and resources for large-scale commercial cultivation. However, further studies were needed to evaluate the use of alternative and environment friendly growth media and other natural products.

ACKNOWLEDGEMENTS

This research was supported by the grant from the Institute of Biological Sciences, Faculty of Science, and University of Malaya. We are grateful to authority of the culture collection centre of the University of Texas, Austin, USA, for providing the algal strain. We thank the Department of Biological Sciences and Chemistry, College of Arts and Sciences, University of Nizwa for encouragement. We thank the University of Nizwa Writing Center for proof reading the English of this manuscript.

REFERENCES

Abu GO, Ogbonda KH, Aminigo RE (2007). Optimization studies of biomass production and protein biosynthesis in a *Spirulina* sp. isolated from polluted flame pit in the Niger Delta. Afr. J. Biotechnol. 6(22): 2550-2554.

Belay A, Ota Y, Miyakawa K, Shimamatsu H (1993). Current knowledge on potential health benefits of *Spirulina*. J. Appl. Phycol. 5: 235-241.

Bezerra RP, Matsudo MC, Converti A, Sato S, Carvalho JCM (2008). Influence of ammonia chloride feeding time and light intensity on the cultivation of *Spirulina* (*Arthrospira*) *platensis*. Biotechnol. Bioeng. 100: 297-305.

Borowitzka M (1999). Commercial production of microalgae: ponds, tanks, tubes and fermenters. J. Biotechnol. 70: 313-321.

Boussiba S, Sandbank E, Shelef G, Cohen Z, Vonshak A, Ben-Amotz A, Arad S, Richmond A (1988). Outdoor cultivation of the marine microalga *Isochrysis galbana* in open raceways. Aquaculture72: 247-253.

Canizares V, Dominguez A, Cruz M, Rios-Leal E (1995). Chemical composition of cyanobacteria grown in diluted, aerated swine wastewater. Biores. Technol. 51: 111-116.

Celekli A, Dönmez G (2006). Effect of pH, light intensity, salt and nitrogen concentration on growth and β-carotenes accumulation by a new isolate *Dunaliella* sp.. World J. Microbiol. Biotechnol. 22: 183-189.

Celekli A, Yavuzatmaca M, Bozkurt H (2009). Modeling of biomass production by *Spirulina platensis* as function of phosphate concentrations and pH regimes. Bioresour. Technol. 100: 3625-3629.

Colla LM, Reinehr CO, Reichert C, Costa JAV (2007). Production of biomass and nutracenutical compounds by *Spirulina platensis* under different temperature and nitrogen sources. Bioresour. Technol. 98: 1489-1493.

Converti A, Scapazzoni S, Lodi A, Carvalho JCM (2006). Ammonia and urea removed by *Spirulina platensis*. J. Ind. Microbiol. Biotechnol. 33: 6-16.

Danesi EDG, Rangel-Yagui CO, Carvalho JCM, Sato S (2002). An investigation of the effect of replacing nitrate by urea in the growth and production of chlorophyll by *Spirulina platensis*. Biomass Bioenergy 23: 261-269.

Danesi EDG, Rangel-Yagui CO, Sato S, de Carvalho JCM (2011). Growth and content of *Spirulina platensis* biomass chlorophyll cultivated at different values of light intensity and temperature using different nitrogen sources. Brazil. J. Microbiol. 42(1): 362-373.

Fagiri YMA, Salleh A, El-Nagerabi SAF (2013). Influence of chemical and environmental factors on the growth performance of *Spirulina platensis* strain SZ100. J. Algal Biomass Utln. 4 (2): 7-15.

Jensen S, Knusten G (1993). Influence of light and temperature on photoinhibition of photosynthesis in *Spirulina platensis*. J. Phycol. 5: 495-504.

Jitendra M, Priyanka S, Madhulika J, Mohsina S, Komal M, Neha K (2012). Impacts of different physical and chemical environment for mass production of *Spirulina platensis*- An immunity promoter. Int. Res. J. Biol. Sci. 1(6): 49-56.

Kamat GM (1995). Gamma linolenic acid production from *Spirulina platensis*. Appl. Microbiol. Biotechnol. 43: 466-469.

Kim MK, Park JW, Kim SJ, Jeune KH, Chang MU, Acreman J (2007). Enhanced production of *Scenedesmus* sp. (green microalgae) using a new medium containing fermented swine wastewater. Bioresour. Technol. 98: 2220-2228.

Madkour FF, Kamil A, Nasr HS (2012). Production and nutritive value of *Spirulina platensis* in reduced cost media. Egypt. J. Aquat. Res. 38: 51-57.

Manabe E, Hirano M, Takano H, Ishikawa-Doi N, Sode K, Matsunaga T (1992). Influence of ammonium chloride on growth and fatty acid production by *Spirulina platensis*. Appl. Biochem. Biotechnol. 34/35: 273-281.

Ogbonda JC, Tanaka H (2000). Light requirement and photosynthetic cell cultivation- development of process for efficient light utilization in photobioreactors. J. Appl. Phycol. 12: 207-218.

Ogbonda KH, Aminigo RE, Abu GO (2007). Influence of temperature and pH on biomass production and protein biosynthesis in a putative *Spirulina* sp. Bioresour. Technol. 98: 2207-2211.

Olgiun E, Galicia S, Angulo-Guerrero O, Hern.nedz E (2001). The effect of low light flux and nitrogen deficiency on chemical composition of *Spirulina* sp. (*Athrospira*) grown on digested pig waste. Bioresour. Technol. 77: 19-24.

Pandey JP, Neerajpathak P, Amit T (2010). Standardization of pH and light intensity for the biomass production of *Spirulina platensis*. J. Algal Biomass Utln. 1(2): 93-102.

Pandey JP, Tiwari A (2010). Optimization of biomass production by *Spirulina platensis*. J. Algal Biomass Utln. 1(2): 20-32.

Paoletti C, Pushparaj B, Tomaselli Feroci L (1975). Ricerche sulla nutrizione minerale di *Spirulina platensis*. In: Atti Cong. Naz. Soc. Ital. Microbiol. 17: 845-853.

Piorreck M, Baasch K, Pohl P (1983). Biomass production, total protein, chlorophylls, lipids and fatty acids of freshwater green and blue green algae under different nitrogen regimes. Phytochem. 23: 207-216.

Pulz O, Gross W (2004). Valuable products from biotechnology of microalgae. Appl. Microbiol. Biotechnol. 65: 635-648.

Rafiqul IM, Jalal KCA, Alam MZ (2005). Environmental factors for optimization of *Spirulina* biomass in laboratory culture. Biotechnol. 4(1): 19-22.

Richmond A (1986). Microalgal culture. CRC Crit. Rev. Biotechnol. 4(4): 369-438.

Richmond A (1986a). Microalgae of economic potentials. In Richmond A (Ed.) CRC Handbook of Algal Mass Culture. CRC Press, Florida, USA, pp. 199-243.

Richmond A (1986b). *Spirulina*. In: Borowitzka M, Borowitzka L (Eds.), Microalgal Biotechnology, Cambridge University Press, London, UK.

Richmond A, Grobbelaar JU (1986). Factors affecting the output rate of *Spirulina platensis* with reference to mass cultivation. Biomass 10(4), 253-264.

Richmond A, Lichtenberg E, Stahl B, Vonshak A (1990). Quantitative assessment of the major limitations of productivity of *Spirulina plantensis* in open raceways. Appl. Phycol. 2: 195-206.

Rodrigues MS, Ferreira LS, Converti A, Sato S (2011). Influence of ammonia sulphate feeding time on fed-batch *Arthrospira* (*Spirulina*) *platensis* cultivation and biomass composition with and without pH control. Bioresour. Technol. 102: 6587-6592.

Rodrigues MS, Ferreira LS, Converti A, Sato S, Carvalho JCM (2010). Fed-batch cultivation of *Arthrospira* (*Spirulina*) *platensis*: potassium nitrate and ammonium chloride as simultaneous nitrogen sources. Bioresour. Technol. 101: 4491-4498.

Samuelsson G, Lönneborg A, Rosenqvist E, Gustafsson P, Öquist G (1985). Photoinhibition and reactivation of photosynthesis in the cyanobacterim *Anacystis nidulans*. Plant Physiol. 79: 992-995.

Soletto O, Binaghi L, Ferrari L, Lodi A (2008). Effects of carbon dioxide feeding rate and light intensity on the fed-batch pulse-feeding cultivation of *Spirulina platensis* in helical photobioreactor. Biochem. Eng. J. 39: 369-375.

Soundarapandian P, Vasanthi B (2008). Effects of chemical parameters on *Spirulina platensis* biomass production: Optimized method for phycocyanin extraction. Int. J. Zool. Res. 4: 1-11.

Stanca D, Popovici E (1996). Urea as nitrogen source in modified Zarrouk medium. Rev. Roum. Biol. Ser. Biol. Veg. 41: 25-31.

Tamiya HE, Hase E, Shibata K, Mituya A, Iwamura T (1953). Kinetic of growth of *Chlorella*, with special reference to its dependence on quality of available light and on temperatures. Algae culture: From laboratory to pilot plant, growth of algae in mass culture. 205-232.

Thirumala M (2012). Optimization of growth of *Spirulina platensis* LN1 for production of carotenoids. Int. J. Life Sci. Biotechnol. Pharm. Res. 1(2): 152-157.

Torzillo G, Pushparaj B, Bocci F, Balloni W, Materassi R, Florenzano G (1986). Production of *Spirulina* biomass in closed photobioreactors. Biomass 11: 61-74.

Torzillo G, Sacchi A, Materassi R, Richmond A (1991). Effect of temperature on yield and night biomass loss in *Spirulina platensis* grown outdoors in tubular photobioreactors. J. Appl. Phycol. 3: 103-109.

Torzillo G, Vonshak A (1994). Effect of light and temperature on the photosynthetic activity of the cyanobacterium *Spirulina platensis*. Biomass Bioenergy (5): 399-403.

Venkataraman LV, Ramanathan PK, Nigam BP (1981). Simplified production technology of blue green alga *Spirulina platensis* for feed application in India. Biotechnol. 3 (2): 619-622.

Volkmann H, Imianovsky U, Oliveira JLB, Santa'Anna ESS (2008). Cultivation of *Arthrospira* (*Spirulina*) *platensis* in desalinator wastewater and salinated synthetic medium. Protein content and amino acid profile. Brazil. J. Microbiol. 39: 98-101.

Vonshak A (1997). *Spirulina*: growth, physiology and Biochemistry. In: Vonshak, A., (Eds): *Spirulina platensis* (*Arthrospira*): Physiology, Cell Biology and Biotechnology. Taylor and Francis, London, UK. pp. 43-65.

Vonshak A, Abeliovick A, Boussimba S, Arad S, Richmond A (1982). Production of *Spirulina* biomass: effects of environmental factors and population density. Biomass 2: 175-185.

Wang CY, Fu CC, Liu YC (2007). Effects of using light-emitting diodes on the cultivation of *Spirulina platensis*. Biochem. Eng. J. 37: 21-25.

Yuan X, Kumar A, Sahu AK, Ergas SJ (2011). Impact of ammonia concentration on *Spirulina platensis* growth in an airlift photobioreactor. Bioresour. Technol. 102: 3234-3239.

Remediation of lead (Pb) by a novel *Klebsiella sp.* isolated from tannery effluent of Ranipet, Vellore district

Anish Saini[1], Rohini Kumar M.[1], Karthik Senan[1], Shakti Sagar[1], Vivekanandan K. E.[2] and Jabez Osborne W.[1]

[1]School of Biosciences and Technology, VIT University, Vellore-632014, Tamil Nadu, India.
[2]CAS in Marine Biology, Faculty of Marine Sciences, Annamalai University, Parangipettai - 608502, Tamil Nadu, India.

Lead is found to be one of the most toxic heavy metal according to American public health association (APHA). Vellore district is one of the most polluted sites in the world. It is more common for lead poisoning to build up slowly over time. Over time, even low levels of lead exposure can harm a child's mental development. Therefore new resources for the removal of lead are the need of the hour. Soil and effluent samples were obtained from common effluent treatment plant; Ranipet, Vellore district. The concentration of heavy metal was also assessed in the collected samples and then isolated lead tolerant bacteria over lead containing mineral salt medium. The isolated desired bacteria was also tested for their ability to remediate other heavy metals like chromium (Cr), iron (Fe), zinc (Zn), cadmium (Cd) which are present in the tannery effluent. The one with good bioremediation activity was further characterized by sequencing 16S rRNA gene and it was found to be a novel species of *Klebsiella* genus.

Key words: Lead tolerant bacteria, *Klebsiella*, heavy metal remediation.

INTRODUCTION

Microorganisms have been used in various biological treatment processes for metals remediation (Thompson et al., 1987; Macaskie et al., 1989; Smith et al., 1994). The microbe-based technologies provide not only an alternative to conventional methods but are also are environmentally safe (Gadd, 1992). They can affect the metal ions directly or indirectly (pH changes, bioaccumulation, or biosorption) (Al-Shahwani et al., 1984; Beveridge et al., 1992; Vesper et al., 1996). Microorganisms may be used as pure as well as mixed cultures for the effective detoxification process (Smith et al., 1994), which also works over a wide pH range and temperature conditions; is cheaper, and also minimizes waste created by the detoxification process (Churchill et al., 1995). This interaction includes both bioaccumulation and biosorption. Microorganisms can remove Pb through biosorption (Tsezos, 1985), and bioaccumulation (Flemming et al., 1990; Thompson et al., 1987), formation of an exopolymer (Pradhan et al., 1995), or as biomethylate Pb (Gadd, 1992). Few heavy metals are important for proper functioning of biological systems but their deficiency or excess could lead to a number of disorders (Ward, 1995). The European Union has specified a level of 300 mg/kg for sewage sludge and 120 mg/kg for soil. In general, it is harder to characterize national standards for soil than for air and water. In addition to differences between nations, many nations have standards that vary according to geographic region, and soil type (for example, pH or organic material content). Moreover standards for cleaning contaminated soil are generally less stringent than are normative standards for uncontaminated soils.

These heavy metals have a harmful effect on humans and the environment as a whole. Heavy metal pollution has arisen due to rapid urbanization, evolution of metal based industries and other developmental activities. Some of the heavy metal acts as harmful environmental pollutants. The sources of these metals include metal plating, mining by-product, pesticide wastes, chemical wastes, coal based wastes, industrial wastes, gasoline, nuclear wastes and mineral leaching (Zayed et al., 2003; Mohan et al., 2006).

Tannery industries are one of the oldest industries in the world. In old days, tanning was performed to fulfill the demands of leather footwear, drums and musical instruments. But recently, due to the increasing demand of leather and its products, tanning industry has become commercially large. The effluent of Tanning industries are known to have higher concentration of heavy metals (Rao et al., 1982). Ranipet in Vellore district is one of the lead polluted places in the country, as several Tannery industries are located in these regions (Sundar et al., 2011). Lead being the least studied heavy metal present in the tannery waste made us choose this topic for our research.

Cadmium occurs in high concentration in the wastes from electroplating, paints, dyes, chrome tanning, paper industries among others (Malarkodi et al., 2007). Lead is also a toxic element and is obtained from industrial wastewater such as printing, dyeing and oil refineries. Hence, the need for a biological method of removal of heavy metals has become essential. The objective of this study was to isolate and characterize these heavy metal sorbing bacteria from tannery sludge.

MATERIALS AND METHODS

Sampling

The tannery effluent and effluent contaminated soil were collected from a common effluent treatment plant (CETP) at Ranipet, Vellore dist. Tamil Nadu, India. The tannery effluent was collected in sterile bottles. The effluent contaminated soil was also collected in a sterile polythene bags from nearby canals. The samples were transported to the laboratory on the same day itself. The microbiological analysis (fungi, bacteria and actinomycetes) of the samples were carried out on reaching the laboratory, as per the American public health association (APHA) (Clesceri et al., 2005). Inductively coupled plasma-optical emission spectrometry (ICP-OES) tests for tannery sludge were carried out to obtain the presence of heavy metal within the effluent.

Isolation of lead tolerant bacteria

The tannery effluent and effluent contaminated soil were serially diluted and then spread plate technique was performed on a mineral salt medium (MSM) media containing lead acetate at the concentration of 500 mg/L. The MSM media consists of (in g/L) Na_2HPO_4, 4.0; KH_2PO_4, 1.5; NH_4Cl 1.0; $MgSO4{\cdot}7H_2O$, 0.2; $C_6H_8O_7FeNH_3$, 0.05; along with modified Hoagland trace element solution (in g/3.6 L) BH_3, 11.0; $MnCl_2{\cdot}4H_2O$, 7.0; $AlCl_3$, 1.0; $CoCl_2$, 1.0; $CuCl_2$, 1.0; KI, 1.0; $NiCl_2$, 1.0; $ZnCl_2$, 1.0; $BaCl_2$, 0.5; KBr, 0.5; LiCl, 0.5; Na_2MoO_4, 0.5; $SeCl_4$, 0.5; $SnCl_2{\cdot}2H_2O$, 0.5; $NaVO_3{\cdot}H_2O$, 0.1; pH 7.0). After incubation, morphologically different colonies were isolated and purified by repeated streaking on agar plates. The isolates were then stored in agar slants for further studies. The strain with higher lead bioremediation was selected for further biochemical (Table 2) and molecular characterization.

Phenotypic characterization

The bacterial cultures were grouped to various genera based on their morphological and biochemical characters as given in Bergey's Manual of Determinative Bacteriology. The bacterial isolates were morphologically characterized by colony morphology, staining techniques like Gram staining and capsule staining. Motility of the organism was tested by hanging drop test. The bacterial isolates were then characterized by various biochemical tests including triple sugar iron agar (TSI), Indole, methyl red, Voges Proskauer, citrate utilization and catalase tests. The tests were carried out on MSM media (Himedia pvt. Ltd). For the comparison of the best lead degrading strain with the standard strains, type cultures were obtained from MTCC, Chandigarh, India.

Minimum inhibitory concentration (MIC) Test

It is the lowest concentration of an antimicrobial agent that inhibits the visible growth of any microorganism after an overnight incubation. In this case, we used lead acetate of concentrations 500, 750, 1000, 1250 and 1500 mg/L, which was added to the MSM broth, stored for 24 h at 25°C, along with the inoculated microorganism (Waranusantigul et al., 2011).

Removal of Pb at different pH

The bacterium was tested for the activity of bioremediation of lead at pH range of 3 to 8. The cells were inoculated into six 100 mL MSM broth containing lead acetate at the concentration of 1250 mg/L with the pH adjusted to 3 to 8 at progression of pH 1 in each of the six broths, respectively. The lead acetate concentration was chosen to be 1250 mg/L as this was the highest concentration at which VITKAS-2 showed growth. The pH was adjusted using sodium hydroxide and acetic acid. The setup was stored for 24 h at 25°C and shaken at 150 rpm. To confirm that the Pb had been taken up by the bacteria, the MSM culture supernatant was subjected to atomic absorption spectrophotometer mentioned by Al-Momani (2007). Briefly, 25 mL of MSM broth medium was supplemented with lead acetate, the bacterial isolates were inoculated and incubated at 37°C for 24 h and shaken at 150 rpm. After the incubation period, the bacterial cultures were centrifuged at 5000 rpm for 10 min, and from the supernatant, 5 mL aliquot was taken for AAS analysis. The differences between the initial and the

Table 1. Inductively coupled plasma optical emission spectroscopy (ICP-OES) results.

Analyte	Mean concentration (mg/L)
Aluminium (Al)	329.0
Boron (B)	313.0
Cadmium (Cd)	13.60
Cobalt (Co)	56.80
Chromium (Cr)	786.0
Copper (Cu)	16.90
Iron (Fe)	283.9
Magnesium (Mg)	451.7
Manganese (Mn)	74.80
Nickel (Ni)	18.90
Lead (Pb)	558.0
Zinc (Zn)	13.8

Table 2. Results of biochemical characterization.

Test	VITKAS-2	*K. pneumoniae*	*K. aerogenes*
Capsule	+	+	+
Catalase Test	+	+	+
Motility	-	-	-
TSI	A/K	K/K	K/K
Indole	-	-	-
Methyl Red	+	-	-
Voges Proskauer	-	+	+
Citrate Utilization	+	+	+
Gas from Lactose	-	-	-
Dextrose	+	+	+
Lactose	+	+	+
Malonate	-	+	+
L-Sorbose	-	+	-

* A/A, Acidic slant and acidic butt; A/K, acidic slant and alkaline butt; K/A, alkaline slant and acidic butt; K/K, alkaline slant and alkaline butt.

final concentrations indicated the amount of heavy metals uptake by the bacterial isolates.

Molecular characterization and phylogenetic analysis

The isolate showing best heavy metal bioremediation property was further characterized by sequencing 16S rRNA gene sequence. The 27F (5' - AGA GTT TGA TCC TGG CTC AG - 3') and 1492R (5' - GGT TAC CTT GTT ACG ACT T - 3') universal bacterialprimers were used to amplify the 16S rRNA gene from the genomic DNA of the isolate. The 16S rDNA molecule was sequenced using an automated DNA sequencer (ABI Prism 310 Genetic Analyzer, Tokyo, Japan). The FASTA sequence of the VITKAS-2 was aligned along with various type culture FASTA sequences belonging to the family Enterobacteriaceae. These aligned sequences were used for construction of phylogenetic tree using MEGA v5.04 package (Tamura et al., 2011).

RESULTS AND DISCUSSION

The effluent was tested for the heavy metal composition by the ICP-OES test, which clearly showed the presence of heavy metals like lead, iron, cadmium, chromium, nickel and zinc, in toxic concentrations (Table 1).

Isolation of lead tolerant bacteria

The growth of bacterial isolates obtained in MSM media supplemented with lead acetate, showed that these isolates were lead tolerant. From the gram staining techniques, the isolates were found to be gram negative rods. After morphological characterization, the colonies

Table 3. Minimum inhibitory concentration of lead tolerant isolate.

Concentration of lead acetate (mg/L)	VITKAS-1	VITKAS-2	VITKAS-3
500	+	+	+
750	+	+	-
1000	-	+	-
1250	-	+	-
1500	-	-	-

+, Growth of colonies; -, no growth of colonies.

Table 4. Effect of different pH values on remediation of lead by VITKAS-2.

pH value	Lead concentration (mg/L)	
	Initial	Final
3.0	1250	1181 ± 2.08
4.0	1250	993 ± 2.65
5.0	1250	765 ± 3.15
6.0	1250	840 ± 4.01
7.0	1250	891 ± 2.50
8.0	1250	1028 ± 5.00

were observed to be non-motile, irregular, raised and smooth in nature.

Minimum inhibitory concentration (MIC) test

The growth of bacteria, VITKAS-2, was seen in all the tubes from 500 to 1250 mg/L of lead acetate containing MSM broth, except at 1500 mg/L of the same (Table 3) whereas, both, VITKAS-1 and VITKAS-3 were found to have lower tolerance level to the higher concentrations of lead acetate compared to VITKAS-2. From the results obtained, we can safely assume that bacteria growing in lead acetate medium can absorbed/accumulate lead present in the medium. We know that some bacteria are able to immobilize the Pb into various less toxic forms. Therefore the lead can be actively removed from the soil, leading to effective reduction of soil pollution.

Bioremediation of Pb at different pH

The VITKAS-2, was then checked in different pH levels for the variation in bioremediation activity, since the pH can be a crucial parameter in the bioremediation of these heavy metals (Sundar et al., 2011; Tamura et al., 2011; Thompson and Watling, 1987; Tsezos, 1985; Vesper et al, 1996; Waranusantigul et al., 2011). VITKAS-2 shows highest removal of lead at pH 5 (Table 4). Compared to *K. pneumonia* (Ward, 1995) at pH 5 (17%), VITKAS-2 reduced 48% of lead (Figure 1). Though, *K. pneumonia* showed highest activity at pH 4 in their study, the influence of acidic pH cannot be ignored.

Characterization and phylogenetic analysis

The 16S rRNA gene sequence results suggested that the culture VITKAS-2 belongs to genus *Klebsiella*, a member of the family Enterobacteriaceae. The nearest neighbour VITKAS-2 was *K.* pneumoniae DSM 30104, Acc. No. X87277 with a similarity of 93.75%. Since, the bacteria with 16S ribosomal RNAs (rRNAs) that are ≤98.7% identical to their nearest neighbor, are always members of different species (Stackebrandt et al., 2006), as such strong differences in rRNA correlate with <70% DNA-DNA similarity. Therefore, *Klebsiella* sp. VITKAS-2 is a potential candidate for novel species. The 16S rRNA gene FASTA sequence has been submitted in GenBank with the Acc. No. JQ398846. The phylogenetic tree further supports the divergence of VITKAS-2 from the *Klebsiella* pneumonia type strain (Figure 2).

Conclusion

Use of a novel *Klebsiella* sp. growing on lead acetate medium was obtained and can serve as a cost effective means of bioremediation of heavy metals present in tannery sludge (Waranusantigul et al., 2011). This can serve as a potent and safe method to curb soil and water pollution from tannery sludge, without producing any

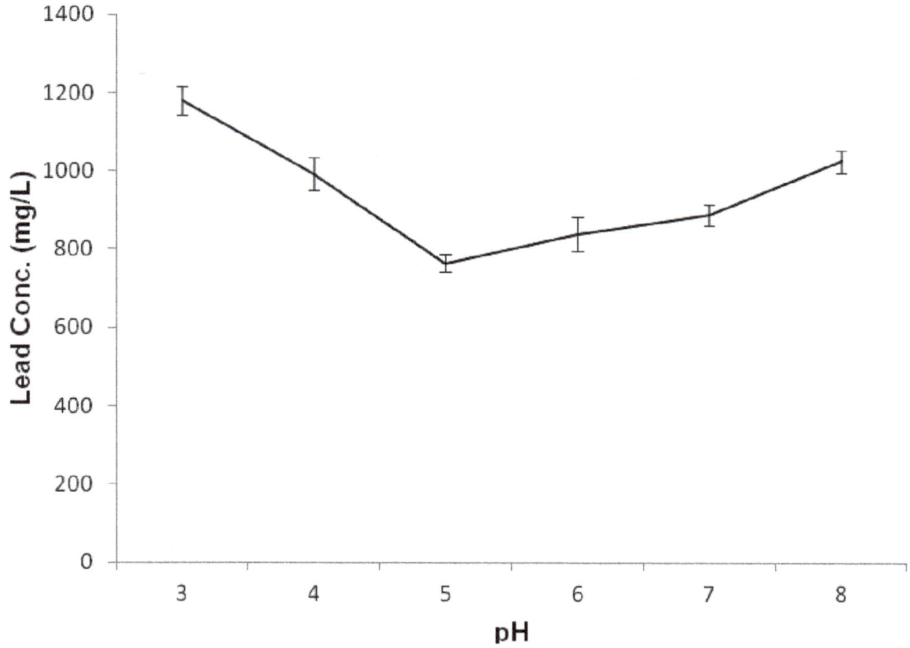

Figure 1. Effect of different pH values on remediation of lead by VITKAS-2.

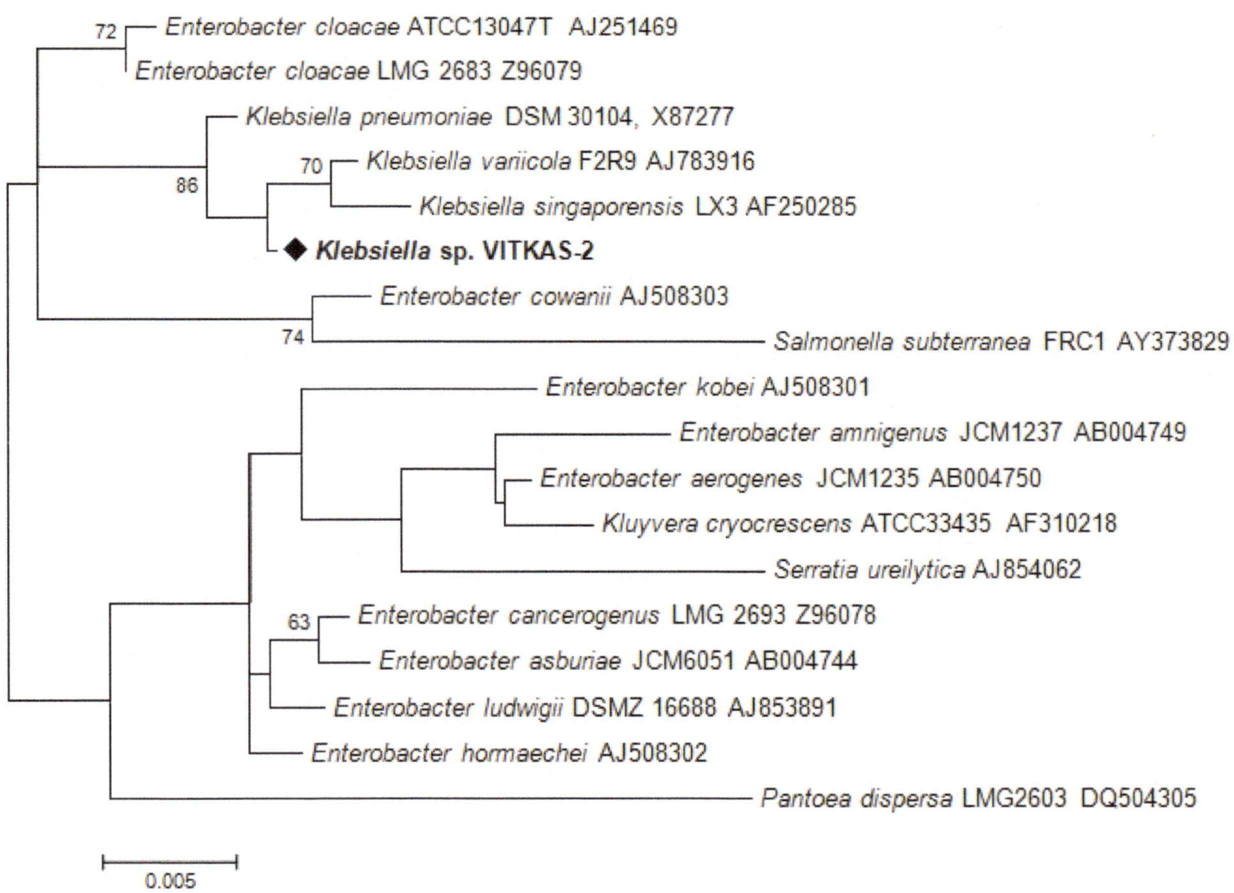

Figure 2. Neighbour joining method based phylogenetic tree for VITKAS-2. Phylogenetic tree based on 16S rDNA gene sequence of VITKAS-2 with various type culture sequences belonging to the family *Enterobacteriaceae*. The numbers at the nodes indicate the levels of the bootstrap support based on a neighbour-joining analysis of 1,000 re-sampled data sets. The bootstrap values below 50% were not indicated. Bar 0.005 substitutions per site.

effective by-products which in turn may lead to soil and water contamination.

ACKNOWLEDGMENTS

We would like to thank the management and laboratory staff of Vellore Institute of Technology, University for providing the facilities to aid our project work.

REFERENCES

Al-Momani FA, Massadeh AM, Hadad YA (2007). Removal of Lead and Cadmium by Halophilic Bacteria Isolated from the Dead Sea Shore, Jordan. Biol. Trace Element Res. 115:291-300.

Al-Shahwani MF, Jazrawi SF, Al-Rawi EH, Ayar NS (1984). Growth and heavy metal removal by *Klebsiella aerogenes* at different pH and temperature. - J. Environ. Sci. Hlth. A 19(4):445-457.

Beveridge T, Doyle RJ (1992). Metal Ions and Bacteria. John Wiley and Sons, New York, NY.

Churchill SA, Walters JV, Churchill PF (1995). Sorption of heavy metal by prepared bacterial cell surfaces. J. Environ. Engr. 121:706-711.

Clesceri LS, Greenberg AE, Eaton AD, Franson MAH (2005). Standard methods for the examination of water and wastewater, 20th ed., American Public Health Association (APHA), Washington, DC, USA.

Flemming CA, Ferris FG, Beveridge TJ, Bailey GW (1990). Remobilization of toxic heavy metals adsorbed to bacterial wall-clay composites. Appl. Environ. Microbiol. 56:3191-3196.

Gadd GM (1992). Microbial Control of Environmental Pollution, Cambridge University Press, Cambridge, MA.

Macaskie LE, Dean A (1989). Use of immobilized biofilm of *Citrobacter* sp. for the removal of uranium and lead from aqueous flows. - Enzyme Microbiol. Technol. 9:2-7.

Malarkodi M, Krishnasamy R, Kumaraperumal R, Chitdeshwari T (2007). Characterization of Heavy Metal Contaminated Soils of Coimbatore District in Tamil Nadu. J. Agron. 6:147-151.

Mohan D, Pittman CU (2006). Activated carbons and low cost adsorbents for remediation of tri- and hexavalent chromium from water. J. Hazard. Mater. 37:762-811.

Pradhan AA, Levine AD (1995). Microbial biosorption of copper and lead from aqueous systems. - Sci. Total Environ. 170:209-220.

Rao MG, Kumar NVN (1982). The tannery industrial effluent effect on succinate dehydrogenase activity pattern in a freshwater snail, *Pila globosa*. Proceedings: Anim. Sci. 91:427-431.

Smith LA, Alleman BC, Copley-Graves L (1994). Biological treatment options. In: Emerging Technology for Bioremediation of Metals. J. L. Means and R. E. Hinchee, ed. Lewis Publishers, New York, NY.

Stackebrandt E, Ebers J (2006). Taxonomic parameters revisited: tarnished gold standards. - Microbiol Today 33:152-155.

Sundar K, Sadiq IM, Mukherjee A, Chandrasekaran N (2011). Bioremoval of trivalent chromium using *Bacillus* biofilms through continuous flow reactor. J. Hazard. Mater. 196:44-51.

Tamura K, Peterson D, Peterson N, Stecher G, Nei M, Kumar S (2011). MEGA5: Molecular Evolutionary Genetics Analysis using Maximum Likelihood, Evolutionary Distance, and Maximum Parsimony Methods. Mol. Biol. Evol. 28:2731-2739.

Thompson GA, Watling RJ (1987). Bioaccumulation potential of heterotrophic bacteria for lead, selenium and arsenic. Bull. - Environ. Contam. Toxicol. 38:1049-1054.

Tsezos M (1985). The selective extraction of metals from solution by microorganisms. - Can. Met. Q. 24:141-144.

Vesper SJ, Donovan-Brand R, Pete-Paris K, Al-Abed SR, Ryan JA, Davis-Hoover WJ (1996). Microbial removal of lead from solid media and soil. - Water, Air Soil Pollut. 86:207-219.

Waranusantigul P, Lee H, Kruatrachue M, Pokethitiyook P, Auesukaree C (2011). Evaluation of Pb Phytoremediation Potential in *Buddleja asiatica* and *B. paniculata*. - Chemosphere. 5:584-590.

Ward NI (1995). In Environmental analytical chemistry-Trace elements. (Eds.F.W. Fifield, and P.J.Haines.), Blackie Academic and Professional, UK. pp. 320-328.

Zayed AM, Terry N (2003). Chromium in the environment: factors affecting biological remediation. Plant Soil 249:139-156.

Permissions

All chapters in this book were first published in AJB, by Academic Journals; hereby published with permission under the Creative Commons Attribution License or equivalent. Every chapter published in this book has been scrutinized by our experts. Their significance has been extensively debated. The topics covered herein carry significant findings which will fuel the growth of the discipline. They may even be implemented as practical applications or may be referred to as a beginning point for another development.

The contributors of this book come from diverse backgrounds, making this book a truly international effort. This book will bring forth new frontiers with its revolutionizing research information and detailed analysis of the nascent developments around the world.

We would like to thank all the contributing authors for lending their expertise to make the book truly unique. They have played a crucial role in the development of this book. Without their invaluable contributions this book wouldn't have been possible. They have made vital efforts to compile up to date information on the varied aspects of this subject to make this book a valuable addition to the collection of many professionals and students.

This book was conceptualized with the vision of imparting up-to-date information and advanced data in this field. To ensure the same, a matchless editorial board was set up. Every individual on the board went through rigorous rounds of assessment to prove their worth. After which they invested a large part of their time researching and compiling the most relevant data for our readers.

The editorial board has been involved in producing this book since its inception. They have spent rigorous hours researching and exploring the diverse topics which have resulted in the successful publishing of this book. They have passed on their knowledge of decades through this book. To expedite this challenging task, the publisher supported the team at every step. A small team of assistant editors was also appointed to further simplify the editing procedure and attain best results for the readers.

Apart from the editorial board, the designing team has also invested a significant amount of their time in understanding the subject and creating the most relevant covers. They scrutinized every image to scout for the most suitable representation of the subject and create an appropriate cover for the book.

The publishing team has been an ardent support to the editorial, designing and production team. Their endless efforts to recruit the best for this project, has resulted in the accomplishment of this book. They are a veteran in the field of academics and their pool of knowledge is as vast as their experience in printing. Their expertise and guidance has proved useful at every step. Their uncompromising quality standards have made this book an exceptional effort. Their encouragement from time to time has been an inspiration for everyone.

The publisher and the editorial board hope that this book will prove to be a valuable piece of knowledge for researchers, students, practitioners and scholars across the globe.

List of Contributors

Marcos Ribeiro da Silva Vieira
Universidade Federal Rural de Pernambuco, Unidade Acadêmica de Serra Talhada, CEP: 59909-460, Serra Talhada, PE, Brasil

Fernando Jose Freire
Departamento de Agronomia, CEP: 52171-900, Recife, PE, Brasil'

Damiana Cleuma de Medeiros
Instituição: UFRN - Unidade Acadêmica Especializada em Ciências Agrárias, Departamento de Agropecuária, Macaíba, RN, Brasil

Fabiana Bezerra Moura
Universidade Federal Rural de Pernambuco, Unidade Acadêmica de Serra Talhada, CEP: 59909-460, Serra Talhada, PE, Brasil

Thialla Laranjeira Amorim
Universidade Federal Rural de Pernambuco, Unidade Acadêmica de Serra Talhada, CEP: 59909-460, Serra Talhada, PE, Brasil

Adriano Nascimento dos Simões
Universidade Federal Rural de Pernambuco, Unidade Acadêmica de Serra Talhada, CEP: 59909-460, Serra Talhada, PE, Brasil

G. I. Ameh
Department of Applied Biology and Biotechnology, Enugu State University of Science and Technology, Enugu, Enugu State, Nigeria

S. C. Eze
Department of Crop Science, University of Nigeria, Nsukka, Enugu State, Nigeria

F. U. Omeje
Department of Applied Biology and Biotechnology, Enugu State University of Science and Technology, Enugu, Enugu State, Nigeria

Karun Thongprajukaew
Interdisciplinary Graduate Program in Bioscience, Faculty of Science, Kasetsart University, Bangkok 10900, Thailand
Biochemical Research Unit for Feed Utilization Assessment, Faculty of Science, Kasetsart University, Bangkok 10900, Thailand
Department of Applied Science, Faculty of Science, Prince of Songkla University, Songkhla 90112, Thailand

Uthaiwan Kovitvadhi
Biochemical Research Unit for Feed Utilization Assessment, Faculty of Science, Kasetsart University, Bangkok 10900, Thailand
Department of Zoology, Faculty of Science, Kasetsart University, Bangkok 10900, Thailand

Satit Kovitvadhi
Department of Agriculture, Faculty of Science and Technology, Bansomdejchaopraya Rajabhat University, Bangkok 10600, Thailand

Arunee Engkagul
Biochemical Research Unit for Feed Utilization Assessment, Faculty of Science, Kasetsart University, Bangkok 10900, Thailand
Department of Biochemistry, Faculty of Science, Kasetsart University, Bangkok 10900, Thailand

Krisna Rungruangsak-Torrissen
Biochemical Research Unit for Feed Utilization Assessment, Faculty of Science, Kasetsart University, Bangkok 10900, Thailand
Institute of Marine Research, Ecosystem Processes Research Group, Matre Research Station, N-5984 Matredal, Norway

Akmal A. Sakr
Botany Department, Faculty of Science, Zagazig University, Zagazig, Egypt
Conservation Department, Civilization Museum, Cairo, Egypt

M. F. Ghaly
Botany Department, Faculty of Science, Zagazig University, Zagazig, Egypt

M. F. Ali
Conservation Department, Faculty of Archaeology, Cairo University, Cairo, Egypt

M. E. F. Abdel-Haliem
Botany Department, Faculty of Science, Zagazig University, Zagazig, Egypt

KiBeom Lee
Biotechnology and Business Center, Incheon Technopark, 7-50 Songdo-Dong, Yeonsu-Gu, Incheon 406-840, South Korea

Sang-Kee Kang
School of Agricultural Biotechnology, Seoul National University, Seoul 151-742, South Korea

Yun Jaie Choi
School of Agricultural Biotechnology, Seoul National University, Seoul 151-742, South Korea

Kyung Hee Han
Department of Horticulture, Hankyong National University, 67 Seokjeong-dong, Ansung 456-749, South Korea

Yu Jin Jung
Institute of Genetic Engineering, Hankyong National University, Ansung 456-749, Korea
Clone Company, 67 Seokjeong-dong, Ansung 456-749, South Korea

Uuganchimeg Bayarsaikhan
Department of Horticulture, Hankyong National University, 67 Seokjeong-dong, Ansung 456-749, South Korea

In Hye Lee
Department of Horticulture, Hankyong National University, 67 Seokjeong-dong, Ansung 456-749, South Korea

Jang Sun Choi
National Institute of Horticultural and Herbal Science, Suwon, 440-746, Korea

Ill Sup Nou
Department of Horticulture, Sunchon National University, Suncheon 540-742, Korea

Yong Gu Cho
Department of Crop Science, Chungbuk National University, Cheongju, Korea

Kwon Kyoo Kang
Department of Horticulture, Hankyong National University, 67 Seokjeong-dong, Ansung 456-749, South Korea
Institute of Genetic Engineering, Hankyong National University, Ansung 456-749, Korea
Clone Company, 67 Seokjeong-dong, Ansung 456-749, South Korea

Raoudha Abdellaoui
Laboratoire d'Ecologie Pastorale, Institut des Régions Arides, 4119 Médenine, Tunisie

Aymen Souid
Laboratoire d'Ecologie Pastorale, Institut des Régions Arides, 4119 Médenine, Tunisie
Département de Biologie, Faculté des Sciences de Tunis, 1060 Tunis, Tunisie

Dhikra Zayoud
Laboratoire d'Ecologie Pastorale, Institut des Régions Arides, 4119 Médenine, Tunisie
Département de Biologie, Faculté des Sciences de Tunis, 1060 Tunis, Tunisie

Mohamed Neffati
Laboratoire d'Ecologie Pastorale, Institut des Régions Arides, 4119 Médenine, Tunisie

A. Shiva Reddy
Analytical Chemistry Division, Indian Institute of Chemical Technology, Hyderabad - 500007, India

P. Sita Devi
Analytical Chemistry Division, Indian Institute of Chemical Technology, Hyderabad - 500007, India

S. Ravi Kiran
Department of Biochemistry and Biotechnology, Aurora's Degree & PG College, Chikkadpally, Hyderabad - 500020, India

Ali Ellafi
Laboratory of Analysis, Treatment and Valorization of Environment Polluants and Product, Faculty of Pharmacy, Monastir (Tunisia), Tunisia

Besma Harbi
Laboratory of Analysis, Treatment and Valorization of Environment Polluants and Product, Faculty of Pharmacy, Monastir (Tunisia), Tunisia

Rihab Lagha
Laboratory of Analysis, Treatment and Valorization of Environment Polluants and Product, Faculty of Pharmacy, Monastir (Tunisia), Tunisia

Amina Bakhrouf
Laboratory of Analysis, Treatment and Valorization of Environment Polluants and Product, Faculty of Pharmacy, Monastir (Tunisia), Tunisia

Regina Ngozi Ugbaja
Department of Biochemistry, College of Natural Sciences, Federal University of Agriculture, Abeokuta, Nigeria

Beno Okechukwu Onunkwor
Department of Biochemistry, College of Natural Sciences, Federal University of Agriculture, Abeokuta, Nigeria

Demilade Akinbola Omoniyi
Department of Biochemistry, College of Natural Sciences, Federal University of Agriculture, Abeokuta, Nigeria

S. Moyo
Department of Livestock and Pasture Science, Faculty of Science and Agriculture, University of Fort Hare, P/Bag X 1314, Alice 5700, South Africa

P. J. Masika
Agricultural and Rural Development Research Institute (ARDRI), Faculty of Science and Agriculture, University of Fort Hare, P/Bag X 1314, Alice 5700, South Africa

R. I. Udegbunam
Department of Veterinary Surgery, Faculty of Veterinary Medicine, University of Nigeria, Nsukka, Enugu State, Nigeria

S. O. Udegbunam
Department of Veterinary Surgery, Faculty of Veterinary Medicine, University of Nigeria, Nsukka, Enugu State, Nigeria

G. N. Anosa
Department of Veterinary Medicine, Faculty of Veterinary Medicine, University of Nigeria, Nsukka, Enugu State, Nigeria

Lola Ajala
Food Science and Technology Department, Rufus Giwa Polytechnic P.M.B 1019, Owo, Ondo State, Nigeria

M. O. Ologunde
Ladoke Akintola University of Technology, Ogbomosho, Oyo State, Nigeria

F. O. Adetuyi
Ondo State University of Technology, Okitipupa, Ondo State, Nigeria

Supalak Sattayasamitsathit
Department of Industrial Biotechnology, Faculty of Agro-Industry, Prince of Songkla University, Hatyai 90112 Thailand

Poonsuk Prasertsan
Department of Industrial Biotechnology, Faculty of Agro-Industry, Prince of Songkla University, Hatyai 90112 Thailand
Palm Oil Product and Technology Research Center, Faculty of Agro-Industry, Prince of Songkla University, Hatyai 90112 Thailand

F. A. G. Gonçalves
Faculdade de Farmácia, Universidade Federal de Minas Gerais, Belo Horizonte, MG, 31270-901, Brazil

G. Colen
Faculdade de Farmácia, Universidade Federal de Minas Gerais, Belo Horizonte, MG, 31270-901, Brazil

J. A. Takahashi
Departamento de Química, Instituto de Ciências Exatas, Universidade Federal de Minas Gerais, Belo Horizonte, MG, 31270-901, Brazil

J. O. AYO
Department of Veterinary Physiology, Ahmadu Bello University, Zaria, Nigeria

F. H. OLAIFA
Department of Veterinary Physiology, Ahmadu Bello University, Zaria, Nigeria

A. S. AKE
Department of Veterinary Physiology, Ahmadu Bello University, Zaria, Nigeria

F. K. Onwu
Department of Chemistry, Michael Okpara University of Agriculture Umudike, P.M.B 7267, Umuahia, Abia State, Nigeria

S. P. I. Ogah
Department of Industrial Chemistry, Ebonyi State University, P.M.B 053, Abakaliki, Ebonyi State, Nigeria

S. O. Ngele
Department of Industrial Chemistry, Ebonyi State University, P.M.B 053, Abakaliki, Ebonyi State, Nigeria

Arturo Silva-Ríos
Facultad de Ciencias Agronómicas, Universidad Autónoma del Estado de México, El Cerrillo Piedras Blancas, Toluca, Edo. de México. CP 50090, México

Octavio Dublán-García
Facultad de Química, Universidad Autónoma del Estado de México, El Cerrillo Piedras Blancas, Toluca, Edo. de México. CP 50090, México

Baciliza Quintero-Salazar
Facultad de Turismo y Gastronomía, Campus "el Rosedal", Universidad Autónoma del Estado de México, Carretera Toluca-Atlacomulco Km. 14.5, Toluca, C.P. 50200 México

Aurelio Dominguez-Lopez
Facultad de Ciencias Agronómicas, Universidad Autónoma del Estado de México, El Cerrillo Piedras Blancas, Toluca, Edo. de México. CP 50090, México

Leobardo Manuel Gómez-Oliván
Facultad de Química, Universidad Autónoma del Estado de México, El Cerrillo Piedras Blancas, Toluca, Edo. de México. CP 50090, México

Leticia Xochitl López-Martínez
Facultad de Química, Universidad Autónoma del Estado de México, El Cerrillo Piedras Blancas, Toluca, Edo. de México. CP 50090, México

Abdel-Fattah Z. M. Salem
Facultad de Medicina Veterinaria y Zootecnia, Universidad Autónoma del Estado de México, El Cerrillo Piedras Blancas, Toluca, Edo. Mexico. CP 50090, Mexico
Faculty of Agriculture (El-Shatby), Alexandria University, Egypt

Darina Ibrahim
Department of Nutrition Science, Kulliyyah of Allied Health Sciences, International Islamic University Malaysia, Jalan Istana, Bandar Indera Mahkota, 25200 Kuantan, Pahang, Malaysia

Norazlanshah Hazali
Department of Nutrition Science, Kulliyyah of Allied Health Sciences, International Islamic University Malaysia, Jalan Istana, Bandar Indera Mahkota, 25200 Kuantan, Pahang, Malaysia

NurhazniKhir Jauhari
Department of Nutrition Science, Kulliyyah of Allied Health Sciences, International Islamic University Malaysia, Jalan Istana, Bandar Indera Mahkota, 25200 Kuantan, Pahang, Malaysia

Muhammad Nor Omar
Department of Biotechnology, Kuliyyah of Science, International Islamic University Malaysia, Jalan Istana, Bandar Indera Mahkota, 25200 Kuantan, Pahang, Malaysia

Mohd Noor Adros Yahya
Malaysian Agriculture Research and Development Institute (MARDI), Sarawak, Malaysia

Idris Adewale Ahmed
Department of Nutrition Science, Kulliyyah of Allied Health Sciences, International Islamic University Malaysia, Jalan Istana, Bandar Indera Mahkota, 25200 Kuantan, Pahang, Malaysia

Maryam Abimbola Mikail
Department of Nutrition Science, Kulliyyah of Allied Health Sciences, International Islamic University Malaysia, Jalan Istana, Bandar Indera Mahkota, 25200 Kuantan, Pahang, Malaysia

Muhammad Ibrahim
Department of Nutrition Science, Kulliyyah of Allied Health Sciences, International Islamic University Malaysia, Jalan Istana, Bandar Indera Mahkota, 25200 Kuantan, Pahang, Malaysia

O. C. Enechi
Department of Biochemistry, University of Nigeria, Nsukka, Enugu State, Nigeria

N. Lewis Manyawo
Department of Biochemistry, University of Nigeria, Nsukka, Enugu State, Nigeria

P. C. Ugwu Okechukwu
Department of Biochemistry, University of Nigeria, Nsukka, Enugu State, Nigeria

C. E. Umebese
Department of Botany, P.M.B 1029 UNILAG Post Office, University of Lagos, Akoka-Yaba, Lagos, Nigeria

A. E. Bankole
Department of Botany, P.M.B 1029 UNILAG Post Office, University of Lagos, Akoka-Yaba, Lagos, Nigeria

H. Zidane
Laboratory of Physical Chemistry of the Natural Resources and Environment, University Mohammed Premier, BP 717, 60000 Oujda, Morocco

M. Elmiz
Laboratory of Physical Chemistry of the Natural Resources and Environment, University Mohammed Premier, BP 717, 60000 Oujda, Morocco

F. Aouinti
Laboratory of Physical Chemistry of the Natural Resources and Environment, University Mohammed Premier, BP 717, 60000 Oujda, Morocco

A. Tahani
Laboratory of Physical Chemistry of the Natural Resources and Environment, University Mohammed Premier, BP 717, 60000 Oujda, Morocco

J. Wathelet
Unity of General and Organic Chemistry, Faculty of the Agronomic Sciences of Gembloux, Passage of the Prisoners 2, B-5030 Gembloux, Belgium

M. Sindic
Unity of General and Organic Chemistry, Faculty of the Agronomic Sciences of Gembloux, Passage of the Prisoners 2, B-5030 Gembloux, Belgium

A. Elbachiri
Laboratory of Physical Chemistry of the Natural Resources and Environment, University Mohammed Premier, BP 717, 60000 Oujda, Morocco

Jinglan Han
Hubei Provincial Key Laboratory for Protection and Application of Special Plants in Wuling Area of China, College of Life Science, South-Central University for Nationalities, Wuhan 430074, China
College of Life Science, South China Agricultural University, Guangzhou 510642, China

Xiangming Zhang
Hubei Provincial Key Laboratory for Protection and Application of Special Plants in Wuling Area of China, College of Life Science, South-Central University for Nationalities, Wuhan 430074, China

Dan Zhang
Hubei Provincial Key Laboratory for Protection and Application of Special Plants in Wuling Area of China, College of Life Science, South-Central University for Nationalities, Wuhan 430074, China

Xuequn Liu
Hubei Provincial Key Laboratory for Protection and Application of Special Plants in Wuling Area of China, College of Life Science, South-Central University for Nationalities, Wuhan 430074, China

Chuntai Wang
Hubei Provincial Key Laboratory for Protection and Application of Special Plants in Wuling Area of China, College of Life Science, South-Central University for Nationalities, Wuhan 430074, China

Xinqiong Liu
Hubei Provincial Key Laboratory for Protection and Application of Special Plants in Wuling Area of China, College of Life Science, South-Central University for Nationalities, Wuhan 430074, China

A. A. Newairy
Department of Biochemistry, Faculty of Science, Moharam Bik. Alexandria University, Egypt

H. M. Abdou
Department of Zoology, Faulty of Science, Moharam Bik. Alexandria University, Egypt

Xiao-Hong Huang,
Institute of Animal Health, College of Animal Science, Fujian Agriculture and Forestry University, Fuzhou 350002, China

Yan-Dong Jin
Institute of Animal Health, College of Animal Science, Fujian Agriculture and Forestry University, Fuzhou 350002, China

Yi-Fan Huang
Institute of Animal Health, College of Animal Science, Fujian Agriculture and Forestry University, Fuzhou 350002, China

Wei-Chao Han
Institute of Animal Health, College of Animal Science, Fujian Agriculture and Forestry University, Fuzhou 350002, China

Wen Zhang
Institute of Animal Health, College of Animal Science, Fujian Agriculture and Forestry University, Fuzhou 350002, China

E. Kolawole Sunday
National Open University of Nigeria, Benin Study Centre, Benin City, Nigeria

O. Obueh Henrietta
Department of Biology, College of Education, P.M.B 1144, Benin City, Nigeria

Hanan A. Said
Botany Department, Faculty of Science, Fayoum University, El Fayoum, Egypt

Wafaa M. Abdel-Aziz
Botany Department, Faculty of Science, Alexandria University, Alexandria, Egypt

Hala M. Taha
Botany Department, Faculty of Science, Alexandria University, Alexandria, Egypt

Yooheon Park
Department of Food and Nutrition, Korea University, Seoul 136-703, Republic of Korea

Hyun Jung Lee
Department of Food and Nutrition, Korea University, Seoul 136-703, Republic of Korea

Jang Won Choi
Department of Bioindustry, College of Life and Environment, Daegu University, Gyeongsan 712-714, Republic of Korea

Song Hwan Bae
Department of Food and Biotechnology, Hankyong National University Anseong, Anseong 456-749, Republic of Korea

Hyung Joo Suh
Department of Food and Nutrition, Korea University, Seoul 136-703, Republic of Korea

N. Djemel Amina
Laboratoire de Génétique, Biochimie et Biotechnologies Végétales University Mentouri of Contantine, Algeria

Benmati Mahbouba
Laboratoire de Génétique, Biochimie et Biotechnologies Végétales University Mentouri of Contantine, Algeria

Ykhlef Nadia
Laboratoire de Génétique, Biochimie et Biotechnologies Végétales University Mentouri of Contantine, Algeria

Belbekri Nadir
Laboratoire de Génétique, Biochimie et Biotechnologies Végétales University Mentouri of Contantine, Algeria

Djekoun Abdelhamid
Laboratoire de Génétique, Biochimie et Biotechnologies Végétales University Mentouri of Contantine, Algeria

Mingjun Wu
Institute of Blood Transfusion, Chinese Academy of Medical Science, Chengdu 610081, Sichuan Province, China
Key Laboratory of Bio-resources and Eco-environment (Ministry of Education, College of Life Sciences, Sichuan University, Chengdu 610064, Sichuan Province, China

Qiang Chen
Institute of Blood Transfusion, Chinese Academy of Medical Science, Chengdu 610081, Sichuan Province, China

Wei Wu
Institute of Blood Transfusion, Chinese Academy of Medical Science, Chengdu 610081, Sichuan Province, China

Zhongwei Zheng
Institute of Blood Transfusion, Chinese Academy of Medical Science, Chengdu 610081, Sichuan Province, China

Bisong Yue
Key Laboratory of Bio-resources and Eco-environment (Ministry of Education, College of Life Sciences, Sichuan University, Chengdu 610064, Sichuan Province, China

Zhiqiang Wu
Laboratory of Molecular Biology, Institute of Basic Medicine, Chinese People's Liberation Army General Hospital, Beijing 100853, China

Yali Zhao
Laboratory of Molecular Biology, Institute of Basic Medicine, Chinese People's Liberation Army General Hospital, Beijing 100853, China

Weidong Han
Laboratory of Molecular Biology, Institute of Basic Medicine, Chinese People's Liberation Army General Hospital, Beijing 100853, China

Xiaobing Fu
Laboratory of Molecular Biology, Institute of Basic Medicine, Chinese People's Liberation Army General Hospital, Beijing 100853, China

Yahia Mustafa A. Fagiri
Institute of Biological Sciences, Faculty of Science, University of Malaya, 50603 Kuala Lumpur, Malaysia

Aisha Salleh
Institute of Biological Sciences, Faculty of Science, University of Malaya, 50603 Kuala Lumpur, Malaysia

Saifeldin Ahmed F. El-Nagerabi
Department of Biological Sciences and Chemistry, College of Arts and Sciences, University of Nizwa, P.O. Box 33, Postal Code 616, Birkat Al Mouz, Nizwa, Oman

Anish Saini
School of Biosciences and Technology, VIT University, Vellore-632014, Tamil Nadu, India

M. Rohini Kumar
School of Biosciences and Technology, VIT University, Vellore-632014, Tamil Nadu, India

Karthik Senan
School of Biosciences and Technology, VIT University, Vellore-632014, Tamil Nadu, India

Shakti Sagar
School of Biosciences and Technology, VIT University, Vellore-632014, Tamil Nadu, India

K. E. Vivekanandan
CAS in Marine Biology, Faculty of Marine Sciences, Annamalai University, Parangipettai - 608502, Tamil Nadu, India

W. Jabez Osborne
School of Biosciences and Technology, VIT University, Vellore-632014, Tamil Nadu, India